T0363922

Methods in Cell Biology

VOLUME 94

Primary Cilia

Series Editors

Leslie Wilson
Department of Molecular, Cellular and Developmental Biology
University of California
Santa Barbara, California

Paul Matsudaira
Department of Biological Sciences
National University of Singapore
Singapore

Methods in Cell Biology

VOLUME 94

Primary Cilia

Edited by

Roger D. Sloboda

Department of Biological Sciences
Dartmouth College
Hanover, New Hampshire

AMSTERDAM • BOSTON • HEIDELBERG • LONDON
NEW YORK • OXFORD • PARIS • SAN DIEGO
SAN FRANCISCO • SINGAPORE • SYDNEY • TOKYO
Academic Press is an imprint of Elsevier

Academic Press is an imprint of Elsevier
30 Corporate Drive, Suite 400, Burlington, MA 01803, USA
525 B Street, Suite 1900, San Diego, CA 92101-4495, USA
32, Jamestown Road, London NW1 7BY, UK
Linacre House, Jordan Hill, Oxford OX2 8DP, UK

First edition 2009

ISBN–13: 978-0-12-375024-2
ISSN: 0091-679X

For information on all Academic Press publications
visit our website at elsevierdirect.com

CONTENTS

SECTION II Cell Biology/Biochemistry

SECTION III Function

CONTRIBUTORS

Numbers in parentheses indicate the pages on which the authors' contributions begin.

Mark C. Alliegro (53), Bay Paul Center for Comparative Molecular Biology and Evolution, Marine Biological Laboratory, Woods Hole, Massachusetts 02543

Kathryn V. Anderson (199), Developmental Biology Program, Sloan-Kettering Institute, 1275 York Avenue, New York, New York 10065

Nicolas F. Berbari (163), Department of Cell Biology, University of Alabama at Birmingham Medical School, Birmingham, Alabama 35294

Robert A. Bloodgood (1), Department of Cell Biology, University of Virginia School of Medicine, Charlottesville, Virginia 22908-0732

Muqing Cao (333), School of Life Sciences, Tsinghua University, Beijing 100084, China

Anna Capasso (223), University of Naples, Yale University, New Haven, Connecticut 06520

Michael J. Caplan (223), Department of Cell Biology and Department of Cellular and Molecular Physiology, Yale University, New Haven, Connecticut 06520

Hannah C. Chapin (223), Department of Cell Biology, Yale University, New Haven, Connecticut 06520

Søren T. Christensen (67, 181), Department of Biology, Section of Cell and Developmental Biology, University of Copenhagen, Universitetsparken 13, DK-2100 Copenhagen OE, Denmark

Christian A. Clement (181), Department of Biology, Section of Cell and Developmental Biology, University of Copenhagen, DK-2100 Copenhagen OE, Denmark

Dusanka Deretic (241), Department of Surgery, Division of Ophthalmology, and Department Cell Biology and Physiology, University of New Mexico, Albuquerque, New Mexico 87131

Eve Donnelly (117), Mineralized Tissues Laboratory, Hospital for Special Surgery, New York, New York 10021

Iain A. Drummond (317), Department of Medicine and Genetics, Harvard Medical School and Nephrology Division, Massachusetts General Hospital, Charlestown, Massachusetts 02129

Cornelia E. Farnum (117), Department of Biomedical Sciences, Cornell University, Ithaca, New York 14865

Gregory G. Germino (273), Department of Medicine, Johns Hopkins University School of Medicine, Baltimore, Maryland 21205 and National Institute of Diabetes and Digestive and Kidney Diseases, National Institutes of Health, Bethesda, Maryland 20892

Sarah C. Goetz (199), Developmental Biology Program, Sloan-Kettering Institute, 1275 York Avenue, New York, New York 10065

Erica A. Golemis (137), Program in Molecular and Translational Medicine, Fox Chase Cancer Center, Philadelphia, Pennsylvania 19111

Bing Huang (103), Miles and Shirley Fiterman Center for Digestive Diseases, Division of Gastroenterology and Hepatology, Mayo Clinic College of Medicine, Rochester, Minnesota 55905

Robert A. Kesterson (163), Department of Genetics, University of Alabama at Birmingham Medical School, Birmingham, Alabama 35294

Lars A. Larsen (181), Wilhelm Johannsen Centre for Functional Genome Research, Department of Cellular and Molecular Medicine, University of Copenhagen, Copenhagen, Denmark

Nicholas LaRusso (103), Miles and Shirley Fiterman Center for Digestive Diseases, Division of Gastroenterology and Hepatology, Mayo Clinic College of Medicine, Rochester, Minnesota 55905

Guihua Li (333), School of Life Sciences, Tsinghua University, Beijing 100084, China

Tatyana Masyuk (103), Miles and Shirley Fiterman Center for Digestive Diseases, Division of Gastroenterology and Hepatology, Mayo Clinic College of Medicine, Rochester, Minnesota 55905

Jana Mazelova (241), Department of Surgery, Division of Ophthalmology, University of New Mexico, Albuquerque, New Mexico 87131

Luis F. Menezes (273), Department of Medicine, Johns Hopkins University School of Medicine, Baltimore, Maryland 21205

Kimberly A.P. Mitchell (87), Department of Biology and Chemistry, Liberty University, Lynchburg, Virginia 24502

Maxence V. Nachury (299), Department of Molecular and Cellular Physiology, Stanford University, Stanford, California 94305

Polloneal J.R. Ocbina (199), Developmental Biology Program, Sloan-Kettering Institute, 1275 York Avenue, New York, New York 10065 and Neuroscience Program, Weill Graduate School of Medical Sciences, Cornell University, New York, New York 10065

Angela de S. Otero (87), Department of Molecular Physiology and Biological Physics, University of Virginia School of Medicine, Charlottesville, Virginia 22903

Junmin Pan (333), School of Life Sciences, Tsinghua University, Beijing 100084, China

Raymond C. Pasek (163), Department of Cell Biology, University of Alabama at Birmingham Medical School, Birmingham, Alabama 35294

Narendra H. Pathak (317), Department of Medicine and Genetics, Harvard Medical School and Nephrology Division, Massachusetts General Hospital, Charlestown, Massachusetts 02129

Lotte B. Pedersen (67), Department of Biology, Section of Cell and Developmental Biology, University of Copenhagen, Universitetsparken 13, DK-2100 Copenhagen OE, Denmark

Olga V. Plotnikova (137), Program in Molecular and Translational Medicine, Fox Chase Cancer Center, Philadelphia, Pennsylvania 19111 and Department of Molecular Biology and Medical Biotechnology, Russian State Medical University, Moscow, Russia

Elena N. Pugacheva (137), Mary Babb Randolph Cancer Center, West Virginia University, Morgantown, West Virginia 26506

Vanathy Rajendran (223), Department of Cellular and Molecular Physiology, Yale University, New Haven, Connecticut 06520

Peter Satir (53), Department of Anatomy and Structural Biology, Albert Einstein College of Medicine, Bronx, New York 10461

Tina Sedmak (259), Department of Cell and Matrix Biology, Institute of Zoology, Johannes Gutenberg University, Mainz D-55099, Germany

E. Scott Seeley (299), Department of Molecular and Cellular Physiology, Stanford University, Stanford, California 94305 and Department of Pathology, Stanford University, Stanford, California 94305

Elisabeth Sehn (259), Department of Cell and Matrix Biology, Institute of Zoology, Johannes Gutenberg University, Mainz D-55099, Germany

Roger D. Sloboda (347), Biological Sciences, Dartmouth College, Hanover, New Hampshire 03755 and The Marine Biological Laboratory, Woods Hole, Massachusetts 02543

Gabor Szabo (87), Department of Molecular Physiology and Biological Physics, University of Virginia School of Medicine, Charlottesville, Virginia 22903

Rikke I. Thorsteinsson (67), Department of Biology, Section of Cell and Developmental Biology, University of Copenhagen, Universitetsparken 13, DK-2100 Copenhagen OE, Denmark

Rebecca M. Williams (117), Department of Biomedical Engineering, Cornell University, Ithaca, New York 14853

Uwe Wolfrum (259) Department of Cell and Matrix Biology, Institute of Zoology, Johannes Gutenberg University, Mainz D-55099, Germany

Bradley K. Yoder (163) Department of Cell Biology, University of Alabama at Birmingham Medical School, Birmingham, Alabama 35294

PREFACE

As an undergraduate at SUNY Albany, I took a course in cell biology with Bob Allen as my professor. I recall, vaguely, at the time Bob impressing upon us the importance of live cell imaging by the then relatively new technique of differential interference contrast (DIC) microscopy. Being the late 1960s, however, I had other things to occupy my mind. A few years later in graduate school at Rensselaer Polytechnic Institute, one of my first courses was Graduate Cytology, taught by Roland Walker. During our first lab session, in which we observed various cells, Prof. Walker chastised me for getting up to leave (at about 6 PM) after only five hours of viewing the cells with the light microscope. He asked me if I had seen everything, to which I answered yes. Then he asked again, did you see *everything*?

It was not until I was an assistant professor at Dartmouth (with Bob Allen now my department chair) that I began to understand Prof. Walker's point of some 10 years before. Bob and I would give graduate students, on whose oral exam committees we sat, his copy of E. B. Wilson's *The Cell in Development and Heredity*. We would ask the students to read the text in the area related to their thesis research and then be prepared to comment during the oral exam on how their area of cell biology had progressed over the past 50 years. This was my first encounter with Wilson's book, and I recall being stunned by the drawings depicting mitosis and cell division, which were about as detailed as the photographic images of isolated mitotic apparatuses my students and I were obtaining with our new DIC microscope; such optics were of course unavailable to Wilson and his contemporaries. For amazing examples of the powers of observation possessed by these early workers, be sure to read Bob Bloodgood's extremely thorough and scholarly review of research on the primary cilium in Chapter 1 in this volume. As with the observations on mitosis cited in Wilson's book, the initial observations of the primary cilium and the speculations about its functions were a century or more ahead of their time.

Widespread interest in the primary cilium, however, dates to relatively recently, in fact, only to about the beginning of this century. The foundation of this interest was laid perhaps 16 years ago when Joel Rosenbaum and his student, Keith Kozminski, first observed intraflagellar transport (IFT) in *Chlamydomonas* using Paul Forscher's video-enhanced DIC microscope. By studying IFT in this model organism, connections between it and the primary cilium of mammalian cells were soon made. The pages of this volume, generously contributed by my colleagues, report contemporary observations of, and the methods used to study, the primary cilium. Such methods include not only microscopic techniques but also those from the fields of biochemistry, genetics, and molecular biology. All of these approaches are of course essential components of the modern cell biologist's tool box, and combined they allow investigators to "see" the primary cilium as never before. Professors Allen and Walker would be thrilled to read this volume.

Roger D. Sloboda
Hanover, New Hampshire
and
Woods Hole, Massachusetts

SECTION I

Background

CHAPTER 1

From Central to Rudimentary to Primary: The History of an Underappreciated Organelle Whose Time Has Come. The Primary Cilium

Robert A. Bloodgood

Department of Cell Biology, University of Virginia School of Medicine, Charlottesville, Virginia 22908-0732

Abstract

For the first time, the history of the central flagellum/primary cilium has been explored systematically and in depth. It is a long and informative story about the course of scientific discovery, memory loss and rediscovery. The progress of our story is saltatory, pushed onward by innovations in technology and retarded by socio-scientific issues of linguistic and temporal chauvinism. Over one hundred and fifty years passed between the discovery of this organelle and full appreciation of its important functions. The main character in our

978-0-12-375024-2
DOI: 10.1016/S0091-679X(08)94001-2

story is an organelle that was relegated to a very minor role in the cellular opera for a very long time, until its rather sudden promotion to a central role in orchestrating many of the sensory and signaling events of the cell. Although early investigators speculated on just such a role for the primary cilium as early as 1898, it was over one hundred years before proof for this hypothesis was forthcoming.

I. Introduction

Today, we know that eukaryotic cilia and flagella come in various flavors; one way to classify cilia and flagella is by whether they have a 9 + 2 or a 9 + 0 arrangement of axonemal microtubules; another way is to classify cilia and flagella by the presence or absence of motility. While, generally, 9 + 2 structures are motile and 9 + 0 structures are not motile, this is hardly a hard and fast rule. For instance, nonmotile 9 + 2 cilia can be found in some sensory receptors, such as the mammalian olfactory epithelium (McEwen *et al.*, 2008) and the hair cells of the vestibular apparatus in some vertebrates (Flock and Duvall, 1965). When a cell has many cilia, they are generally of the 9 + 2 variety and are generally motile; when a cell has a single cilium, it can be of the 9 + 2 variety and motile (in the case of spermatozoa and some fungal zoospores and certain protistan cells). However, in vertebrates, most cases of a cell with a single cilium involve the 9 + 0 immotile variety (often called a monocilium in this context) although there are exceptions (spermatozoa, nodal cilia, kinocilia on hair cells of the inner ear). Generally, 9 + 0 cilia and nonmotile 9 + 2 cilia (such as are found in the olfactory epithelium and some of the hair cells of the inner ear) lack dynein arms. However, there are examples of 9 + 0 cilia that have dynein arms and exhibit motility, albeit unusual motility (insect chordotonal organ cilia; male gametes of centric diatoms; eel spermatozoa) (Boo and Richards, 1975; Corbiere-Tichane, 1971; Gibbons *et al.*, 1983; Manton and von Stosch, 1966). The mammalian embryonic node represents an interesting, still controversial and perhaps unique situation; it may involve a mixture of motile and nonmotile monocilia (McGrath *et al.*, 2003) and it may have a mixture of 9 + 2 and 9 + 0 monocilia (Caspary *et al.*, 2007), although it has not been possible to correlate, at the individual cilium level, the presence or absence of central pair with the presence or absence of motility in this unique structure, essential for establishment of left–right asymmetry in the vertebrate body plan (Basu and Brueckner, 2008). Cilia and flagella can also differ greatly in the organization of the membrane (and hence their overall morphology) (Silverman and Laroux, 2009); the unusually shaped cilia tend to perform specialized sensory roles (such as the connecting cilium associated with photoreceptor cells). Note that it is only recently becoming clear that sensory functions are found within all of the categories of cilia and flagella described above (i.e., is a property of all cilia and flagella). In this chapter, I will be dealing with the history of a special subset of cilia (today called primary cilia) that have a typical ciliary/flagellar morphology, a 9 + 0 axonemal arrangement, no dynein arms and are widely distributed among the tissues of vertebrates. Although they are sometimes referred to as primary cilia, I will not be focusing much attention on the morphologically specialized versions of sensory cilia, such as the connecting cilium of the photoreceptor (Silverman and Leroux, 2009).

Cilia and flagella have a long and rich research history. They were probably first observed by Antoni van Leeuwenhoek in 1674–1675; in a letter to the Royal Society of London (Leeuwenhoek, 1677), he described what were most likely ciliate protozoa "provided with divers incredibly thin little feet, or little legs, which were moved very nimbly" (as quoted by Dobell, 1932). The term cilium (meaning hair or eye lash) was probably first used by Otto Friedrich Muller in 1786 (Muller, 1786). The term flagellum (meaning whip) was introduced much later and is generally attributed to Dujardin (1841). Certainly, by the mid-19th century, the use of the two terms had become rather codified, flagellum being used when a cell had a single (or very small number of) organelle(s) and cilium being used when a cell had many similar organelles. Note, for the sake of our discussion on the history of primary cilia that the term "ciliated cell" or "ciliated epithelium" (wimperepithel; flimmerepithel) refers to cells that have many, motile cilia on their apical surface, whereas "flagellated cell" (geisselzelle) or "flagellated epithelium" (geisselepithel) refers to a situation where each cell has a single organelle on its apical free surface (what we now know as a primary cilium). As light microscope technology improved, so did the attention paid to cilia and flagella, although the focus during most of the 19th century was almost totally on motility, the only postulated function for cilia and flagella and, indeed, a defining feature for identifying them. This is testified to by the huge number of papers in the 19th century German scientific journals whose title included the term "wimperbewegung" or later "flimmerbewegung" (ciliary motility). The first comprehensive reviews of cilia and flagella were published in 1835 (Purkinje and Valentin, 1835; Sharpey, 1835). Indeed, it was Purkinje and Valentin (1834) who were the first to describe ciliary motility in mammals.

The history of the primary cilium cannot be divorced from that of the centrosome/centriole/basal body. The centrosome was discovered and described by Flemming (1875) and independently by Van Beneden (1876). This initiated a 30-year period of intense interest in the centrosome (also called the central body, centrosphere, or attraction sphere in the early literature), the pair of centrioles found within the centrosome, spindle poles, basal bodies, and their various relationships. Theodore Boveri confirmed van Beneden's theory that the centrosome was crucial to cell division and it was Boveri who introduced the term "centrosome" in 1888 (Boveri, 1888) and "centriole" in 1895 (Boveri, 1895). Because of its important role in cell division, Boveri regarded the centrosome as the "dynamic center of the cell" and Flemming (1891) later went on to claim that the discovery of the centrosome marked as important an era in the history of biological science as did the discovery of the nucleus itself. Wheatley (1982) and Gall (2004) provide more detailed history of the centrosome and centriole. The basal body was named by Engelmann (1880), one of the first scientists to recognize and study this organelle. It was in the context of these intense studies on centrosomes, centrioles, and basal bodies and the possible relationships among them that the primary cilium was discovered. In return, the primary cilium would go on to contribute to the resolution of a major controversy in regards to the relationship among centrosomes, centrioles, and basal bodies. Keep in mind that many of the early cell biologists used the terms centrosome and centriole interchangeably, creating some real confusion, especially when they referred to two centrosomes. The term diplosome came into use as a way to refer to the pair of centrioles found at the center of a centrosome, at the center of a spindle pole or associated with the base of a primary cilium.

One of the key observations made very early on was that central flagella were defined by and distinguished from cilia on (multi)ciliated cells by their association with a pair of centrioles (a diplosome) (Zimmermann, 1898). It was also shown early on (Joseph, 1903), often using nonmammalian examples, that an epithelium could have a mixture of central flagellum cells (centralgeisselzellen) and (multi)ciliated cells (flimmerzellen). Lack of motility had become an accepted feature of primary cilia by the time of the Heidenhain (1907) textbook. Since it was understood early on (Zimmermann, 1898) that the primary cilium was associated with the distal centriole of the diplosome (belonging to the centrosome), it was implicit that primary cilia would need to be eliminated at the time of cell division in order to liberate the centrioles to function in mitosis (Section IX). Walter (1929) may have been the first investigator to explicitly report that the central flagella disappeared during cell division, although the figures in Meves (1899) clearly show that in the kidney epithelium.

II. Technical Issues in the Discovery and Study of Primary Cilia

It should be noted that phase contrast microscopy was not commercially available until the early 1940s (Zernicke, 1955); even if it had been available much earlier, it would have been of very limited value in identifying a structure like the primary cilium, which is found on the apical surface of epithelia of hollow organs and glands, on neurons, and in a host of other difficult-to-image sites. Hence the primary technical approach available to most students of animal structure in the latter half of the 19th century was to fix, embed, section, stain, and observe stained sections in the bright field microscope. In 1889, the Carl Zeiss Jena company, due to the efforts of Ernst Abbe (a physicist) and Otto Schott (a glass chemist) brought out a dramatically improved bright field light microscope that addressed the problems of spherical and chromatic aberration as well as introducing the Abbe condenser. The critical technical development that made possible the discovery of primary cilia in vertebrates, as well as many of the early findings on centrioles and basal bodies, was the development of the iron–hematoxylin stain. Hematoxylin, derived from the logwood tree, had been around a long time as a textile stain (Titford, 2005). In order to function as a useful histological stain, hematoxylin must be oxidized and used in combination with a heavy metal mordant. Although earlier workers used aluminum, Benda (1886) introduced the use of a ferric ammonium sulfate mordant and this iron–hematoxylin stain was further refined by Heidenhain (1892), whose formulation rapidly became the stain of choice (Clark and Kasten, 1983). Even with this technical advance, it was not easy to observe primary cilia, as pointed out by Joseph (1903): "Anyone who has been involved in studying the central flagellum will agree with Zimmermann's dictum, that 'this structure is among the most optically difficult' to study. In my opinion, it is largely a matter of luck to be able to see this fine little thread, which due to its great fineness also stains quite faintly." In referring to the Heidenhain (1892) iron–hematoxylin method, Joseph (1903) goes on to say that he "saw things that I would never have been in a position to demonstrate [otherwise]."

III. Origins: 1844–1910

The early history of the primary cilium is mostly a story about epithelia (including glandular epithelia). Although Zimmermann (1898) is always the researcher who is cited for the discovery of primary cilia (central flagella), it is clear that other workers had observed primary cilia well before Zimmermann. Ecker (1844) described the epithelium of the semicircular canals of the ear of the sea lamprey (*Petromyzon marinus*), what would come to be known as "Ecker's epithelium" (Langerhans, 1876), and noted that each cell possessed one cilium. Kolliker (1854), in his *Mikroskopische Anatomie* textbook, when describing the arrangement of cilia on epithelial cells, pointed out that they can occur "even, as it is said, singly on a cell."

Kowalevsky (1867) reported that, in the gastrula stage of *Amphioxus*, every cell of the external seminal leaf is covered with numerous cilia and these only disappear at a later point, to be replaced by one flagellum on each cell; he shows several figures that clearly depict primary cilia (Fig. 1). Kowalevsky (also called Kovaleskii) continued these observations, with many additional figures of primary cilia in a later publication (Kowalevsky, 1877). Another striking example, that clearly reports on primary cilia, is the paper by Langerhans (1876), who, 22 years before Zimmermann (1898), described what are clearly primary cilia on a wide variety of epithelia in *Amphioxus*, including those on the epithelium lining the oral cavity. Numerous clear drawings of epithelial cells with primary cilia accompany his paper (Fig. 2); he calls these "flagellum cells" and points out that "one never detects any movement on their part." Paul Langerhans (1847–1888) (Fig. 3) was a scientific Wunderkind, who became famous for two eponymous discoveries made before he had finished his medical training: the

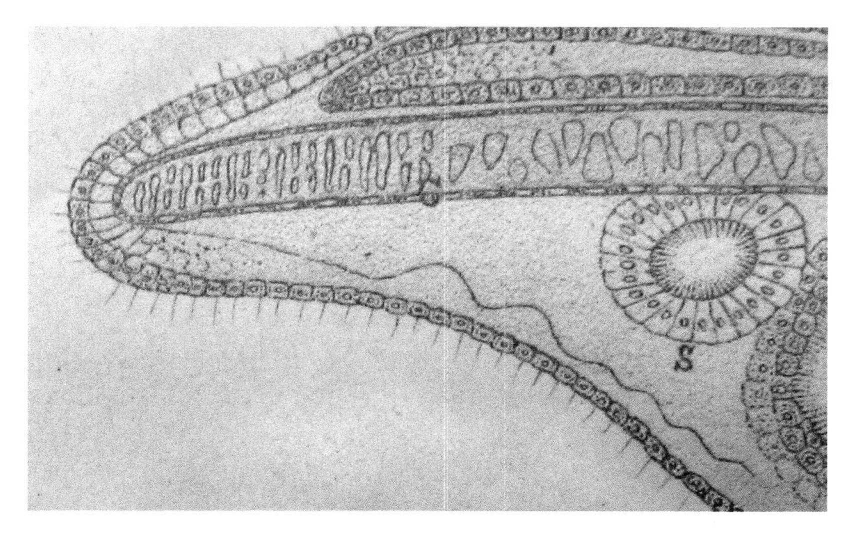

Fig. 1 Portion of Fig. 24 from Kowalevsky (1867) showing a portion of a 30-h-old *Amphioxus* embryo. The epithelium of the surface ectoderm consists of simple cuboidal cells, each of which possesses a single flagellum.

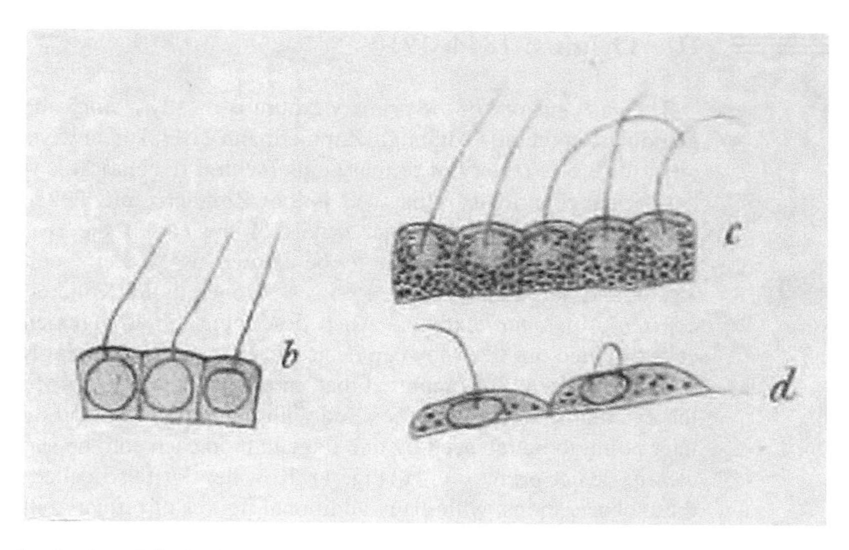

Fig. 2 Portion of Fig. 30 from Langerhans (1876) showing peritoneal epithelium from the liver (b), the hindgut (c), and the ovary (d) of adult *Amphioxus lanceolatus*. Note the single central flagellum/primary cilium on each cell of each of these epithelia. Reproduced with permission of Springer.

Fig. 3 Photograph of Paul Langerhans taken in 1878 in Funchal, Portugal. Taken from "Die Inseln des Paul Langerhans" by Bjoern M. Hausen.

Langerhans cells of the skin (dendritic cells of the immune system) and the Islets of Langerhans, comprising the endocrine tissue of the pancreas. His detailed study of the anatomy and histology of *Amphioxus* was written at the age of 29, 2 years after he had contracted tuberculosis. Joseph (1903) and Erhard (1911) appear to be the only papers that have ever mentioned Langerhans' descriptions of primary cilia, although Joseph (1903) did not provide a citation for the work. The fact that this work remained overlooked by Zimmermann and all subsequent cilia and flagella workers (except the two mentioned above) remains a mystery, given that Langerhans published his work in a mainstream scientific journal of his day and the paper has been cited in a number of papers and books on *Amphioxus* and comparative anatomy, though always in other contexts. It is pretty remarkable the wide range of *Amphioxus* tissues in which Kowalevsky and Langerhans saw primary cilia, although they did not provide much insight or speculation about these structures.

While it is now known that Zimmermann (1898) was not the first to observe primary cilia (as is universally claimed in the primary cilia literature), it was Karl Wilhelm Zimmermann (1861–1935) (Fig. 4) who, while working at the Institute of Anatomy of the University of Berne (Switzerland), first recognized primary cilia in mammals (including humans) using Heidenhain's iron–hematoxylin stain, identified them as a special class of cilia/flagella distinct from the motile cilia previously observed on protists and multiciliated epithelial cells (flimmerzellen), gave this class of cilia/flagella a new and unique name, Central Flagella (Centralgeissel or Zentralgeissel), and, most importantly, correctly predicted their sensory function. Dr. Zimmermann, who spent his entire scientific career (1894–1933) at the Institute of Anatomy in Berne, was a very accomplished scientist whose career involved a series of significant discoveries related to the kidney, especially in regard to mesangial cells and the juxtaglomerular apparatus (Hintzsche, 1936; Reubi: http://www.soc-nephrologie.org/enephro/publications/hier/4_1.htm). The first description of central flagella by Zimmermann (on epithelia from human ureters and rabbit kidney), and the first use of the term central flagellum (Centralgeissel) are actually found in Zimmermann (1894), a very brief report (with no illustrations) from the Anatomical Meeting held in Strasbourg, May 13–16, 1894 (what we would call a meeting abstract today but was called a "Demonstration" then). In a huge paper on mammalian epithelia and glands, Zimmermann (1898) described central flagella on rabbit kidney tubule epithelial cells, human pancreatic duct epithelial cells, human seminal vesicle epithelial cells, uterine fundus epithelial cells, and thyroid gland epithelial cells (Fig. 5 and Table I). Zimmermann particularly pointed out, in the case of the rabbit kidney epithelial cells, that every cell had a single central flagellum. He noted that the central flagellum is always associated with a pair of centrioles (the diplosome) and that the central cilium was always associated with the distal centriole, the one furthest from the nucleus and closest to the plasma membrane (what we now know to be the older centriole). Together with a filament (of somewhat questionable reality) that extended from the diplosome into the cytoplasm (called the "innenfaden"), these structures constituted what Zimmermann called the Central Flagellum Apparatus (Centralgeisselapparat).

Since Zimmermann was working from fixed and stained preparations, he was not able to determine whether his central flagella were motile organelles. While Zimmermann

Fig. 4 Photograph of Karl Wilhelm Zimmermann in his laboratory at the Anatomical Institute of the University of Berne, Switzerland. Courtesy of Professor Urs Boschung, MD, director of the Institute for the History of Medicine of the University of Berne, Switzerland.

emphasized that Meves (1897) and others had, prior to him, discovered “axial filaments” (flagella) on developing spermatids that were identical in appearance to the central flagella he was reporting on tubular and glandular epithelial cells (and clearly associated with the diplosome of the cell), it is not appropriate to give credit to these reports for the identification of primary cilia since the structures Meves studied in his 1897 paper were destined to become motile sperm flagella (and were later shown to be 9 + 2 structures). Zimmermann states, as one of the main conclusions from his paper, that “the microcenter is the motoric center—the ‘Kinetic Center’— of the cell (in contrast to the nucleus as the ‘Chemical Center’).” He was probably referring both to the role of the centrosome in mitosis (Boveri’s “dynamic center of the cell”) as well as a possible role of the centrosome in bringing about central flagella motility, as he did not know whether this structure he had named was motile or not. A number of early students of ciliary motility thought that the basal body was responsible for the motility of cilia (reviewed in Gray, 1928).

While Zimmermann (1898) was a remarkably accurate observer in many respects, not all of his observations on central flagella and the central flagellum apparatus would prove to be entirely accurate. In particular, Zimmermann (1898) reported a knob-shaped thickening or swelling at the ends of central flagella, something that would be quickly discredited by Joseph (1903) and later by Alverdes (1927). Joseph (1903) astutely

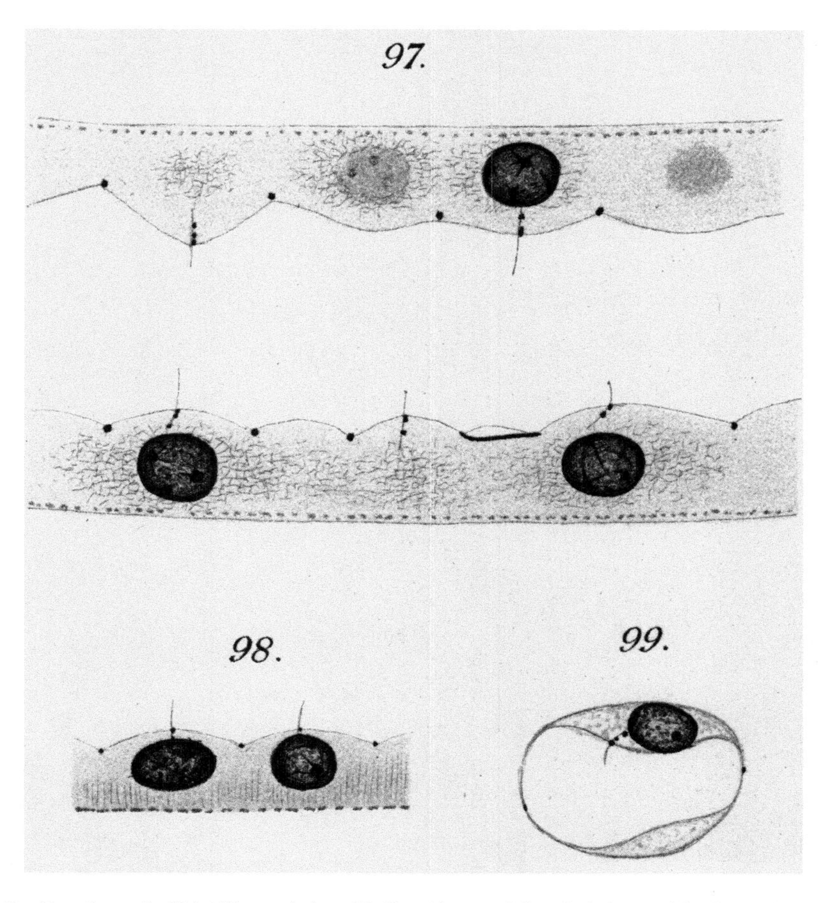

Fig. 5 Drawings of rabbit kidney tubule epithelia with central flagella/primary cilia, from Zimmermann (1898). The darkly staining dots that are not associated with central flagella/primary cilia represent junctional complexes between neighboring cells. Reproduced with permission of Springer.

attributes this to an optical artifact due "to the twisting of the peripheral end of the filament, in that the optical cross-section of the filament may be superimposed here." A more persistent observation, but one of questionable validity, concerns the large "inner filament" (Innenfaden) described by Zimmermann (1898) as an integral part of the central flagellum apparatus ("centralgeisselapparat"). This structure, shown in many drawings to be as thick as and often almost as long as the primary cilium itself (Joseph, 1903, 1905; Zimmermann, 1898) (see Figs. 5, 7, 8, 9) will go on, in later reports, to be greatly altered in size and appearance (Alverdes, 1927; Meves, 1899) (see Figs. 6 and 10) or to be absent altogether (Pollister, 1933; Walter, 1926). Alverdes (1927), while using the same iron–hematoxylin stain used by the early workers, reported that

Table I
Early Locations Where Vertebrate Primary Cilia were Discovered

	Zimmermann (1898)	Meves (1899)	Gurwitsch (1901)	Benda (1901)	Joseph (1903)	Cowdry (1921)	Walter (1926)	Alverdes (1927)	Walter (1929)	Pollister (1933)
Kidney tubules	X	X			X		X			X
Bowman's capsule					X					
Ureter					X					
Pancreatic duct	X				X					X
Thyroid gland	X					X				
Liver bile ducts					X		X			
Otic vesicle										X
Inner ear cochlea					X					
Iris							X			
Uterine fundus	X									
Ovarian cyst									X	
Oviduct			X							
Choroid plexus			X							
Ependyma				X						
Seminal vesicles	X									
Rete testis								X		
Spermatic ducts					X					
Epididymis							X			
Mesothelium of serosa							X			

an inner filament could not be found. Yet he reported on a similar structure, depicted often as two or three branches associated with each diplosome and each branch containing multiple fine filaments (Fig. 10), which he called the "inner fascicle" after a term used by Joseph (1905). Later workers would go on to refer to a flagellar rootlet (wurzeln) and, indeed, perhaps the innenfaden really represents an exaggerated version of what we know today as ciliary rootlet fibers, due, in part, to the overstaining properties of iron–hematoxylin, the very property that made it valuable in the earliest studies of centrioles, basal bodies and primary cilia.

By far, the most striking aspect of Zimmermann's (1898) paper, besides the spectacular drawings of epithelial cells with central flagella (Fig. 5), which obviously benefited from his early training as an artist, lies in his speculations on the function of the central flagella: "One could also imagine that this delicate flagellum, whose movement may have no significant impact upon the secretions that are found in the glandular lumen, may work as a kind of sensory organ, that is, changes in the configuration of the secretions flowing inside the glandular lumen might have a stimulating effect upon the flagellum, whereby the secretory function might be qualitatively or quantitatively affected." In this brilliant piece of intuition, Zimmermann not only correctly predicted the sensory role of primary cilia in general, he was over 100 years ahead of his time in proposing that the primary cilia in the kidney tubule were flow sensors. In this, he was 100% correct but it was left to Praetorius and Spring (2001, 2002) to experimentally demonstrate this. Although numerous papers on primary cilia published in the subsequent 110 years cite Zimmermann's precedence in describing them (not quite true as we saw above), as far as I can determine, the only sources that have ever quoted his eloquent prediction of their sensory function are Heidenhain (1907) in his very well-known cell biology monograph Plasma und Zelle (p. 275) and Erhard (1911) in his review of the Henneguy–Lenhossek Hypothesis (see Section IV). However, several other (albeit it surprisingly few) workers (Alverdes, 1927; Pollister, 1933; Walter, 1929; Wilson and McWhorter, 1963) have alluded to Zimmermann (1898) having postulated a sensory role for central flagella. E. B. Wilson, as late as the 3rd edition of his landmark textbook *The Cell in Development and Heredity* (Wilson, 1934) fails to make any mention of central flagella or of the Zimmermann (1898) paper, probably a contributing factor as to why so many subsequent English-speaking authors appear to be totally unaware of the central flagellum (centralgeissel) name that Zimmermann gave these organelles or of his prediction that primary cilia were sensory organelles (and specifically flow detectors). The only portion of Zimmermann's prescient statement quoted above that has not yet been demonstrated is his prediction that the flow-induced sensory reception mediated by the primary cilia in the kidney tubules regulated secretion from these cells. Sottiurai and Malvin (1972), in a transmission electron microscopy (TEM) report of primary cilia on macula densa cells of the kidney speculated that they could regulate renin secretion or sodium absorption. It is worth noting that it was actually Zimmermann (1933) who gave the macula densa cells their name. Despite the fact that Zimmermann (1898) has an extensive and spectacular collection of drawings (sample shown in Fig. 5) of a wide range of cell types (all

Fig. 6 Drawing of a single epithelial cell from a kidney tubule of a salamander larva, showing a central flagellum/primary cilium; Fig. 1 from Meves (1899).

from mammalian epithelia) possessing primary cilia, the only paper, chapter, or book that ever reproduced any figure from his paper was Wheatley (1982). Even Heidenhain (1907) chose to use illustrations from Joseph (1903) and Meves (1899) for the section of his book on central flagella.

The next report of primary cilia was that of Meves (1899), who, while also using the iron–hematoxylin staining technique of Heidenhain (1892), reported the presence of what he called external filaments ("aussenfaden") on "the cells I studied from the second section of the salamander larval kidney, which are equivalent to the convoluted tubules in mammals" (Fig. 6). Although he was clearly seeing Zimmermann's central flagella and cited Zimmermann (1898) for observing these structures in the rabbit kidney, he never used the term "central flagella" in his paper, although he referred to the structure as a flagellum (geissel) in one footnote (p. 59). He emphasized that these structures were "at least as long as the (secretory) cell was tall" and emphasized that they were longer than those reported by Zimmermann (1898). Meves (1899) shows only an extremely reduced version of Zimmermann's "innenfaden" (Fig. 6). Joseph (1903) pointed out that the Meves (1899) work showed "that the central flagellum cells of the kidney reproduce in a karyokinetic manner, a process where the central flagellum diplosome comes into play." What neither Meves (1899) nor Joseph (1903) pointed out (although Joseph inferred it in his statement quoted above) is that Meves' very clear drawings of mitosis in these kidney epithelial cells shows that the central flagellum is absent during mitosis. While these are the first data showing absence (and hence presumably loss) of central flagella during mitosis, this result is predicted by the observations of Zimmermann (1898) that central flagella are associated with the cell's centriole pair (diplosome) (see Section IX).

Gurwitsch (1901) showed clear illustrations of central flagella in both the epithelium of the "Tela choroidia" (choroid plexus) of the salamander larva and the epithelium of the fimbria of the rabbit oviduct. Benda (1901) observed central flagella in epithelia from human cadaver tissues; in particular, he reported primary cilia from ependymal cells from cerebral ventricles.

Joseph (1903), in what was the most detailed and analytical study of central flagella to date, observed the central flagellum apparatus (centralgeisselapparat), meaning the pair of centrioles/basal bodies (diplosome) plus the central flagellum, in the kidneys of *Torpedo* fish, salamander larvae, and rabbits, as well as in pancreatic ducts and liver "excretory ducts" (bile ducts) in salamander larvae, in various locations in the guinea pig inner ear, in ectodermal epithelium of *Amphioxus*, and lizard embryo epithelia and expressed the feeling that central flagella were widespread (Fig. 7). Although he usually used Zimmermann's term central flagellum (centralgeissel), Joseph (1903) also used the term "flagellum-like (geisselartiger) filament." As had Zimmermann (1898), Joseph (1903) pointed out that the central flagellum is associated with the

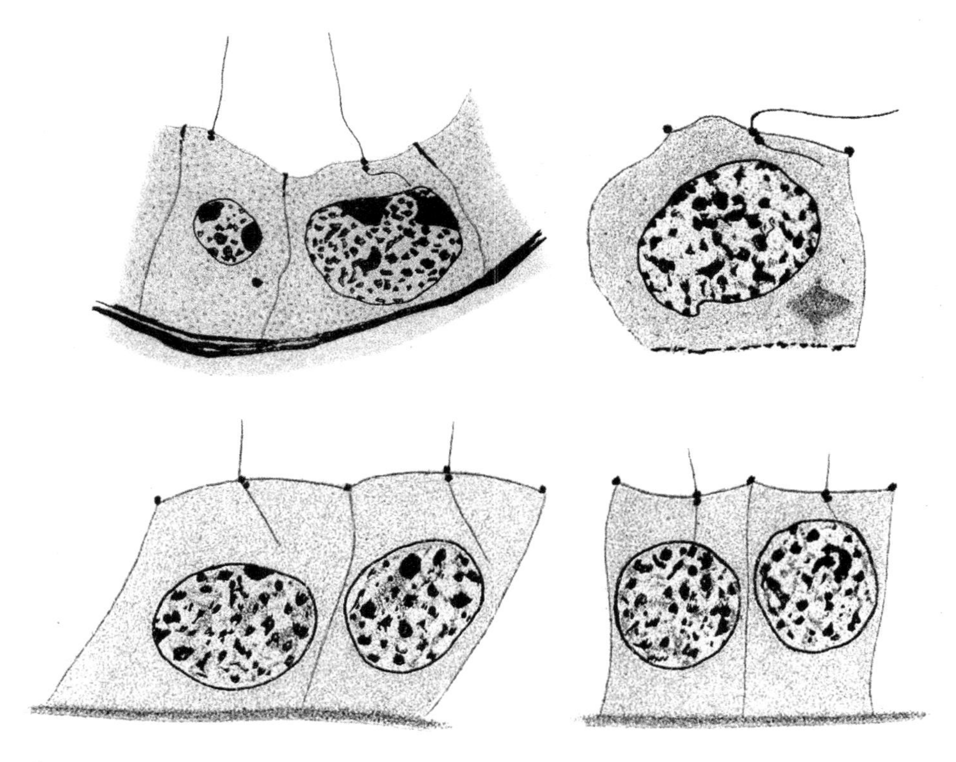

Fig. 7 Drawings from Joseph (1903) showing epithelial cells containing the "central flagella apparatus." The figures in the top row are from the kidney of the *Torpedo* fish while the figures in the bottom row are from the sulcus spiralis internus of the cochlea from the inner ear of the guinea pig embryo.

distal centriole/basal body of the diplosome (which tends to be oriented perpendicular to the cell surface in cells possessing primary cilia) and predicted that all central flagella will be found associated with a diplosome. Joseph (1903) was one of the first to report (in the kidney tubules of the *Torpedo* fish) the presence of both central flagellum cells and multiciliated cells within the same epithelium (Fig. 8) and to hypothesize that only the central flagellum cells are capable of cell division.

Fig. 8 Drawings from Joseph (1903) showing mixed epithelia containing both "central flagella" cells (centralgeisselzellen) and ciliated cells (flimmerzellen). The drawing in the upper left shows epithelial cells from the pancreatic duct of the salamander larva. The drawing in the upper right shows epithelial cells from a kidney tubule from the *Torpedo* fish. The drawing in the bottom of this figure shows epithelial cells from the glomerular (Bowman's) capsule from the kidney of the *Torpedo* fish.

Joseph (1903) hypothesized that cells with primary cilia are in transition between an unciliated state and a (multi)ciliated state and that central flagella cells can divide, unlike ciliated cells. While this idea that the cell with a primary cilium is a temporary transitional state on the way to a true ciliated cell is not generally thought to be the case today, there are exceptions that appear to fit the situation envisioned by Joseph (1903). Banizs *et al.* (2005) report that, during mouse development, the ependymal cells start out only with primary cilia and then later in development become ciliated cells (meaning cells which each possess many motile cilia). A similar situation may exist in the lung (Sorokin, 1968). This theme proposed by Joseph (1903) would reappear as late as 1968, when Sorokin (1968) referred to "the rudimentary, or abortive, cilia that are known to occur in a wide variety of cells" and which "have only a transitory existence" and used this as justification for creating the name "primary cilia" for this organelle.

Another issue raised by the Joseph (1903) paper is whether epithelial cells containing a brush border can contain a central flagellum. While known today to be composed of actin-based microvilli, in his day some observers mistakenly thought that the brush border was composed of cilia. Joseph (1903) reports that he "never saw a central flagellum apparatus in a brush border cell, nor was I able to detect a brush border in a central flagellum cell" which he felt would create an insurmountable challenge to the Henneguy–Lenhossek Hypothesis (Section IV) on the interconvertibility of centrioles and basal bodies "because today only a few people doubt that the brush border cell is totally homologous to a ciliated cell." Actually, Meves (1899) had previously reported central flagella on kidney epithelial cells with a brush border, Gurwitsch (1901) appears to have previously shown epithelial cells with both a central flagellum and a brush border and Walter (1929) would go on to report the same thing. Ironically, in a subsequent publication, Joseph (1905) while attempting a classification of various categories of central flagellum cells (Fig. 9), shows several examples of central flagella cells containing a brush border. For some time to come, the brush border (or striated border) would continue to create confusion [see discussion in Chase (1923) and Newell and Baxter (1936)] because some subsequent observers, like Joseph (1903), thought that it was composed of short cilia and/or a row of basal bodies and this would cloud the picture for a while in terms of both the Henneguy–Lenhossek Hypothesis (see Section IV) and the Exclusion Principle (Wheatley, 1982, 2005). The latter states that any one cell would have either motile cilia or a primary cilium (and never both).

The first monograph/textbook on cell biology to mention central flagella was *Morphologie und Biologie der Zelle* (Gurwitsch, 1904) which mentioned central flagella very briefly at two points in this large book and reproduced two drawings from Joseph (1903) of cells containing central flagella (Fig. 206 in Gurwitsch, 1904).

Kupelwieser (1906) clearly showed (in his Fig. 3) epithelial cells in the larvae of the bryozoan *Cyphonautes* that contained one cilium per cell and suggested that these "Ein-Cilien-Zellen" (one cilium cells) may have a sensory function. Kupelwieser (1906) does not cite Zimmermann (1898) or any of the other previous central flagellum papers.

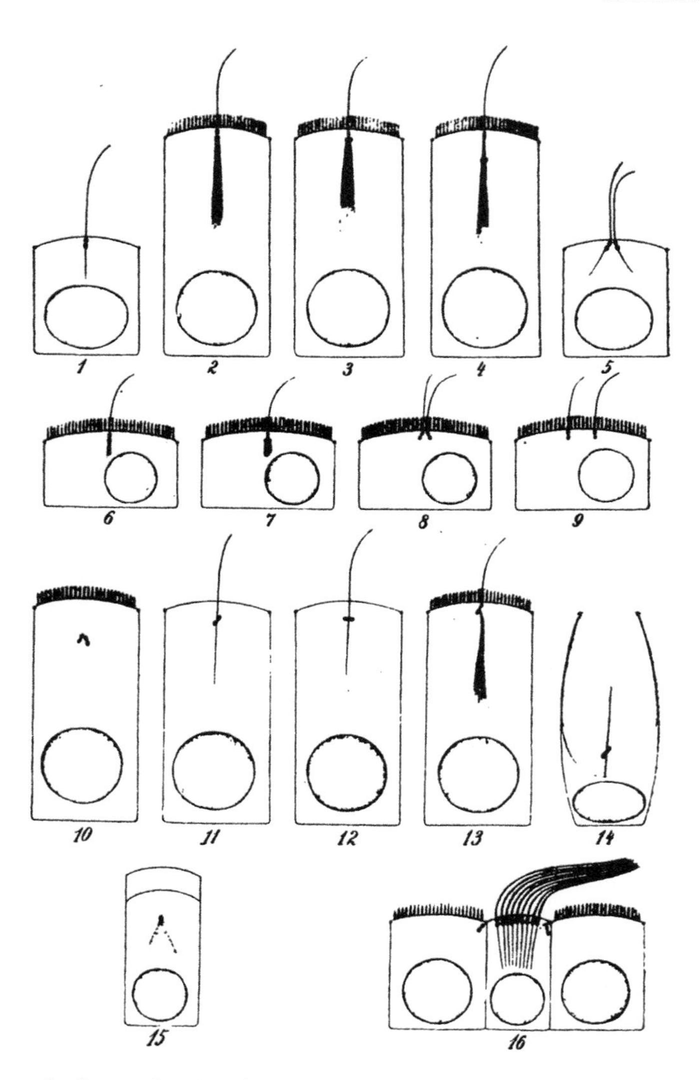

Fig. 9 Composite diagram from Josephs (1905) summarizing the appearance of central flagella/primary cilia cells from a wide range of sources. Note that quite a number of these cell types exhibit brush borders and some are shown to have two central flagella, something only rarely reported in the literature. Reproduced with permission of Elsevier.

It is appropriate to end this Classic Period in the history of primary cilia with the first review devoted to central flagella, a small but separately labeled section in the classic cell biology monograph by Martin Heidenhain (Heidenhain, 1907) entitled *The Central Flagellum Apparatus*. He defined the central flagellum apparatus, as did Zimmermann (1898), to include the central flagellum, the

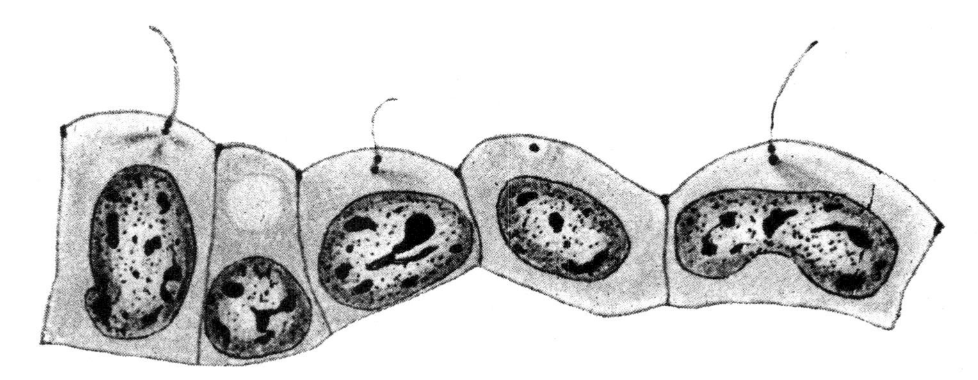

Fig. 10 Drawing of epithelial cells from the human rete testis, some of which exhibit central flagella. Note that the left most cell shows a tripartite "inner fascicle," probably related to the "innenfaden" of Zimmermann (1898); Fig. 1 from Alverdes (1927).

centriole pair (diplosome), and the inner filament (which would become the one controversial element). Heidenhain points out that "its [referring to the central flagellum] preservation and observation belong to the most difficult tasks in microscopy." He states that no one to date had claimed that central flagella were motile and that their solitary presence on each epithelial cell makes it very unlikely that they could "accomplish the forward movement of secretions." Heindenhain feels that this leaves only one "functional explanation" and that is Zimmermann's striking suggestion of a sensory role.

IV. The Henneguy–Lenhossek Hypothesis and the Role of Primary Cilia

1898 was a special year in the history of cell biology. In addition to the Zimmermann (1898) paper on central flagella, two papers (Henneguy, 1898; Lenhossek, 1898b) were published that together crystallized a hypothesis (The Henneguy–Lenhossek Hypothesis) already deeply rooted piecemeal in the literature and one that would generate a great deal of controversy that would last into the 1920s, even though, in retrospect, its validity seems to have been well established from the very beginning. I will devote some space to the discussion of the Henneguy–Lenhossek Hypothesis because of the important role that central flagella (primary flagella) played in supporting it. The basic hypothesis is that centrosomes (meaning the pair of centrioles within what we today call a centrosome) are basically the same organelles as basal bodies. Joseph (1903) includes an interesting diagram (see Fig. 11) illustrating the idea of the interconvertibility of the centrosomes in primary cilia cells for two purposes (assembly and

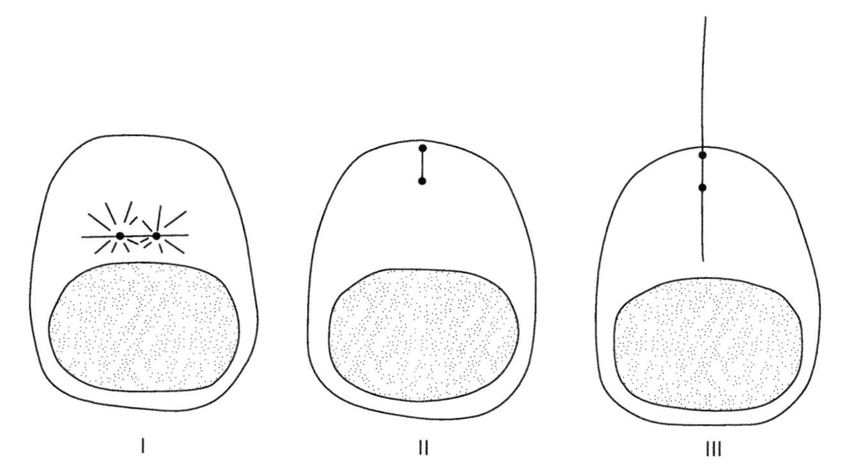

Fig. 11 Diagram from Joseph (1903) in which he illustrates the interconvertability of the centrioles with the basal body of the central flagellum/primary cilium, which is the basis of the Henneguy–Lenhossek Hypothesis.

maintenance of a central flagellum vs mitosis). The corollary expressly implied in this hypothesis is that the basal bodies in ciliated cells are derived from the centrosome through a special process of multiplication that uses up the centrosome *per se*. The consequence/prediction of this is that ciliated cells lack any centrosome besides the basal bodies associated with the cilia in a one-to-one association and, because of this, ciliated cells cannot divide. Note here that, whenever we refer to a ciliated cell, we are referring to a cell with many motile cilia. While most workers took very strong sides one way or the other, Heidenhain (1907) does an excellent job of evaluating the arguments for and against the Henneguy–Lenhossek Hypothesis in a very objective manner and, indeed, his concluding remark is that "we are not going to try to come to a decision in favor of one side or the other."

Aside from some of the weaker arguments for the homology of centrioles and basal bodies [similar size and shape, similar staining with iron–hematoxylin, similar appearance in polarization microscopy, similar position in the apical portion of mixed epithelia (epithelia containing both ciliated cells and central flagellum cells)], there are several major arguments that were put forth in support of the Henneguy–Lenhossek Hypothesis:

1. The major early reports of the central flagellar apparatus (centralgeisselapparat) (Joseph, 1903; Zimmermann, 1898), all utilizing epithelia, clearly showed that the interphase cell's pair of centrioles was intimately associated with the central flagellum. Indeed, it came to be recognized early on that a defining feature of central flagella was the association with a pair of centrioles (a diplosome) and that the cell was capable of future cell division, which required mobilizing this

pair of centrioles to become spindle poles. As Joseph (1903) put it: "What I especially want to emphasize here is the irrefutable fact that centrosomes undoubtedly can function as basal bodies for cilia, as shown in the central flagellum cells, and further, that the central flagellum cells are capable of karyokinetic reproduction" and illustrated this point with a diagram (Fig. 11). Joseph (1903) goes on to point out that the controversy over the Henneguy–Lenhossek Hypothesis greatly increased interest in primary cilia. "It is quite clear that after the emergence of the Henneguy-Lenhossek Hypothesis, the central flagellum in particular has become an object of great interest and has been employed in support of this hypothesis." One interesting prediction of this is that epithelial central flagella cells would have to lose their flagella during cell division in order to release the diplosome so that it could migrate to the proper position to serve as centrioles for spindle poles (see Section IX), although this was not explicitly discussed until Walter (1929). Not surprisingly, most of the early students of central flagella were supporters of the Henneguy–Lenhossek Hypothesis. In general, those who opposed the Henneguy–Lenhossek Hypothesis accepted the claim that the central flagellum was associated with a true centriole pair (or centrosome as they referred to it). Although he was severely skeptical of the Henneguy–Lenhossek Hypothesis, Erhard (1911) states that "It is demonstrated with certainty that [the central flagellum] possesses a centrosome." However, there was one major holdout (Gurwitsch, 1901) who challenged this major pillar of the Henneguy–Lenhossek Theory (the centrosomal nature of the central flagella diplosome). Joseph (1903) responded by saying that he "must confront today the most emphatic of the withering judgments with which Gurwitsch (1901) denies the centrosomal character of epithelial cell diplosomes, and must characterize this approach as utterly arbitrary."

2. During spermiogenesis, the centrosome gives rise to the sperm tail flagellum. This observation was well established (Hermann, 1889; Lenhossek, 1898a; Meves, 1897; Moore, 1895) by the time of the Henneguy–Lenhossek Hypothesis and none of the parties in the subsequent controversy denied the accuracy of this important observation. The only controversy would relate to whether a spermatozoan tail was equivalent to a cilium on an epithelial cell.
3. The most dramatic and irrefutable observation in support of the Henneguy–Lenhossek Hypothesis comes from observations of insect spermatocytes that possess flagella (claimed to be nonmotile) during the course of meiotic divisions [Henneguy, 1898; Meves, 1897, 1903; confirmed by electron microscopy by Friedlander and Wahrman (1970) and LaFountain (1976), neither of whom cited either of these original authors]. In these meiotic cells, a very unusual circumstance occurs in which the centrosomes/centrioles simultaneously serve as both spindle poles (hence as centrioles) and as basal bodies, the very definition of the Henneguy–Lenhossek Hypothesis (Fig. 12)! Although there have been no reports that this can happen in epithelial cells with primary flagella, there have been a number of reports of this in various flagellate protozoa, such as *Trichomonas* (Fig. 13).

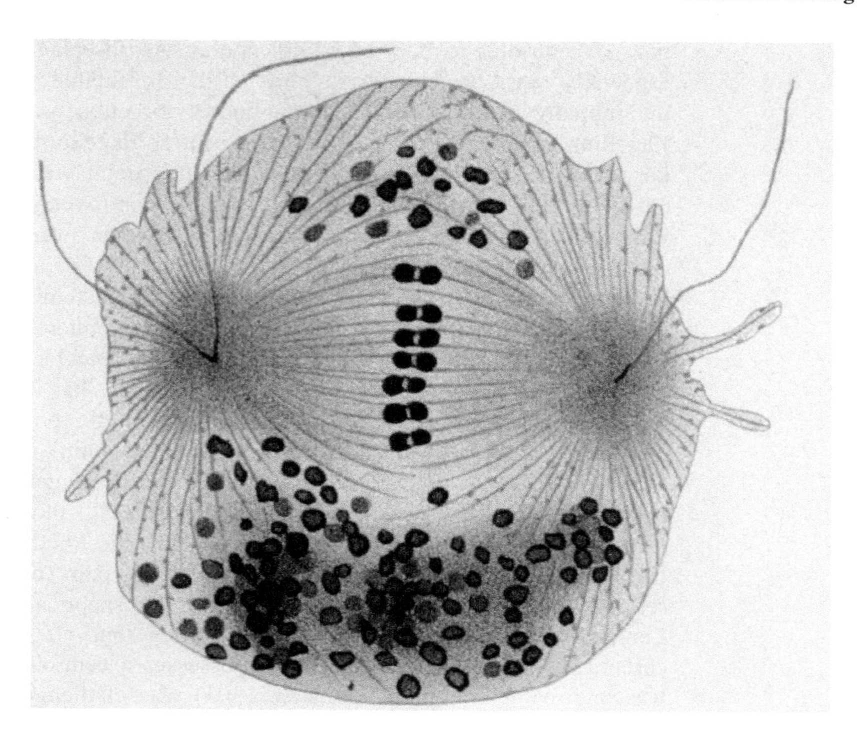

Fig. 12 Drawing (Fig. 55 from Meves, 1900) showing a spermatocyte from the testis of the moth *Pygaera bucephala*. During meiosis, the centrioles are serving simultaneously as spindle poles and basal bodies. Reproduced with permission of Springer.

4. Observations that ciliated epithelial cells lacked any centrosomes or centrioles separate from the basal bodies associated with the cilia led to the conclusion that the centrioles/centrosome must "have all been confiscated to become basal bodies" during the process of ciliation (Lenhossek, 1898b). It is interesting that, during the subsequent long-lived debates over the Henneguy–Lenhossek Hypothesis, it was felt by the opponents of the theory that finding independent centrosomes in ciliated epithelial cells (something that would turn out not to be the case) was a death knell for the Henneguy–Lenhossek Theory, even though Lenhossek (1898b) himself pointed out that "the demonstration of definite centrosomes located somewhere else in the cell would not exclude the possibility that basal bodies originate from centrosomes."

Given the array of supporting evidence, it is curious, in hindsight, that there was such vigorous and long-lived opposition to the Henneguy–Lenhossek Hypothesis (Erhard, 1910, 1911; Gurwitsch, 1902; Maier, 1903; Prowazek, 1901; Studnicka, 1899; Wallengren, 1905; Saguchi, 1917; Sharp, 1921). Benda (1901) points out that K. W. Zimmermann himself had problems with the hypothesis. Two of the most scholarly and objective reviewers of this debate (Heidenhain, 1907; Wilson, 1934)

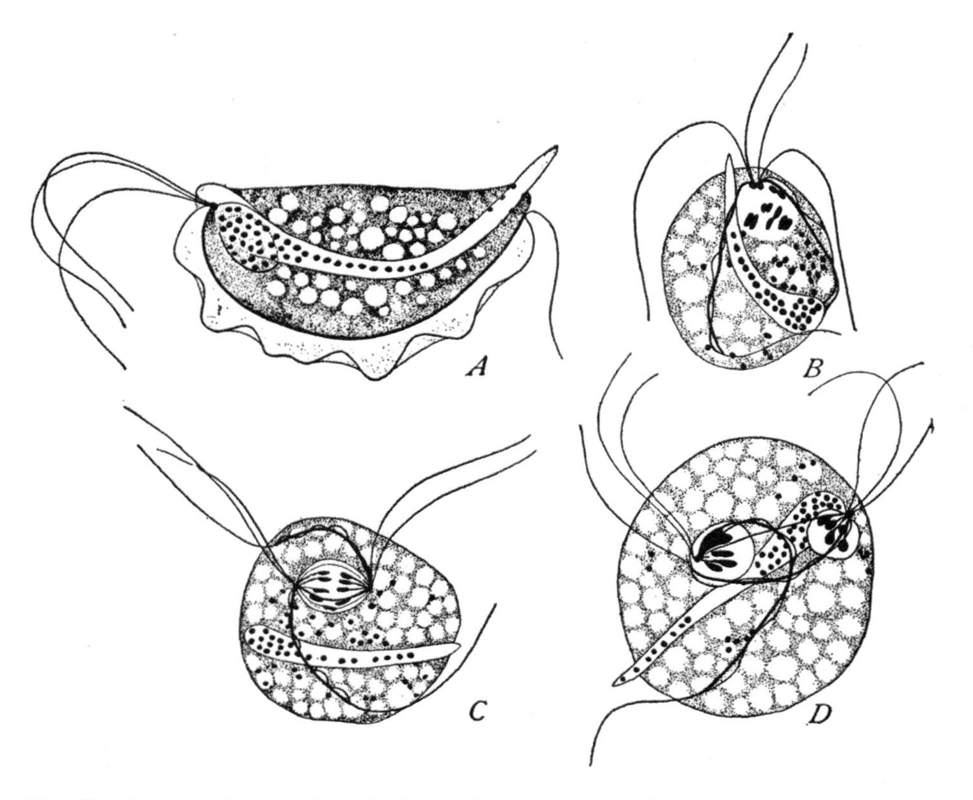

Fig. 13 Diagrams showing mitosis in the flagellate protozoan *Trichomonas augusta*. Adapted from Kofoid and Swezy (1915). A, interphase; B, prophase; C, metaphase; D, telophase. Note that the flagella remain associated with the centrioles during mitosis.

both felt that the jury was still out. The only dedicated review article published on the Henneguy–Lenhossek Hypothesis (Erhard, 1911, citing 146 references) came down against the hypothesis ("We must reject the Henneguy-Lenhossek Hypothesis in its strict formulation as presented by its originators"). Erhard (1911) gives quite a biased account of the literature and its interpretation; indeed, he was already a biased party to the debate, having published a paper the previous year arguing against the hypothesis (Erhard, 1910). It has to be appreciated that there was much confusion at this time (reflected in the Heidenhain, 1907 text book, pp. 286–287) about the identity of the "brush border" found in some kidney epithelial cells, epididymal epithelium, and epithelial cells of the small intestine. Many authors thought that these structures were cilia and hence must have basal bodies and this contributed to some of the mistaken data (such as some of the examples of mitosis in ciliated cells) cited (Erhard, 1911; Heidenhain, 1907) as inconsistent with the Henneguy–Lenhossek Hypothesis.

So what were the major arguments used against the Henneguy–Lenhossek Hypothesis? The major battle lines were drawn over the interrelated questions of whether ciliated epithelial cells had a centrosome independent of the cilium-associated structures (basal bodies) and whether ciliated epithelial cells exhibited cell division. While Henneguy (1898), Lenhossek (1898b), Joseph (1903), and Walter (1929), among many others, were adamant that they had never seen mitosis in ciliated epithelial cells, Erhard (1911), in his extensive review, mustered quite a number of references to authors who claimed to have observed mitosis in ciliated cells; at least some of these were cases of superimposition of dividing and ciliated cells in thick sections and cases of mistaken identity (for instance, mitosis in a brush border cell thought to be a ciliated cell or mitosis of a basal cell in a ciliated pseudostratified columnar epithelium). Indeed, Heidenhain (1907), who found fault with some of the arguments for the Henneguy–Lenhossek Hypothesis (including the idea that ciliated cells lacked independent centrosomes) was quite clear that the literature came down against mitosis in ciliated cells. He was quite emphatic that workers who saw mitotic figures in ciliated epithelia saw the mitoses in nonciliated cells (such as the basal cells in the respiratory epithelium). Even for those workers who believed in the absence of independent centrosomes/centrioles and of mitosis in ciliated epithelial cells, the question remained as to how a single pair of centrioles could give rise to the large numbers of basal bodies necessary for a ciliated cell to have one basal body per cilium. Benda (1901), Ikeda (1906), Luban (1918), Renyi (1924), and Walter (1929) provided support for the Henneguy–Lenhossek Theory by showing what they interpreted to be stages in the development of basal bodies from centrioles/centrosomes in ciliated epithelial cells (meaning epithelial cells with multiple motile cilia). Benda (1901) and Ikeda (1906) described, in the place of a traditional centrosome, the presence of "Centralkorperballen" in the apical cytoplasm of epithelial cells in the process of forming multiple cilia. These large granular aggregates (Fig. 14) contained structures that looked like both centrioles and basal bodies to early light microscopists (description in Walter, 1929). These structures appear (in the static images) to be migrating to the apical surface of the cell and to be morphing into basal bodies, as the cilia appear. Walter (1929) noted that these structures were never seen in central flagella cells. With the advent of TEM, things would get more complicated, in the sense that TEM studies showed evidence for both a centriole-dependent and a centriole-independent pathway for formation of basal bodies in ciliated epithelial cells (Anderson and Brenner, 1971; Dirksen, 1971; Sorokin, 1968; reviewed in Dawe *et al.*, 2007), although both involved the disappearance of the traditional centrosome/diplosome from the ciliated cell. On this one point, both supporters and opponents of the Henneguy–Lenhossek Theory can claim victory on this aspect of the argument (whether centrioles directly give rise to basal bodies). Sorokin (1968) and Dawe *et al.* (2007) both show, using TEM, a structure that is likely to be equivalent to the "centralkorperballen" of Benda (1901), Ikeda (1906) and Walter (1929) (Fig. 14).

There were more esoteric objections raised against the Henneguy–Lenhossek Hypothesis, ones that were more difficult for supporters to deal with (Erhard, 1911; Heidenhain, 1907). These included arguments related to the division of ciliate

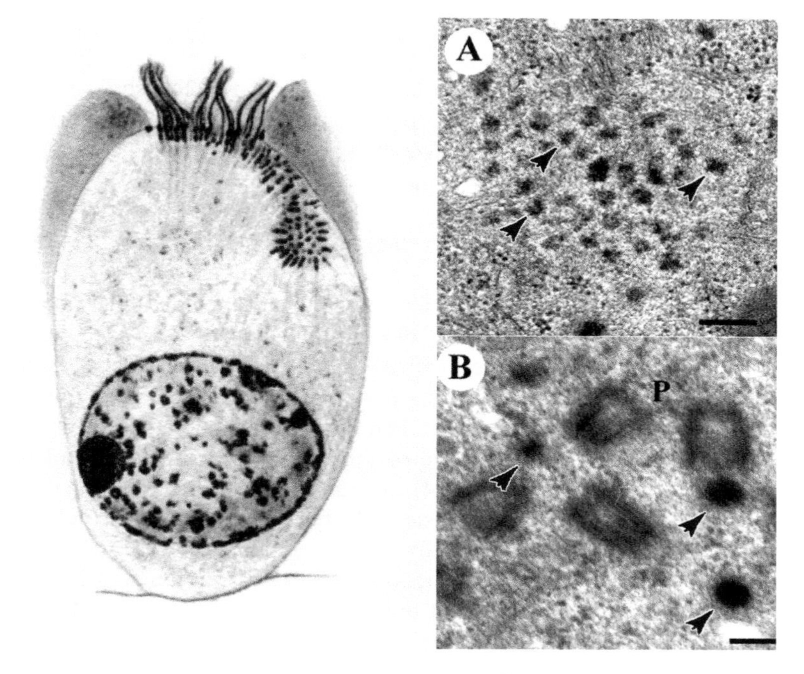

Fig. 14 Image on the left is a drawing (Fig. 6 from Walter, 1929) illustrating the process of ciliation (cilia formation) on the epithelial cells of an ovarian cyst. Note the cluster of dark spots in the apical cytoplasm, allegedly representing the proliferation and migration of future basal bodies ("Centralkorperballen") from the cell's diplosome. The TEM images on the right (from Dawe *et al.*, 2007) show what may be the ultrastructural equivalent of the "Centralkorperballen." Figure A shows the appearance of dense granules which precede the appearance of procentrioles (shown in B) which can go on to become basal bodies. This is one of two possible pathways for ciliogenesis discussed by Dawe *et al.* (2007). Reproduced with permission of The Company of Biologists and John Wiley and Sons.

protozoans, the appearance *de novo* of centrioles/basal bodies in ameboflagellate protozoa, and the absence of centrosomes in plants that gave rise to multiflagellated sperm. While active opposition to the Henneguy–Lenhossek Theory would certainly linger on (for instance, Kindred, 1927) and be confused by bogus issues like amitosis in epithelial cells (Jordan, 1913), the Henneguy–Lenhossek Hypothesis would survive its many early critics, endure, and turn out to be basically correct, if not totally universal. Soon after the advent of thin section transmission electron microscopy, Burgos and Fawcett (1955) and deHarven and Bernhard (1956) would demonstrate the essential structural identity of basal bodies and centrioles and Heidemann and Kirschner (1975) would later show that purified basal bodies can substitute for centrioles in aster formation when injected into *Xenopus* eggs. In an exciting recent study, Dammermann *et al.* (2009) showed that a single protein is sufficient to shift centrioles between their roles in cell division and ciliogenesis in *Caenorhabditis*. While the Henneguy–Lenhossek Hypothesis is generally accepted implicitly today (although few have heard of the theory by name), it is surprising how little we still know about some of the issues surrounding the debate over the hypothesis, such as the migration of centrioles to and from the cell surface in primary cilia cells and the process by which large numbers of basal bodies are produced during the differentiation of ciliated cells (Dawe *et al.*, 2007).

V. The Middle Period (1910–1935)

After a decade of inactivity that coincided with World War I, reports of central flagella picked up and continued at a steady, though very modest pace; some correction or refinement of the early observations on central flagella was already occurring during this period. While most of these authors were aware of at least some of the earlier literature, we are already starting to see the beginnings of a reduction in the use of the central flagellum terminology.

Heidenhain (1914), a usually very reliable scientist, reported on central cilia in gustatory cells in taste buds, but subsequent TEM (deLorenzo, 1958; Trufillo-Cenoz, 1957) studies have shown this not to be the case.

Cowdry (1921) reported in detail on the presence of "flagella" on the apical (luminal) surface of thyroid follicle cells of the dogfish. He reported that every cell had one flagellum, regardless of whether the follicle cells were in an active (cuboidal) configuration or in the inactive (flattened) state. Although he was unable to observe the flagella in living thyroid follicles, Cowdry performed an interesting experiment (probably the first experiment ever performed with primary cilia) in order to ask if they were motile. He introduced dye particles into the colloid filling the thyroid follicle and observed that "the granules are stationery, exhibiting no currents or eddies which one might expect flagellar action to produce." Once he shifts from the reporting of data to discussion and speculation, his paper becomes a study in contradiction. He states that "it is almost inconceivable that such an elaborate mechanism should not serve some useful purpose ... although I have not actually observed movement of the flagella, they have all the distinctive characters of a typical motor apparatus ...," this despite the fact that he reports in this paper an experiment arguing against motility. This reflects how difficult it was for most scientists of the 1930s to conceive of any other function for cilia and flagella than motility, even in the face of clear data. Cowdry was apparently unaware of the earlier papers that postulated a sensory role for central flagella as he cited not a single paper on central flagella in his bibliography (and hence the reason he did not use that term) and was unaware that Zimmermann (1898) had already observed central flagella on thyroid cells 23 years earlier. Cowdry concluded his paper with one further contradiction. After stating that "I do not venture to suggest that these thyroid cells have developed flagella merely because they acquired the habit years ago," he then switches course and states in the summary to his paper that "the development of flagella in the thyroid gland of the dogfish is ... without apparent adaptive value," a sentiment that would continue to be voiced by workers in the field of primary flagella as late as the 1960s and 1970s.

Walter (1926) provided a detailed account of central flagella on a wide variety of mammalian serosal epithelial cells (what we call mesothelium today but what he referred to as endothelium) from mesenteries, pleura, and peritoneum. He noted that every cell had a flagellum (he alternated between using the terms flagellum and central flagellum) associated with the distal centriole of a diplosome. He noted in passing that he also observed central flagella on epithelial cells from all segments of human kidney tubules, the bile duct, the iris, and the appendix of the epididymis. He insisted that central flagella were lacking from corneal epithelium and the endothelium of blood vessels, although both of those sources have subsequently been shown to have primary cilia (Bystrevskaya *et al.*,1988; Gallagher, 1980; Iomini *et al.*, 2004). Walter (1926) wrote that he would "forego expressing an opinion regarding the presumable physiological role of the endothelial flagellum."

Alverdes (1927) reported in detail on central flagella on the epithelium of the human rete testis (Fig. 10) and once more confirmed their association with the diplosome and connection to the distal centriole. While acknowledging the inability to know from sectioned material whether the central flagellum is motile, he expressed a preference for the term central flagellum over Walter's (1926) use of "flagellum" because he felt the latter implied motility. Building on Zimmermann's (1898) postulate of a sensory function, Alverdes (1927) more specifically suggests that "the central flagellum apparatus might in some way represent a cellular receptor, which communicates fluctuations in the physicochemical composition of the lumen contents to the cell and thereby influences its activity," a striking prediction of what would be found, some 75 years later, about the critical roles of primary cilia in signaling mechanisms, such as the hedgehog signaling pathway (Wong and Reiter, 2008) (see Section X).

Walter (1929), in a very important paper in terms of the Henneguy–Lenhossek Hypothesis (see Section IV), reported on the presence of central flagella on epithelial cells in an ovarian cyst. These central flagella cells were mixed with ciliated cells in a simple cuboidal epithelium, similar to the situation reported by Joseph (1903). In the last sentence of his paper, Walter (1929) noted that, during cell division, the central flagellum cell loses its central flagellum and the centrioles move away from the apical cell surface descending into the cell, one of the first actual observations of something predicted by Henneguy (1898), Lenhossek (1898b), Meves (1899), and Joseph (1903) and something that would go on to become an important topic of primary cilia experimentation, but only in the 1960s (see Section VIII).

Pollister (1933), in an important study of centrioles in various amphibian (*Ambystoma*) larval cells, noted the presence of central flagella (though he never used the term) in epithelial cells of the pancreatic duct, the pronephric and mesonephric tubules (future kidney), the segmental duct, and the otic vesicle (future ear). Pollister reports that he is unable to find flagella in the adult *Necturus* kidney. While acknowledging that "Zimmermann suggested that the flagella were actually sensory in function, serving to detect the rate of passage of material along the tubule into which they project," Pollister goes on to conclude that he agrees "with Joseph in his view that these flagella in vertebrate cells are merely vestigial structures reminiscent of a remote ancestral type of animal which had flagellated cells throughout extensive areas of epithelium."

Given this group of papers that we have just discussed, it remains somewhat puzzling why the terminology of central flagella did not penetrate the subsequent (post-World War II) literature. However, a look at some of the major monographs reinforces the lack of awareness of central flagella (or at least the lack of a feeling that these were important organelles). As we have already mentioned, Martin Heidenhain's excellent book *Plasma und Zelle* (Heidenhain, 1907) does have a designated section on "Der Centralgeisselapparat" (the central flagella apparatus) and may well be the last monograph to do so until the Wheatley (1982) volume on the centriole. Oscar Hertwig in his 1906 cell biology textbook *Allgemeine Biologie* (Hertwig, 1906) cites Zimmermann (1898) but makes no mention of central flagella. Sharp's 1921 textbook entitled *An Introduction to Cytology* also makes no mention of central flagella. The first book devoted entirely to cilia and flagella in English, Gray (1928), makes no mention of central flagella and makes only one brief comment about any possible sensory role for cilia or flagella (p. 32, in connection with the protozoan, *Heteromastix*). E. B. Wilson, as late as the 1934 printing of the third edition of *The Cell in Development and Heredity*, makes no mention of central flagella; as the most influential cell biology textbook in English for the first half of the 20th century, this may be a true predictor of the upcoming dark age of primary cilia. The rather comprehensive book by Sleigh (1962) on the biology of cilia and flagella contains no mention of the central flagellum (primary cilium) although he does have a section entitled "Sensory Cilia" in which he discusses the very specialized sensory cilia of the retina, cochlea, insect mechanoreceptors, and the very unusual cilia of the *saccus vasculosus* of fish. Sleigh does report on the generalization that sensory cilia tend to have a 9 + 0 arrangement of microtubules.

VI. Origin of Photoreceptor Connecting Cilia from Primary Cilia

Although this history focuses primarily on the traditional primary cilium (central flagellum) rather than the highly specialized sensory cilia, it is necessary to point out the developmental origins of the photoreceptor connecting cilium from more traditional primary cilia on cells of the early neuroepithelium. Studnicka (1898) believed that photoreceptors arose from flagellated ependymal cells. The outer segment of visual cells differentiates "... exactly like any flagellum. It starts as a filament of centrosomic material rooted in a diplosome or dumb-bell shaped centriole embedded in the future inner segment ...." Salzmann (1912), in discussing the development of the human retina, pointed out that "at the fifth month, the rods appear as small caps projecting over the membrane limitans externa; the diplosomes lie in them, and at each cell a fine thread (the outer thread) goes into the pigment epithelium from the diplosome." In their atlas of human eye development, Bach and Seefelder (1911, 1912, 1914) showed beautiful drawings of the development of human retinal cone outer segments from very traditional primary cilia (Fig. 15). In the early stages, these cells

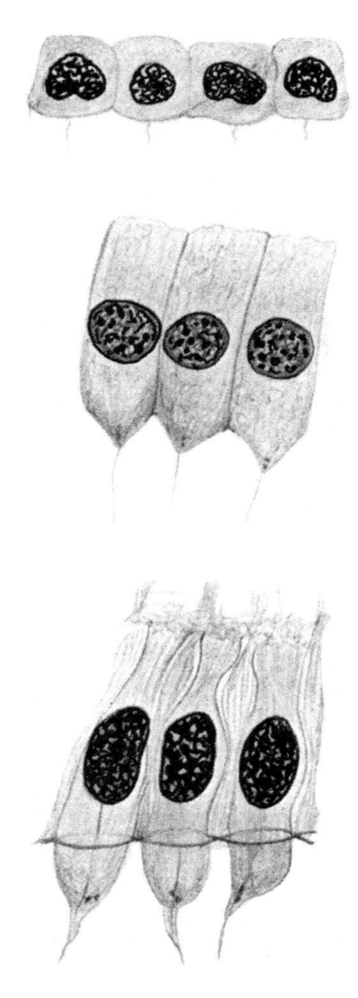

Fig. 15 Drawings adapted from Bach and Seefelder (1911, 1912, 1914) showing the early prenatal development of cone photoreceptor cells in the human retina. The cells of the developing photoreceptor layer first develop what look to be typical primary cilia prior to morphogenesis of these primary cilia into cone outer segments and the connecting cilium.

appear to possess a very typical primary cilium, which was confirmed much later by electron microscopy (Olney, 1968; Weidman and Kuwabara, 1968). The first TEM of the mature photoreceptor connecting cilium (in the mature differentiated photoreceptor) was published by Sjostrand (1953) (Fig. 16) and probably represents the first TEM image of any sensory or primary cilium, followed shortly thereafter by TEM images of the very unusual sensory cilia seen in the saccus vasculosus of fish (Bargmann and Knoop, 1955).

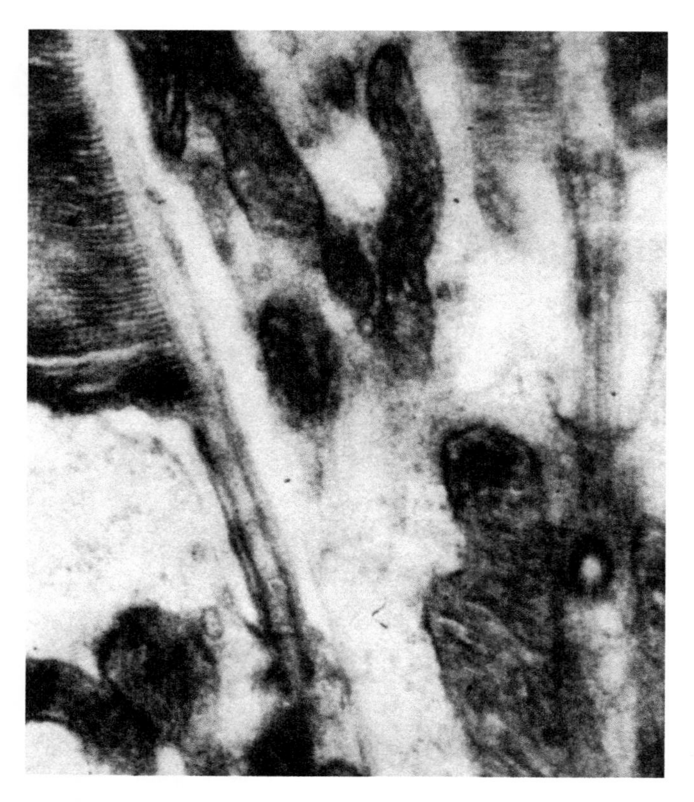

Fig. 16 This is the first published TEM of the photoreceptor connecting cilium (Fig. 6 from Sjostrand, 1953) and shows rod photoreceptor cells from the guinea pig retina. Reproduced with permission of John Wiley and Sons.

VII. The Dark Ages of Primary Cilia (1935–1955)

During the years from 1935 to 1955, there was a remarkable drought in terms of scientific progress in the study of cilia, flagella, centrosomes, centrioles, and basal bodies in general, and especially so in terms of primary cilia. This is certainly reflected in the dearth of papers on primary cilia; in Bowser's web bibliography of over 600 primary cilia papers (http://www.bowserlab.org/primarycilia/ciliumref.html), he lists only three papers published in this period and one of those is a TEM report of the connecting cilium in photoreceptor cells (Sjostrand, 1953). It is not unreasonable to speculate that this period of publication drought contributed to the "loss" of the old name and hence the eventual renaming (in 1968) of the organelle. While it appears that cell biology studies in general were going through a low point during these years (which included the Great Depression and WWII), other fields of science and engineering, including nuclear physics, particle physics, radar, radio, and TV were blossoming. One product of this period of technological innovation, of particular relevance to the history of primary cilia, was the electron microscope (Rasmussen, 1997). While the electron microscope was invented by Ernst Ruska and Max Knott in 1931, the first commercial electron microscopes became available only at the beginning of WWII (Siemens, 1938 and Radio Corporation of America, 1941). Wide distribution of these instruments only began after the end of the war and their application to biological tissues had to await the later development of reliable techniques for fixation (buffered osmium tetroxide, Palade, 1952), embedding (development of methacrylate; Newman *et al.*, 1949), and thin sectioning (Porter–Blum MT1 microtome; Porter and Blum,1953).

VIII. The 1950s and 1960s: Transmission Electron Microscopy and the "Renaissance" of the Primary Cilium

The development of TEM and its application to the study of cilia and flagella (as well as to centrioles and basal bodies) gave an enormous boost to what had become a rather stagnant field of study. Grigg and Hodge (1949) and Manton and Clark (1952), using whole-mount preparations, were among the first groups to make effective use of TEM to understand the organization of cilia and flagella. The first really informative TEM study of sectioned material was the study of ciliated epithelia by Fawcett and Porter (1954). While not having quite as immediate an impact on the study of primary cilia, the explosive application of TEM to all sorts of cell and tissue types did lead to the "rediscovery" of primary cilia and to an appreciation (actually a reappreciation; see Table I) of their widespread distribution. While this initial spate of reports of primary cilia in sectioned material did lead to the discovery of primary cilia in new places (such as neurons and lung) where they had not previously been described, TEM studies *per se* did not serve to clarify the function of the primary cilium and many researchers, unaware of the extensive pre-WWII literature on central flagella still thought of them as "rudimentary," "primary," "abortive," or "vestigial" organelles.

Leaving aside the first TEM observations of the connecting cilium in the retinal photoreceptor cells (Sjostrand, 1953) and other specialized sensory cilia, such as those found in the saccus vasculosus of fish (Bargmann and Knoop, 1955), the first TEM observation of what had been called "central flagella" before WWII and would only much later (Sorokin, 1968) be called "primary cilia" was that of deHarven and Bernhard (1956). Although this paper focused on the TEM structure of centrioles, one figure (Fig. 17 in this chapter) shows a primary cilium in a chicken embryo cell but the authors did not appear to recognize it as such and referred to this primary cilium as "une structure fusiforme, osmiophile, et presentant une fine striation longitudinal" (an osmiophilic, fusiform-shaped structure with thin longitudinal striations). The authors noted that this structure had the same diameter as the centriole with which it was associated. These same authors went on to describe primary cilia in a subsequent paper presented in 1958 at the 4th International Conference on Electron Microscopy in Berlin, but only published in 1960 (Bernhard and deHarven, 1960). They refer to finding these structures, which they now identify as cilia, in the embryonic chicken

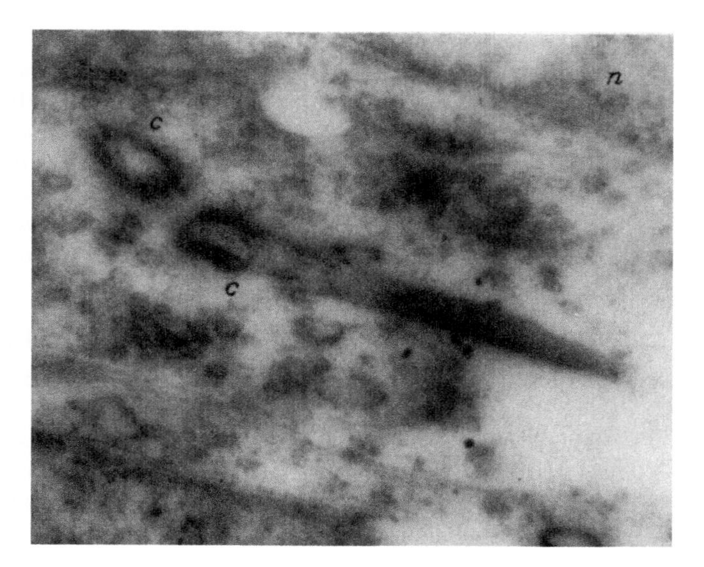

Fig. 17 This is probably the first published TEM picture of a "typical" primary cilium (Fig. 6 from deHarven and Bernhard, 1956) and shows a cell from the spleen of a chicken embryo. Reproduced with permission of Springer.

spleen and a mouse ovarian tumor. They state that, to their knowledge, no such structures had been shown by light microscopy, but then later in the same paragraph suggest that their "petit cili" may be related to Zimmermann's "centralgeissel." They demonstrate that these primary cilia are associated with the distal centriole of a pair of centrioles (diplosome) as had been observed by Zimmermann (1898) but give him no credit for this observation. More importantly, they are the first scientists to report that the microtubules ("tubules") of the distal (older) centriole are continuous with the microtubules (which they variously refer to as "tubules," "fibers," or "canalicules") of the primary cilium. Duncan (1957) used TEM to show solitary cilia on the luminal surface of chick neural tube cells and this is the first report of primary cilia in neural tissues. Sotelo and Trujillo-Cenoz (1958) went on to describe in much more detail, again using TEM, the primary cilia in this same neural epithelium of the developing chick; they noted the postmitotic migration of the pair of centrioles to the plasma membrane prior to initiation of primary cilia assembly and further noted that there was one cilium/cell with a 9 + 0 structure. However, the first reports of primary cilia in adult neurons were by Palay (1960, 1961) in secretory neurons of the preoptic nucleus of the goldfish and Taxi (1961) in sympathetic neurons of the frog. Munger (1958) showed TEM of primary cilia in embryonic and adult pancreatic beta cells and referred to them simply as "cilia." He postulated three possible roles: (1) Embryonic remnants that, at best, might be merely signposts for newly formed β cells, (2) If motile, they might serve to stir the extracellular fluid, and (3) Chemoreceptors, perhaps for glucose, in analogy with known sensory cilia in photoreceptors and hair cells. Leeson (1960) was the first TEM report of primary cilia in kidney tubules (distal tubules in newborn hamsters), to be followed by Latta *et al.* (1961) who described primary cilia in all segments of the kidney tubule system. Noting that "it is difficult to suggest a function for such rare cilia in the mammalian kidney," Latta suggests two possibilities: (1) That they create turbulence and prevent laminar flow of tubular fluid, thereby facilitating the resorption of solutes or (2) They may be an evolutionary remnant because they are found more frequently in the excretory ducts of lower animals. Just as Zimmermann (1898) had, Latta *et al.* (1961) notes that "the cilia of the mammalian kidney could hardly propel much fluid." DeRobertis and Sabatini (1960) described primary cilia in chromaffin cells of the adrenal medulla. Barnes (1961) gave a detailed ultrastructural description of primary cilia in anterior pituitary cells and reported that these were 9 + 0 cilia, each associated with a basal body (proximal centriole) and a distal centriole (Zimmermann's "Central geisselapparat" diplosome). Interestingly, she pointed out that the pars distalis (anterior pituitary) is derived from a traditional ciliated epithelium, suggesting a reverse transformation of a cell from a ciliated epithelium to a primary cilium phenotype, reminiscent of what was reported by Kowalevsky (1867) in *Amphioxus*, but opposite to what Joseph (1903), Sorokin (1968), and Banizs *et al.* (2005) report (transformation of primary cilia epithelium into a true ciliated epithelium). Barnes (1961) went on to review the TEM literature to date (but cited no paper before the 1950s) and proposed three general criteria for primary cilia: (1) lack of motility, (2) 9 + 0 arrangement of microtubules, and (3) association with a pair of centrioles (one functioning as a basal body), although she cited exceptions to the last criterion. Arguing from analogy with known sensory structures like the photoreceptor connecting cilium (already known to be a 9 + 0 structure), she argued that primary cilia were probably performing a sensory role (with no credit to Zimmermann, 1898). After these pioneering TEM reports of primary cilia, reports began to proliferate and are far too numerous to discuss here (see Table 6.2 in Wheatley, 1982, and the Bowser list of cell types reported to have primary cilia at: http://www.bowserlab.org/primarycilia/cilialist.html). Leeson (1962), who used TEM to observe primary cilia on the epithelium of the rete testes was one of the very few post-WWII papers to use Zimmermann's term Central Flagellum, consistent with the fact that he cited the Alverdes (1927) report on central flagella in the rete testis. He suggested, after Munger (1958), that cilia in the rete testes were chemoreceptive. By the end of the 1960s, it was clear that primary cilia were widely distributed, were associated with both embryonic and adult cell types, were found on cells of both rapidly cycling tissues and nondividing tissues, were associated with the distal centriole of a centriole pair (diplosome), contained a 9 + 0 structural organization, and were nonmotile. There was little consensus (and almost no data whatsoever) regarding their function. The fact that the 9 + 0 arrangement of microtubules was found both in primary cilia and in a cilium clearly thought to be associated with a sensory function (the connecting cilium of retinal photoreceptors) led a number of workers to hypothesize a sensory role for primary cilia, by analogy.

Moving beyond descriptive reports of the occurrence of primary cilia in various cell types using TEM, the next major advance in our knowledge about primary cilia involves TEM reports of the process of primary cilia assembly (Sorokin, 1962, 1968; Sotelo and

Trujillo-Cenoz, 1958) and a comparison with "ciliogenesis," the assembly of multiple motile cilia on a cell (Sorokin, 1968). The Sorokin (1968) paper is worthy of special attention for two reasons: (1) This represents the best and most extensive TEM study (using mammalian lung tissue) comparing the ciliogenesis of primary cilia and motile cilia, representing a prodigious amount of work. (2) Sorokin is the researcher who initiated the use of the currently accepted name (primary cilium) and effectively renamed an organelle which had already been given a perfectly workable and probably more appropriate name (central flagellum) in 1898. This happened because of a total lack of knowledge of the pre-1950s literature by Sorokin and most other 1950s and 1960s students of primary cilia, probably compounded by the "Dark Ages" from 1935 to 1950 during which there were almost no reports on primary cilia, perhaps exacerbated by a post-WWII decline in the study of the German language by US scientists. While Sorokin (in both his 1962 and 1968 papers) cites the Zimmermann (1898) paper, it is clear that he had not read it because he is clearly unaware of the "central flagellum" name applied by Zimmermann, and widely used by pre-WWII students of primary cilia, and because he is unaware that Zimmermann had shown primary cilia as a normal structure in adult mammalian epithelia and persists in seeing primary cilia as rudimentary structures. Indeed, in his 1968 paper, he alternates between calling them primary cilia and rudimentary cilia with an occasional "abortive" cilia term thrown in. The point of view that "these cilia are considered to have only a transitory existence" may have come from his choice of experimental material (lung airway epithelium), where it appears that the epithelium does transition from one with all primary cilia to the true ciliated (motile) epithelium of the adult state (as also occurs on the ependyma, Banizs *et al.*, 2005). However, Sorokin (1968) states that the primary cilia in the lung "resemble closely the rudimentary, or abortive, cilia that are known to occur in a wide variety of cells present in other organs" and cites many of the earlier TEM studies, as well as Zimmermann (1898) again testifying to his lack of knowledge of that paper. Sorokin (1968) concludes that "the function of primary cilia remains unknown," a statement which will be echoed in many subsequent papers over the next 25 years. Sorokin's term "primary cilia" and its incorrect implication that these organelles are always transitory intermediates caught on quickly and soon was used by many subsequent papers in the field, which continued to accumulate observations documenting the broad distribution of primary cilia in adult, as well as developing, tissues.

In summary, the 1960s ended with a clear characterization of the primary cilium as a 9 + 0, nonmotile structure associated with a pair of centrioles (diplosome) and an appreciation of the extremely widespread (but not quite ubiquitous) distribution of primary cilia among vertebrate cell types. The function of the organelle remained mysterious and controversial.

IX. Primary Cilia and the Cell Cycle

An additional important trend in the primary cilia literature involves the relationship of primary cilia to the cell cycle (see Chapter 7 by Plotnikova, Pugacheva and Golemis,

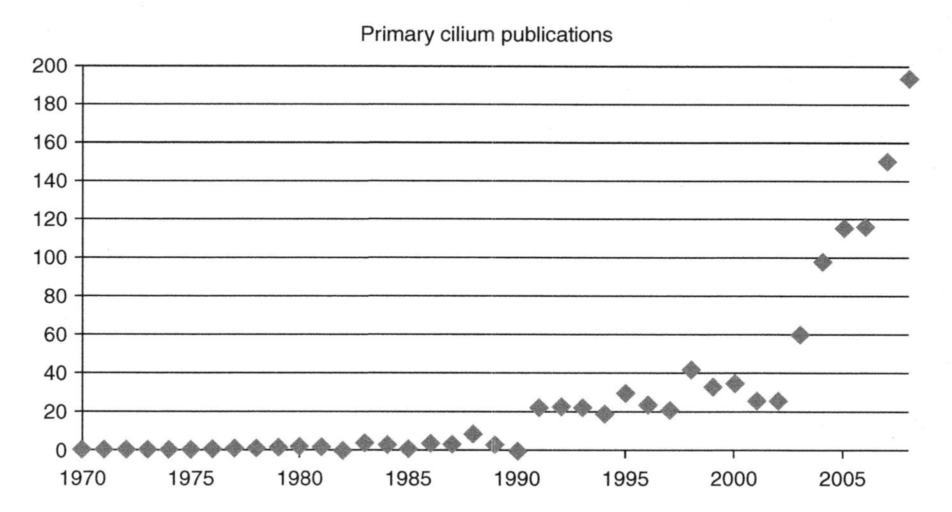

Fig. 18 This graph shows the kinetics of publication of primary cilia papers from 1970 to the present. The data were obtained from the Web of Science database and papers were selected for matching "primary cilia" or "primary cilium."

this volume). The Henneguy (1898) and Lenhossek (1898b) papers formulating the Henneguy–Lenhossek Hypothesis (Section IV) carried the implicit prediction that central flagella cells could divide while true ciliated cells (cells containing multiple, motile cilia) were incapable of future cell division. The Henneguy–Lenhossek Hypothesis also clearly predicted that cells would lose their primary cilia during mitosis, when the centriole pair (diplosome) associated with the primary cilium would need to migrate to a different position within the cell in order to assume its other role as an organizer of the mitotic spindle. Joseph (1903) states that central flagella cells can divide (citing Meves, 1899, as well as his own Fig. 18). The illustrations accompanying the Meves (1899) paper clearly show that, in salamander larval kidney tubule epithelium, the primary cilia disappear during mitosis. Joseph (1903) states that "unfortunately, we know nothing about the fate of the central flagellum apparatus [during cell division], knowledge that would be of great significance."

Walter (1929) noted that ciliated cells never divided but felt that the other population of cells (the central flagella cells) did divide: "I could never find 'Centralgeisselaparat' on the surface of the dividing cell, although the centrioles were always seen in the position characteristic of mitosis ... as the process of division ensues the flagellum disappears, and the granules [centrioles] descend to the inner part of the cell."

In the 1960s, there emerged a controversy over whether primary cilia were "labels" for immature or developmental cells, were labels for differentiating or differentiated, postmitotic cells, or were labels for the G_1 stage of the cell cycle. Going beyond that, some thought them capable of actively inhibiting mitosis. In a rare argument for the exact opposite, Wilson and McWhorter (1963), while studying primary cilia on basal cells in human epidermis, hypothesized that "flagella of epidermal cells might conceivably play a part in the initiation of mitosis", citing a prior observation (Pinkus, 1951) that repeated stripping of the superficial cornified layers of the skin induced mitosis in the basal cell layer.

With the exception of Cowdry's (Cowdry, 1921) unique experiment concerning central flagellar motility in thyroid follicles, experimental studies on primary cilia really only began once primary cilia were identified on cultured cell lines and addressed the issue of the fate of primary cilia during the cell cycle. Although Stubblefield and Brinkley (1966) documented an absence of primary cilia on the Don/C fibroblast cell line when blocked in metaphase using colcemid and the appearance of primary cilia after washing out the colcemid, they misinterpreted their data as showing induction of primary cilia by colcemid because they mistakenly thought that this cell line normally lacked primary cilia. Wheatley (1969) clearly showed the normal presence of primary cilia in interphase cells of the Don/C fibroblast cell line.

While Wheatley (1967) noted that "it is of interest to know what happens to a cilium when the cell bearing it enters division," it is only with the demonstration of primary cilia on cultured cell lines (primarily cell lines derived from fibroblasts) (Stubblefield and Brinkley, 1966; Wheatley, 1969) that Wheatley was able to experimentally address the problem in an important paper published in 1971 (Archer and Wheatley, 1971). Using a BHK 21/C13 fibroblast cell from hamster kidney, Archer and Wheatley (1971) showed a complete disappearance of primary cilia during mitosis, regardless of whether the mitotic cells had been obtained naturally or by use of a drug such as colcemid or vinblastine. Further, they showed reappearance of primary cilia after washing out the mitotic drugs. Archer and Wheatley (1971) argued that their data (and that of others) did not support a hypothesis that primary cilia could regulate the frequency of mitosis in cell populations. They pointed out that primary cilia (though they did not use that term) are found in rapidly dividing populations of cells as well as in highly differentiated cell lines exhibiting little or no mitosis (i.e., neurons) and they noted that cultured cell lines that do not exhibit primary cilia divide at about the same rate as cultured cell lines with high cilium indexes.

Boquist (1968) reported that, following alloxan treatment of Chinese hamsters, regenerating pancreatic tissue exhibited increased numbers of cilia associated with duct cells and islet β-cells, although he also, paradoxically, reported that "unequivocal signs of mitotic cell division were not found." At least one of his TEM pictures showed a duct cell with multiple $9 + 2$ cilia, although other images appear to show primary cilia. Boquist resurrects the long dead ghost of "amitosis" and speculates "that the cilia may register changes in the intercellular milieu and initiate cell division and new formation of cilia when there is need for formation of cells." Dingemans (1969) showed a rough correlation between agents that are known to stimulate proliferation of cells in the adenohypophysis (anterior pituitary) and a decline in the frequency of primary cilia, though he did not quantitate numbers of mitotic cells in either the control or experimental samples.

Rash *et al.* (1969) observed large numbers (~700) of primary cilia on myoblasts and fibroblasts in developing chick embryos, as well as on fibroblasts on adult mammalian hearts but never observed primary cilia on dividing cells ($p = .001$). These workers doubted a sensory role for primary cilia in muscle and connective tissues and suggested that primary cilia may arrest mitosis in differentiated cells by tying up the centrioles so they cannot participate in mitosis. Subsequently, Myklebust *et al.* (1977) showed that primary cilia were present in both the embryonic and adult human heart on what were probably fibroblasts and noted that "no cilia were seen in cells going into mitotic division."

Fonte *et al.* (1971) used TEM to examine cells in the mesenchyme of the limb bud of stage 19 chick embryos. In this very confusing and poorly reasoned article, these authors end up basically arguing that actively cycling embryonic cells can have primary cilia and that primary cilia are no indication of the state of differentiation. Although they start out implying that their data will refute the proposed "reciprocal relationship between the presence of cilia on cells and mitotic activity" dating back to Henneguy (1898) and Lenhossek (1898b), all their data really end up arguing is that, statistically, some S phase cells may have primary cilia, which is in no way incompatible with the "reciprocal relationship."

In one of the first TEM studies of primary cilia to utilize high-voltage electron microscopy and thick sections, Rieder *et al.* (1979) directly showed that PtK1 cultured rat kangaroo kidney epithelial cells resorb their cilia during the early stages of spindle formation (prophase and prometaphase). By observing the process of resorption in a mammalian cell type, this constitutes an important paper that demonstrates directly that a cell with a primary cilium has the capacity to enter into mitosis, arguing against models of primary cilia function (such as that of Rash *et al.*, 1969) that suggest that primary cilia interfere with mitosis (prevent a cell from going through cell division). Tucker *et al.* (1979) showed that serum stimulation of growth-arrested Swiss 3T3 cultured fibroblasts resulted in loss of primary cilia as cells synchronously entered the cell cycle. However, unlike Rieder *et al.* (1979), Tucker *et al.* (1979) argued that cells lost their primary cilia at or soon after entering the S phase of the cell cycle, a time when centriole replication usually occurs.

By the end of the 1970s (reviewed by Wheatley, 1982), the generalization was pretty clearly established that $9 + 0$ primary cilia (at least in vertebrate systems) were resorbed just prior to or during mitosis and reassembled soon after entry of the daughter cells into G_1, an observation that was predicted by and clearly supported the Henneguy–Lenhossek Hypothesis (see Section IV). However, confusion would persist right to the present (even with the clear demonstration of the sensory function of primary cilia in 2001) as to the significance of this observation. Is it merely that primary cilia must be eliminated at the G_2–M boundary of the cell cycle in order to release the centriole pair (diplosome) to migrate within the cell in order to participate in mitosis or does the primary cilium play some sort of active (stimulatory or inhibitory) role in regulating the cell cycle. Although protein kinase-associated signaling pathways (Christensen *et al.*, 2008; Pugacheva *et al.*, 2007) are being shown to coordinate primary cilia resorption with cell cycle progression, it is still unclear whether primary cilia must be eliminated primarily because they are an impediment to entry into mitosis or whether primary cilia are more active players in regulating the cell cycle (Sharma *et al.*, 2008). Wheatley *et al.* (1996) suggested that, in cultured cells, the expression of primary cilia may be more dependent on cell density (and hence perhaps cell contact) than on cell cycle activity. However, cell density and cell cycle are not totally independent variables and confluent cells cease cell division and hence would be expected to have more primary cilia than rapidly cycling cells. In studying a rat liver cell line, Mori *et al.* (1979) noted that no primary cilia were seen until 5 days after confluence of the cultures.

Although not strictly the purview of this review, it should be pointed out that the situation with regard to motile 9 + 2 cilia and flagella during mitosis and meiosis is a bit more complicated and messier than that related to 9 + 0 primary cilia. Despite all the early controversy on this point (critical to the arguments of the opponents of the Henneguy–Lenhossek Hypothesis), it is now well established that vertebrate ciliated epithelial cells cannot divide (Dawe *et al.*, 2007) as predicted by the Henneguy–Lenhossek Theory and as observed by early workers like Joseph (1903).

As mentioned earlier (see Fig. 12) when discussing the Henneguy–Lenhossek Hypothesis, insect spermatocytes can maintain their flagella (9 + 2 flagella destined to become motile sperm tails) during meiotic cell divisions (Friedlander and Wahrman, 1970; Henneguy, 1898; LaFountain, 1976; Meves, 1897, 1900) in a clear demonstration that the same structure can function (even at the same time) as a centriole and a basal body, a dramatic piece of support for the Henneguy–Lenhossek Hypothesis.

Many parasitic and symbiotic flagellate protozoa, such as *Trichomonas* (Kofoid and Swezy, 1915; Wenrich, 1921) (Fig. 13), *Trypanosoma* (Sherwin and Gull, 1989), and *Oxymonas* (Cleveland, 1950) exhibit flagellar association with the spindle poles during mitosis/meiosis, again providing potent support for the Henneguy–Lenhossek Hypothesis. Many ciliate protozoa maintain their many cortical cilia during cell division; in general, nuclear division in ciliates does not involve centrioles (Friedlander and Wahrman, 1970; Jenkins, 1967). On the other hand, there are clearly protistan flagellates, like the biflagellate green algal cell *Chlamydomonas*, that do resorb their flagella before cell division (Bloodgood, 1974; Ringo, 1967) releasing their basal bodies/centrioles to migrate to the spindle poles. Although *Chlamydomonas* has long been the paradigm for predivision resorption of 9 + 2 flagella, it may well be in the minority among protistan organisms, containing 9 + 2 organelles, in regard to exhibiting predivision resorption.

There is also a small but interesting literature on treatments and conditions that may alter the assembly state of primary cilia in what are presumably nondividing cells. Doughty (1998) reviews a number of studies showing that primary cilia on endothelial cells of the cornea increase in number and length after explanting the cornea and incubating under various buffer and fluid flow conditions. Iomini *et al.* (2004) showed that laminar shear stress (fluid flow) induced disassembly of all primary cilia on cultured human umbilical vein endothelial cells and presented some evidence that this effect was mediated through inhibition of intraflagellar transport (IFT). Subsequent work has shown the situation with vascular endothelium to be quite complicated. Primary cilium formation is stimulated in areas of low and disturbed/oscillating flow, while primary cilium formation is inhibited in areas of continuous high flow (Van der Heiden *et al.* 2006, 2008). Verghese *et al.* (2008) reported that renal ischemia-reperfusion injury resulted in a decrease in length of the primary cilia in the kidney tubules and collecting ducts and later repair was associated with cilia elongation. Verghese *et al.* (2009) showed that renal tubular necrosis occurring during transplantation resulted in more than a doubling in length of the primary cilia all along the nephron and collecting duct and that the ciliary lengths normalized after repair of tissue damage. Lack of contact with another cell or with a substrate of some kind may be incompatible with expression of primary cilia. Wheatley (1969) cites unpublished evidence that a BHK21/C13 tissue culture cell line has a high instance of ciliation under confluent culture conditions, but when adapted to growth in suspension, no primary cilia are seen. This is consistent with the observation that white blood cells are one of the few cell types that lack primary cilia.

X. The Golden Age of Primary Cilia (2000–Onward)

The 1980s and 1990s represented a period of relative stasis in the field of primary cilia research, despite Denys Wheatley's best efforts to keep primary cilia on the radar of the cell biology community (Wheatley, 1995; Wheatley *et al.*, 1996). Indeed, the field was self-limiting as long as it was the purview primarily of electron microscopists. However, the entire field of cilia and flagella received a kick-start in 1992, when Keith Kozminski (while a graduate student in Joel Rosenbaum's laboratory) discovered (on January 24, 1992) intraflagellar transport (IFT) in *Chlamydomonas* flagella, with the imaging assistance of Paul Forscher (Kozminski *et al.*, 1993), an observation that would initiate a chain of events leading to the enormous expansion of our knowledge of ciliopathies (see below). This discovery was critically dependent upon new technical advances in light microscopy (video microscopy using new improvements in differential interference contrast optics). Subsequently, it was shown that IFT was necessary for the assembly and maintenance of almost all cilia and flagella (reviewed in Rosenbaum and Witman, 2002; see Han *et al.*, 2003 for an interesting exception), leading to a resurgence in interest in the assembly, disassembly, and turnover of cilia and flagella. This also signaled a move away from heavy dependence on TEM among students of cilia and flagella and a return to the use of novel light microscopic techniques in the study of cilia and flagella (the technical domain where the study of primary cilia began over a hundred years earlier).

During the period 2000–2002, a remarkable conjunction of events occurred that would rapidly lead to an incredible expansion of interest in cilia and flagella, including primary cilia, and this interest would spread way beyond the cell biology community into the physiology, developmental biology, cell signaling, and medical research communities. These discoveries gave cilia and flagella a level of respect and interest and a breadth of audience they had never had before and that could scarcely have been imagined prior to 2000. Contributing factors to this Big Bang for the Universe of cilia and flagella included the direct demonstration of a sensory role (flow detection) for kidney tubule primary cilia, the discovery that the Oak Ridge polycystic kidney disease (ORPK) mouse had a ciliary defect, the discovery of the central role of specialized cilia in nodal signaling and the establishment of L–R asymmetry in vertebrates, the localization of transient receptor potential (TRP) ion channels (such as polycystins) and membrane receptors involved in sensory signaling in the ciliary/flagellar membrane, and the association of primary cilia with major signaling systems in mammals, in particular, hedgehog signaling. This is a remarkable conjunction of events, all happening within the period of a few years, and all focusing dramatically increased attention on cilia and flagella, especially primary cilia. These events also served to focus attention on ciliary and flagellar membranes (Bloodgood, 1990; Pazour and Bloodgood, 2008), the long neglected stepchild to the axoneme. The use of a combination of model organisms (*Chlamydomonas, Caenorhabditis*, and the mouse) exhibiting motile and primary cilia as well as new technical approaches (including immunofluorescent light microscopic techniques, molecular genetic techniques, and clever micromanipulation techniques) led to some fundamental observations that would forever change the landscape for cilia and flagella.

Prior to 2000, there were only two well-established ciliopathies (human diseases clearly due to ciliary or flagellar defects). The classic ciliopathy was Kartagener's syndrome (Kartagener, 1933; Siewert, 1904) (also subsequently called Siewert's/Kartagener's syndrome, immotile cilia syndrome, and primary ciliary dyskinesia). While the clinical description of Siewert's and Kartagener's triad (sinusitis, bronchiectasis, and situs inversus; only later expanded to include male infertility) was published in 1904 by Siewert and 1933 by Kartagener, it was only in 1976 that Bjorn Afzelius, using TEM of sperm and nasal biopsies, made the connection between Kartagener's syndrome and a motility defect in cilia and flagella throughout the human body due to lack of dynein arms (Afzelius, 1976). Since Kartagener's syndrome results from a defect in ciliary and flagellar motility, and not from the absence of cilia, it does not impact on the function of primary cilia, depending on one's definition of nodal cilia (Basu and Brueckner, 2008) associated with the situs inversus aspect of Kartagener's syndrome, which may be a mixture of 9 + 2 motile and 9 + 0 immotile sensory cilia). The second ciliopathy to be identified involved cilium defects in patients with certain kinds of retinitis pigmentosa, in particular, Usher syndrome. Usher Syndrome 1B results from a mutation in the gene coding for myosin VIIa (Well, 1995) which localizes to the connecting cilium in photoreceptors and is thought to be involved in transport of proteins from the photoreceptor inner segment to the photoreceptor outer segment (Wolfrum, 2003). As with Kartagener's syndrome, this ciliopathy does not affect primary cilia.

It was the ORPK disease mouse (Lehman *et al.*, 2008), in collaboration with the algal cell *Chlamydomonas*, that was to prove to be the gateway into the many human ciliopathies that affect primary cilia and to bring primary cilia to the attention of the wider medical research community. The ORPK mouse was first described in 1994 (Moyer *et al.*, 1994) and resulted from a large-scale insertional mutagenesis screen. Mice die within a few weeks of birth from what resembles human autosomal recessive polycystic kidney disease. Although the dominant defect in these mice is polycystic kidney disease, they have been shown to have a wide range of additional defects including hepatic and pancreatic ductal abnormalities and cysts, retinal degeneration, skeletal defects, cerebellar hypoplasia, hydrocephalus, situs inversus, and skin and hair abnormalities (Lehman *et al.*, 2008). The ORPK mouse has a hypomorphic mutation in a gene called Tg737 and the original name given the gene product was Polaris and, for many years, the function of Polaris was unknown. This brings us back to the discovery of IFT in *Chlamydomonas* (Kozminski *et al.*, 1993). Subsequent work in *Chlamydomonas* and *Caenorhabditis* and other systems led to the characterization of the protein components of the IFT particles that are translocated along the length of the flagellum and the microtubule-associated motor proteins responsible for the anterograde and retrograde movements of the IFT particles and the demonstration that IFT was necessary for ciliary and flagellar assembly and maintenance (reviewed by Rosenbaum and Witman, 2002; Scholey, 2003). The key paper that first demonstrated that the ORPK mouse had an underlying ciliary defect was published by Pazour *et al.* (2000), who showed that a *Chlamydomonas* insertional mutant in the gene encoding an IFT particle protein (IFT88) was unable to assemble flagella and that the mouse homolog of the IFT88 gene was Tg737, the gene that is defective in the ORPK mouse. They also showed that the primary cilia in the kidney tubules of the ORPK mouse are significantly shorter than in wild type mice. Finally, these authors predicted very accurately that defects in IFT were likely to play a role in many disease states. In the years subsequent to 2000, there would rapidly emerge many ciliopathies (Fliegauf *et al.*, 2007; Sharma *et al.*, 2008; Tobin and Beales, 2009) and papers on primary cilia and ciliopathies would become a staple of medical journals (see the kinetics of primary cilia publications in Fig. 18).

Although a sensory role for primary cilia had been hypothesized by Zimmermann (1898) in connection with his study of primary cilia in renal tubules, it was over 100 years before a sensory role for primary cilia was directed demonstrated, again using primary cilia on renal tubular epithelium. Although Schwartz *et al.* (1997) showed that primary cilia on kidney epithelial cells bent in response to fluid shear, it was Praetorius and Spring (2001) who first showed that bending these primary cilia with a micropipette or through increased flow rate of a perfusate resulted in an increase in cytoplasmic free calcium. They further showed that loss of the primary cilium resulted in a loss of this flow sensing (Praetorius and Spring, 2002). Later studies by other investigators would show that primary cilia were involved in flow sensing in other systems, including the vascular endothelium (Nauli *et al.*, 2008), cardiac endothelium (Hierck *et al.*, 2008), osteoblasts and osteocytes in bone (Malone *et al.*, 2007), and cholangiocytes in the bile duct epithelium (Masyuk *et al.*, 2006). Coming immediately in time after the Pazour *et al.* (2000) revelation of the ORPK mouse as having a ciliopathy and the Praetorius and Spring (2001, 2002) demonstration of the mechanosensitivity of kidney primary cilia came the first observations on mechanoreceptive ion channels (TRP channels) in the membranes of mammalian primary cilia (again using kidney tubules). In 2002, TRPP polycystins (products of the PKD1 and PKD2 genes) were localized to kidney tubule primary cilia (Pazour *et al.*, 2002; Yoder *et al.*, 2002). Soon after, a number of additional TRP family mechanosensitive channels (particularly TRPV) would be found in the membranes of insect (Kernan, 2007) and worm (Kahn-Kirby and Bargmann, 2006) sensory cilia. This group of 2000–2002 papers served as real gateway papers for the soon to come enormous expansion of work on primary cilia.

The latest chapter in the story of sensory reception by cilia and flagella is the recent widespread appreciation that mammalian motile cilia also function in sensory reception and this is mediated via TRPV4 channels in the ciliary membrane (Andrade *et al.*, 2005; Fernandes *et al.*, 2008; Lorenzo *et al.*, 2008; Shah *et al.*, 2009; Teilmann *et al.*, 2005, 2006). In retrospect, it should not be surprising that motile cilia are also sensory; Jennings (1906) described the avoidance response in *Paramecium*, in which the motile cilia on *Paramecium* serve a mechanosensory role. The sensory role of motile cilia can also be indirect, by virtue of their motility creating gradients of fluid flow and/or extracellular signaling molecules, as in the cases of nodal cilia (Basu and Brueckner, 2008; Nonaka *et al.*, 2002) and ependymal cilia (Sawamoto *et al.*, 2006).

This leaves one major event in the history (up to now) of primary cilia to arrive on the scene and that was the discovery that some of the major signaling pathways used during vertebrate development and in the adult organism utilize primary cilia. The first big splash in this area came when it was appreciated that primary cilia are necessary for mammalian Hedgehog signaling (Huangfu and Anderson, 2005; Wong and Reiter, 2008; Chapter 10 by Goetz, Ocbina and Anderson, this volume). While the situation is far less clear, there is emerging evidence for primary cilia involvement in certain cases of Wnt signaling, both in suppression of canonical Wnt signaling as well as transmission of noncanonical Wnt signaling (Gerdes and Katsanis, 2008). A platelet-derived growth factor receptor (PDGFRα) has been localized to the primary cilium in several cell types, including fibroblasts and stem cells and it has been shown that the primary cilium can coordinate PDGFRα signaling associated with cell cycle control and directional cell migration (Christensen *et al.*, 2008). These primary cilia-associated signaling pathways are now serving to define new ciliopathies and help characterize existing ones and further expand the range of disciplines and scientists to acquire an interest in primary cilia. Today, primary cilia have come to justify the name that Zimmermann (1894, 1898) originally gave them—central flagella—for they have come to occupy a central position in the life of the cell, serving as a focus for sensory reception and for the hedgehog signaling pathway.

XI. Conclusions: The Morals of this Story

The history of primary cilia can be seen as an enormous success story, a neglected organelle rising from its modest beginnings to become a superstar on the modern scientific stage. It appeared with a bit of a splash at the very end of the 19th century, in particular because of its close ties to the controversial Henneguy–Lenhossek Hypothesis. But this was a premature stardom that was fated to leave the organelle, for the want of a clear function, in obscurity for the next full century before conditions (both technical and intellectual) were favorable for the appreciation of the important roles that it plays, particularly in the sensory and signaling domains. In all likelihood, we still do not appreciate its full worth and there will be surprises yet to come.

One lesson that can surely be learned is that no organelle should be written off as without function (vestigial) as so many students of primary cilia (particularly in the 1950s and 1960s) were willing to do. Indeed, the scientists of the late 19th century (such as Zimmermann, 1898) were much more imaginative in their thoughts about primary cilia than many of these later workers of the early TEM era. The moral is clear: do not write off any organelle as being without function. More attention should have been paid to the intermittent suggestions, occurring throughout the entire history of primary cilia, that primary cilia had a sensory role; these were based on a combination of brilliant intuition and argument from analogy with more obviously sensory cilia, particularly the connecting cilium of retinal photoreceptors.

From a somewhat different point of view, the story of primary cilia can be seen as a disappointing tale from which we might learn additional lessons. Because of a failure of scientific memory, so much about primary cilia (their widespread distribution, their putative function, even the name) had to be reinvented after WWII and after the development of biological TEM. There are two dangerous trends in post-WWII US graduate education in the sciences that conspire to wall off students from the messages of the past and the wisdom of our scientific pioneers. The first is English language chauvinism that is reflected in the decline in foreign language training in US colleges and its near disappearance in US graduate science education; this closes off access to the foreign literature (which generally means the older literature). Prior to WWII, so much of the best biological literature was published in German, as is well illustrated by the primary cilia literature and German was an important part of a US graduate education in the sciences at that time. The second trend can be referred to as temporal chauvinism and it is reflected in the feeling among young scientists today that it is neither necessary nor cool to study and to cite the older literature and older, in this context, can be as recent as a decade ago. Granted, part of this trend can be ascribed to the enormous explosion in the biological literature and the enormity of the task of keeping up with the current literature in even a relatively narrow field. Still one senses a decline in the value of scholarship, as opposed to the high premium on the pursuit of laboratory science *per se*, and this is driven by scientific journals that limit space, reward brevity and set rigid constraints on the nature of scientific writing, things not felt by scientists like Zimmermann (1898) and Joseph (1903), whose papers were 150 and 80 pages long! One further observation, related to the language barrier issue, is the widespread citation of older articles by authors who had not read the papers being cited. This is clearly the case with Zimmermann (1898) where there is clear and repeated evidence that citing authors were totally unaware of the contents of his paper.

Primary cilia also reinforce that important rule of science that study of one thing can often lead one into an unexpected discovery of something very different. The scientists who discovered primary cilia were focused on the study of the centrosome and centrioles, which were, at that time, thought to be primarily involved in cell division. Their fortuitous discovery of the primary cilium then led them to reverse engineer new insights about the relationship between centrioles and basal bodies and their interconvertibility (the Henneguy–Lenhossek

Hypothesis). Another example of this may relate to the ciliary/flagellar membrane. The early and long-term emphasis on the study of motile cilia and flagella lead to a focus on the axoneme; it was only with the increased emphasis on primary cilia and the identification of their sensory functions that much more attention was shifted to the ciliary/flagellar membrane.

Like so many advances in the biological sciences, another way to look at primary cilia is to view their history through the lens of technology. It was probably advances in light microscope technology and sectioning technology, as well as a critical advance in stain technology (the iron–hematoxylin stain) that led to the initial discoveries about primary cilia. The rebirth of primary cilia after WWII was clearly driven by the development of TEM and the associated preparative techniques that made this technique useful for the study of tissues and organs. Later still, it was technical advances in light microscopy (driven, in part, by advances in computer and camera technology), molecular biology, and molecular genetics that gave rise to the spectacular increase in our knowledge about primary cilia that started about 2000. An additional factor in the maturation of the primary cilia field was the development of appropriate model organisms; yeast and *Drosophila* lacked cilia while the refinement of the *Chlamydomonas, Caenorhabditis*, and mouse model systems provided great experimental opportunities, as we have seen. As an aside, the first model organism that was key to the study of primary cilia was *Amphioxus* (Joseph, 1903; Kowalevsky, 1867, 1877; Langerhans, 1876), organisms now referred to as the lancelets, the modern representatives of the subphylum cephalochordata. The entire polycystic kidney disease story shows how the special advantages of one model organism (*Chlamydomonas*) can inform the situation in a less experimentally amenable but more clinically relevant model system (the mouse).

It is particularly remarkable to see how much of the story of primary cilia is connected to a single organ, the kidney. The kidney tubule epithelium shows up at every single step in the history of primary cilia, starting with Zimmermann (1898), Meves (1899), and Joseph (1903). Zimmermann (1898) emphasized that "the most typical conformation of the central flagellum occurs in the renal tubules" (p. 693). It is interesting to note that Zimmermann went on to devote much of his later career to the kidney and, in 1933, named the "macula densa" (Zimmermann, 1933), a key component of the sensory machinery in the kidney. The Leeson (1960) report on primary cilia in the distal tubule epithelial cells of the kidney was among the first half-dozen TEM reports of primary cilia. The first careful TEM study of the resorption of primary cilia at mitosis (Rieder *et al.*, 1979) utilized cultured kidney cells. The initial direct demonstration of a sensory role for primary cilia (Praetorius and Spring, 2001, 2002) and the first demonstration of TRP calcium channels in primary cilia (Pazour *et al.*, 2002; Yoder *et al.*, 2002) both utilized kidney tubule epithelial cells (and both were part of that critical gateway cluster of papers in 2000–2002). The rapid expansion of the field of ciliopathies began with polycystic kidney disease and the ORPK mouse.

The biggest message to be gained from this exploration of the history of primary cilia lies in the growing appreciation of the complexity and sophistication of this long underappreciated organelle. Truly, we are in a Golden Age for cilia and flagella (with no end in sight) and the present volume documents much of the intellectual excitement in this field, as well as many of the emerging tools for its study.

Acknowledgments

My thanks go to the very helpful staff of the University of Virginia Health Sciences Library, the Stanford University Lane Medical Library, the Dartmouth College Dana Biomedical Library, the MBL/WHOI Library, and the National Library of Medicine. Google Scholar, Google Books, and the Biodiversity Heritage Library (http://www.biodiversitylibrary.org/) served as additional valuable resources. Thanks go to Roger Sloboda (Dartmouth College) and Dennis Diener (Yale University) for assistance in obtaining access to certain papers. Thanks go to Dr Paul Hamburg (Chestnut Hill, Massachusetts) for his nuanced translations of the German scientific papers, as well as his interest in the entire project. The photograph of Dr. K.W. Zimmermann was kindly provided by Professor Urs Boschung, MD, director of the Institute for the History of Medicine of the University of Berne (Switzerland). The photograph of Paul Langerhans was kindly provided by Dr. Nikolaus Romani of the Innsbruck Medical University. The value of the primary cilia bibliography (http://www.bowserlab.org/primarycilia/ciliumref.html) and the annotated listing of cell types with primary cilia (http://www.bowserlab.org/primarycilia/cilialist.html) created by Samuel Bowser (Wadsworth Center, New York State Department of Health, Albany, New York) with assistance from Denys N. Wheatley (Aberdeen University) as an entry point into the literature, cannot be underestimated. A special tenacity award goes to Denys Wheatley for his sustained efforts to keep the visibility of primary cilia (often barely) above the horizon in those dark (and unenlightened) days when there was very limited interest in this organelle. The preparation of this chapter was supported by a Harrison Distinguished Teaching Award from the University of Virginia School of Medicine.

References

Afzelius, B. A. (1976). A human syndrome caused by immotile cilia. *Science* **193**, 317–319.

Alverdes, K. (1927). Der Zentralgeisselapparat der Epithelzellen im Rete testis des Menschen. *Zeitschrift fur Mikroskopisch-anatomische Forschung* **11**, 172–180.

Anderson, R. G., and Brenner, R. M. (1971). The formation of basal bodies (centrioles) in the Rhesus monkey oviduct. *J. Cell Biol.* **50**, 10–34.

Andrade, Y. N., Fernandes, J., Valquez, E., Fernandez-Fernandez, J. N., Arniges, M., Sanchez, T. M., Villalon, M., and Valverde, M. A. (2005). TRPV4 channel is involved in the coupling of fluid viscosity changes to epithelial ciliary activity. *J. Cell Biol.* **168**, 869–874.

Archer, F. L., and Wheatley, D. N. (1971). Cilia in cell-cultured fibroblasts II. Incidence in mitotic and post-mitotic BHK 21/C13. *J. Anat.* **109**, 277–292.

Bach, L., and Seefelder, R. (1911, 1912, 1914). "Atlas zur Entwicklungsgeschichte des menschlichen Auges, Parts 1–3." Wilhelm Engelmann, Leipzig.

Banizs, B., Pike, M. M., Millican, C. L., Ferguson, W. B., Komlosi, P., Sheetz, J., Bell, P. D., Schwiebert, E. M., and Yoder, B. K. (2005). Dysfunctional cilia lead to altered ependyma and choroid plexus function, and result in the formation of hydrocephalus. *Development* **132**, 5329–5339.

Bargmann, W., and Knoop, A. (1955). Elektronmikroskopische Untersuchung der Kronchenzellen des Saccus vasculosus. *Z. Zellforsch.* **43**, 184–194.

Barnes, B. G. (1961). Ciliated secretory cells in the pars distalis of the mouse hypophysis. *J. Ultrastruct. Res.* **5**, 453–467.

Basu, B. and Brueckner, M. (2008). Cilia: Multifunctional organelles at the center of vertebrate left-right asymmetry. *Curr. Top. Dev. Biol.* **85**, 151–174.

Benda, C. (1886). Uber eine neue Farbemethode des Centralnervensystems, und Theoretisches uber Haematoxylinfarbungen. *Arch. Anat. Physiol. (Phys. Abth.)* 562–564.

Benda, C. (1901). Ueber neue Darstellungsmethoden der Centralkorperchen und die Verwandtschaft der Basalkorper der Cilien mit Centralkorperchen. *Arch. Anat. Physiol. (Phys. Abth.)*, 147–157.

Bernhard, W., and deHarven, E. (1960). L'ultrastructure du centriole et d'autres elements de l'appareil achromatique. *In* "Proceedings of the 4th International Congress Electron Microscopy" (W. Bargmann, D. Peters, and C. Wolpers, eds.), Vol. 2, pp. 217–227. Springer-Verlag, Berlin.

Bloodgood, R. A. (1974). Resorption of organelles containing microtubules. *Cytobios* **9**, 142–161.

Bloodgood, R. A. (1990). "Ciliary and Flagellar Membranes." Plenum Press, New York.

Boo, K. S., and Richards, A. G. (1975). Fine structure of scolopidia in Johnston's organ of female *Aedes aegypti* compared with that of the male. *J. Insect Physiol.* **21**, 1129–1139.

Boquist, L. (1968). Cilia in normal and regenerating islet tissue. An ultrastructural study in the Chinese hamster with particular reference to the β-cells and the ductular epithelium. *Z. Zellforsch.* **89**, 519–532.

Boveri, T. (1888). Zellen-Studien II. Die Befruchtung und Teilung des Eies von *Ascaris megalocephala. Jena. Z. Naturwiss.* **22**, 685–882.

Boveri, T. (1895). Ueber die Befruchtungs und Entwickelungsfahigkeit kernloser Seeigeleier und uber die Moglichkeit ihrer Bastardierung. *Arch. Entwicklungsmech. Org. (Wilhelm Roux)* **2**, 394–443.

Burgos, M. H., and Fawcett, D. W. (1955). Studies on the fine structure of mammalian testes: 1. Differentiation of the spermatids in the cat (*Felis domestica*). *J. Biochem. Biophys. Cytol.* **1**, 287–300.

Bystrevskaya V. B., Lichkun, V. V., Antonov, A. S., and Perov, N. A. (1988). An ultrastructural study of centriolar complexes in adult and embryonic human aortic endothelial cells. *Tissue Cell* **20**, 493–503.

Caspary, T., Larkins, C. E., and Anderson, K. V. (2007). The graded response to sonic hedgehog depends on cilia architecture. *Dev. Cell* **12**, 767–778.

Chase, S. W. (1923). The mesonephros and urogenital ducts of *Necturus maculosus*, rafinesque. *J. Morphol.* **37**, 457–531.

Christensen, S. T., Pedersen, S. F., Satir, P., Veland, I. R., and Schneider, L. (2008). The primary cilium coordinates signaling pathways in cell cycle control and migration during development and tissue repair. *Curr. Top. Dev. Biol.* **85**, 261–301.

Clark, G., and Kasten, F. H. (1983). "The History of Staining." 3rd edn. Williams and Wilkins, Baltimore.

Cleveland, L. R. (1950). Hormone-induced sexual cycles of flagellates II. Gametogenesis, fertilization, and one-division meiosis in Oxymonas. *J. Morphol.* **86**, 185–213.

Corbiere-Tichane, G. (1971). Ultrastructure de l'equipement sensoriel de la mandibule chez la larve de *Speophyes lucidulus* Delar. (Coleoptere Cavernicole de la souis-famille des Bathysciinae). *Z. Zellforsch. Mikrosk. Anat.* **112**, 129–138.

Cowdry, E. V. (1921). Flagellated thyroid cells in the dogfish (*Mustelus canis*). *Anat. Rec.* **22**, 289–299.

Dammermann, A., Pemble, H., Mitchell, B. J., McLeod, I., Yates, J. R., Kintner, C., Desai, A. B., and Oegema, K. (2009). The hydrolethalus syndrome protein HYLS-1 links core centriole structure to cilia formation. *Genes Dev.* **23**, 2046–2059.

Dawe, H. R., Farr, H., and Gull, K. (2007). Centriole/basal body morphogenesis and migration during ciliogenesis in animal cells. *J. Cell Sci.* **120**, 7–15.

deHarven, E., and Bernhard, W. (1956). Etude au microscope de l'ultrastructure du centriole chez les vertebres. *Z. Zellforsch. Mikrosk. Anat.* **45**, 378–398.

deLorenzo, A. J. (1958). Electron microscopic observations on the taste buds of the rabbit. *J. Biophys. Biochem. Cytol.* **4**, 143–150.

deRobertis, E. D.P., and Sabatini, D. D. (1960). Submicroscopic analysis of the secretory process in the adrenal medulla. *Fed. Proc.* **19**(Suppl. 5), 70–78.

Dingemans, K. P. (1969). The relation between cilia and mitoses in the mouse adenohypophysis. *J. Cell Biol.* **43**, 361–367.

Dirksen, E. R. (1971). Centriole morphogenesis in developing ciliated epithelium of the mouse oviduct. *J. Cell Biol.* **51**, 286–302.

Dobell, C. (1932). "Antony van Leeuwenhoek and his 'Little Animals'." p. 435. Harcourt, Brace and Co., New York.

Doughty, M. J. (1998). Changes in cell surface primary cilia and microvilli concurrent with measurements of fluid flow across the rabbit corneal endothelium *ex vivo*. *Tissue Cell* **30**, 634–643.

DuJardin, F. (1841). "Histoire Naturelle des Zoophytes (Infusoires)." Roret, Paris.

Duncan, D. (1957). Electron microscope study of the embryonic neural tube and notochord. *Tex. Rep. Biol. Med.* **15**, 367–377.

Ecker, A. (1844). Flimmerbewegung im Gehörorgan von *Petromyzon marinus*. *Arch. Anat. Physiol. Wiss. Med. (Müller's Archiv)* 520–521.

Engelmann, T. W. (1880). Zur Anatomie und Physiologie der Flimmerzellen. *Pflugers Arch.* **23**, 505–535.

Erhard, H. (1910). Studien uber Flimmerzellen. *Arch. Zellforsch.* **4**, 309–442.

Erhard, H. (1911). Die Henneguy-Lenhosseksche Theorie. *Ergeb. Anat. Entwickelungsgesch.* **19**, 893–929.

Fawcett, D. W., and Porter, K. R. (1954). A study of the fine structure of ciliated epithelium. *J. Morphol.* **94**, 221–281.

Fernandes, J., Lorenzo, I. M., Andrade, Y. N., Garcia-Elias, A., Serra, S. A., Fernandez-Fernandez, J. M., and Valverde, M. A. (2008). IP_3 sensitizes TRPV4 channel to the mechano- and osmotransducing messenger 5–6-epoxyeicosatrienoic acid. *J. Cell Biol.* **181**, 143–155.

Flemming, W. (1875). Studien uber die Entwicklungsgeschichte der Najaden. *Sitzungsgeber. Akad. Wiss. Wien* **71**, 81–147.

Flemming, W. (1891). Ueber Zellteilung. *Verh. Anat. Ges. Versamml. Munchen* **8**, 125–143.

Fliegauf, M., Benzing, T., and Omran, H. (2007). Mechanisms of disease—When cilia go bad: Cilia defects and ciliopathies. *Nat Rev. Mol. Cell Biol.* **8**, 880–893.

Flock, A. and Duvall, A. J. (1965). The ultrastructure of the kinocilium of the sensory cells in the inner ear and lateral line organs. *J. Cell Biol.* **25**, 1–8.

Fonte, V. G., Searls, R. L., and Hilfer, S. R. (1971). The relationship of cilia with cell division and differentiation. *J. Cell Biol.* **49**, 226–229.

Friedlander, M. and Wahrman, J. (1970). The spindle as a basal body distributor: A study in the meiosis of the male silkworm moth, *Bombyx mori*. *J. Cell Sci.* **7**, 65–89.

Gall, J. G. (2004). Early studies on centrioles and centrosomes. *In* "Centrosomes in Development and Disease" (E. A. Nigg, ed.), pp. 3–15. Wiley-VCH, Weinheim.

Gallagher, B. C. (1980). Primary cilia of the corneal epithelium. *Am. J. Anat.* **159**, 475–484.

Gerdes, J. M., and Katsanis, N. (2008). Ciliary function and Wnt signal modulation. *Curr. Top. Dev. Biol.* **85**, 175–195.

Gibbons, B. H., Gibbons, I. R., and Baccetti, B. (1983). Structure and motility of the $9+0$ flagellum of eel spermatozoa. *J. Submicrosc. Cytol.* **15**, 15–20.

Gray, J. (1928). "Ciliary Movement." The MacMillan Company, New York.

Grigg, G. W. and Hodge, A. J. (1949). Electron microscopic studies of spermatozoa. *Aust. J. Sci. Res. (B)* **2**, 271–286.

Gurwitsch, A. (1901). Studien uber Flimmerzellen. Theil I. Histogenese der Flimmerzellen. *Arch. Mikrosk. Anat.* **57**, 184–229.

Gurwitsch, A. (1902). Der Haarbuschel der Epithelzellen im Vas epididymis des Menschen. Zugleich ein Beitrag zur Centralkoperfrage in den Epithelien. *Arch. Mikrosk. Anat.* **59**, 32–62.

Gurwitsch, A. (1904). "Morphologie und Biologie der Zelle." Gustav Fischer, Jena.

Han, Y-G, Kwok, B. H., and Kernan, M. J. (2003). Intraflagellar transport is required in *Drosophila* to differentiate sensory cilia but not sperm. *Curr. Biol.* **13**, 1679–1686.

Heidemann, S. R. and Kirschner, M. W. (1975). Aster formation in eggs of *Xenopus leavis*. Induction by basal bodies. *J. Cell Biol.* **67**, 105–117.

Heidenhain, M. (1892). Uber Kern and Protoplasma. *In* "Festschrift zur 50 jahr. Doctorjubilaum von Geheimrat A. V. Kolliker." pp. 109–166. W. Englemann, Leipzig.

Heidenhain, M. (1907). "Plasma und Zelle. Erste Lieferung: Die Grundlagen der Mikroskopischen Anatomie, die Kerne, die Centren und die Granulalehre." 506 pp. Gustav Fischer, Jena.

Heidenhain, M. (1914). Uber die Sinnesfelder und die Geschmacksknosphen der Papilla foliate des Kaninchens. *Arch. Mikrosk. Anat.* **85**, 365–479.

Henneguy, L. F. (1898). Sur les rapports des cils vibratiles avec les centrosomes. *Archives d'Anatomie Microscopique* **1**, 481–496.

Hermann, F. (1889). Beitrage zur Histologie des Hodens. *Arc. Mikrosk. Anat.* **34**, 58–106.

Hertwig, O. (1906). "Allgemeine Biologie." Fischer, Jena.

Hierck, B. P., Van der Heiden, K., Alkemade, F. E., Van de Pas, S., Van Thienen, J. V., Groenendijk, B. C. W., Bax, W. H., Van der Laarse, A., DeRuiter, M. C., Horrevoets, A. J.G. and Poelmann, R. E. (2008). Primary cilia sensitize endothelial cells for fluid shear stress. *Dev. Dyn.* **237**, 725–735.

Hintzsche, E. (1936). Karl Wilhelm Zimmermann zum Gedachtnis. Anatomischer Anzeiger **82**, 300–311.

Huangfu, D., and Anderson, K. V. (2005). Cilia and hedgehog responsiveness in the mouse. *Proc. Natl. Acad. Sci. USA* **102**, 11325–11330.

Ikeda, R. (1906). Uber das Epithel im Nebenhoden des Menschen. *Anat. Anz.* **29**, 1–14, 76–82.

Iomini, C., Tejada, K., Mo, W., Vaananen, H., and Piperno, G. (2004). Primary cilia of human endothelial cells disassemble under laminar shear stress. *J. Cell Biol.* **164**, 811–817.

Jenkins, R. A. (1967). Fine structure of division in ciliate protozoa I. Micronuclear mitosis in *Blepharisma. J. Cell Biol.* **34**, 463–481.

Jennings, H. S. (1906). "Behavior of the Lower Organisms." pp. 47–49. Columbia University Press, New York.

Jordan, H. E. (1913). Amitosis in the epididymis of the mouse. *Anat. Anz.* **43**, 598–612.

Joseph, H. (1903). Beitrage zur Flimmerzellen-und Centrosomafrage. *Arbeiten aus den Zoologischen Instituten der Universität Wien und der Zoologischen Station in Triest* **14**, 1–80.

Joseph, H. (1905). Uber die Zentralkorper der Nierenzelle. *Verh. Anat. Ges.* **19**, 178–187.

Kahn-Kirby, A. H., and Bargmann, C. I. (2006). TRP channels in *C. elegans. Annu. Rev. Physiol.* **68**, 719–736.

Kartagener, M. (1933). Zur Pathogenese der Bronchiektasien. I. Mitteilung: Bronchiektasien bei Situs viscerum inversus. *Beitr. Klin. Erforsch. Tuberk Lungenkr.* **83**, 489–501.

Kernan, M. J. (2007). Mechanotransduction and auditory transduction in *Drosophila. Pflugers Arch. Eur. J. Physiol.* **454**, 703–720.

Kindred, J. E. (1927). Cell division and ciliogenesis in the ciliated epithelium of the pharynx and esophagus of the tadpole of the green frog, *Rana clamitans. J. Morphol.* **43**, 267–297.

Kofoid, C. A., and Swezy, O. (1915). Mitosis in trichomonas. *Proc. Natl. Acad. Sci. USA* **1**, 315–321.

Kolliker, A. (1854). "Manual of Human Microscopical Anatomy" (translation by George Busk of Mikroskopische Anatomie). Lippincott, Grambo and Company, Philadelphia.

Kowalevsky. A. (1867). Entwickelungsgeschichte des *Amphioxus lanceolatus. Mém. Acad. Imp. Sci. St.-Pétersbourg (Ser VII)* **11**(4), 1–17 + Tafel I–III.

Kowalevsky, A. (1877). Weitere Studien über die Entwicklungsgeschichte des *Amphioxus lanceolatus*, nebst einem Beitrage zur Homologie des Nervensystems der Würmer und Wirbelthiere. *Arch. Mikrosk. Anat.* **13**, 181–204 + Tafel XV and XVI.

Kozminski, K. G., Johnson, K. A., Forscher, P., and Rosenbaum, J. L. (1993). A motility in the eukaryotic flagellum unrelated to flagellar beating. *Proc. Natl. Acad. Sci. USA* **90**, 5519–5523.

Kupelwieser, H. (1906). Untersuchungen uber den feineren Bau und die Metamorphose des Cyphonautes. *Zool. Heft* 47, **19**, 1–50.

LaFountain J. R., Jr. (1976). Analysis of birefringence and ultrastructure of spindles in primary spermatocytes of *Nephrotoma suturalis* during anaphase. *J. Ultrastruct. Res.* **54**, 333–346.

Langerhans, P. (1876). Zur Anatomie des *Amphioxus. Arch. Mikrok. Anat.* **12**, 290–348.

Latta, H., Maunsbach, A. B., and Madden, S. C. (1961). Cilia in different segments of the rat nephron. *J. Biophys. Biochem. Cytol.* **11**, 248–252.

Leeson, T. S. (1960). Electron microscope studies of newborn hamster kidney. *Norelco Rep.* **7**, 45–47.

Leeson, T. S. (1962). Electron microscopy of the rete testis of the rat. *Anat. Rec.* **144**, 57–67.

Leeuwenhoek, A. (1677). Concerning little animals observed in rain-, well-, sea- and snow-water; as also in water wherein pepper had lain infused. *Philos. Trans. Lond.* **12**, 821–31.

Lehman, J. M., Michaud, E. J., Schoeb, T. R., Aydin-Son, Y., Miller, M., and Yoder, B. K. (2008). The Oak Ridge polycystic kidney mouse: Modeling ciliopathies of mice and men. *Dev. Dyn.* **237**, 1960–1971.

Lenhossek, M. von (1898a). Untersuchungen uber Spermatogenese. *Arch. Mikrosk. Anat.* **51**, 215–318.

Lenhossek, M. von (1898b). Ueber Flimmerzellen. *Verh. Anat. Ges.* **12**, 106–128.

Lorenzo I. M., Liedtke, W., Sanderson, M. J., and Valverde, M. A. (2008). TRPV4 channel participates in receptor-operated calcium entry and ciliary beat frequency regulation in mouse airway epithelial cells. *Proc. Natl. Acad. Sci. USA* **105**, 12611–12616.

Luban, S. (1918). Uber Eigentumliche vorgange in den Flimmerzellen des Menschlichen Uteruskorpers. *Anat. Hefte. Beitrage und Referate zur Anatomie und Entwickelungsgeschichte* **56**, 269–303.

Maier, H. N. (1903). Ueber den feineren Bau der Wimperapparate der Infusorien. *Arch. Protistenkunde* **2**, 73–179.

Malone, A. M.D., Anderson, C. T., Padmaja, T., Kwon, R. Y., Johnson, T. R., Sterns, T., and Jacobs, C. R. (2007). Primary cilia mediate mechanosensing in bone cells by a calcium-independent mechanism. *Proc. Natl. Acad. Sci. USA* **104**, 13325–13330.

Manton, I., and Clarke, B. (1952). An electron microscopic study of the spermatozoid of *Sphagnum*. *J. Exp. Bot.* **3**, 265–275.

Manton, I., and Stosch, H. A. von (1966). Observations on the fine structure of the male gamete of the marine centric diatom *Lithodesmium undulatum*. *J. R. Microsc. Soc.* **85**, 119–13.

Masyuk, A. I., Masyuk, T. V., Splinter, P. L., Huang, B. Q., Stroope, A. J., and LaRusso, N. F. (2006). Cholangiocyte cilia detect changes in luminal fluid flow and transmit them into intracellular Ca^{2+} and cAMP signaling. *Gastroenterology* **131**, 911–920.

McEwen, D. P., Jenkins, P. M., and Martens, J. R. (2008). Olfactory cilia: Our direct neuronal connection to the outside world. *Curr. Top. Dev. Biol.* **85**, 333–370.

McGrath, J., Somlo, S., Makova, S., Tian, X., and Brueckner, M. (2003). Two populations of node monocilia initiate left-right asymmetry in the mouse. *Cell* **114**, 61–73.

Meves, F. (1897). Ueber Struktur und Histogenese der Samenfaden von *Salamandra maculosa*. *Arch. Mikrosk. Anat.* **50**, 110–141.

Meves, F. (1899). Ueber den Einfluss der Zellteilung auf den Sekretionsvorgang, nach Beobachtungen an der Niere der Salamanderlarve. *In* "Festschrift zum Siebenzigsten Geburtstag von Carl von Kupffer." pp. 57–62. Gustav Fischer, Jena.

Meves, F. (1900). Ueber den von la Valette St. George entdeckten Nebenkern (Mitochondrienkorper) der Samenzellen. *Arch. Mikrosk. Anat.* **56**, 553–606.

Meves, F. (1903). Ueber oligopyrene und apyrene Spermien und uber ihre Entstehung, nach Beobachtungen an Paludina und Pygaera. *Arch. Mikrosk. Anat.* **61**, 1–84.

Moore J. E.S. (1895). Structural changes in the reproductive cells during spermatogenesis of elasmobranchs. *Q. J. Microsc. Sci.* **38**, 275–313.

Mori, Y., Akedo, H., Tanigaki, Y., Tanaka, K., and Okada, M. (1979). Ciliogenesis in tissue-cultured cells by the increased density of cell populations. *Exp. Cell Res.* **12**, 435–439.

Moyer, J. H., Lee-Tischler, M. J., Kwon, H. Y., Schrick, J. J., Avner, E. D., Sweeney, W. E., Godfrey, V. L., Cacheiro, N. L., Wilkinson, J. E., and Woychik, R. P. (1994). Candidate gene associated with a mutation causing recessive polycystic kidney disease in mice. *Science* **264**, 1329–1333.

Muller, O. F. (1786). "Animalcula infusoria; fluvia tilia et marina, que detexit, systematice descripsit et ad vivum delineari curavit." Molleri, Havniae.

Munger, B. L. (1958). A light and electron microscopic study of cellular differentiation in the pancreatic islets of the mouse. *Am. J. Anat.* **103**, 275–297.

Myklebust, R., Engedal, H., Saetersdal, T. S., and Ulstein, M. (1977). Primary 9 + 0 cilia in the embryonic and the adult human heart. *Anat. Embryol.* **151**, 127–139.

Nauli, S. M., Kawanabe, Y., Kaminski, K. K., Pearce, W. J., Ingbar, D. E., and Zhou, J. (2008). Endothelial cilia are fluid shear sensors that regulate calcium signaling and nitric oxide production through polycystin-1. *Circulation* **117**, 1161–1171.

Newell, G. E., and Baxter, E. W. (1936). On the nature of the free cell-border of certain mid-gut epithelia. *Q. J. Microsc. Sci.* **s2–79**, 123–150.

Newman, S., Borysko, E., and Swerdlow, M. (1949). A new sectioning technique for light and electron microscopy. *Science* **110**, 66–68.

Nonaka, S., Shiratori, H., Saijoh, Y., and Hamada, H. (2002). Determination of left right patterning of the mouse embryo by artificial nodal flow. *Nature* **418**, 96–99.

Olney, J. W. (1968). An electron microscopic study of synapse formation, receptor outer segment development, and other aspects of developing mouse retina. *Invest. Ophthalmol.* **7**, 250–268.

Palade, G. E. (1952). A study of fixation for electron microscopy. *J. Exp. Med.* **95**, 285–298.

Palay, S. I. (1960). The fine structure of secretory neurons in the preoptic nucleus of the goldfish (*Carassias auratus*). *Anat. Rec.* **138**, 417–443.

Palay, S. I. (1961). Structural peculiarities of the neurosecretory cells in the pre-optic nucleus of the goldfish *Carassias auratus*. *Anat. Rec.* **139**, 262 (Abstract).

Pazour, G. J., and Bloodgood, R. A. (2008). Targeting proteins to the ciliary membrane. *Curr. Top. Dev. Biol.* **85**, 115–149.

Pazour, G. J., Dickert, B. L., Vucica, Y., Seeley, E. S., Rosenbaum, J. L., Witman, G. B., and Cole, D. G. (2000). *Chlamydomonas* IFT88 and its mouse homologue, polycystic kidney disease gene Tg737, are required for assembly of cilia and flagella. *J. Cell Biol.* **151**, 709–718.

Pazour, G. J., San Augustin, J. T., Follit, J. A., Rosenbaum, J. L., and Witman, G. B. (2002). Polycystin-2 localizes to kidney cilia and the ciliary level is elevated in orpk mice with polycystic kidney disease. *Curr. Biol.* **12**, R378–R380.

Pinkus, H. (1951). Examination of the epidermis by the strip method of removing horny layers. I. Observations on thickness of the horny layer, and on mitotic activity after stripping. *J. Invest. Dermatol.* **16**, 383–386.

Pollister, A. W. (1933). Notes on the centrioles of amphibian tissue cells. *Biol. Bull. (Woods Hole)* **65**, 529–545.

Porter, K. R., and Blum, J. (1953). A study in microtomy for electron microscopy. *Anat. Rec.* **117**, 685–710.

Praetorius, H. A., and Spring, K. R. (2001). Bending the MCDK cell primary cilium increases intracellular calcium. *J. Membr. Biol.* **184**, 71–79.

Praetorius, H. A., and Spring, K. R. (2002). Removal of the MDCK cell primary cilium abolished flow sensing. *J. Membr. Biol.* **191**, 69–76.

Prowazek, S. von (1901). Spermatologische Studien. I. Spermatogenese der Weinbergschnecke (Helix pomatia L.). *Arbeiten aus den Zoologischen Instituten der Universität Wien und der Zoologischen Station in Triest* **13**, 197–222.

Pugacheva, E. N., Jablonski, S. A., Hartmann, T. R., Henske, E. P., and Golemis, E. A. (2007). HEF1-dependent Aurora A activation induces disassembly of the primary cilium. *Cell* **129**, 1351–1363.

Purkinje, J. E., and Valentin, G. G. (1834). Entdeckung continuerlicher durch Wimperhaare erzeugter Flimmerbewegungen. *Arch. Anat. Physiol. Wiss. Med.*, 391–400.

Purkinje, J. E., and Valentin, G. G. (1835). De phaenomeno generali et fundamentali motus vibratorii continui in membranis cum externis tum internis animalium plurimorum et superiorum et inferiorum ordinum obvii. *In* "Opera Omnia" (J. E. Purkinje, ed.), Vol. 1, pp. 277–371. Vratislaviae, Bratislava.

Rash, J. E., Shay, J. W., and Biesele, J. J. (1969). Cilia in cardiac differentiation. J. *Ultrastruct. Res.* **29**, 470–484.

Rasmussen, N. (1997). "Picture Control: The Electron Microscope and the Transformation of Biology in America, 1940–1960." p. 436. Stanford University Press, Stanford.

Renyi, G. (1924). Untersuchungen uber Flimmerzellen. *Z. Anat. Entwicklungsgesch.* **73**, 338–357.

Rieder, C. L., Jensen, C. G., and Jensen, L. C.W. (1979). The resorption of primary cilium during mitosis in a vertebrate (PtK1) cell line. *J. Ultrastruct. Res.* **68**, 173–185.

Ringo, D. L. (1967). Flagellar motion and find structure of the flagellar apparatus in *Chlamydomonas*. *J. Cell Biol.* **38**, 543–571.

Rosenbaum, J. L., and Witman, G. B. (2002). Intraflagellar transport. *Nat. Rev. Mol. Cell Biol.* **3**, 813–825.

Saguchi, S. (1917). Studies on ciliated cells. *J. Morphol.* **29**, 217–279.

Salzmann, M. (1912). "The Anatomy and Histology of the Human Eyeball in its Normal State, its Development and Senescence." University of Chicago Press, Chicago.

Sawamoto, K., Wichterle, H., Gonzalez-Perez, O., Cholfin, J. A., Yamada, M., Spassky, N., Murcia, N. S., Garcia-Verdugo, J. M., Marin, O., Rubenstein, J. L.R., Tessier-Lavigne, M., Okano, H., et al. (2006). New neurons follow the flow of cerebrospinal fluid in the adult brain. *Science* **311**, 629–632.

Scholey, J. M. (2003). Intraflagellar transport. *Annu. Rev. Cell Dev. Biol.* **19**, 423–443.

Schwartz, E. A., Leonard, M. L., Bizios, R., and Bowser, S. S. (1997). Analysis and modeling of the primary cilium bending response to fluid shear. *Am. J. Physiol.* **272**, F132–F138.

Shah, A. S., Ben-Shahar, Y., Moninger, T. O., Kline, J. N., and Welsh, M. J. (2009). Motile cilia of human airway epithelia are chemosensory. *Science* **325**, 1131–1134.

Sharma, N., Berbari N. F., and Yoder, B. K. (2008). Ciliary dysfunction in developmental abnormalities and diseases. *Curr. Top. Dev. Biol.* **85**, 371–427.

Sharp, L. W. (1921). "An Introduction to Cytology." p. 452. McGraw-Hill Book Company, New York.

Sharpey, W. (1835). Cilia. *In* "Cyclopoedia of Anatomy and Physiology" (R. B. Todd, ed.), Vol. 1, pp. 606–638. Longman, Brown, Green, Longman and Roberts, London.

Sherwin, T., and Gull, K. (1989). The cell division cycle of *Trypanosoma brucei*: Timing of event markers and cytoskeletal modulations. *Phil. Trans. R. Soc. Lond. B* **323**, 573–588.

Siewert, A. K. (1904). Über einen Fall von Bronchiectasie bei einem Patienten mit situs inversus viscerum. *Berl. Klin. Wochenschr.* **41**, 139–141.

Silverman, M. A., and Leroux, M. R. (2009). Intraflagellar transport and the generation of dynamic, structurally and functionally diverse cilia. *Trends Cell Biol.* **19**, 306–316.

Sjostrand, F. S. (1953). The ultrastructure of the inner segments of the retinal rods of the guinea pig eye as revealed by electron microscopy. *J. Cell. Comp. Physiol.* **42**, 45–70.

Sleigh, M. A. (1962). "The Biology of Cilia and Flagella." The MacMillan Company, New York.

Sorokin, S. (1962). Centrioles and the formation of rudimentary cilia by fibroblasts and smooth muscle cells. *J. Cell Biol.* **15**, 363–377.

Sorokin, S. P. (1968). Reconstructions of centriole formation and ciliogenesis in mammalian lungs. *J. Cell Sci.* **3**, 207–230.

Sotelo, J. R., and Trujillo-Cenoz, O. (1958). Electron microscope study on the development of ciliary components of the neural epithelium of the chick embryo. *Z. Zellforsch. Mikrosk. Anat.* **49**, 1–12.

Sottiurai, V., and Malvin, R. L. (1972). The demonstration of cilia in canine macula densa cells. *Am. J. Anat.* **135**, 281–286.

Stubblefield, E., and Brinkley, B. R. (1966). Cilia formation in Chinese hamster fibroblasts *in vitro* as a response to colcemid treatment. *J. Cell Biol.* **30**, 645–652.

Studnicka, F. K. (1898). Untersuchungen uber den Bau der Sehnerven der Wirbeltiere. *Jena. Z. Naturwiss.* **31**, 1–28.

Studnicka, F. K. (1899). Ueber Flimmer-und Cuticularzellen mit besonderer Berucksichtigung der Centrosomenfrage. *Sitzungsberichte Konigl. Bohmischen Gesellschaft der Wissenschaften Math.-Naturwiss. Classe* **35**, 1–22 (Juni).

Taxi, J. (1961). Sur l'existence de neurons cilies dans les ganglions sympathetiques de certains vertebres. *C. R. Soc. Biol. (Paris)* **155**, 1860–1863.

Teilmann, S. C., Byskov, A. G., Pedersen, P. A., Wheatley, D. N., Pazour, G. J., and Christensen, S. T. (2005). Localization of transient receptor potential ion channels in primary and motile cilia of the female murine reproductive organs. *Mol. Reprod. Dev.* **71**, 444–52.

Teilmann, S. C., Clement, C. A., Thorup, J., Byskov, A. G., and Christensen, S. T. (2006). Expression and localization of the progesterone receptor in mouse and human reproductive organs. *J. Endocrinol.* **191**, 525–535.

Titford, M. (2005). The long history of hematoxylin. *Biotech. Histochem.* **80**, 73–78.

Tobin, J. L., and Beales, P. L. (2009). The nonmotile ciliopathies. *Genet. Med.* **11**, 386–402.

Trufillo-Cenoz, D. (1957). Electron microscopic studies of the rabbit gustatory bud. *Z. Zellforsch. Mikrosk. Anat.* **46**, 272–280.

Tucker, R. W., Pardee, A. B., and Fujiwara, K. (1979). Centriole ciliation is related to quiescence and DNA synthesis in 3T3 cells. *Cell* **17**, 527–535.

Van Beneden, E. (1876). Contribution a l'histoire de la vesiculaire germinative et du premier noyau embryonnaire. *Bull. Acad. R. Belg (2me series)* **42**, 35–97.

Van der Heiden, K., Groenendijk, B. C., Hierck, B. P., Hogers, B., Koerten, H. K., Mommaas, A. M., Gittenberger-de Groot, A. C., and Poelmann, R. E. (2006). Monocilia on chicken embryonic endocardium in low stress areas. *Dev. Dyn.* **235**, 19–28.

Van der Heiden, K., Hierck, B. P., Krams, R., de Crom, R., Cheng, C., Baiker, M., Pourquie, M. J.B. M., Alkemade, F. E., DeRuiter, M. C., Gittenberger-de Groot, A. C., and Poelmann, R. E. (2008). Endothelial primary cilia in areas of disturbed flow are at the base of atherosclerosis. *Atherosclerosis* **196**, 542–550.

Verghese, E., Ricardo, S. D., Weidenfeld, R., Zhuang, J., Hill, P. A., Langham, R. G., and Deane, J. A. (2009). Renal primary cilia lengthen after acute tubular necrosis. *J. Am. Soc. Nephrol.* **20**, 2147–2153.

Verghese, E., Weidenfeld, R., Bertram, J. F., Ricardo, S. D., and Deane, J. A. (2008). Renal cilia display length alterations following tubular injury and are present early in epithelial repair. *Nephrol. Dial. Transplant.* **23**, 834–41.

Wallengren, H. (1905). Zur Kenntnis der Flimmerzellen. *Z. Allg. Physiol.* **5**, 351–414.

Walter, L. (1926). Das flagellum der endothelzellen. Zeitschrift fur die gesamte. *Anatomie* **81**, 738–741.

Walter, L. (1929). Ciliogenesis in a cystically dilated hydatid of Morgagni. *Anat. Rec.* **42**, 177–187.

Weidman, T. A., and Kuwabara, T. (1968). Postnatal development of the rat retina. *Arch. Ophthalmol.* **79**, 470–484.

Well, D., Blanchard, S., Kaplan, J., Guilford, P., Gibson, F., Walsh, J., Mburu, P., Varela, A., Levilliers, J., Weston, M. D., Kelley, P. M., Kimberling, W. J., et al. (1995). Defective myosin VIIA gene responsible for Usher syndrome type 1B. *Nature* **374**, 60–61.

Wenrich, D. H. (1921). The structure and division of *Trichomonas muris* (Hartmann). *J. Morphol.* **36**, 119–155.

Wheatley, D. N. (1967). Cilia and centrioles in the rat adrenal cortex. *J. Anat.* **101**, 223–237.

Wheatley, D. N. (1969). Cilia in cell-cultured fibroblasts. I. On their occurrence and relative frequencies in primary cultures and established cell lines. *J. Anat.* **105**, 351–362.

Wheatley, D. N. (1982). "The Centriole, a Central Enigma of Cell Biology." Elsevier Biomedical Press, Amsterdam.

Wheatley, D. N. (1995). Primary cilia in normal and pathological tissues. *Pathobiology* **63**, 222–238.

Wheatley, D. N. (2005). Landmarks in the first hundred years of primary (9 + 0) cilium research. *Cell Biol. Int.*, **29**, 333–339.

Wheatley, D. N., Wang, A. M., and Strugnell, G. E. (1996). Expression of primary cilia in mammalian cells. *Cell Biol. Int.* **20**, 73–81.

Wilson, E. B. (1934). "The Cell in Development and Heredity." 3rd edn., with corrections. The Macmillan Company, New York.

Wilson, R. B., and McWhorter, C. A. (1963). Isolated flagella in human skin. *Lab. Invest.* **12**, 242–249.

Wolfrum, U. (2003). The cellular function of the usher gene product myosin VIIa is specified by its ligands. *Adv. Exp. Med. Biol.* **533**, 133–42.

Wong, S. Y., and Reiter, J. F. (2008). The Primary Cilium: At the crossroads of mammalian hedgehog signaling. *Curr. Top. Dev. Biol.* **85**, 225–260.

Yoder, B. K., Hou, X., Guay-Woodford, L. M. (2002). The polycystic kidney disease proteins, polycystin-1, polycystin-2, Polaris, and cystin, are co-localized in renal cilia. *J. Am. Soc. Nephrol.* **13**, 2508–2516.

Zernicke, F. (1955). How I discovered phase contrast. *Science* **121**, 345–349.

Zimmermann, K. W. (1894). Demonstration: Plastische reconstruction des hirnrohres; Schnittserie, Kaninchenembryo; Photogramm; Praparate von Uterus, Nebenhoden, Darm, Ureter, Niere, Thranendruse. *Verhandlungen der Anatomischen Gesellschaft auf der achten Versammlung zu Strassburg, vom 13–16 Mai*, **8**, 244–245.

Zimmermann, K. W. (1898). Beitrage zur Kenntniss einiger Drusen und Epithelien. *Arch. Mikrosk. Anat.* **52**, 552–706.

Zimmermann, K. W. (1933). Uber den Bau des Glomerulus der Saugerniere. *Z. Mikrosk. Anat. Forsch.* **32**, 176–278.

CHAPTER 2

Origin of the Cilium: Novel Approaches to Examine a Centriolar Evolution Hypothesis

Mark C. Alliegro* *and* **Peter Satir†**

*Bay Paul Center for Comparative Molecular Biology and Evolution, Marine Biological Laboratory, Woods Hole, Massachusetts 02543

†Department of Anatomy and Structural Biology, Albert Einstein College of Medicine, Bronx, New York 10461

Abstract

Recently, a new hypothesis was proposed regarding the evolution of the cilium from an enveloped RNA virus (Satir *et al.*, 2007, *Cell Motil. Cytoskeleton* **64,** 906). The hypothesis predicts that there may be specific centriolar or basal body RNAs with sequences reminiscent of retroviruses, and/or that the nuclear genes for certain centriole-specific proteins would have viral origins. Four independent

978-0-12-375024-2
DOI: 10.1016/S0091-679X(08)94002-4

laboratories have reported the existence of centrosomal RNA (cnRNA). Methods for studying cnRNA are described. We analyzed evidence of relatedness of known full-length cnRNAs to extant viral molecules. Out of 14 cnRNAs studied, 12 have similarity to entries in viral databases, all but one of these with E-values of $\leq 1e^{-4}$. Some centrosomal, and possibly uniquely centriolar, proteins also have relatives in viral databases that meet the criteria accepted to indicate a relationship by descent. Nine general cytoskeleton proteins exhibited no significant similarity to viral proteins. The speculation that centrioles are invaders of RNA viral origin in the evolving eukaryotic cell is strengthened by these findings.

I. Introduction

Recently, a new detailed hypothesis was proposed regarding the evolution of the cilium from an enveloped RNA virus whose core was a primitive basal body (Satir *et al.*, 2007). The hypothesis contrasted in many important ways with the two previous scenarios of ciliary evolution: (1) from a spirochete (Margulis, 1981) or (2) autogenously, from a bundling of spindle or cytoplasmic microtubules via motor and membrane interactions (Jékely and Arendt, 2006; Mitchell, 2007). Important features of the autogenous and viral hypothesis have been summarized and compared in a new review (Satir *et al.*, 2009).

The viral hypothesis proposes that the invasion occurred in a proto-eukaryotic cytoplasm that was already capable of endocytosis and exocytosis, and membrane movement in general, in part along a microtubule cytoskeleton. The invasive RNA, released from the core after invasion, was processed by reverse transcriptase as a retrovirus that initially incorporated into the developing eukaryotic nucleus. Transcription of viral genes led to the production of several novel proteins so that the assembly produced multiple protocentriolar-like cores in the cytoplasm (Fig. 1). At maturity these protocentrioles reincorporated viral RNA and bound to the cell membrane, which encapsulated the virus as it exited the cell.

The basal body first evolved attached to the cell membrane when such full encapsulation and exit were blocked. The membrane above the protocentriolar attachment could then incorporate specific signaling receptors, which permitted directional detection of ligands and acted as a sensory bulge. These features would seem to be advantageous for the cell and would be selected for. Selection would also favor a reduction in number of the attached protocentrioles to defined positions on the cell surface, so that directional detection could be enhanced. Elongation of the basal body (comparable to length mutations in bacteriophage—Kellenberger, 1966) would then extend the bulge as a primitive 9 + 0 sensory cilium. Motility would develop later as axonemal dynein isoforms arose from cytoplasmic dynein (Asai and Koonce, 2001) and a central apparatus appeared. The motile 9 + 2 axoneme would convey further selective advantage for behavioral response to attractants or repellants.

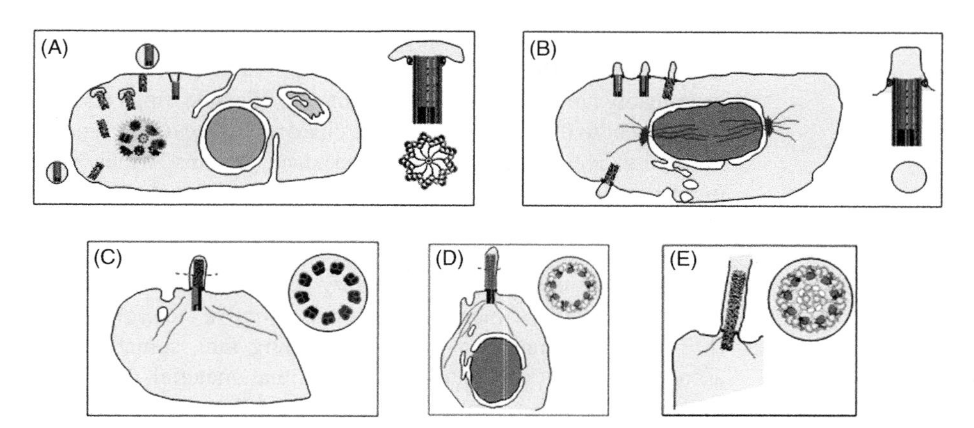

Fig. 1 The viral hypothesis of ciliary evolution. (A) Viral invasion and reproduction in the protoeukayotic cytoplasm. The centriole-like viral core arises in a cytoplasmic factory. Each emerging RNA virus attaches to a vesicle and is eventually enveloped by membrane and shed. Cross section shows the cartwheel. (B) The centriole-like virus no longer becomes completely enveloped and shed; instead it remains attached to the cell membrane. A sensory bulge with membrane receptors delivered by IFT and capable of positional signaling develops. The cell is drawn to emphasize that intranuclear division occurs on microtubules unrelated to the centriolar virus. (C) Signaling leads to restriction in organelle number. An elongation mutation extends the capsid into the sensory bulge giving rise to a 9 + 0 axoneme, shown in cross section. (D) Cytoplasmic dynein diversification and permanent attachment of novel dyneins to axonemal doublet microtubules give rise to motility (E). Central pair develops, completing 9 + 2 axoneme with efficient signaling and motility. Images from Satir *et al.* (2007). Courtesy *Cell Motil. Cytoskeleton.*

A. Features of the Basal Body/Centriole Explained by the Viral Hypothesis

As discussed previously (Satir *et al.*, 2009), the viral hypothesis offers relatively easy explanation of several remarkable features of ciliary morphogenesis:

1. Explosive and *de novo* assembly, with many basal bodies, sometimes hundreds, arising at once in a cytoplasmic factory (reminiscent of a viral factory), is found in many phyla. Some examples are anarchic oral field formation in ciliates, gametogenesis in *Gingko* or *Marsilea* (Mizukami and Gall, 1966), and ciliogenesis in mammalian respiratory epithelia (Dirksen, 1991). In the ameboflagellate *Naegleria* and related genera, normally only two to four cilia form, but ciliogenesis is *de novo* (Fulton and Dingle, 1971). Are these situations, rather than templated mother–daughter assembly, the primitive conditions? (for further discussion see Marshall, 2009; Nigg, 2007; Salisbury, 2007; Kuriyama, 2009)
2. In viral assembly, symmetry and length are often rigidly defined by self-assembly of subunit proteins. Such stereotypic self-assembly with triplet microtubules of defined length polymerizing on a basal plate (the cartwheel), defined ninefold symmetry and a single enantiomorphic form, is characteristic of basal bodies and centrioles.

3. Basal bodies contain unique proteins including unusual tubulin isoforms, tektins, and ribbon proteins (Setter *et al.*, 2006) not present in cytoplasmic microtubules, which presumably are necessary for the formation of stable triplet and doublet microtubules. Attachment to the membrane via the ciliary necklace (Gilula and Satir, 1972) involves novel, relatively uncharacterized, necklace proteins, which probably interact with intraflagellar transport (IFT) particles. Some additional proteins that are required for centriole formation are described in Leidel and Gönczy (2005).
4. The only microtubules that grow directly from basal body (centriolar) microtubules are the outer doublets of the ciliary axoneme. Other cytoplasmic microtubules including spindle microtubules arise from a microtubule organizing center (MTOC) consisting of γ-tubulin-containing foci, sometimes (but not always) in association with pericentriolar centrosomal material (Gould and Borisy, 1977). These observations imply that the ciliary doublet microtubules uniquely arise on a centriolar template and not vice versa.
5. Centrioles are necessary for cilia formation, but not for mitosis. A case in point is *Naegleria* (Fulton and Dingle, 1971), which has cilia in its flagellate state, but does not have centrioles as a dividing ameba. In a particularly informative experiment, Basto *et al.* (2006) ablated centrioles in *Drosophila* cells. Flies with a normal overall body plan and tissue differentiation developed, but they were missing all sensory cilia. These observations, as well as the existence of many mitotic cell types that completely lack centrioles, uncouple centriole and cilia evolution from cell division, suggesting that centriolar duplication connected to the cell cycle is a secondary event.

B. Predictions of the Viral Hypothesis

Three specific predictions of the viral hypothesis as stated are :

1. The sensory cilium originates before the motile cilium.
2. The first ciliated cells were neither unikonts nor bikonts, but rather multiciliated, as might be expected from an abortive viral budding process. This prediction is quite different from those made from the autogenous hypothesis and might be testable via extensive genomic analysis of primitive protists.
3. As originally described by Alliegro *et al.* (2006) (see below), there may be specific centriolar or basal body RNAs with sequences reminiscent of retroviruses sometimes found within the organelle. Although the centriolar genome is now completely or largely nuclear and dispersed, during morphogensis of the organelle, centriolar genes must be turned on as cassettes in specific sequences. These centriolar genes, some of which must code for the specific unique centriolar proteins mentioned above, could have features that link them to the original centriolar virus. This prediction, if correct, would provide strong support for the viral hypothesis (or if the sequences were related to spirochetes to that hypothesis and by contrast they would weaken an autogenous explanation). This chapter examines recent approaches and observations that may provide evidence for or against this prediction in greater detail.

C. Status of Centrosomal/Centriolar RNA Work

Until recently, the evidence for RNA in centrosomes was tantalizing but largely indirect. In the past few years, however, reports have been published by four independent laboratories to substantiate the existence of centrosomal RNA (cnRNA). Unlike earlier studies that relied on less specific biochemical and histochemical approaches, the more recent reports identify specific sequences and provide direct evidence of localization. The level of resolution does not permit a distinction between centriolar and centrosomal localization, but coupled with ultrastructural studies in the 1970s and 1980s, it seems likely that RNA is associated with the inside and outside of the centriolar barrel, as well as the pericentriolar matrix (PCM).

Alliegro *et al.* (2006) and Alliegro and Alliegro (2008) demonstrated a population of RNAs that coisolated with surf clam (*Spisula*) oocyte centrosomes and were localized in and around the centrosome by *in situ* hybridization (Fig. 2). *Spisula* cnRNAs are not simply cytoplasmic mRNAs adsorbed to or shuttling through the centrosome; few have cognates in protein and nucleotide databases and none of the abundant cytoplasmic mRNAs are represented among them. The ontological profile of sequences that are identifiable is heavily weighted toward nucleic acid binding and metabolism and includes several reverse transcriptase-like sequences. *Spisula* cnRNAs also appear to be set apart from cytoplasmic RNAs by the low intron content of their corresponding

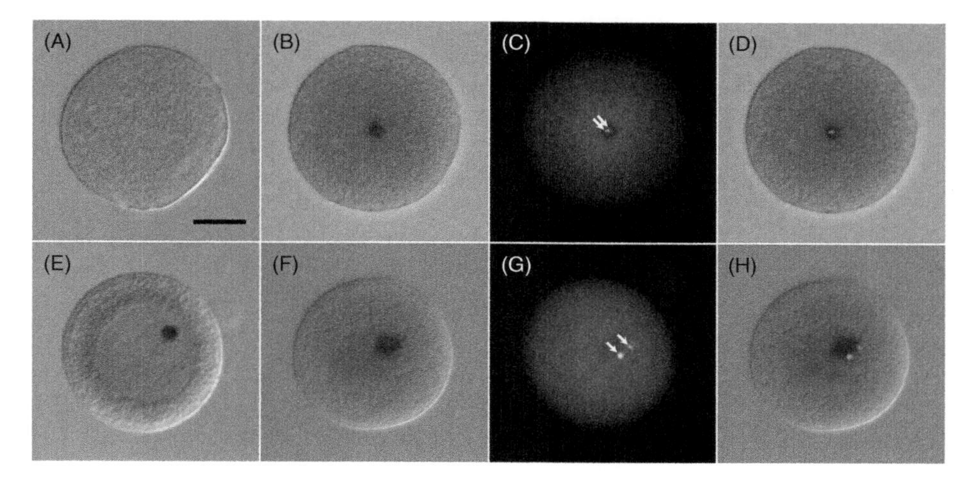

Fig. 2 Localization of cnRNAs by *in situ* hybridization. (A) DIC image of a *Spisula* oocyte fixed 7 min after parthenogenetic activation with 0.5 M KCl and labeled with sense RNA probe (negative control) for cnRNA239. (B) seven minutes postactivation oocyte labeled with antisense cnRNA239 shows a discrete cytoplasmic patch. (C) Immunofluorescence image of centrosomes (arrows) localized in the same oocyte with anti-γ-tubulin antibody. (D) Overlay of DIC and fluorescent images. (F–H) is a similar series showing localization of cnRNA65 at 7 min postactivation (F, DIC; G, anti-γ-tubulin immunofluorescence; H, overlay). Unlike cnRNA239, cnRNA65 is present in unactivated oocytes (E), appearing as a crisp, spherical patch within the oocyte germinal vesicle (GV). Images from Alliegro *et al.* (2006). Courtesy *Proc. Natl. Acad. Sci. USA*.

nuclear genes. This seems to be the case for the cnRNA homologues identified in the sea urchin (*Strongylocentrotus purpuratus*) genome database as well (unpublished observations).

Lécuyer *et al.* (2007) localized six RNAs to the centrosome in *Drosophila* embryos by *in situ* hybridization and speculated on the role that localized RNAs may play in assembly of the centrosome and other organelles by serving as a scaffold upon which organelle-specific proteins may assemble. The authors' statistical analysis predicts a total of 24 *Drosophila* cnRNAs genome-wide. Among these is the mRNA encoding CP309, which is Drosophila pericentrin (PCNT). This is an important finding with implications for centrosome architecture and function, as PCNT itself is an integral component of the PCM. The protein contains a series of coiled coil domains and a highly conserved PCM-targeting motif (PACT domain) in its C-terminus. PCNT binds protein kinase A to the centrosome, binds calmodulin and γ-tubulin, and is thus important for a range of fundamental centrosome functions, including microtubule nucleation and cell cycle progression.

The findings of Kingsley *et al.* (2007) and Bluwer *et al.* (2007) confirm the presence of RNA at the centrosome and spindle pole in the snail *Ilyanassa* and *Xenopus*, respectively. There are interesting similarities between the findings of these two groups and those reported by Alliegro *et al.* (2006) and Alliegro and Alliegro (2008), including a significant number of sequences unmatched in protein or nucleotide databases and, of those that are identifiable, a preponderance of molecules associated with nucleotide binding and metabolism. These studies support the notion that cnRNAs are a unique set of molecules and that they are present in the centrosome as more than just visitors. Both reports are focused on the potential functional significance of cnRNAs in their own experimental systems and present clear and intriguing evidence in support of the investigators' respective hypotheses, but neither addresses the potential implications for the ontogenic or evolutionary origin of the centriole/centrosome system (Alliegro, 2008).

II. Methods

The methods described below follow Alliegro *et al.* (2006) to which the reader is referred for additional detail.

A. cnRNA Library Development

Centrosomes from activated *Spisula solidissima* oocytes were isolated as previously described (Palazzo and Vogel, 1999). These centrosomes are functionally competent to form asters and duplicate, and have been extensively characterized. Total [non-oligo(dT)-selected] RNA was extracted from the peak sucrose density gradient fraction using Qiagen (Valencia, CA) RNeasy reagents, reverse transcribed (*Retroscript,* Ambion, Austin, TX) and amplified by polymerase chain reaction (PCR) using the random primers and methods described by Froussard (1992) and Von Eggeling and Spielvogel (1995), and blunt-cloned into the PCR-Script plasmid (Stratagene, Valencia, CA). A nonreverse-transcribed control was employed to guard against the potential amplification and subsequent cloning of genomic DNA contaminants. When this procedure was followed, nonreverse-transcribed cnRNA yielded no detectable products from random PCR nor, subsequently, cloned inserts.

B. PCR Screening for Centrosome Enrichment

Alliegro *et al.* (2006) were unable to detect cnRNAs by northern analysis in preliminary experiments, suggesting very low copy number. A PCR approach was therefore used to screen for biochemical enrichment in centrosomes. Equal quantities of RNA isolated from whole, lysed oocytes and RNA isolated from peak centrosome fractions were reverse transcribed and used as template. Internal controls, primer pairs for known *Spisula* cytoplasmic RNAs, included poly(A)-binding protein, ribonucleotide reductase, and 18s rRNA were also used. Quantitative PCR is not necessary, especially when looking for strong enrichment; semiquantitative reaction conditions can readily be determined empirically to generate reaction product sensitive to template input and within the linear range of yield. Alliegro *et al.* used 20 ng template, 300–500 bp amplification products, and 30 cycles of 94°C × 30 s, 55°C × 30 s, and 72° × 1 min.

C. *In Situ* Localization of cnRNA11

Gravid *S. solidissima* are obtained from the Marine Resources Department at the Marine Biological Laboratory, Woods Hole, MA. Oocytes and zygotes are fixed for *in situ* hybridization for 2 h at 4°C in 4% paraformaldehyde in 3-(*N*-morpholino propanesulfonic acid (MOPS) buffer, pH7.4. They were subsequently washed by settling and resuspension twice in MOPS buffer and once in 0.5 M NaCl, dehydrated in a graded series of ethanols to 70%, and stored at –20°C until use. Two *in situ* hybridization methods to localize cnRNAs with high sensitivity and low background were developed. One is a modification of methods described by Lee *et al.* (2003) and employs 1% sodium dodecyl sulfate (SDS) in the prehybridization and hybridization solutions. Chromatin structure is damaged using this protocol and endogenous protein antigenicity is greatly reduced, especially given the use of relatively high temperatures and proteinase K to expose target. The second method for *in situ* hybridization anticipated the possibility that targets remained masked, either by components of the centrosome itself or by interaction between sense and antisense RNA strands. The protocol, referred to as *fluorescent extreme touchdown* in situ *hybridization* (FETISH), was therefore designed with exaggerated protease treatment to ameliorate persistent protein masking, and the temperature was raised to 94°C for 10 min to melt duplexes. Since the window available for hybridization was expected to close as the temperature was lowered to a range typical for *in situ* hybridization, the temperature was then reduced to final hybridization temperature in 5°C steps of 20 minutes each. Although protein antigenicity is destroyed in FETISH, chromatin organization was better preserved than in SDS-treated embryos. Optimal hybridization and development times (for alkaline phosphatase reactions) can vary greatly between cnRNA targets, the former typically 2–3 days and the latter up to 6 days.

III. Viral Relatives of cnRNAs and Other Centriolar and Centrosomal Components

The proposal of a viral origin for the centriole and related structures would be supported by evidence of relatedness of cnRNAs to extant viral molecules. We conducted an analysis to address this question. Only those cnRNAs with strong centrosomal localization, as determined by *in situ* hybridization, and for which full-length sequences are available are considered here. Although some partial sequences reported by Alliegro and Alliegro (2008) and Kingsley *et al.* (2007) do apparently have viral relatives, there is no way to reliably assess false negatives using only fragments. It should be noted that there is no clear guideline for a cutoff E-value to demonstrate relationship by descent. Some investigators accept values of $\leq 1e^{-2}$ as indicating relatedness, others prefer the more rigorous $1e^{-5}$.

The results of the analysis are detailed in Table I. Out of an aggregate 14 cnRNAs reported by three independent laboratories (Alliegro and Alliegro, 2008; Alliegro *et al.*,

Table I
Similarity Between cnRNAs and Viral Sequences

Query cnRNA	Accession	Viral hit	*E*-value
Drosophila Bsg 25D[a]	NM_057568	Human herpes virus protein	$1.0e^{-8}$
CG1962, isoform A[a]	NM_136205	Squirrel pox virus protein	$1.0e^{-4}$
CP309 (pericentrin)[a]	AY373570	Hypothetical protein SGHV062	$5.0e^{-12}$
CG14438, isoform A[a]	NM_132120	Equid herpesvirus envelope glycoprotein	$4.0e^{-3}$
Cyclin B, isoform A[a]	NM_166555	*Ateline* herpesvirus 3 v-cyclin	$1.0e^{-7}$
Rapsynoid[a]	NM_080260	*Saccharopolyspora erythraea* genomic[b]	$2.0e^{-33}$
Nanos-like protein[c]	EU087572	Cyprinid herpesvirus 3 DNA[b]	$6.0e^{-5}$
Ankyrin-like protein[c]	EU087604	*Paramecium* chlorella virus hypothetical protein	$7.0e^{-18}$
Lipoprotein VsaC-like protein[c]	EU087576	Crocodilepox virus virion core protein	$3.0e^{-10}$
Tis11-like protein[c]	EU087577	–	–
RNA binding 1-like protein[c]	EU087580	–	–
Zinc finger family 135 protein[c]	EU087585	Human APM-1 protein	$9.0e^{-13}$
cnRNA11[d]	DQ359732	Baculovirus retrotransposón	$3.0e^{-64}$
cnRNA15[e]	EU069824	Lymphocystis virus RNA-dependent DNA polymerase	$8.0e^{-7}$

[a] Lecuyer *et al.* (2007).
[b] Algorithm used, tBLASTx (all others are BLASTx).
[c] Kingsley *et al.* (2007).
[d] Alliegro *et al.* (2006).
[e] Alliegro and Alliegro (2008).

2006; Kingsley *et al.*, 2007; Lécuyer *et al.*, 2007), 12 have similarity to entries in viral databases, all but one with E-values of $\leq 1e^{-4}$. The values shown in Table I range from approximately e^{-3} to e^{-64}. Viral classes among the top hits include both RNA and DNA viruses, as well as several which are unclassified. Notable hits include cnRNA11 (Alliegro *et al.*, 2006), matching a baculoviral RNA-dependent nucleotide polymerase ($3e^{-64}$), and an RNA with predicted ankyrin-like repeats (Kingsley *et al.*, 2007), matching a *Paramecium bursaria* chlorella viral sequence ($7e^{-18}$).

We extended our analysis to include known centriolar and centrosomal proteins with results that were perhaps even more instructive, since the precise localization and function of these proteins is understood, at least to some degree. Tektins are found in centrioles and the ciliary axoneme associated with the outer doublet microtubules. It has been suggested that tektins may function as molecular rulers to regulate axoneme length (Setter *et al.*, 2006; Yanagisawa and Kamiya, 2004). Satir *et al.* (2007) suggested that tektin could have been brought into the evolving eukaryotic cell by the centriolar virus. In our analysis, we found that horse tektin (XP_001504783) returned a hit ($9e^{-5}$) on a *Streptococcus mutans* bacteriophage M102-predicted protein and bore similarity to HIV pol protein. Mosquito (*Anopheles*)-predicted tektin (EAA04762) also generated a hit on *Lactobacillus* prophage minor tail protein and bore some similarity to *Toscano* virus polyprotein. A predicted tektin-like protein from *Caenorhabditis* (AA96184) showed similarity to varicella virus early protein 0 ($3e^{-4}$). There were a number of other viral hits on tektins and tektin-like proteins in this preliminary analysis, some with highly significant E-values, but these will require further scrutiny. These three examples are highlighted because in each case the tektin protein and its viral hit represented reciprocal best hits (RBHs), indicating orthology.

Centriolin localizes to the maternal centriole and functions in cytokinesis and cell cycle progression (Gromley *et al.*, 2003). This protein was not considered by Satir *et al.* (2007), but in line with that discussion, it could have originally been associated with the centriolar virus. Human, mouse, and horse centriolins (AAP43846, A2AL36, and XP_001501659, respectively) return viral hits in the range of e^{-6}. Porcine (XP_001925810) and sea urchin (XP_001190047) centriolin are confirmed orthologues of viral proteins by RBH analysis.

We also conducted a search in viral databases for proteins with similarity to PCNTs and related eukaryotic centrosomal proteins. As noted above, PCNT is important for MTOC function and cell cycle progression. Sea urchin pericentrin (predicted; accession XP_001178124) returned a hit on, among others, a *P. bursaria* chlorella virus ankyrin repeat protein with an E-value of $1e^{-15}$. Human chromodomain helicase DNA-binding protein (CHD; NP_005843) returned hits on several viral proteins, including *Plutella xylostella* multiple nucleopolyhedrovirus global transactivator ($5e^{-31}$). CHD3 and CHD4 form complexes with pericentrin, and this complex is required for centrosomal anchoring of pericentrin/γ-tubulin and for centrosome integrity. Mammalian AKAP-9 [A-kinase anchoring protein-9, a family that includes PCNT; e.g., chimp (XP_527814), human (Q99996), and cow (XP_599034)] comes back with hits in the range of e^{-5} on *Bacillus* phage TMP (tape measure protein) repeat protein 1. TMP is thought to regulate viral tail assembly and length. All three of these

AKAPs represent RBHs, as do other pairings between viral and fungal (*Neurospora* and *Coprinopsis*) PCNTs.

Finally, to assess whether similar results would be obtained with any cytoskeletal proteins used to interrogate the viral database, we performed Basic Local Alignment Search Tool for proteins (BLASTp) alignments with 10 arbitrarily chosen cytoskeleton-related proteins. Of these 10, only α-1 actin returned a significant hit—on the feline sarcoma viral oncogene, tyrosine-protein kinase-transforming protein (Fgr). The *Fgr* gene product is a known fusion protein including a portion of actin and a tyrosine-specific protein kinase (Naharro *et al.*, 1984). Together with the observation that Fgr was the only protein listed with significant similarity to actin, this suggests a late lateral gene transfer (the other two proteins listed in the BLAST report exhibited E-values of 5.8 and 7.6). The nine other cytoskeleton-associated proteins used in this search exhibited no significant similarity to viral proteins, even at the low-stringency cut off of e^{-2}. These included α-tubulin, lamin B1, myosin heavy chain, keratin, spectrin, dynein, kinesin, desmin, and calmodulin.

Given the range of E-values discussed above as well as the orthologous relationships revealed by RBH analysis, some centrosomal and centriolar molecules clearly have relatives among the viruses that meet or exceed criteria conventionally accepted to indicate a relationship by descent. The list, even from our preliminary study, is far from complete and only a few proteins have been examined. Proteins specifically assigned to the centriole in the studies of Leidel and Gönczy (2005) and others have not yet been tested. Additional cnRNA sequences may also be revealing. The large-scale analysis now underway will undoubtedly reveal additional orthologous relationships and hopefully permit a phylogenetic analysis of these proteins and cnRNAs reaching back into the viruses. In this way, the relationship of cnRNA to specific viral RNA and of true centrosomal and basal body protein sequences from basal protists to the viral protein sequences may be confirmed.

IV. Conclusions

cnRNA is a reality, but whether this represents centriolar RNA remains undetermined. The generality of RNA localized at the centrosome is also unclear, that is, whether cnRNAs are only found in cells with true centrosomes and centrioles, and whether they play a direct role in centrosome function or duplication. There is enough data to suggest that some cnRNAs are viral-like and that some centrosomal, and possibly uniquely centriolar, proteins also have protein relatives in viral databases. The speculation that centrioles are invaders of RNA viral origin in the evolving eukaryotic cell is strengthened by these findings. Neither the hypothesis of autogenous origin of the centriole/cilium nor the older spirochete hypothesis makes the predictions that centrosomal proteins and RNAs have viral cognates, but there could be other explanations, for example, late transfer and incorporation of cnRNAs into modern RNA viruses. One potential function of the centriole, the determination of left/right axes within the cell (see Bell *et al.*, 2008 for discussion), could arise with the initial viral invasion. In this regard, it would be interesting to examine genes involved in chirality determination in snails and planar cell polarity (PCP) genes (Okumura *et al.*, 2008) to see whether they too have viral cognates.

Finally, the following three lines of investigation will be helpful in further strengthening the hypothesis that the centriole has evolved from a viral precursor. These would be (a) determining if proteins unique to centrioles or specifically involved in basal body morphogenesis are represented by viral precursor proteins, (b) defining the relationship of these proteins to cnRNA, and (c) generating a detailed phylogeny of cnRNAs.

Acknowledgments

Supported by NIH (GM075163) and NSF (0843092) grants to MCA. PS is partially supported by NIDDK.

References

Alliegro, M.C. (2008). The implications of centrosomal RNA. *RNA Biol.* **5**, 198–201.

Alliegro M.C., and Alliegro, M.A. (2008). Centrosomal RNA correlates with intron-poor nuclear genes in Spisula oocytes. *Proc. Natl. Acad. Sci. USA* **105**, 6993–6997.

Alliegro, M.C., Alliegro, M.A., and Palazzo, R.E. (2006). Centrosome-associated RNA in surf clam oocytes. *Proc. Natl. Acad. Sci. USA* **103**, 9034–9038.

Asai, D.J., and Koonce, M.P. (2001). The dynein heavy chain: Structure, mechanics, and evolution. *Trends Cell Biol.* **11**, 196–202.

Basto, R., Lau, J., Vingradova, T., Gardiol, A., Woods, C.G., Khodjakov, A., and Raff, J.W. (2006). Flies without centrioles. *Cell* **125**, 1375–1386.

Bell, A.J., Satir, P., and Grimes, G.W. (2008). Mirror-imaged doublets of *Tetmemena*: Implications for the development of left-right symmetry. *Dev. Biol.* **314**, 150–160.

Bluwer, M., Feric, E., Weis, K., and Heald, R. (2007). Genome-wide analysis demonstrates conserved localization of messenger RNAs to mitotic microtubules. *J. Cell. Biol.* **179**, 1365–1373.

Dirksen, E.R. (1991). Centriole and basal body formation during ciliogenesis revisited. *Biol. Cell.* **72**, 31–38.

Froussard, P. (1992). A random-PCR method (rPCR) to construct whole cDNA library from low amounts of RNA. *Nucleic Acids Res.* **20**, 2900.

Fulton, C., and Dingle, D.A. (1971). Basal bodies, but not centrioles, in *Naegleria*. *J. Cell Biol.* **51**, 826–836.

Gilula N.B., and Satir, P (1972). The ciliary necklace: A ciliary membrane specialization. *J. Cell Biol.* **53**, 494–509.

Gould, R.R., and Borisy, G.G. (1977). The pericentriolar material in Chinese hamster ovary cells nucleates microtubule formation. *J. Cell Biol.* **73**, 601–615.

Gromley, A., Jurczk, A., Sillibourne, J., Halilovic, E., Mogensen, M., Groisman, I., Blomberg, M., and Doxsey, S. (2003). *J. Cell Biol.* **161**, 535–545.

Jékely, G., and Arendt, D. (2006). Evolution of intraflagellar transport from coated vesicles and autogenous origin of the eukaryotic cilium. *Bioessays* **28**, 191–198.

Kellenberger, E. (1966). Control mechanisms in bacteriophage morphopoiesis. *In* "Principles of Biomolecular Organization" (G.E.W. Wolstenholme and M. O'Conner, eds.), CIBA Symposium, pp. 192–226. Little Brown, Boston.

Kuriyama, R. (2009). Centriole assembly in CHO cells expressing Plk4/SAS6/SAS4 is similar to centriologenesis in ciliated epithelial cells. *Cell Motil. Cytoskeleton* **66**, 588–596.

Kingsley, E.P., Chen, X.Y., Duan, Y., and Lumbert, J. (2007). Widespread RNA localization in the spiralian embryo. *Evol. Dev.* **9**, 527–539.

Lécuyer, E., Yoshida, H., Parthasarathy, N., Alm, C., Babak, T., Cerovina, T., Hughes, T.R., Tomancak, P., and Krause, H.M. (2007). Global analysis of mRNA localization reveals a prominent role in organizing cellular architecture and function. *Cell* **131**, 174–187.

Lee, P.N., Callaerts, P., De Couet, H.G., and Martindale, M.Q. (2003). Cephalopod *Hox* genes and the origin of morphological novelties. *Nature* **424**, 1061–1065.

Leidel, S., and Gönczy, P. (2005). Centrosome duplication and nematodes: Recent insights from an old relationship. *Dev. Cell* **9**, 317–325.

Margulis, L. (1981). "Symbiosis in Cell Evolution. Life and its Environment on the Early Earth." W. H. Freeman, San Francisco.

Marshall, W.F. (2009). Centriole evolution. *Curr. Opinion Cell Biol.* **21**, 14–19.

Mitchell, D.R. (2007). The evolution of eukaryotic cilia and flagella as motile and sensory organelles. *Adv. Exp. Med. Biol.* **607**, 130–140.

Mizukami, I., and Gall, J. (1966). Centriole replication. II. Sperm formation in the fern *Marsilea*, and the cycad. *Zamia. J. Cell Biol.* **29**, 97–111.

Naharro, G., Robbins, K.C., and Reddy, E.P. (1984). Gene product of v-fgr onc: Hybrid protein containing a portion of actin and a tyrosine-specific protein kinase. *Science* **223**, 63–66.

Nigg, E.A. (2007). Centrosome duplication: of rules and licenses. *Trends Cell Biol.* **17**, 215–221.

Okumura,T., Utsunn, H., Kuroda, J., Gittenberger, E., Asami, T., and Matsuno, K. (2008). The development and evolution of left-right asymmetry in invertebrates: Lessons from *Drosophila* and snails. *Dev. Dyn.* **237**, 3497–3515.

Palazzo, R.E., and Vogel, J.M. (1999). Isolation of centrosomes from *Spisula solidissima* oocytes. *Methods Cell Biol.* **61**, 35–56.

Salisbury, J.L. (2007). A mechanistic view of the origin for centrin-based control of centriole duplication. *J. Cell. Physiol.* **213**, 420–428.

Satir, P., Guerra, C., and Bell, A.J. (2007). Evolution and persistence of the cilium. *Cell Motil. Cytoskeleton.* **64**, 906–913.

Satir, P., Mitchell, D.R., and Jékely, G. (2009). How did the cilium evolve? *Curr. Top. Dev. Biol.* **85**, 63–82.

Setter, P.W., Malvey-Dorn, E., Steffen, W., Stephens, R.E., and Linck R.W. (2006). Tektin interactions and a model for molecular functions. *Exp. Cell Res.* **312**, 2880–2896.

Von Eggeling, F., and Spielvogel, H. (1995). Applications of random PCR. *Cell. Mol. Biol.* **41**, 653–670.

Yanagisawa, H.A., and Kamiya, R. (2004). A tektin homologue is decreased in *Chlamydomonas* mutants lacking an axonemal inner-arm dynein. *Mol. Biol. Cell* **15**, 2105–2115.

SECTION II

Cell Biology/Biochemistry

CHAPTER 3

Using quantitative PCR to Identify Kinesin-3 Genes that are Upregulated During Growth Arrest in Mouse NIH3T3 Cells

Rikke I. Thorsteinsson, Søren T. Christensen, *and* Lotte B. Pedersen

Department of Biology, Section of Cell and Developmental Biology, University of Copenhagen, Universitetsparken 13, DK-2100 Copenhagen OE, Denmark

METHODS IN CELL BIOLOGY, VOL. 94

978-0-12-375024-2
DOI: 10.1016/S0091-679X(08)94003-6

Abstract

Most cells in our body form a single primary cilium when entering growth arrest. During the past decade, a number of studies have revealed a key role for primary cilia in coordinating a variety of signaling pathways that control important cellular and developmental processes. Consequently, significant effort has been directed toward the identification of genes involved in ciliary assembly and function. Many candidate ciliary genes and proteins have been identified using large-scale "omics" approaches, including proteomics, transcriptomics, and comparative genomics. Although such large-scale approaches can be extremely informative, additional validation of candidate ciliary genes using alternative "small-scale" approaches is often necessary. Here we describe a quantitative PCR-based method that can be used to screen groups of genes for those that are upregulated during growth arrest in cultured mouse NIH3T3 cells and those that might have cilia-related functions. We employed this method to specifically search for mouse kinesin-3 genes that are upregulated during growth arrest and identified three such genes (*Kif13A, Kif13B*, and *Kif16A*). In principle, however, the method can be extended to identify other genes or gene families that are upregulated during growth arrest.

I. Introduction

Nonmotile primary cilia are found in most quiescent/growth-arrested cells in our body and consist of a microtubule (MT)-based axoneme surrounded by a bilayer lipid membrane. The ciliary membrane is rich in receptors, ion channels, and other signaling proteins that enable the cilia to receive signals from the extracellular environment and transmit these signals to the cell to control cell growth, migration, and differentiation (Christensen *et al.*, 2007). Lack of normal functioning cilia causes various diseases (ciliopathies), including polycystic kidney disease, infertility, respiratory diseases, blindness, developmental defects, and probably cancer (Mans *et al.*, 2008; Marshall, 2008).

The formation of primary cilia is initiated in G1/G0 and involves the addition of axonemal precursors to the distal end of the mother centriole/basal body. During ciliary elongation, precursor proteins are transported from the cytoplasm toward the distal tip of the cilium by intraflagellar transport (IFT), which is a bidirectional MT-based transport system required for assembly and maintenance of almost all cilia and flagella. Motor proteins belonging to the kinesin-2 family mediate anterograde IFT whereas cytoplasmic dynein-2 is responsible for retrograde IFT (Rosenbaum and Witman, 2002; Scholey, 2003). In addition to kinesin-2, motor proteins belonging to other kinesin families may contribute to ciliary structural and functional diversity [reviewed in (Scholey, 2008)]. For example, in *Caenorhabditis elegans* the kinesin-3 family protein Klp6 has been implicated in localization of polycystin-2 to the membrane of sensory cilia (Peden and Barr, 2005). In addition, *in silico* screens for *C. elegans* and *Drosophila* genes containing consensus binding sites (*x*-boxes) for cilia-related

Rfx-type transcription factors identified the genes *unc-104* and *klp98A*, respectively, which both encode kinesin-3 family proteins, among the positive hits (Blacque *et al.*, 2005; Laurencon *et al.*, 2007), and KLP98A was also suggested to be cilia-related by comparative genomics (Avidor-Reiss *et al.*, 2004; Li *et al.*, 2004). Finally, a homolog of a human kinesin-3 family member (GI 27549391) was identified in the ciliome of *Tetrahymena thermophila* (Smith *et al.*, 2005). Genomic analysis shows that the mouse genome contains eight different kinesin-3 genes (*Kif1A, Kif1B, Kif1C, Kif13A, Kif13B, Kif14, Kif16A*, and *Kif16B*) whereas the corresponding number for the human genome is seven, owing to the presence of only one *KIF16* gene in humans (Miki *et al.*, 2005; Wickstead and Gull, 2006). However, so far there are no reports that directly link kinesin-3 family proteins to primary cilia in mammals.

Identification of genes involved in cilia assembly or function is essential for understanding the molecular basis of cilia-related diseases. Several studies have approached the identification of ciliary genes from different angles. These include comparative genomics analyses of ciliated and nonciliated organisms (Avidor-Reiss *et al.*, 2004; Li *et al.*, 2004), genome-wide screens to identify promoters with *x*-box motifs (Blacque *et al.*, 2005; Chen *et al.*, 2006; Efimenko *et al.*, 2005; Laurencon *et al.*, 2007), and transcriptome analyses to identify genes that are upregulated during ciliogenesis (Chhin *et al.*, 2008; Stolc *et al.*, 2005). Furthermore, large-scale proteomics has been performed for various organisms and ciliary compartments, including isolated *Chlamydomonas reinhardtii* flagella and centrioles (Keller *et al.*, 2005; Pazour *et al.*, 2005), human ciliary axonemes (Ostrowski *et al.*, 2002), rat olfactory ciliary membranes (Mayer *et al.*, 2008), photoreceptors from mice (Liu *et al.*, 2007), and isolated human centrosomes (Andersen *et al.*, 2003). To date two ciliary and one centrosomal database have been established based on a number of genomics, transcriptomics, and proteomics studies performed in different organisms (Gherman *et al.*, 2006; Inglis *et al.*, 2006; Nogales-Cadenas *et al.*, 2009). Indeed, comparing results obtained using different approaches is probably the most reliable method for obtaining a global picture of genes involved in ciliary assembly or function, as each method for identifying the "ciliome" of a cell or organism has its advantages and disadvantages. For example, in proteomics studies, low-abundance proteins or proteins that are difficult to separate by gel electrophoresis may not be detected and, depending on the isolation method, contamination by nonciliary proteins can be an issue. In addition, ciliary proteomics studies are designed to only reveal proteins found in the cilium, and potentially also the coisolated centrioles, whereas the advantage of comparative genomics or transcriptomics studies is that they will likely identify genes coding for proteins that themselves do not localize to the cilium, but which may nevertheless play a role in cilia assembly or function, for example, by promoting cytosolic transport of cilia precursors to the basal bodies or by regulating cytoplasmic events in response to cilia-mediated signaling. On the other hand, transcriptome studies might miss genes that are only moderately upregulated during ciliogenesis, due to the experimental threshold set up, and comparative genomics may not identify ciliary proteins that also have nonciliary roles, such as tubulin. In any case, for all of the methods mentioned above, validation of candidate ciliary genes is essential.

II. Rationale

Here we provide a method for validating candidate ciliary genes identified, for example, by comparative genomics or transcriptomics approaches. Since cilia-related genes are often upregulated during growth arrest in mammalian cells [e.g., Gas8 (Yeh *et al.*, 2002) and PDGFRα (Lih *et al.*, 1996; Schneider *et al.*, 2005)] we used a quantitative PCR (qPCR) approach to screen a family of genes for those that are upregulated during growth arrest. Specifically, we used qPCR to identify mouse kinesin-3 family genes that are upregulated during growth arrest, since cilia-related functions for kinesin-3 family members are suggested by studies in *C. elegans* (Peden and Barr, 2005), *Drosophila* (Avidor-Reiss *et al.*, 2004; Chen *et al.*, 2006; Efimenko *et al.*, 2005; Laurencon *et al.*, 2007; Li *et al.*, 2004), and *Tetrahymena* (Smith *et al.*, 2005). The benefit of the qPCR method for expression profiling is that it is very simple and sensitive, so if one or more candidate ciliary genes identified by large-scale "omics" approaches need to be validated, expression profiling by qPCR is a convenient method for initial validation. The qPCR analysis presented here can be extended to other families of genes that are suggested to be ciliary on the basis of results obtained for one or more orthologous genes in other organisms.

III. Materials

A. Cell Culture Reagents and Growth Media

T75 flasks, 10-cm diameter Petri dishes, 6-well plates for cell culturing
Phosphate-buffered saline (PBS)
Growth medium: Dulbecco's modified Eagle's medium with glutamax and penicillin/streptomycin (100 U/ml) plus 10% heat-inactivated fetal bovine serum (FBS)
Starvation medium: same as growth medium but without FBS
Trypsin mix: 1% trypsin-EDTA in PBS

B. Molecular Biology Reagents

Kit for RNA preparation (e.g., NucleoSpin®RNAII from Macherey-Nagel, Duren, Germany which includes a DNase step)
Reverse transcriptase (e.g., SuperScript® II from Invitrogen, Carlsbad, CA, USA)
PCR primers (Table I)
dNTP mix, 10 mM of each dNTP
RNAse inhibitor (e.g., rRNasin from Promega, Madison, WI, USA)
PCR enzyme (e.g., Herculase® II from Stratagene, La Jolla, CA, USA)
PCR-grade nuclease-free water
SYBR® *premix Ex Taq*™ (TaKaRa, Kyoto, Japan)
Agarose, ethidium bromide, TAE buffer (4.84 g Tris base, 1.14 ml glacial acetic acid, 2 ml 0.5 M Na_2 EDTA pH 8.0, ddH_2O up to 1 l), nucleic acid molecular weight marker

Table I
PCR Primers Used in This Study

Gene name[a]	GenBank Acc.No.	Forward Primer	Reverse Primer	PCR prod. size (bp)	Tm (°C)[b]
Kif1A	BC062891.1	5′-GCCTCTGTGAAG GTGGCGGTG-3′	5′-GTTCCCTGAGTCCATG AGGTCCTG-3′	581	60
Kif1C	NM_153103.2	5′-CCCGTGAGACCA GCCAGGATG-3′	5′-CATGTTGGTGGCAGCC ACGGTTC-3′	580	60
Kif3A	NM_008443.3	5′-GTCGGAGAAGCC GGAAAGCTGC-3′	5′-GTCATGATTCTATCCA TGTCGTCAGCG-3′	584	60
Kif13A	AB037923.2	5′-CCCATGAACCGA AGAGAGCTGG-3′	5′-CCCCAGGACCTTATGC TCTCGG-3′	500	59
Kif13B	NM_001081177.1	5′-GTCCGGCCAATG AACCGACGAG-3′	5′-CTTTGGGGTCAAGAAG GTCTCGGAC-3′	462	60
Kif14	NM_001081258.1	5′-CAACATCGGAAG CCTGGAGGTTTC-3′	5′-CGTCAACCCGGTGTCG TTGGAG-3′	507	59
Kif16A	AB001425.1	5′-GCGTATGGGCAG ACAGGCTCTG-3′	5′-GGTCGACAAGATTGAT CTTGCTGGC-3′	460	60
Kif16B	NM_001081133.2	5′-CTGGATCTGGAA AGTCCTACACCATG-3′	5′-CCTGAGTGAACTTGAT GGTGAAGATGG-3′	589	60
KifC3	NM_010631.2	5′-GCTTACGGCCAG ACAGGTGCC-3′	5′-CCTGTACCACCATAAG GGTCTTGC-3′	657	60
B2M	NM_009735.3	5′-ATTTTCAGTGGC TGCTACTCG-3′	5′-ATTTTTTTCCCGTTCT TCAGC-3′	248	51

[a] All genes listed are from mouse.
[b] Melting temperatures (Tm) were calculated using software available at http://www.basic.northwestern.edu/biotools/oligocalc.html

C. Reagents and Solutions for Immunofluorescence Microscopy

Microscope slides and glass coverslips (12-mm diameter)
Concentrated HCl
Ethanol (96% v/v and 70% v/v)
Paraformaldehyde (PFA), 4% solution in PBS
Blocking buffer: 2% bovine serum albumin (BSA) in PBS
Permeabilization buffer: 0.2% v/v Triton X-100, 1% w/v BSA in PBS
Mouse anti-acetylated α-tubulin antibody (Sigma, St. Louis, MO, USA, catalog number T6793)
Rabbit anti-detyrosinated α-tubulin antibody (Chemicon, Billerica, MA, USA, catalog number AB3201)
Alexa Flour 568 Goat anti-rabbit antibodies (Molecular Probes, Carlsbad, CA, USA, catalog number 21069)
Alexa Fluor 488 Donkey anti-mouse (DAM) antibodies (Molecular Probes, catalog number 21202)
DAPI (4′,6-diamidino-2-phenylindole, dihydrochloride)
Mounting reagent (5 ml): 4.5 ml glycerol, 0.5 ml 10× PBS, 0.1 g *N*-propylgallate. The solution is stirred 1–2 h until the *N*-propylgallate is dissolved
Nail polish
Humidity chamber

IV. Methods

A. Introductory Notes and Experimental Outline

Most mammalian cells form a single primary cilium when entering growth arrest/ quiescence. For many types of cultured mammalian cells, such as mouse NIH3T3 fibroblasts, growth arrest can be induced by growing the cells to near confluence and/or by serum deprivation (Schneider *et al.*, 2005). Thus by examining the expression level of target genes in cell populations grown in the presence or absence of serum for variable amounts of time, one can identify genes that are upregulated during growth arrest and therefore might have potential roles in cilia assembly or function. Here we use mouse NIH3T3 cells as a model system to screen for kinesin-3 family genes that are upregulated during growth arrest, but with a few adjustments the procedures should be applicable to most kinds of adherent mammalian cells and target genes. Previous work showed that in cultures of NIH3T3 cells that have been serum starved for 12 and 24 h, respectively, approximately 40 and 90% of the cells are ciliated (Schneider *et al.*, 2005). Therefore, to correlate the expression level of selected kinesin genes with ciliogenesis, samples were collected from NIH3T3 cells that were either (1) non-starved, (2) serum starved for 12 h, (3) serum starved for 24 h, or (4) serum starved for 24 h followed by 24 h of serum supply to make the cells reenter the cell cycle

(Section IV.B). Samples of cells grown on coverslips were used for immunofluorescence microscopy (IFM) to assess the number of ciliated cells present under the various culture conditions (Section IV.C). In parallel, samples were collected for preparation of mRNA, which was translated to cDNA by reverse transcriptase (Section IV.D) and subsequently subjected to an initial PCR screen for expression of the kinesin genes of interest (Section IV.E). Finally, positive hits from the initial PCR screen were subjected to qPCR analysis (Sections IV.F and IV.G).

B. Cell Culture

Cells were cultured at 37°C, 5% CO_2, 95% humidity in T75 flasks in 15-ml growth medium to ca. 80% confluence. After a brief 37°C wash in PBS the cells were trypsinized (1–2 ml trypsin mix per flask), re-suspended in 8 ml 37°C fresh growth medium, and transferred to new T75 flasks (0.5 ml) or prepared for experiments as follows: For cilia quantification by IFM, approximately 5.5×10^4 cells/ml were grown on coverslips placed in a 6-well culture dish (2 ml cell suspension per well). Prior to use the coverslips were acid-washed by immersion in concentrated HCl for 1 h under gentle agitation followed by 10–15 washes in ddH_2O. The acid-washed coverslips were then sterilized by rinsing 10–15 times in 96% ethanol, stored in 70% ethanol, and air dried immediately before use. For RNA purification the same concentration of cells was plated on Petri dishes (10 ml cell suspension per dish). To obtain four different growth situations corresponding to different cell cycle stages (i.e., nonstarved, serum starved for 12 h, serum starved for 24 h, and serum starved for 24 h followed by 24 h of serum supply) three sets of cell cultures were grown to ca. 80% confluence in growth medium and then grown in starvation medium for 0, 12, and 24 h, respectively. One additional set of cell cultures was grown to ca. 30% confluence in growth medium and then serum starved for 24 h to enter growth arrest, followed by 24 h of growth in normal growth medium to re-enter the cell cycle. For all samples, it is very important to avoid confluence of more than 80% because this leads to ciliogenesis despite the presence of serum.

C. Immunofluorescence Microscopy and Quantification of Cilia

Unless otherwise stated, all steps were performed at room temperature. The cells grown on coverslips (see Section IV.B) were washed once in cold (4°C) PBS and then fixed in 4% PFA solution for 15 min directly in the 6-well culture dish. After briefly washing with PBS the coverslips were incubated in permeabilization buffer for 12 min and then transferred with forceps to a humidity chamber. To prepare a humidity chamber, one can place a moist piece of Whatman paper in an empty Petri dish and put a piece of parafilm on top onto which the coverslips are placed, cells facing up. Incubations with various solutions are achieved by simply placing a drop of liquid on top of each coverslip. The lid of the Petri dish should be closed during all incubations to prevent evaporation. Immediately following transfer to the humidity chamber, cells were incubated in 100-μl blocking buffer for 30 min. The cells were then incubated in blocking buffer with primary antibodies against acetylated α-tubulin (1:400) and detyrosinated α-tubulin (GluTub; 1:500) for 1–2 h. After three washes, 5 min each, in blocking buffer cells were incubated with flourochrome-conjugated secondary antibodies (1:600) for 45 min followed by 5 s staining with DAPI (1:1000 in PBS) and three washes, 5 min each, in PBS. Coverslips were mounted by placing them, cells facing down, in a droplet of mounting reagent on microscope slide cleaned with 96% ethanol. After removing excess mounting reagent with a Kimwipe the edges of the coverslips were sealed with nail polish. Fluorescence was visualized on an Eclipse E600 microscope (Nikon, Tokyo, Japan) with EPI-FL3 filters, and images acquired using a MagnaFire cooled CCD camera (Optronics, Goleta, CA, USA). Digital images were processed using Adobe Photoshop and the percentage of ciliated cells in each sample quantified (Figs. 1 and 2).

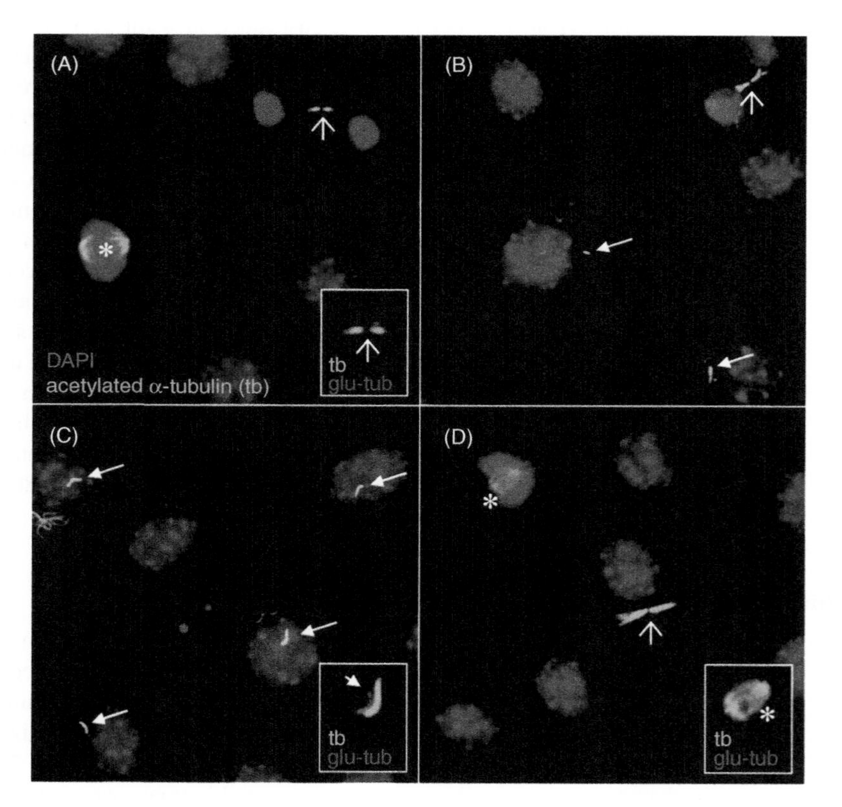

Fig. 1 Immunoflourescence microscopy of NIH3T3 cells. Overview of cells that were either (A) nonstarved, (B) serum-starved for 12 h, (C) serum-starved for 24 h, or (D) serum-starved for 24 h followed by 24 h of serum supply. The cells were processed for IFM and stained with antibodies against acetylated α-tubulin (green) and detyrosinated α-tubulin (GluTub; red), and nuclei were stained with DAPI (blue). Open arrows: midbodies; closed arrows: primary cilia; asterisks: mitotic cells. The insets show shifted overlays of selected regions of stained cells highlighting the colocalization of the antiacetylated and antidetyrosinated α-tubulin antibodies in midbodies (A, inset), primary cilia (C, inset), and spindle poles (D, inset). (See Plate no. 1 in the Color Plate Section.)

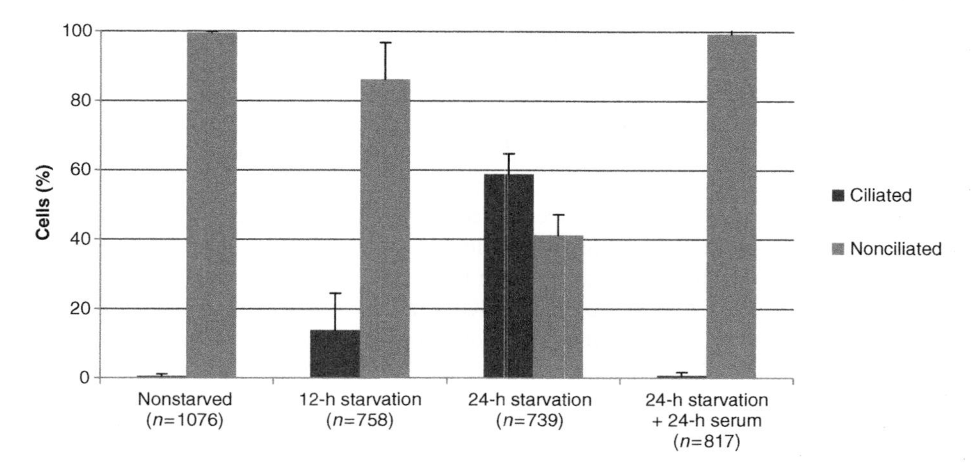

Fig. 2 Quantification of ciliated NIH3T3 cells by IFM. Images of cells similar to those shown in Fig. 1 were used for quantification of cilia. Three independent experiments were carried out. Error bars indicate SD.

D. RNA Extraction and cDNA Preparation

The cells in Petri dishes were harvested and total RNA extracted and purified according to the NucleoSpin®RNAII protocol from Macherey-Nagel <http://www.macherey-nagel.com/>, which includes a DNase step to eliminate contamination with genomic DNA. The total RNA concentration was quantified by measuring the optical density at 260 nm and Eq. (1): C (μg/ml) = $A_{260}/0.025$. RNA was converted to single-stranded cDNA using random primers (oligodeoxyribonucleotides, mostly hexamers) and SuperScript II® Reverse Transcriptase from Invitrogen, essentially as described in the protocol from the manufacturer. Briefly, 500-ng random primers, 2 μl dNTP mix (10 mM each), 1 μg RNA, and H_2O to 24 μl were mixed in a nuclease-free centrifuge tube. The tube was incubated at 65°C for 5 min and quickly chilled on ice. After a short spin 8 μl 5× first-strand buffer, 4 μl 0.1 M DTT and 2 μl RNAse inhibitor were added, and the tube incubated at 25°C for 2 min. Next, 2 μl SuperScript II RT enzyme was added and the mixture incubated at 25°C for 10 min followed by 50 min incubation at 42°C and 15 min incubation at 70°C to inactivate the reaction. The cDNA was stored at –20°C until use. The absence of genomic DNA contamination in the purified RNA samples was confirmed by including a control reaction without SuperScript® II enzyme.

E. Initial PCR Screen for Kinesin Gene Expression

Prior to the qPCR analysis an initial PCR screen for expression of selected kinesin-3 genes in NIH3T3 cells was performed for two reasons: first, to confirm that the kinesin genes of interest are actually expressed in these cells, and second, to optimize the annealing temperature of the PCR to avoid the formation of nonspecific PCR products in the subsequent qPCR analysis. To this end, random-primed cDNA prepared from cells subjected to 0 or 24 h of serum starvation was used as template (see Sections IV.B and IV.D).

As a starting point, the analysis was carried out on seven different mouse kinesin-3 genes: *Kif1A, Kif1C, Kif13A, Kif13B, Kif14, Kif16A*, and *Kif16B* (Miki *et al.*, 2005; Wickstead and Gull, 2006). *Kif14* expression was used as a control, because it is known to be upregulated in mitotic cells and associated with the developing spindle poles in early mitotic cells and the midbody during cytokinesis (Carleton *et al.*, 2006; Gruneberg *et al.*, 2006). Therefore we expected *Kif14* to be downregulated during growth arrest. The Kif1 subfamily is well investigated and Kif1A, Kif1B, and Kif1C are all known to participate in intracellular transport of, for example, mitochondria, synaptic vesicle precursors, and membrane trafficking from Golgi to the ER (Dorner *et al.*, 1998, 1999; Nangaku *et al.*, 1994; Okada *et al.*, 1995). Kif1C also seems to be involved in podosome dynamics (Kopp *et al.*, 2006). Kif16B is known to transport PI_3P-containing endosomes (Hoepfner *et al.*, 2005), whereas the role for Kif16A is unclear. Kif13A associates with the β1-adaptin subunit of the adaptor complex adaptor protein 1 (AP-1) and transports vesicles containing the mannose-6-phosphate receptor from the trans-Golgi network to the plasma membrane (Nakagawa *et al.*, 2000). In *C. elegans* AP-1 has been implicated in ciliary transport of calcium channels and odorant receptors (Dwyer *et al.*, 2001). *Kif13A* is ubiquitously expressed in the mouse and humans including most subregions of the nervous system and to a high degree in regions where sensory neurons are developed (Jamain *et al.*, 2001; Nakagawa *et al.*, 2000). *Kif13B* (also known as GAKIN) is ubiquitously expressed but shows tissue- and cell-dependent variations with relatively high abundance in human kidney, pancreas, brain, and testis (Hanada *et al.*, 2000). Kif13B has at least two binding partners: human discs large (hDlg) tumor suppressor and a PIP_3-binding protein termed PIP_3BP or centarurin-α_1, which both are implicated in regulation of cell polarity pathways (Hanada *et al.*, 2000; Horiguchi *et al.*, 2006). In addition to these kinesin-3 genes, we also included *KifC3* and *Kif3A* in the analysis. *KifC3* encodes a minus end-directed kinesin motor of the kinesin-14 family that is known to be ubiquitously expressed (Yang *et al.*, 2001) and is required for Golgi positioning and integration (Xu *et al.*, 2002; Yang *et al.*, 2001) as well as for anchoring of MT minus ends at adherens junctions (Meng *et al.*, 2008). Chromosome mapping has located the *KifC3* gene near a locus associated with Bardet–Biedl syndrome (Hoang *et al.*, 1998; Yang *et al.*, 2001), but *KifC3* null mice exhibit no apparent cilia-related defects and develop normally (Hoang *et al.*, 1998; Yang *et al.*, 2001), suggesting that KifC3 is not important for cilia assembly or function. It is possible that the reported linkage of *KifC3* with Bardet–Biedl syndrome (Hoang *et al.*, 1998; Yang *et al.*, 2001) is due to the nearby *CNGB1* gene, mutations in which have been associated with retinitis pigmentosa in humans (Bareil *et al.*, 2001). *Kif3A* codes for one of the motor subunits of the well-studied heterotrimeric kinesin-2, which is involved in anterograde IFT (reviewed in Rosenbaum and Witman, 2002; Scholey, 2003), although nonciliary functions have also been reported for this kinesin (e.g., Corbit *et al.*, 2008; Haraguchi *et al.*, 2006; Tuma *et al.*, 1998). However, the *Chlamydomonas Kif3A* ortholog *Fla10* is moderately upregulated during flagellar regeneration (Stolc *et al.*, 2005) and therefore we expected to see similar upregulation of *Kif3A* in mammalian cells during ciliogenesis.

All primers used for PCR are listed in Table I. The primers were designed to recognize a sequence in the motor domain-encoding region of each kinesin gene, because this is the most conserved part of the genes and therefore alternative splicing of transcripts within this region is unlikely. The PCR was carried out in a total reaction volume of 25 μl essentially as described in the Herculase® II manual (Stratagene). One microliter of cDNA was used for template in each reaction and the PCR program was as follows (Ta = annealing temperature): 95°C for 2 min, 35 cycles at 95°C for 30 s, Ta for 30 s, 72°C for 1 min, and 72°C for 10 min. For each primer set three different reactions were carried out with Ta ranging from minus to plus 3°C of the theoretical melting temperatures (Tm) calculated for each primer set (Table I). Following PCR, 10 μl of each PCR product were mixed with loading dye and analyzed by agarose gel electrophoresis using standard procedures. For all the genes analyzed PCR products of the expected sizes were identified, but *Kif1A* and *Kif16B* were excluded from the proceeding qPCR due to the presence of additional unspecific bands.

F. Quantitative PCR: Theoretical Considerations

Following the initial PCR analysis (Section IV.E), a qPCR analysis was performed in order to assess the relative expression levels of selected kinesin genes during growth and quiescence. By converting isolated RNA to cDNA (Section IV.D) and then performing a qPCR where the increase in PCR product for each cycle is measured, it is possible to directly determine the amplification of PCR product after each cycle and hence quantify the relative expression level of specific target genes (Kubista *et al.*, 2006). To measure the increase in PCR product we used the SYBR green method. SYBR green is an asymmetric cyanine dye with two aromatic systems, which has no fluorescence when free in solution, but has a bright signal when bound to double-stranded PCR product. The fluorescence signal is used to generate amplification curves as illustrated in Fig. 3A. To compare expression levels between different samples, a

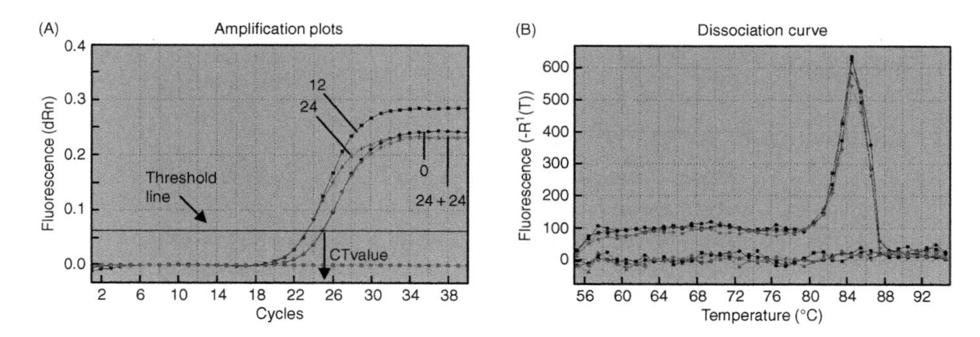

Fig. 3 qPCR plots exemplified by a *Kif13B* experiment. Screen prints. (A) Amplification curve where the number of cycles is depicted on the *x*-axis and the normalized fluorescence (Rn) on the y-axis. Each curve is the mean of a triplicate, SD < 0.3. The CT value is the point where the amplification curve crosses the threshold line and shows that the samples from cells subjected to 12- and 24-h starvation (12 and 24, respectively) have similar and low CT values compared to the samples from cells subjected to 0-h starvation or 24-h starvation followed by 24 h serum supply (0 and 24 + 24, respectively). (B) Dissociation curve for the same experiment; temperature is indicated on the *x*-axis and [–Rn′(T)], which is the first derivative of the normalized fluorescence reading multiplied by –1, on the *y*-axis.

fluorescence threshold line is set at the level where the amplification curves are parallel, which means that the amplification is in the exponential growth phase and the reactions are equally effective (Fig. 3A). The crossing point with the amplification curve and the threshold line is called the CT value. The CT value reflects the number of cycles necessary to reach a certain amount of PCR product and thereby makes it possible to calculate the initial amount of cDNA or compare the relative amount of expression of a gene between two samples (Kubista *et al.*, 2006). We used the comparative threshold method where the expression of a target gene is normalized to a reference gene to correct for general differences in mRNA levels between samples. The normalized expression value for each experimental sample is compared to that of a control sample or "starting point" (in our example the nonstarved cells) to provide a relative measurement of the expression level as a function of, for example, cell cycle stage (Pfaffl, 2001).

Various reference genes are suggested in the literature depending on the type of study or cell lines used. It is recommended to try a few, and it is important to select one that is constitutively expressed throughout the cell cycle. Furthermore, it is recommended to choose genes whose expression is not much higher than that of the target genes. According to these criteria we found that the gene encoding beta-2-microglobulin (*B2m*), which is the beta-chain of major histocompatibility complex class I molecules, is a suitable reference gene for the qPCR analysis described here.

Because the SYBR green method is based on fluorescence emitted from double-stranded DNAi is important that the PCR reactions take place without formation of unspecific PCR product or primer dimers. This is tested by integrating a dissociation curve as part of the PCR program. After a PCR of 40 cycles, a dissociation curve is generated by heating the samples to 95°C. The drop in fluorescence appearing when the PCR products denature is measured; if all PCR products in the sample are of similar size, the curve is a well-defined peak confirming the absence of unspecific PCR products (Fig. 3B).

G. Quantitative PCR: Experimental Design

Based on results from the initial PCR expression screen (Section IV.E), we chose to analyze the expression, by qPCR using *B2m* as reference gene, of six different target genes coding for various kinesin family members (*Kif3A, Kif13A, Kif13B, Kif14, Kif16A*, and *KifC3*). For each target gene three independent experiments were carried out. The qPCR was performed in 20-μl reaction volumes using SYBR® *premix Ex Taq*™ (TaKaRa) and 200 nM of each primer (Table I). The reactions were performed according to the protocol provided by the supplier (TaKaRa) using the following program: 95°C for 10 min, 40 cycles of (95°C for 30 s, Ta for 1 min, 72°C for 30 s), 95°C for 1 min. Ta varies between the different genes and was chosen based on the initial PCR (Section IV.E). To confirm the absence of unspecific PCR products, a dissociation curve was subsequently generated by cooling the samples to 55°C, leaving them at 55°C for 30 s, and then heating to 95°C. For each primer set five different reactions were carried out: four reactions with 1 μl cDNA template from each of the four cell growth conditions and a negative control with 1 μl PCR-grade water. Also, for each primer set a series of reactions using 10^{-1}–10^{-7} dilutions of the template were performed to generate a standard curve. All reactions were performed in triplicate. If the standard deviation (SD) of the triplicate was larger than 0.3, the diverging sample was excluded. Calculations were carried out as described by Pfaffl (2001) using Eq. (2), resulting in a measure of the ratio of target gene (*Kif*) expression in a specific sample (e.g., serum starved for 12 h) to the nonstarved control. The results are presented in Fig. 4.

Eq. (2):

$$\text{ratio} = \frac{(E_{\text{target}})^{\Delta CP_{\text{target}}^{(\text{control}-\text{sample})}}}{(E_{\text{ref}})^{\Delta CP_{\text{target}}^{(\text{control}-\text{sample})}}}$$

E is the efficiency of the primers calculated from a standard curve of 10^{-1}–10^{-7} dilutions of the template and given by $E = 10^{(-1/\text{slope})}$. $\Delta CP^{(\text{control}-\text{sample})}$ is the difference in CT value between the "control" and the "sample." "Target" is the *Kif* gene of interest and "ref" is the reference gene (*B2m*).

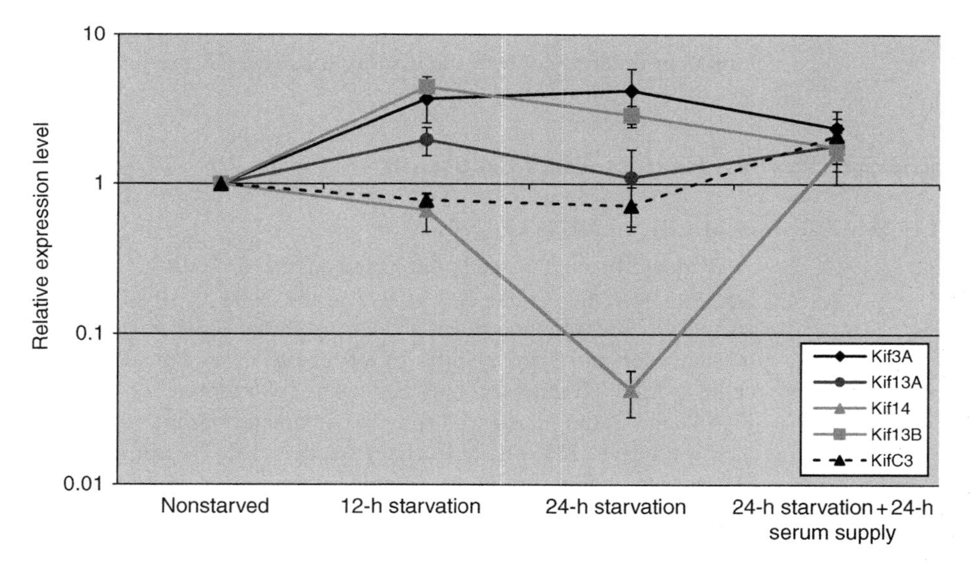

Fig. 4 Quantitative expression of *Kif3A, Kif13A, Kif13B, Kif14*, and *KifC3* in NIH3T3 cells. NIH3T3 cells were grown under the various conditions listed on the *x*-axis (see also the legend to Fig. 1), total mRNA isolated from the cells, and RT-qPCR was performed using primers specific for each kinesin gene, as indicated (primers are listed in Table I). The expression level for each gene under the different growth conditions is shown relative to the expression of the nonstarved cells. Each experiment was performed in triplicate at least three times using independently prepared RNA samples. Each data point thus represents the average value from three different experiments (see Section IV.G for details). Error bars indicate the standard error of the mean.

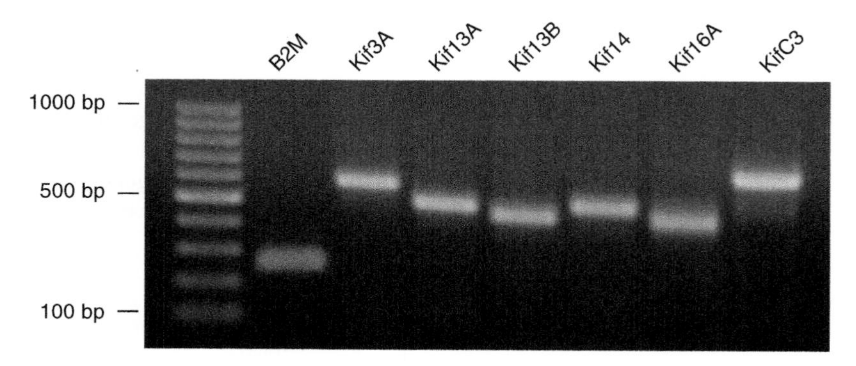

Fig. 5 Verification of qPCR products by band size. The qPCR products from individual genes were pooled and analyzed by agarose gel electrophoresis and staining with ethidium bromide. The apparent band sizes for all qPCR products were in accordance with those expected. Expected band sizes were as follows: *B2M*, 248 bp; *Kif3A*, 589 bp; *Kif13A*, 500 bp; *Kif13B*, 462 bp; *Kif14*, 507 bp; *Kif16A*, 460 bp; *KifC3*, 657 bp. To further confirm the identity of the qPCR products, all qPCR products were purified and subjected to sequence analysis using the same primers as used for PCR.

Following completion of qPCR, all PCR products were analyzed by agarose gel electrophoresis (Fig. 5) as well as by sequencing <http://www.eurofinsdna.com/home.html> in order to confirm the identity of each PCR product.

V. Results and Discussion

A. Quantification of Cilia in NIH3T3 Cells

To quantify cilia in cells subjected to the four different growth conditions (see Section IV.A and IV.B), we analyzed the cells by IFM using antibodies against acetylated α-tubulin and detyrosinated α-tubulin (GluTub), which are highly enriched in centrioles and primary cilia of vertebrate cells, but also present in other stable cellular MTs (Gundersen and Bulinski, 1986; Poole *et al.*, 2001). As shown in Figs. 1 and 2, the nonstarved sample contained both interphase and mitotic cells, but hardly any (ca. 2%) ciliated cells were detected. Dividing cells in different mitotic phases were recognized in this sample by staining of the mitotic spindle, whereas cells in the last step of cytokinesis were detected by staining of the midbodies (Figs. 1A and 2). After 12 h of serum starvation, ca. 20% of the cells had cilia and a few dividing cells, primarily in the last step of cytokinesis, were also observed (Figs. 1B and 2). After 24 h of starvation, the majority of cells (ca. 60%) were ciliated and no dividing cells were seen (Figs. 1C and 2). In contrast, cells that had been serum-starved for 24 h and subsequently resupplied with serum for 24 h displayed no cilia, and the appearance of mitotic cells further indicated that virtually all cells grown under these conditions had reentered the cell cycle (Figs. 1D and 2). In summary the IFM results indicated that the highest proportion of ciliated cells was found in the cells that had been serum-starved for 12 or 24 h (ca. 20 and 60% ciliated cells, respectively), whereas almost no cilia were found in the nonstarved or serum-supplied cultures. By growing the cells in the absence or presence of serum for variable amounts of time it was thus possible to induce the formation or resorption/disassembly of primary cilia. We noted that the percentage of cells displaying cilia after 24 h of serum starvation (ca. 60%) was lower than what we previously observed for NIH3T3 cells (ca. 90%; Schneider *et al.*, 2005). This discrepancy is likely due to differences in the specific batch and passage of cell cultures used.

B. Expression Profile of Kinesin-3 Genes

Having established conditions to induce the formation or resorption/disassembly of primary cilia in cultured NIH3T3 cells we next sought to correlate the expression profile of selected kinesin genes with ciliogenesis using qPCR. Specifically, the relative expression levels of five kinesin-3 genes (*Kif1C, Kif13A, Kif13B, Kif14*, and *Kif16A*) as well as the kinesin-14 gene *KifC3* and the kinesin-2 gene *Kif3A* were analyzed. Total RNA from cells subjected to different culture conditions was extracted and converted to cDNA using reverse transcriptase, and after confirming the expression of specific kinesin genes by PCR and agarose gel electrophoresis, their relative expression levels were determined by qPCR using *B2m* as a reference gene. We also tested the gene encoding β-actin as a potential reference gene for qPCR, but found that this gene is dramatically downregulated during growth arrest and therefore not suitable as reference gene for the type of analysis described here.

In Fig. 4 the expression levels of each kinesin gene under the different culture conditions are shown relative to the nonstarved sample. Consistent with previous results (Carleton *et al.*, 2006; Gruneberg *et al.*, 2006), we found that the expression of *Kif14* is upregulated during cell growth and downregulated during growth arrest. A similar expression profile was observed for *KifC3* although the extent of downregulation during growth arrest (24 h serum starvation) was less prominent. In contrast, *Kif3A* was ~4-fold upregulated during growth arrest (12 and 24 h of serum starvation), consistent with results reported for the *Chlamydomonas Kif3A* ortholog *Fla10* (Stolc *et al.*, 2005), and in line with the known role of Kif3A in IFT and cilia assembly (Rosenbaum and Witman, 2002; Scholey, 2003). Interestingly, among the remaining genes analyzed *Kif13B* showed a dramatic (~6-fold) and statistically significant (*t*-test: $p < 0.05$) upregulation during growth arrest (12 or 24 h serum starvation) compared to the nonstarved cells, whereas *Kif13A* and *Kif16A* showed moderate (about 1- to 3-fold) upregulation during growth arrest relative to cell growth (Fig. 4 and data not shown). These results indicate that Kif13B and perhaps also Kif13A and Kif16A might have cilia-related functions, although additional analyses, such as IFM using antibodies against the native proteins or by GFP- or epitope-tagging, will be required to test this. Such analyses are currently being performed in our laboratory.

Because the motor domain-encoding region for many of the kinesin genes examined above are highly homologous, it was important to verify that the products of the qPCR indeed corresponded to the expected kinesin genes. The qPCR products were therefore analyzed by agarose gel electrophoresis and actual band sizes compared to the expected sizes. As shown in Fig. 5 all qPCR products had the expected band sizes. To further confirm the identity of the qPCR products, all the PCR products were sequenced and the sequences found to be identical to those expected.

C. Evaluation of the Technique

The qPCR method described here is a straightforward and useful way to identify a limited number of growth arrest-specific and potentially cilia-related genes in mammalian cells, and can be used to validate candidate cilia-related genes identified initially by proteomics-, transcriptomics-, or bioinformatics-based studies. Indeed, qPCR has been used previously by others to validate candidate ciliary genes identified, for example, in *Chlamydomonas* by comparative genomics (Li *et al.*, 2004) or proteomics (Keller *et al.*, 2005; Pazour *et al.*, 2005), or by comparative transcriptomics of ciliated and nonciliated human airway epithelial (Chhin *et al.*, 2008). Here we extend the use of qPCR to screen a family of genes, that is, mouse kinesin-3 genes, to identify those that are upregulated during growth arrest/ciliogenesis in cultured mouse NIH3T3 cells, and which we suspected might be ciliary based on previous results from *C. elegans, Drosophila*, and *Tetrahymena* (Avidor-Reiss *et al.*, 2004; Chen *et al.*, 2006; Efimenko *et al.*, 2005; Laurencon *et al.*, 2007; Li *et al.*, 2004; Peden and Barr, 2005; Smith *et al.*, 2005). Among the mouse kinesin-3 genes tested, we found that at least three genes (*Kif13A, Kif13B*, and *Kif16A*) displayed moderate (1- to 3-fold) to high (6-fold) upregulation during growth arrest, while others (*Kif14*) were downregulated during growth arrest, consistent with previous reports (Carleton *et al.*, 2006; Gruneberg *et al.*, 2006). The latter result, as well as the observation that the kinesin-2 gene *Kif3A* was upregulated during growth arrest, as expected, indicates that the qPCR method described here is a valid method for identifying growth arrest-specific (*gas*) genes in cultured mammalian cells. Since ciliogenesis is strongly correlated with growth arrest in cultured NIH3T3 cells, some of the *gas* genes identified here (i.e., *Kif13A, Kif13B*, and *Kif16A*) may have cilia-related functions. However, additional experiments will be required to test this further as some *gas* genes may be involved exclusively in noncilia-related processes. Another caveat of the method is that all cilia-related genes are not necessarily upregulated during growth arrest, but could be regulated, for example, by alternative splicing. Indeed, alternative splicing was shown to be involved in generating a cilia-specific isoform of the Crumbs3 protein, CRB3-CLPI, involved in regulating cilia and centrosome biogenesis (Fan *et al.*, 2007), and multiple splice variants have also been reported for *Kif13A* (Jamain *et al.*, 2001), which complicates the interpretation of the qPCR results for *Kif13A* presented here. Thus it is possible that some isoforms of Kif13A (and potentially other kinesins as well) are abundant during growth arrest while others are not. Additional qPCR analyses using isoform-specific primers might provide further insight into this issue.

VI. Summary

We have described a qPCR-based method for identifying genes that are upregulated during growth arrest in mouse NIH3T3 cells concomitant with assembly of primary cilia. Using this method, we identified three kinesin-3 genes (i.e., *Kif13A, Kif13B*, and *Kif16A*) that are upregulated specifically during growth arrest, suggesting potential cilia-related functions for these genes. The validity of the method is underscored by the observations that *Kif14* was downregulated during growth arrest/ciliogenesis whereas *Kif3A* was upregulated, as expected. Given that not all genes that are upregulated during growth arrest have cilia-related functions, further validation by, for example, IFM is required to test the potential ciliary association of Kif13A, Kif13B, and Kif16A. In addition to identifying potential cilia-associated kinesin-3 genes, the method can be extended to identify/confirm other genes or gene families that are upregulated during growth arrest.

Acknowledgments

We thank Linda Schneider for assistance with immunofluorescence analysis and Kristian Arild Poulsen for help with qPCR. This work was supported by grants from the Lundbeck Foundation (No. R9-A969), the Novo Nordisk Foundation, and the Danish Natural Science Research Council (No. 272-07-0530 and No. 272-07-0411) to S. T. C. and L. B. P., as well as a Novo Nordisk/Novozymes scholarship to R. I. T.

References

Andersen, J.S., Wilkinson, C.J., Mayor, T., Mortensen, P., Nigg, E.A., and Mann, M. (2003). Proteomic characterization of the human centrosome by protein correlation profiling. *Nature* **426**, 570–574.

Avidor-Reiss, T., Maer, A.M., Koundakjian, E., Polyanovsky, A., Keil, T., Subramaniam, S., and Zuker, C.S. (2004). Decoding cilia function: Defining specialized genes required for compartmentalized cilia biogenesis. *Cell* **117**, 527–539.

Bareil, C., Hamel, C.P., Delague, V., Arnaud, B., Demaille, J., and Claustres, M. (2001). Segregation of a mutation in CNGB1 encoding the beta-subunit of the rod cGMP-gated channel in a family with autosomal recessive retinitis pigmentosa. *Hum. Genet.* **108**, 328–334.

Blacque, O.E., Perens, E.A., Boroevich, K.A., Inglis, P.N., Li, C., Warner, A., Khattra, J., Holt, R.A., Ou, G., Mah, A.K., McKay, S.J., Huang, P., *et al.* (2005). Functional genomics of the cilium, a sensory organelle. *Curr. Biol.* **15**, 935–941.

Carleton, M., Mao, M., Biery, M., Warrener, P., Kim, S., Buser, C., Marshall, C.G., Fernandes, C., Annis, J., and Linsley, P.S. (2006). RNA interference-mediated silencing of mitotic kinesin KIF14 disrupts cell cycle progression and induces cytokinesis failure. *Mol. Cell Biol.* **26**, 3853–3863.

Chen, N., Mah, A., Blacque, O.E., Chu, J., Phgora, K., Bakhoum, M.W., Newbury, C.R., Khattra, J., Chan, S., Go, A., Efimenko, E., Johnsen, R., *et al.* (2006). Identification of ciliary and ciliopathy genes in Caenorhabditis elegans through comparative genomics. *Genome Biol.* **7**, R126.

Chhin, B., Pham, J.T., El, Z.L., Kaiser, K., Merrot, O., and Bouvagnet, P. (2008). Identification of transcripts overexpressed during airway epithelium differentiation. *Eur. Respir. J.* **32**, 121–128.

Christensen, S.T., Pedersen, L.B., Schneider, L., and Satir, P. (2007). Sensory cilia and integration of signal transduction in human health and disease. *Traffic* **8**, 97–109.

Corbit, K.C., Shyer, A.E., Dowdle, W.E., Gaulden, J., Singla, V., Chen, M.H., Chuang, P.T., and Reiter, J.F. (2008). Kif3a constrains beta-catenin-dependent Wnt signalling through dual ciliary and non-ciliary mechanisms. *Nat. Cell Biol.* **10**, 70–76.

Dorner, C., Ciossek, T., Muller, S., Moller, P.H., Ullrich, A., and Lammers, R. (1998). Characterization of KIF1C, a new kinesin-like protein involved in vesicle transport from the Golgi apparatus to the endoplasmic reticulum. *J. Biol. Chem.* **273**, 20267–20275.

Dorner, C., Ullrich, A., Haring, H.U., and Lammers, R. (1999). The kinesin-like motor protein KIF1C occurs in intact cells as a dimer and associates with proteins of the 14-3-3 family. *J. Biol. Chem.* **274**, 33654–33660.

Dwyer, N.D., Adler, C.E., Crump, J.G., L'Etoile, N.D., and Bargmann, C.I. (2001). Polarized dendritic transport and the AP-1 mu1 clathrin adaptor UNC-101 localize odorant receptors to olfactory cilia. *Neuron* **31**, 277–287.

Efimenko, E., Bubb, K., Mak, H.Y., Holzman, T., Leroux, M.R., Ruvkun, G., Thomas, J.H., and Swoboda, P. (2005). Analysis of xbx genes in C. elegans. *Development* **132**, 1923–1934.

Fan, S., Fogg, V., Wang, Q., Chen, X.W., Liu, C.J., and Margolis, B. (2007). A novel Crumbs3 isoform regulates cell division and ciliogenesis via importin beta interactions. *J. Cell Biol.* **178**, 387–398.

Gherman, A., Davis, E.E., and Katsanis, N. (2006). The ciliary proteome database: An integrated community resource for the genetic and functional dissection of cilia. *Nat. Genet.* **38**, 961–962.

Gruneberg, U., Neef, R., Li, X., Chan, E.H., Chalamalasetty, R.B., Nigg, E.A., and Barr, F.A. (2006). KIF14 and citron kinase act together to promote efficient cytokinesis. *J. Cell Biol.* **172**, 363–372.

Gundersen, G.G. and Bulinski, J.C. (1986). Microtubule arrays in differentiated cells contain elevated levels of a post-translationally modified form of tubulin. *Eur. J. Cell Biol.* **42**, 288–294.

Hanada, T., Lin, L., Tibaldi, E.V., Reinherz, E.L., and Chishti, A.H. (2000). GAKIN, a novel kinesin-like protein associates with the human homologue of the Drosophila discs large tumor suppressor in T lymphocytes. *J. Biol. Chem.* **275**, 28774–28784.

Haraguchi, K., Hayashi, T., Jimbo, T., Yamamoto, T., and Akiyama, T. (2006). Role of the kinesin-2 family protein, KIF3, during mitosis. *J. Biol. Chem.* **281**, 4094–4099.

Hoang, E.H., Whitehead, J.L., Dose, A.C., and Burnside, B. (1998). Cloning of a novel C-terminal kinesin (KIFC3) that maps to human chromosome 16q13-q21 and thus is a candidate gene for Bardet-Biedl syndrome. *Genomics* **52**, 219–222.

Hoepfner, S., Severin, F., Cabezas, A., Habermann, B., Runge, A., Gillooly, D., Stenmark, H., and Zerial, M. (2005). Modulation of receptor recycling and degradation by the endosomal kinesin KIF16B. *Cell* **121**, 437–450.

Horiguchi, K., Hanada, T., Fukui, Y., and Chishti, A.H. (2006). Transport of PIP3 by GAKIN, a kinesin-3 family protein, regulates neuronal cell polarity. *J. Cell Biol.* **174**, 425–436.

Inglis, P.N., Boroevich, K.A., and Leroux, M.R. (2006). Piecing together a ciliome. *Trends Genet.* **22**, 491–500.

Jamain, S., Quach, H., Fellous, M., and Bourgeron, T. (2001). Identification of the human KIF13A gene homologous to Drosophila kinesin-73 and candidate for schizophrenia. *Genomics* **74**, 36–44.

Keller, L.C., Romijn, E.P., Zamora, I., Yates, J.R., III, and Marshall, W.F. (2005). Proteomic analysis of isolated chlamydomonas centrioles reveals orthologs of ciliary-disease genes. *Curr. Biol.* **15**, 1090–1098.

Kopp, P., Lammers, R., Aepfelbacher, M., Woehlke, G., Rudel, T., Machuy, N., Steffen, W., and Linder, S. (2006). The kinesin KIF1C and microtubule plus ends regulate podosome dynamics in macrophages. *Mol. Biol. Cell* **17**, 2811–2823.

Kubista, M., Andrade, J.M., Bengtsson, M., Forootan, A., Jonak, J., Lind, K., Sindelka, R., Sjoback, R., Sjogreen, B., Strombom, L., Stahlberg, A., and Zoric, N. (2006). The real-time polymerase chain reaction. *Mol. Aspects Med.* **27**, 95–125.

Laurencon, A., Dubruille, R., Efimenko, E., Grenier, G., Bissett, R., Cortier, E., Rolland, V., Swoboda, P., and Durand, B. (2007). Identification of novel regulatory factor X (RFX) target genes by comparative genomics in Drosophila species. *Genome Biol.* **8**, R195.

Li, J.B., Gerdes, J.M., Haycraft, C.J., Fan, Y., Teslovich, T.M., May-Simera, H., Li, H., Blacque, O.E., Li, L., Leitch, C.C., Lewis, R.A., Green, J.S., *et al.* (2004). Comparative genomics identifies a flagellar and basal body proteome that includes the BBS5 human disease gene. *Cell* **117**, 541–552.

Lih, C.J., Cohen, S.N., Wang, C., and Lin-Chao, S. (1996). The platelet-derived growth factor alpha-receptor is encoded by a growth-arrest-specific (gas) gene. *Proc. Natl. Acad. Sci. USA* **93**, 4617–4622.

Liu, Q., Tan, G., Levenkova, N., Li, T., Pugh, E.N., Jr., Rux, J.J., Speicher, D.W., and Pierce, E.A. (2007). The proteome of the mouse photoreceptor sensory cilium complex. *Mol. Cell Proteomics* **6**, 1299–1317.

Mans, D.A., Voest, E.E., and Giles, R.H. (2008). All along the watchtower: Is the cilium a tumor suppressor organelle? *Biochim. Biophys. Acta* **1786**, 114–125.

Marshall, W.F. (2008). The cell biological basis of ciliary disease. *J. Cell Biol.* **180**(1), 17–21.

Mayer, U., Ungerer, N., Klimmeck, D., Warnken, U., Schnolzer, M., Frings, S., and Mohrlen, F. (2008). Proteomic analysis of a membrane preparation from rat olfactory sensory cilia. Chem. Senses **33**, 145–162.

Meng, W., Mushika, Y., Ichii, T., and Takeichi, M. (2008). Anchorage of microtubule minus ends to adherens junctions regulates epithelial cell–cell contacts. *Cell* **135**, 948–959.

Miki, H., Okada, Y., and Hirokawa, N. (2005). Analysis of the kinesin superfamily: Insights into structure and function. *Trends Cell Biol.* **15**, 467–476.

Nakagawa, T., Setou, M., Seog, D., Ogasawara, K., Dohmae, N., Takio, K., and Hirokawa, N. (2000). A novel motor, KIF13A, transports mannose-6-phosphate receptor to plasma membrane through direct interaction with AP-1 complex. *Cell* **103**, 569–581.

Nangaku, M., Sato-Yoshitake, R., Okada, Y., Noda, Y., Takemura, R., Yamazaki, H., and Hirokawa, N. (1994). KIF1B, a novel microtubule plus end-directed monomeric motor protein for transport of mitochondria. *Cell* **79**, 1209–1220.

Nogales-Cadenas, R., Abascal, F., Diez-Perez, J., Carazo, J.M., and Pascual-Montano, A. (2009). CentrosomeDB: A human centrosomal proteins database. *Nucleic Acids Res.* **37**, D175–D180.

Okada, Y., Yamazaki, H., Sekine-Aizawa, Y., and Hirokawa, N. (1995). The neuron-specific kinesin superfamily protein KIF1A is a unique monomeric motor for anterograde axonal transport of synaptic vesicle precursors. *Cell* **81**, 769–780.

Ostrowski, L.E., Blackburn, K., Radde, K.M., Moyer, M.B., Schlatzer, D.M., Moseley, A., and Boucher, R.C. (2002). A proteomic analysis of human cilia: Identification of novel components. *Mol. Cell Proteomics* **1**, 451–465.

Pazour, G.J., Agrin, N., Leszyk, J., and Witman, G.B. (2005). Proteomic analysis of a eukaryotic cilium. *J. Cell Biol.* **170**, 103–113.

Peden, E.M., and Barr, M.M. (2005). The KLP-6 kinesin is required for male mating behaviors and polycystin localization in Caenorhabditis elegans. *Curr. Biol.* **15**, 394–404.

Pfaffl, M.W. (2001). A new mathematical model for relative quantification in real-time RT-PCR. *Nucleic Acids Res.* **29**, e45.

Poole, C.A., Zhang, Z.J., and Ross, J.M. (2001). The differential distribution of acetylated and detyrosinated alpha-tubulin in the microtubular cytoskeleton and primary cilia of hyaline cartilage chondrocytes. *J. Anat.* **199**, 393–405.

Rosenbaum, J.L., and Witman, G.B. (2002). Intraflagellar transport. *Nat. Rev. Mol. Cell Biol.* **3**, 813–825.

Schneider, L., Clement, C.A., Teilmann, S.C., Pazour, G.J., Hoffmann, E.K., Satir, P., and Christensen, S.T. (2005). PDGFRalphaalpha signaling is regulated through the primary cilium in fibroblasts. *Curr. Biol.* **15**, 1861–1866.

Scholey, J.M. (2003). Intraflagellar transport. *Annu. Rev. Cell Dev. Biol.* **19**, 423–443.

Scholey, J.M. (2008). Intraflagellar transport motors in cilia: Moving along the cell's antenna. *J. Cell Biol.* **180**, 23–29.

Smith, J.C., Northey, J.G., Garg, J., Pearlman, R.E., and Siu, K.W. (2005). Robust method for proteome analysis by MS/MS using an entire translated genome: Demonstration on the ciliome of Tetrahymena thermophila. *J. Proteome. Res.* **4**, 909–919.

Stolc, V., Samanta, M.P., Tongprasit, W., and Marshall, W.F. (2005). Genome-wide transcriptional analysis of flagellar regeneration in Chlamydomonas reinhardtii identifies orthologs of ciliary disease genes. *Proc. Natl. Acad. Sci. USA* **102**, 3703–3707.

Tuma, M.C., Zill, A., Le, B.N., Vernos, I., and Gelfand, V. (1998). Heterotrimeric kinesin II is the microtubule motor protein responsible for pigment dispersion in Xenopus melanophores. *J. Cell Biol.* **143**, 1547–1558.

Wickstead, B., and Gull, K. (2006). A "holistic" kinesin phylogeny reveals new kinesin families and predicts protein functions. *Mol. Biol. Cell* **17**, 1734–1743.

Xu, Y., Takeda, S., Nakata, T., Noda, Y., Tanaka, Y., and Hirokawa, N. (2002). Role of KIFC3 motor protein in Golgi positioning and integration. *J. Cell Biol.* **158**, 293–303.

Yang, Z., Xia, C., Roberts, E.A., Bush, K., Nigam, S.K., and Goldstein, L.S. (2001). Molecular cloning and functional analysis of mouse C-terminal kinesin motor KifC3. *Mol. Cell Biol.* **21**, 765–770.

Yeh, S.D., Chen, Y.J., Chang, A.C., Ray, R., She, B.R., Lee, W.S., Chiang, H.S., Cohen, S.N., and Lin-Chao, S. (2002). Isolation and properties of Gas8, a growth arrest-specific gene regulated during male gametogenesis to produce a protein associated with the sperm motility apparatus. *J. Biol. Chem.* **277**, 6311–6317.

CHAPTER 4

Methods for the Isolation of Sensory and Primary Cilia—An Overview

Kimberly A.P. Mitchell[†], **Gabor Szabo**[*], ***and*** **Angela de S. Otero**[*]

[*]Department of Molecular Physiology and Biological Physics, University of Virginia School of Medicine, Charlottesville, Virginia 22903

[†]Department of Biology and Chemistry, Liberty University, Lynchburg, Virginia 24502

Abstract

Detailed proteomic analyses of mammalian olfactory and rod photoreceptor sensory cilia are now available, providing an inventory of resident ciliary proteins and laying the foundation for future studies of developmental and spatiotemporal changes in the composition of sensory cilia. Cilia purification methods that were elaborated and perfected over several decades were essential for these advances. In contrast, the proteome of primary cilia is yet to be established, because

978-0-12-375024-2
DOI: 10.1016/S0091-679X(08)94004-8

purification procedures for this organelle have been developed only recently. In this chapter, we review current techniques for the purification of olfactory and photoreceptor cilia, and evaluate methods designed for the selective isolation of primary cilia.

I. Introduction

Primary and sensory cilia of higher organisms play essential roles as detectors and transducers of extracellular and intercellular signals (for a recent review, see Berbari *et al.*, 2009). The primary cilium present in the majority of vertebrate cells responds to chemical and mechanical stimuli (reviewed in Gerdes *et al.*, 2009; Praetorius and Spring, 2005; Wheatley, 2008), while modified cilia of specialized cells in the retina and olfactory epithelium sense odorants and light. Multiple olfactory cilia arise from a knob at the extremity of apically oriented dendrites of olfactory receptor neurons (ORNs) and project into the mucous layer on the epithelial surface (reviewed in McEwan *et al.*, 2009; Menco, 1997). The photoreceptor outer segment is the enlarged tip of the connecting cilium, which is anchored to a basal body in the inner segment (Besharse and Horst, 1990; Fig. 1). Dysfunctions of nonmotile cilia are associated with a number of human diseases, referred to as ciliopathies (reviewed in Tobin and Beales, 2009).

The outer segments from retinal photoreceptor cells have been isolated by mechanical agitation followed by fractionation in density gradients for more than 70 years (Saito, 1938), and the techniques for isolation of olfactory and primary cilia rely on classical protocols developed over several decades for the purification of flagella and traditional "9 + 2" cilia from invertebrates and protozoa. The invaluable contribution of these methods to recent and remarkable developments in the understanding of the structure and function of sensory and primary cilia cannot be understated (Gherman *et al.*, 2006; Keller *et al.*, 2005; Liu *et al.*, 2007; Mayer *et al.*, 2009; Ostrowski *et al.*, 2002; Pazour *et al.*, 2005). As pointed out by Pazour *et al.* (2005), comparative genomics and *in silico* approaches provide important information on the identity of ciliary proteins, but cannot substitute for direct proteomic analysis of purified cilia, which is a critical step toward establishing a definitive inventory of the proteins involved in the structure and function of any organelle. Here we recapitulate the techniques developed for the isolation of olfactory and photoreceptor cilia of a purity suitable for proteomics studies, and assess existing methods for isolation of primary cilia for their potential values in future structural and functional studies.

II. Isolation of Olfactory Cilia

Early methods for removing cilia from *Tetrahymena* utilized ethanol treatment, followed by exposure to digitonin, high concentrations of KCl or glycerol (Child, 1959; Child and Mazia, 1956). In 1962, Watson and Hopkins published a method

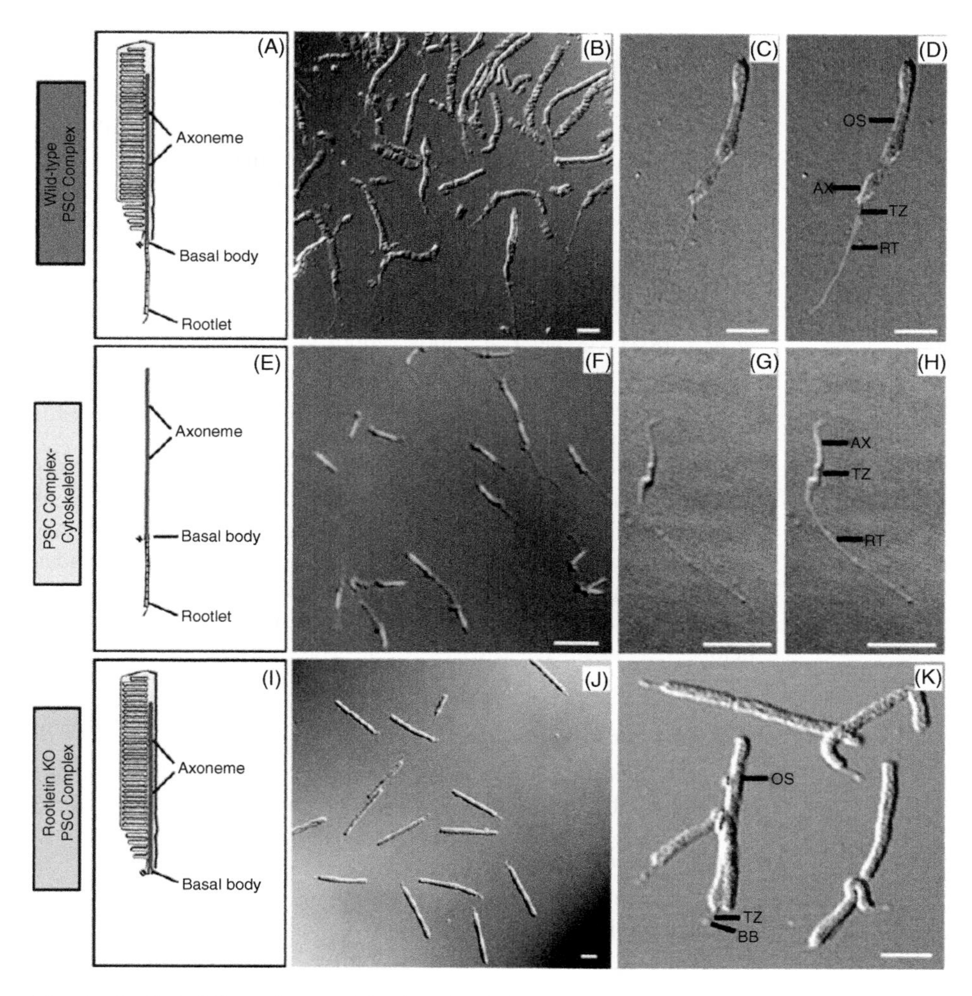

Fig. 1 Purified PSC complexes and fractions. (A) Schematic of a wild-type PSC complex showing that it comprises the outer segment and its cytoskeleton, including the rootlet, basal body, and axoneme. (B–D) wild-type PSC complex preparation viewed with differential interference contrast and fluorescence microscopy. At higher magnification (C and D), it can be seen that the PSC complexes consist of the outer segments with thin extensions at their bases. The extensions are the portions of the cytoplasmic cytoskeleton that were attached to the basal bodies as indicated by staining with antibodies to rootletin (RT, red; B and D). The axoneme in the outer segment is demonstrated by staining with antibodies to Rp1 (AX, green; D). At lower magnification (B), it is evident that the preparation consists of highly enriched PSC complexes with minimal contamination by the other structures noted. (E–H) Drawing of PSC complex-cytoskeleton (E) and isolated PSC complex-cytoskeleton preparation viewed with differential interference contrast (G) and fluorescence microscopy (F and H) using antibodies to rootletin (RT, *red*) and Rp1 (AX, *green*). The cytoskeletons consist of rootlets, basal bodies, and axonemes and are highly purified as illustrated in the images. (I–K) Drawing of a rootletin KO PSC complex (I) and rootletin KO PSC complex preparation (J and K) viewed with differential interference contrast and fluorescence microscopy and stained with antibodies to rootletin (*red*) and Rp1 (*green*). In these PSC complexes, *small dots* of rootletin staining are noted at the basal bodies, but no formed rootlets are present. OS, outer segment; BB, basal body; RT, ciliary rootlet; TZ, transition zone; AX, axoneme. Bars: 5μm. Modified from Liu *et al.* (2007). (See Plate no. 2 in the Color Plate Section.)

for isolating *Tetrahymena* cilia that combined ethanol treatment with high calcium, a procedure that became the basis for deciliation and deflagellation protocols in subsequent years (Gibbons, 1963; Rosenbaum and Child, 1967; Winicur, 1967; Witman *et al.*, 1972; Adoutte *et al.*, 1980). Eventually the use of ethanol was discontinued (Adoutte *et al.*, 1980), and deciliation through incubation in high Ca^{2+}, often combined with mechanical agitation, became known as the Ca^{2+}-shock method, a classic procedure that is still the method of choice for detaching the olfactory cilia from ORNs (Anholt *et al.*, 1986; Chen *et al.*, 1986; Delgado *et al.*, 2003; Mayer *et al.*, 2008, 2009; Schandar *et al.*, 1998). The common event triggered by all these treatments is the entry of calcium into cells, which activates the molecular machinery that results in the severance of cilia or flagella from the cell body (reviewed in Quarmby, 2004).

The isolation of olfactory cilia is not a straightforward process, given that the cilia and the dendritic knob are surrounded by numerous microvilli of supporting cells that can contaminate the final product. For instance, the standard Ca^{2+}-shock method produces a fraction containing significant amounts of nonciliary material (Mayer *et al.*, 2008), a finding that led to modifications of the procedure that yield highly purified olfactory cilia appropriate for proteomic analysis. This method is summarized below.

A. Ca^{2+}/K^{+}-shock and NaBr Treatment (Mayer *et al.*, 2009)

In this modified procedure, Mayer *et al.* (2009) reduce contaminants of the cilia purified by the traditional Ca^{2+}-shock method (Mayer *et al.*, 2008) through the inclusion of K^{+} in the deciliation buffer, an approach previously adopted for the isolation of *Paramecium* cilia (Adoutte *et al.*, 1980).

Briefly, the rat olfactory epithelium is washed for 5 min in a neutral isotonic buffer containing EGTA to remove mucus. Subsequently, the cilia are severed by gentle agitation in the same buffer, this time containing 20 mM $CaCl_2$ and 30 mM KCl—but no EGTA—for 20 min at 4°C. Cells and debris are pelleted by low-speed centrifugation. The detached cilia are collected by ultracentrifugation on a sucrose cushion (45% sucrose in deciliation buffer), and the interface band is diluted and pelleted. The ciliary fraction is washed twice with NaBr (2 M followed by 1 M) by resuspension and ultracentrifugation. The resulting pellet is washed by resuspension in a hypotonic buffer containing EGTA, and the olfactory cilia are recovered through a final ultracentrifugation step.

The yields in the olfactory cilia protein using this procedure are one-fifth of those observed with the traditional Ca^{2+}-shock method, presumably due to the decrease in contaminating cellular proteins and microvilli. However, these modifications increase the specificity of the deciliation process, so that larger amounts of the olfactory cilia are detached from the tissue while contaminating cellular proteins from microvilli and other sources remain behind, leading to an estimated enrichment of ~100-fold in ciliary proteins as assessed by immunoblotting of marker proteins. Examination of the deciliated tissue shows stubs of the ciliary proximal segments, suggesting that the fraction obtained through this method consists chiefly of the distal portions of the olfactory cilia. Analysis of the purified cilia by mass spectrometry led to the identification of 377 proteins, the majority of those being linked to the specialized function of olfactory neurons. Clearly, this is not a complete inventory of the proteins present in the olfactory cilia, but these results give new insights and perspectives regarding the components of the complex molecular machinery involved in olfactory signaling.

Note that the isolation of olfactory cilia by a Ca^{2+}-shock method may inflict damage to cellular processes sensitive to high calcium levels, such as signaling mechanisms, and may promote proteolytic activity. With this caveat in mind, Washburn *et al.* (2002) devised a method for the purification of olfactory cilia membranes in which cilia are detached by mechanical agitation in a neutral, isotonic buffer containing EDTA. The ciliary yields in the supernatant are slightly higher than those obtained with a standard Ca^{2+}-shock method. In addition, the presence of ciliary markers (evidenced by immunoblotting) and the adenylyl cyclase activities of the two preparations are comparable. However promising, this protocol has yet to be utilized specifically for the isolation of intact cilia and has not been characterized in terms of purity and integrity. Until this is done, the Ca^{2+}/K^{+}-shock method described above will likely remain the standard means of isolating the olfactory cilia.

III. Isolation of Rod Photoreceptor Cilia

Intact, sealed photoreceptor sensory complexes (PSCs) which comprise the outer segments and their ciliary axoneme, as well as the basal bodies and ciliary rootlets from the inner segments (Fig. 1), are readily detached from the retina by mechanical agitation and can be separated from the remainder of the photoreceptor cell by centrifugation in density gradients. Versions of this procedure have been used for many years to obtain large amounts of virtually pure PSCs from bovine retinas and were instrumental in building our present understanding of signal transduction in the visual system (Papermaster and Dryer, 1974; Saito, 1938; Schnetkamp and Daemen, 1982; Zimmerman and Godchaux, 1982). Published methods differ in the technique adopted to detach PSCs from retinas (e.g., vortexing or swirling), the composition and osmolarity of buffer solutions, the gradient-forming material (sucrose, Percoll, Ficoll, and metrizamide), and the type of gradient utilized (continuous vs discontinuous) and will not be reviewed here. The ciliary cytoskeleton of PSCs, comprising the axonemes, basal bodies, and rootlets, can be isolated by detergent treatment, which solubilizes the plasma membrane and the outer segment discs (Fleischman and Denisevich, 1979; Horst *et al.*, 1987; Pagh-Roehl and Burnside, 1995; Schmitt and Wolfrum, 2001). Note that the separation of connecting cilia from the remainder of the PSC has not been achieved, since this bridge-like structure is quite small (1 μm long and 0.3 μm in)

Schmitt and Wolfrum, 2001) and is effectively buried between the inner and outer segments.

We summarize below a procedure that was chosen for the purification of rod outer segments (ROS) for proteomic analysis (Liu *et al.*, 2007), providing a wealth of qualitative and quantitative information on the composition of the photoreceptor cilium.

A. Mechanical Shear (Liu *et al.*, 2007)

This procedure is a modification of those of Papermaster and Dryer (1974) and Liu *et al.* (2004). Mouse retinas are gently vortexed for 1 min in a hypertonic buffer containing 50% (w/v) sucrose. The suspension is centrifuged at high speed to pellet debris and tissue. The PSC band at the top is diluted to reduce the sucrose concentration and layered on a 50% sucrose shelf and recentrifuged. Purified PSCs are collected at the interface, diluted, and pelleted. This fraction contains highly purified photoreceptor cilia. The same purification procedure was followed to isolate PSCs from retinas from rootletin knockout (KO) mice, which do not form ciliary rootlets and therefore are essentially free of inner segment components (Fig. 1). Additionally, wild-type PSCs were treated with detergent to allow detection of low-abundance cytoskeletal proteins that otherwise would be obscured by the copious amounts of signaling proteins such as rhodopsin.

Analysis of these three preparations by mass spectrometry led to the identification of ~2000 proteins, along with an estimate of their individual copy numbers. The assignment of the proteins to the outer versus inner segments was accomplished through the assessment of overlaps and divergences between the three datasets. Comparison with the proteomes of cilia from lower organisms shows that numerous proteins are conserved and expressed in the mammalian photoreceptor cilium, in particular those involved in ciliogenesis and maintenance of mature cilia. A significant number of the proteins identified in this work are linked to diseases that impair vision, for example, retinal degeneration, and others are the product of genes linked to ciliary anomalies associated with human ciliopathies such as Bardet–Biedl syndrome (Tobin and Beales, 2009).

IV. Isolation of Primary Cilia

Anyone attempting to isolate primary cilia in amounts compatible with biochemical or functional analysis faces a formidable task. Most vertebrate cells display a single cilium of diminutive proportions, with a diameter of ~0.2 μm and lengths that vary from 1 to 4 μm. These forbidding characteristics stack up against a single encouraging feature: in the majority of cells the primary cilium protrudes from the cell surface, its accessibility being a trait that in principle should simplify the purification process. The situation is somewhat improved in confluent cultures of renal epithelial cells that express long primary cilia, with average lengths of 8–25 μm depending on the specific cell line (Mitchell *et al.*, 2004; Roth *et al.*, 1988; Wheatley and Bowser, 2000; Wheatley *et al.*, 1996;). Still, the ratio of the volume of a primary cilium to that of a typical renal epithelial cell in culture is very low, for example, about 1:3000 in Madin–Darby canine kidney (MDCK) cells for a cilium with a length of 8 μm (Praetorius and Spring, 2002). Because cell volume is proportional to protein content (Erlinger and Saier, 1982), a confluent 10-cm dish of MDCK cells with 3 mg of protein can yield at best 1 μg of pure primary cilium protein. This is a highly optimistic estimate, which assumes 100% of full-length cilia that are harvested by breakage at the ciliary necklace. In reality, primary cilia are very fragile structures prone to fragmentation (Gallagher, 1980) and are easily severed along the axoneme. Indeed, electron microscopy and immunofluorescence microscopy analyses of isolated primary cilia fractions (Mitchell *et al.*, 2004; Raychowdhury *et al.*, 2005) show cilia with shortened shafts and numerous detached tips, indicating that fracture of the ciliary stem often takes place at its distal portion and not at the base. Thus, it is essential to start with generous amounts of cells in order to obtain sufficient material for further analysis. Equally important is to use the gentlest possible deciliation process in order to preserve the integrity of the cilium membrane, retain components of the ciliary matrix, and reduce the odds of contamination by other organelles or cell components.

A. Mechanical Shear (Mitchell *et al.*, 2004)

The rationale behind this procedure is twofold: to harvest large amounts of primary cilia by using a cell line that has long cilia and to prevent damage to the cilium by avoiding the use of anesthetics or extremes of pH, osmolarity, or high calcium. The cells chosen, from the A6 line, are derived from the distal nephron of *Xenopus* and have a high incidence of primary cilia that extend to an average length of 23 μm after 7–10 days of culture in complete medium (Mitchell *et al.*, 2004). Therefore, in principle one could obtain 3 μg of pure primary cilia from a confluent, 150-mm dish of A6 cells. Additionally, when A6 cells are plated at high density, they express sizeable primary cilia within 24 h, a feature that can be exploited to study changes in the composition (Mitchell *et al.*, 2004) and function of cilia during the elongation process. The technique is based on the observation that shear forces can easily lead to deciliation of specimens (Gallagher, 1980). The procedure consists of brief mechanical agitation in phosphate-buffered saline followed by differential centrifugation and can be completed in less than 1 h. With this gentle approach, the cell monolayer and its apical surface suffer a minimum of mechanical damage, reducing contamination of the primary cilium fraction with proteins from the cell body. Examination of cells by immunofluorescence microscopy after agitation shows that the apical microvilli, which are stained by an antibody to nucleoside diphosphate (NDP) kinase, remain intact after deciliation. Note that not all cilia are removed, since the surface of the confluent monolayer is uneven due to dome formation (Moberly and Fanestil, 1988), so the drag force on stems and tips of cilia is not uniform across the culture dish.

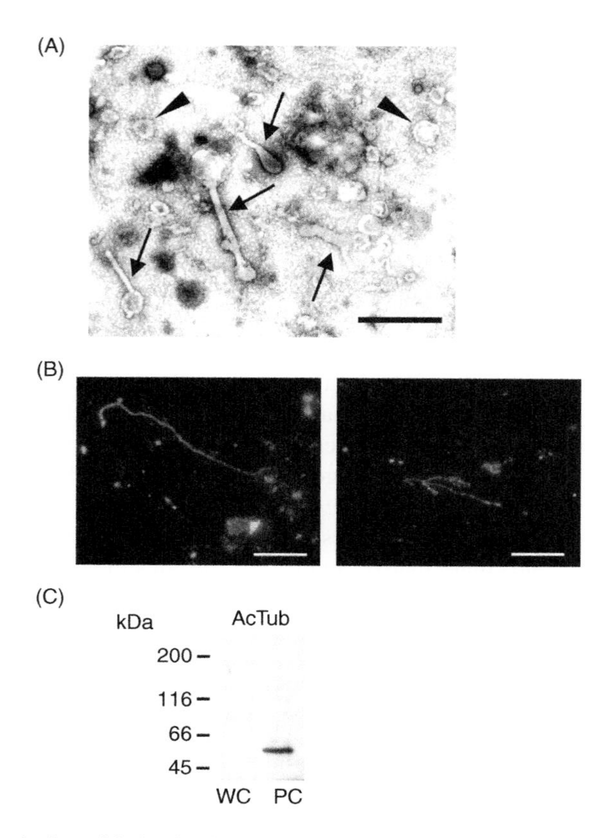

Fig. 2 Characterization of isolated primary cilia. (A) Electron microscopy of isolated primary cilia negatively stained with uranyl acetate. Arrows indicate primary cilia, and arrowheads point to ciliary tip-like structures. Bar: 2.5 μm. (B) Immunofluorescence microscopy of isolated primary cilia doubly stained with antibodies to acetylated α-tubulin and NDP kinase. Red, acetylated α-tubulin; green, NDP kinase; yellow, overlap regions. Bars: 20 μm. (C) Immunoblot analysis of isolated primary cilia: equal amounts of protein from whole-cell (WC) extracts and isolated primary cilia (PC) were resolved by SDS-PAGE and probed with antibodies to acetylated α-tubulin (AcTub). Modified with permission from Mitchell *et al.* (2004). (See Plate no. 3 in the Color Plate Section.)

Electron microscopy of negatively stained specimens (Fig. 2A) reveals structures with the characteristic lollipop shape of primary cilia (Roth *et al.*, 1988; Wheatley and Bowser, 2000). The bulb-like tips often appear as discrete structures. Immunofluorescence analysis of preparations fixed and then stained for both acetylated α-tubulin and NDP kinase, which is present in primary cilia as well as in microvilli (Donowitz *et al.*, 2007; Mitchell *et al.*, 2004), confirms that the preparation contains not only typical cilia, but also round structures of roughly the size of ciliary tips observed in cells stained for acetylated α-tubulin. Structures labeled solely by antibodies to NDP kinase, presumably nonciliary contaminants, constitute an estimated 40–60% of the isolated primary cilia fraction.

Note that the ciliary shafts are significantly longer in samples visualized by immunofluorescence (Fig. 2B) when compared with those seen with negative stain microscopy (Fig. 2A), suggesting that fixation of the specimen helps to preserve the ciliary structure.

Yields from four 150-mm dishes are 6–20 μg of isolated primary cilia, falling within the range of the theoretical number of 12 μg of protein for an ideal, 100% pure preparation of full length, 24-μm cilia, and the procedure is easily scaled up. Analysis of immunoblots (Fig. 2C) shows a considerable enrichment in acetylated α-tubulin in comparison with whole A6 cell extracts, on average 17-fold ($n = 4$). Note that in some preparations a purification factor could not be estimated, because of the absence of a signal for acetylated α-tubulin in the cell extract (e.g., Fig. 2C), as expected from an antigen present at low levels in cells (Piperno *et al.*, 1987).

In brief, this protocol suffers from the usual trade-off: a high degree of enrichment is achieved, but yields are relatively low. The same approach can be utilized to isolate primary cilia from MDCK cells (A. S Otero and K. A. P Mitchell, unpublished results), although one should keep in mind that the process of ciliogenesis in this line is slower than in A6 cells (Praetorius and Spring, 2005).

B. Ca^{2+}-shock (Raychowdhury *et al.*, 2005)

This procedure uses a Ca^{2+}-shock protocol (see Section II.A) to purify cilia from the LLC-PK1 cells, a renal cell line that has well developed primary cilia when grown to confluence for 2–3 weeks (17 μm; Roth *et al.*, 1988). The major modification to the classic Ca^{2+}-shock is a preliminary step in which cells are scraped from the culture dish in Ca^{2+}-free saline. The cells are then pelleted, resuspended in a high Ca^{2+} solution (112 mM NaCl, 3.4 mM KCl, 10 mM $CaCl_2$, 2.4 mM $NaHCO_3$, 2 mM HEPES, pH 7.0), and shaken for 10 min at 4°C. After low-speed centrifugation ($7700 \times g$ for 5 min) to remove debris, the suspension is placed on a 45% sucrose shelf in high Ca^{2+} solution and centrifuged at 100,000 × g for 1 h. The interface band is removed, and cilia are pelleted by centrifugation at 100,000 × g for 1 h after a 10-fold dilution of the band at the saline/sucrose interface. The final pellet is resuspended in normal PBS containing 2.0 mM EGTA and 0.5 mM sucrose. Yields were not reported. This method was successfully adapted by Wang and Brautigan (2008) to purify primary cilia from human retinal epithelial cells.

The possibility of contamination of the preparation by basolateral membranes was explored in a subsequent publication (Raychowdhury *et al.*, 2009) through immunostaining with antibodies to tubulin, type 2 vasopressin receptor (V2R), and Na^+–K^+-ATPase. The results are not shown, but the authors report a negligible contribution of basolateral membranes to the isolated cilia fraction.

The cilia preparation obtained through this approach was characterized by immunofluorescence and electron microscopy, as well as electrophysiological methods. Staining with antibodies to acetylated α-tubulin shows rounded structures and cilia that are short and appear much thicker than the native structures seen in cells, although comparison between native and isolated primary cilia is hampered by the absence of magnification bars (Li *et al.*, 2006; Raychowdhury *et al.*, 2005). Additional immunolabeling experiments show the presence of epithelial sodium channels in cilia isolated from wild-type cells and V2R-*gfp* and adenylyl cyclase in primary cilia isolated from V2R-*gfp* expressing LLC-PK1 cells.

Negative stain electron microscopy shows numerous cilia tips and relatively long shafts of irregular diameter (Raychowdhury *et al.*, 2005). As stated in Raychowdhury *et al.* (2005), this procedure swells the cilia to the extent that electrophysiological recordings can be obtained using a very small, high-resistance (140 mΩ, or ~0.1 μm) patch pipette. However, a true gigaseal (i.e., a pipette-to-membrane electrical seal with a resistance greater than 5 GΩ) could not be established. The poor quality of the membrane–pipette seal may result from membrane damage, the presence of debris on the membrane surface, or, similarly to what happens in spermatozoa (Kirichok *et al.*, 2006), the existence of a tight connection between the plasma membrane and the axoneme preventing seal formation. This leaky seal, in combination with the high resistance of very small patch pipettes, precludes proper voltage clamping of the preparation and accurate measurement of channel properties, since the leak current (~1 pA) is far from negligible when compared to the size of the single-channel currents observed. Within these constraints, two levels of ion currents attributed to nonselective ion channels can be detected in cilium-attached patches, as well as a small Na^+-permeable channel induced by vasopressin. The activity of the latter is increased by bath application of cAMP-dependent protein kinase (PKA) and ATP and is inhibited by amiloride. The mechanism by which PKA gains access to the inner surface of the membrane is not known; it is speculated that the enzyme can cross the leaky membrane and/or diffuse through the open end of the isolated cilia (Raychowdhury *et al.*, 2005). The latter hypothesis is unlikely to be correct, because the ciliary lumen is a congested compartment, mostly filled with the axoneme and attached proteins, and the cilia isolated by Ca^{2+}-shock retain not only axonemal proteins but also cytosolic and membrane proteins such as soluble and membrane-associated adenylyl cyclase, cyclic nucleotide phosphodiesterase (Raychowdhury *et al.*, 2009), the catalytic subunit of protein phosphatase 1 (PP1C), and phosphatase inhibitor-2 (Wang and Brautigan, 2008). Thus, the tears in the ciliary membrane are likely to be larger than 50 Å, since the hydrodynamic radius of PKA complexed with ATP is 24 Å (Yang *et al.*, 2005).

Reconstitution of the cilia preparation onto lipid bilayers indicates the presence of several types of ion channels, including a cation-selective channel inhibited by an antibody to polycystin-2 and an anion-selective channel (Li *et al.*, 2006; Raychowdhury *et al.*, 2005, 2009). Nevertheless, in the absence of a rigorous assessment of the degree of contamination by subcellular components released by the initial scraping step, the origin of these channels remains undefined.

In summary, this approach produces a fraction enriched in primary cilia, but may lead to significant contamination of the final material by proteins from other cell compartments (Mayer *et al.*, 2009). The physical damage to the membrane is also a disadvantage, interfering with the direct electrophysiological recording of ion channel currents *in situ*. Note that scrape-loading experiments have demonstrated that the tears in the plasma membranes of cells are unable to reseal quickly at low calcium levels (McNeil and Steinhardt, 1997), so both problems may be reduced if scraping in a Ca^{2+}-free solution is avoided. An alternative and milder approach would be to apply this protocol to cells lifted with normal saline containing 1 mM EDTA, a technique that preserves primary cilia (Mitchell *et al.*, 2004).

C. Peel-Off and Slide-Pull (Huang *et al.*, 2006)

These two approaches were utilized to isolate primary cilia of medium size (7–10 μm) from cultures of normal mouse and rat cholangiocytes. Basically, a coverslip or small culture dish is coated with 0.1% poly-L-lysine (PLL) and placed atop the upper surface of the cells. After application of pressure (peel-off) or mechanical agitation (slide-pull), the coated surface with attached structures is removed and used for cilia purification. The composition of the solutions utilized during cilia isolation is not reported.

Both methods are derived from the "rip-off" or "deroofing" procedure originally designed for detaching sheets of plasma membrane from the cell surface (Perez *et al.*, 2006; Sanan and Anderson, 1991). This technique has been successfully utilized to isolate plasma membranes with attached endocytic vesicles and cytoskeletal components from the apical surface of normal rat cholangiocytes (Doctor *et al.*, 2002). Indeed, portions of the cholangiocyte apical membrane adhere to the PLL-coated surface and are detached along with the cilia (Huang *et al.*, 2006), contaminating the cilia preparation.

In the peel-off method, the coated coverslip is manually pressed onto the cells for 20 s while the medium is removed and then quickly lifted off with forceps. The preparation is either fixed and analyzed by microscopy or scraped from the coverslip and purified in a sucrose cushion, as described by Raychowdhury *et al.* (2005). A deciliation efficiency of 70% was estimated by staining the coverslips with antiacetylated α-tubulin.

In the slide-pull technique, a 6-cm PLL-coated dish is placed on the top of cells cultured in a 10-cm dish, and this assembly is placed on a sliding shaker at 70–100 rpm. After 5 min the supernatant is removed, vortexed for 1 min, and placed in an ice bath for 20–30 min to settle debris. The cilia are collected and purified as above.

To assess purity, ciliary pellets and extracts from the corresponding deciliated cell extracts were solubilized and analyzed by immunoblotting. Analysis of the isolated primary cilia shows the presence of acetylated α-tubulin, polycystins 1 and 2, and fibrocystin. However, amounts of the ciliary marker acetylated α-tubulin in the final fraction of the peel-off technique are only 50% higher than those measured in the whole-cell lysate. The slide-pull approach leads to a slightly higher enrichment in acetylated α-tubulin: threefold. The yields of ciliary protein from three 10-cm dishes are 35–130 μg and 150–270 μg for the peel-off and slide-pull techniques, respectively. These excessively high yields (see Section IV.A), combined with the modest enrichment in acetylated α-tubulin, imply that both preparations are impure, consisting mostly of nonciliary elements. The ciliary fraction obtained by the peel-off technique is likely to be heavily contaminated by plasma membranes, endocytic vesicles, caveolae, coated pits, and the submembranous cytoskeleton (Heuser, 2000; Huang *et al.*, 1997; Parton and Hancock, 2001). Furthermore, the cilia recovered from the coated coverslip by scraping are likely to be irreparably damaged and thus unsuitable for studies that require a minimum of structural integrity. As to the slide-pull method, movement of the PLL-coated dish may well wound the cells and even result in the detachment of membrane sheets, increasing the amounts of contaminants in the fraction.

D. Perspectives

The emergence of methods to isolate primary cilia is a critical step toward understanding the normal functions of these complex structures and their role in human disease. The approaches described here use traditional techniques based on chemical or mechanical stress to sever primary cilia from cells. Methodological improvements will be required to reduce damage to this delicate organelle during purification. The discovery of selective deciliation agents, which remove motile cilia reversibly, rapidly, and at low concentrations (Semenova *et al.*, 2008), raises hopes that analogous compounds may soon become available for the removal of immotile cilia, yielding preparations of intact and essentially pure primary cilia. Recent advances in patch clamp techniques for the study of ion channels in tiny structures such as the sperm head (Jiménez-González *et al.*, 2007) imply that a gigaseal recording of ion channel activity in isolated primary cilia is an attainable goal. Those are compelling reasons to believe that a comprehensive picture of the composition and function of primary cilia will be a reality in the near future.

References

Adoutte, A., Ramanathan, R., Lewis, R. M., Dute, R. R., Ling, K.-Y., Kung, C., and Nelson, D. L. (1980). Biochemical studies of the excitable membrane of *Paramecium tetraurelia*. III Proteins of cilia and ciliary membranes. *J. Cell Biol.* **84**, 717–738.

Anholt, R. R., Aebi, U., and Snyder, S. H. (1986). A partially purified preparation of isolated chemosensory cilia from the olfactory epithelium of the bullfrog, *Rana catesbeiana*. *J. Neurosci.* **6**, 1962–1969.

Berbari, N. F., O'Connor, A. K., Haycraft, C. J., and Yoder, B. K. (2009). The primary cilium as a complex signaling center. *Curr. Biol.* **19**, R526–R535.

Besharse, J. C., and Horst, C. J. (1990). The photoreceptor connecting cilium. A model for the transition zone. *In* "Ciliary and Flagellar Membranes" (R. A. Bloodgood, ed.), pp. 389–417. Plenum Press, New York.

Chen, Z., Pace, U., Ronen, D., and Lancet, D. (1986). Polypeptide gp95: A unique glycoprotein of olfactory cilia with transmembrane receptor properties. *J. Biol. Chem.* **261**, 1299–1305.

Child, F. M. (1959). The characterization of the cilia of *Tetrahymena pyriformis*. *Exp. Cell Res.* **18**, 258–267.

Child, F. M., and Mazia, D. (1956). A method for the isolation of the parts of ciliates. *Experientia* **12**, 161–162.

Delgado, R., Saavedra, M. V., Schmachtenberg, O., Sierralta, J., and Bacigalupo, J. (2003). Presence of Ca^{2+}-dependent K^{+} channels in chemosensory cilia support a role in odor transduction. *J. Neurophysiol.* **90**, 2022–2028.

Doctor, R. B., Dahl, R., Fouassier, L., Kilic, G., and Fitz, J. G. (2002). Cholangiocytes exhibit dynamic, actin-dependent apical membrane turnover. *Am. J. Physiol. Cell Physiol.* **282**, C1042–C1052.

Donowitz, M., Singh, S., Salahuddin, F. F., Hogema, B. M., Chen, Y., Gucek, M., Cole, R. N., Zachos, N. C., Kovbasnjuk, O., Lapierre, L. A., Broere, N., Goldenring, J., et al. (2007). Proteome of murine jejunal brush border membrane vesicles. *J. Proteome Res.* **6**, 4068–4079.

Erlinger, S., and Saier, M. H., Jr. (1982). Decrease in protein content and cell volume of cultured dog kidney epithelial cells during growth. *In Vitro* **18**, 196–202.

Fleischman, D., and Denisevich, M. (1979). Guanylate cyclase of isolated bovine retinal rod axonemes. *Biochemistry* **18**, 5060–5066.

Gallagher, B. C. (1980). Primary cilia of the corneal endothelium. *Am. J. Anat.* **159**, 475–484.

Gerdes, J. M., Davis., E. E., and Katsanis, N. (2009). The vertebrate primary cilium in development, homeostasis, and disease. *Cell* **137**, 32–45.

Gherman, A., Davis, E. E., and Katsanis, N. (2006). The ciliary proteome database: An integrated community resource for the genetic and functional dissection of cilia. *Nat. Genet.* **38**, 961–962.

Gibbons, I. R. (1963). Studies on the protein components of cilia from *Tetrahymena pyriformis*. *Proc. Natl. Acad. Sci. USA* **50**, 1002–1010.

Heuser, J. (2000). The production of 'cell cortices' for light and electron microscopy. *Traffic* **1**, 545–552.

Horst, C. J., Forestner, D. M., and Besharse, J. C. (1987). Cytoskeletal-membrane interactions: A stable interaction between cell surface glycoconjugates and doublet microtubules of the photoreceptor connecting cilium. *J. Cell Biol.* **105**, 2973–2987.

Huang, B. Q., Masyuk, T. V., Muff, M. A., Tietz, P. S., Masyuk, A. I., and LaRusso, N. F. (2006). Isolation and characterization of cholangiocyte primary cilia. *Am. J. Physiol. Gastrointest. Liver Physiol.* **291**, G500–G509.

Huang, C., Hepler, J. R., Chen, L. T., Gilman, A. G., Anderson, R. G., and Mumby, S. M. (1997). Organization of G proteins and adenylyl cyclase at the plasma membrane. *Mol. Biol. Cell* **8**, 2365–2378.

Jiménez-González, M. C., Gu, Y., Kirkman-Brown, J., Barratt, C. L., and Publicover, S. (2007). Patch-clamp 'mapping' of ion channel activity in human sperm reveals regionalisation and co-localisation into mixed clusters. *J. Cell. Physiol.* **213**, 801–808.

Keller, L. C., Romijn, E. P., Zamora, I., Yates, J. R., III, and Marshall, W. F. (2005). Proteomic analysis of isolated *Chlamydomonas* centrioles reveals orthologs of ciliary-disease genes. *Curr. Biol.* **15**, 1090–1098.

Kirichok, Y., Navarro, B., and Clapham, D. E. (2006). Whole-cell patch clamp measurements of spermatozoa reveal an alkaline-activated Ca^{2+} channel. *Nature* **439**, 737–740.

Li, Q., Montalbetti, N., Wu, Y., Ramos, A., Raychowdhury, M. K., Chen, X. Z., and Cantiello, H. F. (2006). Polycystin-2 cation channel function is under the control of microtubular structures in primary cilia of renal epithelial cells. *J. Biol. Chem.* **281**, 37566–37575.

Liu, Q., Tan, G., Lavenkova, N., Li, T., Pugh, E. N., Jr., Rux, J. J., Speicher, D. W., and Pierce, E. A. (2007). The proteome of the mouse photoreceptor sensory cilium complex. *Mol. Cell. Proteomics* **6**, 1299–1317.

Liu, Q., Zuo, J., and Pierce, E. A. (2004). The retinitis pigmentosa 1 protein is a photoreceptor microtubule-associated protein. *J. Neurosci.* **24**, 6427–6436.

Mayer, U., Kuller, A., Daiber, P. C., Neudorf, I., Warnken, U., Schnolzer, M., Frings, S., and Mohrlen, F. (2009). The proteome of rat olfactory sensory cilia. *Proteomics* **9**, 322–334.

Mayer, U., Ungerer, N., Klimmeck, D., Warnken, U., Schnolzer, M., Frings, S., and Mohrlen, F. (2008). Proteomic analysis of a membrane preparation from rat olfactory sensory cilia. *Chem. Senses* **33**, 145–162.

McEwan, D. P., Jenkins, P. M., and Martens, J. R. (2009). Olfactory cilia: Our direct neuronal connection to the external world. *Curr. Top. Dev. Biol.* **85**, 333–370.

McNeil, P. L., and Steinhardt, R. A. (1997). Loss, restoration and maintenance of plasma membrane integrity. *J. Cell Biol.* **137**, 1–4.

Menco, B. P. (1997). Ultrastructural aspects of olfactory signaling. *Chem. Senses* **22**, 295–311.

Mitchell, K. A.P., Gallagher, B. C., Szabo, G., and Otero, A. S. (2004). NDP kinase moves into developing primary cilia. *Cell Motil. Cytoskeleton* **59**, 62–73.

Moberly, J. B., and Fanestil, D. D. (1988). A monoclonal antibody that recognizes a basolateral membrane protein in A6 epithelial cells. *J. Cell. Physiol.* **135**, 63–70.

Ostrowski, L. E., Blackburn, K., Radde, K. M., Moyer, M. B., Schlatzer, D. M., Moseley, A., and Boucher, R. C. 2002.A proteomic analysis of human cilia: Identification of novel components. *Mol. Cell. Proteomics* **1**, 451–465.

Pagh-Roehl, K., and Burnside, B. (1995). Preparation of teleost rod inner and outer segments. *In* "Methods in Cell Biology" (A. Dentler and G. Witman, eds.), Vol. 47, pp. 83–92. Academic Press, San Diego.

Papermaster, D. S., and Dryer, W. J. (1974). Rhodopsin content in outer segment membranes of bovine and frog retinal rods. *Biochemistry* **13**, 2438–2444.

Parton, R. G., and Hancock, J. F. (2001). Caveolin and Ras function. *Method Enzymol.* **333**, 172–183.

Pazour, G. J., Agrin, N., Leszyk, J., and Witman, G. B. (2005). Proteomic analysis of an eukaryotic cilium. *J. Cell Biol.* **170**, 103–113.

Perez, J.-B., Martinez, K. L., Segura, J.-M., and Vogel, H. (2006). Supported cell-membrane sheets for functional fluorescence imaging of membrane proteins. *Adv. Funct. Mater.* **16**, 306–312.

Piperno, G., LeDizet, M., and Chang, X. J. (1987). Microtubules containing acetylated alpha-tubulin in mammalian cells in culture. *J. Cell Biol.* **104**, 289–302.

Praetorius, H. A., and Spring, K. R. (2002). Removal of the MDCK cell primary cilium abolishes flow sensing. *J. Membr. Biol.* **191**, 69–76.

Praetorius, H. A., and Spring, K. R. (2005). A physiological view of the primary cilium. *Annu. Rev. Physiol.* **67**, 515–529.

Quarmby, L. M. (2004). Cellular deflagellation. *Int. Rev. Cytol.* **233**, 47–91.

Raychowdhury, M. K., McLaughlin, M., Ramos, A. J., Montalbetti, N., Bouley, R., Ausiello, D. A., and Cantiello, H. F. (2005). Characterization of single channel currents from primary cilia of renal epithelial cells. *J. Biol. Chem.* **280**, ,34718–34722.

Raychowdhury, M. K., Ramos, A. J., Zhang, P., McLaughin, M., Dai, X. Q., Chen, X. Z., Montalbetti, N., Del Rocío Cantero, M., Ausiello, D. A., and Cantiello, H. F. (2009). Vasopressin receptor-mediated functional signaling pathway in primary cilia of renal epithelial cells. *Am. J. Physiol. Renal Physiol.* **296**, F87–F97.

Rosenbaum, J. L., and Child, F. M. (1967). Flagellar regeneration in protozoan flagellates. *J. Cell Biol.* **34**, 345–364.

Roth, K. E., Rieder, C. L., and Bowser, S. S. (1988). Flexible substratum technique for viewing from the side: Some *in vivo* properties of primary (9 + 0) cilia in cultures of kidney epithelia. *J. Cell Sci.* **89**, 457–466.

Saito, Z. (1938). Isolierung der Stäbchenaussenglieder und spektrale Untersuchung des daraus hergestellten Sehpurpurextraktes. *Tohoku Y. Exp. Med.* **32**, 432–466.

Sanan, D., and Anderson, R. (1991). Simultaneous visualization of LDL receptor distribution and clathrin lattices on membranes torn from the upper surface of cultured cells. *J. Histochem. Cytochem.* **39**, 1017–1024.

Schandar, M., Laugwitz, K.-L., Boekhoff, I., Kroner, C., Gudermann, T., Schultz, G., and Breer, H. (1998). Odorants selectively activate distinct G protein subtypes in olfactory cilia. *J. Biol. Chem.* **273**, 16669–16677.

Schmitt, A., and Wolfrum, U. (2001). Identification of novel molecular components of the photoreceptor connecting cilium by immunoscreens. *Exp. Eye Res.* **73**, 837–849.

Schnetkamp, P. P.M., and Daemen, F. J.M. (1982). Isolation and characterization of osmotically sealed bovine rod outer segments. *Meth. Enzymol.* **81**, 110–116.

Semenova, M. N., Tsyganov, D. V., Yakubov, A. P., Kiselyov, A. S., and Semenov, V. V. (2008). A synthetic derivative of plant allylpolyalkoxybenzenes induces selective loss of motile cilia in sea urchin embryos. *ACS Chem. Biol.* **3**, 95–100.

Tobin, J. L., and Beales, P. L. (2009). The nonmotile ciliopathies. *Genet Med.* **11**, 386–402.

Wang, W., and Brautigan, D. L. (2008). Phosphatase inhibitor 2 promotes acetylation of tubulin in the primary cilium of human retinal epithelial cells. *BMC Cell Biol.* **9**, 62–73.

Washburn, K. B., Turner, T. T., and Talamo, B. R. (2002). Comparison of mechanical agitation and calcium shock methods for preparation of a membrane fraction enriched in olfactory cilia. *Chem. Senses* **27**, 635–642.

Watson, M. R., and Hopkins, J. M. (1962). Isolated cilia from *Tetrahymena pyriformis*. *Exp. Cell Res.* **28**, 280–295.

Wheatley, D. N. (2008). Nanobiology of the primary cilium-paradigm of a multifunctional nanomachine complex. *In* "Methods in Cell Biology" (B. P. Jena, ed.), Vol. 90, pp. 139–156. Elsevier Inc, San Diego

Wheatley, D. N., and Bowser, S. S. (2000). Length control of primary cilia: Analysis of monociliate and multiciliate cells. *Biol. Cell.* **92**, 573–582.

Wheatley, D. N., Wang, A.-M., and Strugnell, G. E. (1996). Expression of primary cilia in mammalian cells. *Cell Biol. Int.* **20**, 73–81.

Winicur, S. (1967). Reactivation of ethanol-calcium isolated cilia from *Tetrahymena pyriformis*. *J. Cell Biol.* **35**, C7–C9.

Witman, G. B., Carlson, K., Berliner, J., and Rosenbaum, J. L. (1972). *Chlamydomonas* flagella. Isolation and electrophoretic analysis of microtubules, matrix, membranes and mastigonemes. *J. Cell Biol.* **54**, 507–539.

Yang, S., Rogers, K. M., and Johnson, D. A. (2005). MgATP-induced conformational change of the catalytic subunit of cAMP-dependent protein kinase. *Biophys. Chem.* **113**, 193–199.

Zimmerman, W. F., and Godchaux, W. III. (1982). Preparation and characterization of sealed bovine rod cell outer segments. *Meth. Enzymol.* **81**, 52–57.

CHAPTER 5

Isolation of Primary Cilia for Morphological Analysis

Bing Huang, Tatyana Masyuk, *and* Nicholas LaRusso

Miles and Shirley Fiterman Center for Digestive Diseases, Division of Gastroenterology and Hepatology, Mayo Clinic College of Medicine, Rochester, Minnesota 55905

978-0-12-375024-2
DOI: 10.1016/S0091-679X(08)94005-X

Abstract

Primary cilia are present in most mammalian cells and have lately been recognized as important cellular sensors that integrate and transduce extracellular signals into functional responses. Development of approaches to isolate primary cilia of sufficient quantity and quality for biochemical and molecular studies are crucial to understand their roles and functions under normal and pathological conditions. Two separate but complementary techniques (i.e., peel-off and slide pulling) to isolate enriched ciliary fractions from cultured epithelial cells are described. The purity and quantity of isolated cilia is verified by immunofluorescent confocal microscopy, light microscopy, scanning electron microscopy (SEM), and transmission electron microscopy (TEM), and western blot analysis. Examples of detection of ciliary-associated proteins using isolated cilia are shown. These techniques will allow the isolation of primary cilia from cultured epithelial cells and permit further examination of the expression and localization of proteins of interest, helping to elucidate the role of primary cilia in health and disease.

I. Introduction

Cilia are highly conserved organelles present in most mammalian cells. In general, cilia are classified as motile and primary. Structurally, primary cilia consist of the basal body (i.e., mother centriole) and ciliary axoneme that contain nine peripherally located microtubule pairs (i.e., 9 + 0 arrangement) and differ from motile cilia (i.e., 9 + 2 arrangement) by the absence of the central pair of microtubules (D'Angelo and Franco, 2009; Gerdes et al., 2009). Being considered for a long time as vestigial appendages with no functions, primary cilia have recently been recognized as important cell sensors that integrate and transduce extracellular signals into functional responses (Christensen et al., 2007; Gerdes et al., 2009; Masyuk et al., 2008; Veland et al., 2009). Despite progress that has been made in identifying the role of primary cilia in health and disease, their functions are still poorly understood. Current approaches to study ciliary functions and expression of ciliary-associated proteins are mostly restricted to using whole cells in culture or in tissues. Therefore, developing procedures to isolate primary cilia for molecular and biochemical studies are critical to further elucidate the roles and functions of these organelles under normal and pathological conditions. While various deciliation approaches (i.e., chemical and detergent treatments, excessive pH, and calcium shock) have been developed for the isolation of motile cilia in a variety of organisms (Dentler, 1995; Huang et al., 2006; Nelson, 1995), these techniques are not suitable for the isolation of primary cilia because they do not produce sufficient yields of cilia for biochemical analysis. To the best of our knowledge, there are only two reports of primary cilia isolation (Mitchell et al., 2004; Raychowdhury et al., 2005).

Here we describe two separate but complementary approaches to isolate enriched ciliary fractions from cultured cholangiocytes (i.e., epithelial cells lining intrahepatic bile ducts) that should be applicable to any epithelial cell growing in polarized monolayers.

II. Materials and Instrumentations

A. Materials, Solutions, and Instrumentation for Isolation of Primary Cilia

- Two cholangiocyte cell lines were used for primary cilia isolation: (1) simian virus 40-transformed normal mouse cholangiocytes (NMCs, a generous gift from Dr. Ueno) and spontaneously immortalized normal rat cholangiocytes (NRCs). Both NMCs and NRCs grown in DMEM/F-12 culture medium for 10–15 days form polarized monolayers with well-developed apically located primary cilia, suitable electrical resistance, and retain the phenotypic and functional characteristics that define these cells *in vivo* (Muff *et al.*, 2006)
- Phosphate-buffered saline (PBS)
- 0.1% poly-L-lysine solution (Sigma, St Louis, MO, USA P8920)
- 0.01% soybean trypsin inhibitor
- 0.1 mM phenylmethylsulphonyl fluoride (PMSF)
- Glass coverslips or thermanox coverslips (22 mm × 60 mm, sterilized)
- Petri dish (10 cm in diameter, sterilized)
- Petri dish (6 cm in diameter, sterilized)
- Refrigerated centrifuge
- Fisher Marathon 21 K/BK centrifuge (Haverhill, MA, USA)
- AROS 160™ Adjustable reciprocating orbital shaker (Dubuque, IO, USA)
- 50-ml polypropylene centrifuge tubes

B. Materials, Solutions, and Instrumentation for Purification of Isolated Primary Cilia

- Partially purified cilia supernatant (see Section III)
- 45% sucrose–PBS–10 mM $CaCl_2$
- PBS-10 mM $CaCl_2$
- 2× PAGE sample buffer
- Ultracentrifuge with SW 28 and 45 Ti rotor
- Polyallomer tubes appropriate for SW 28 rotor
- Thick-walled polycarbonate tubes appropriate for the 45 Ti rotor
- 10-ml syringe
- 18-G needle

C. Materials, Solutions, and Instrumentation for Immunofluorescence Analysis of Isolated Primary Cilia

- Partially or completely isolated cilia supernatant
- Coverslip with attached isolated cilia and coverslip with remaining cells
- 4% paraformaldehyde (PFA)
- PBS
- 0.2% (v/v) Triton X-100
- Blocking solution [10% fetal calf serum (FCS) in PBS]
- 0.01 M glycine

- Primary antibody (e.g., antiacetylated α-tubulin)
- Appropriate secondary antibody (conjugated with fluorescence)
- Fluorescence microscope or confocal laser scanning microscope

D. Materials, Solutions, and Instrumentation for Scanning and Transmission Electron Microscopy of Isolated Primary Cilia

- Partially or completely isolated cilia supernatant
- Coverslip with isolated cilia and coverslip with remaining cells
- 4% PFA
- 2% glutaraldehyde
- 1% osmium tetroxide
- PBS
- Serial ethanol (50–70–95–100–100–100%)
- Specimen holders for scanning electron microscopy (SEM)

III. Methods

A. Isolation of Cholangiocyte Primary Cilia

Primary cilia are isolated by two techniques: peel-off and slide pulling. Cells are cultured on either sterilized coverslips (for isolation by peel-off technique) or on 10-cm diameter Petri dishes (for isolation by slide-pulling technique) for 7–10 days after cells became confluent. The general idea of primary cilia isolation by two different approaches is depicted in Figs. 1 and 2. The first approach (i.e., the peel-off technique) is designed to pull off the cilia from the apical membrane of the cells by applying poly-L-lysine-coated coverslips to the cultured cells and peeling off primary cilia or ciliary fragments by lifting the coverslip (Fig. 1). The second approach (i.e., the slide pulling) is designed to isolate the primary cilia from the apical membrane of the cells by the slide movement of poly-L-lysine-coated Petri dish cover floating on top of culture medium and the force generated by the shaker (Fig. 2).

B. Preparation of Coverslips and Petri Dishes for Cilia Isolation

- Glass coverslips or thermanox coverslips are sterilized under the ultraviolet light in a tissue culture hood overnight.
- Coat the faceup side of coverslips and the exterior surface of a 6-cm diameter Petri dish cover with 0.1% poly-L-lysine solution (Fig. 3). Use an appropriate volume of poly-L-lysine solution to cover the coverslips and Petri dishes with an even layer, and then air dry them overnight. *Note:* To avoid confusion, the side of coated coverslips and Petri dish cover should always laid faceup.

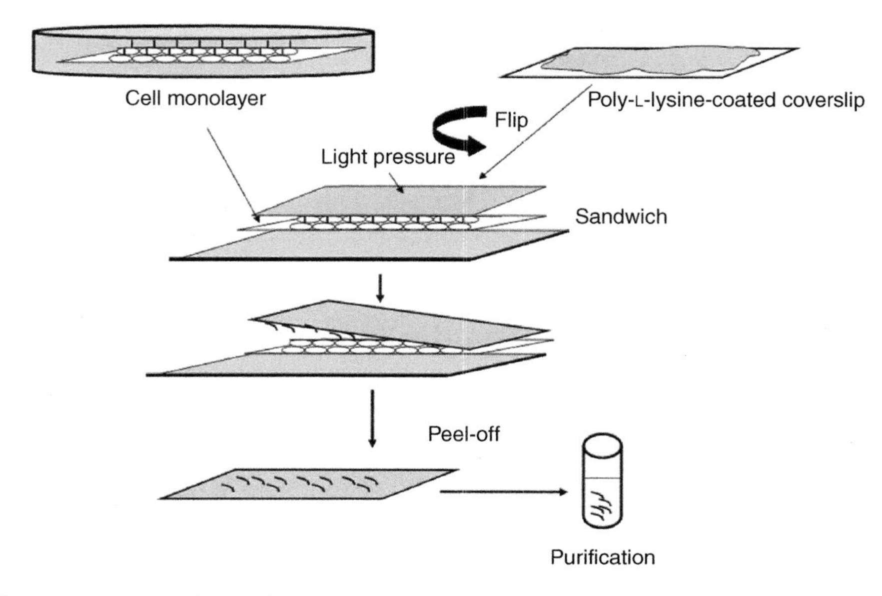

Fig. 1 Isolation of primary cilia by the peel-off technique.

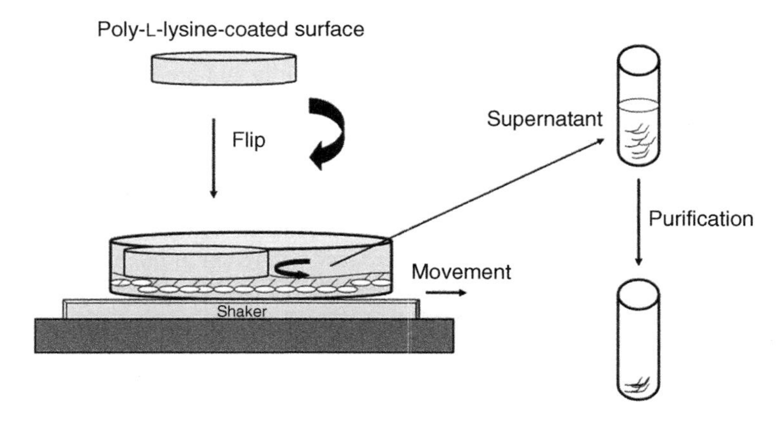

Fig. 2 Isolation of primary cilia by slide-pulling technique.

C. Procedure of Primary Cilia Isolation by Peel-Off Technique

- Take a coverslip with cells from culture medium and drain the medium by filter paper and place this coverslip on a clean slide. *Note:* keep the side of coverslip with monolayer cells faceup.
- Flip a poly-L-lysine-coated coverslip facedown and place on the top of coverslip with cell monolayer.

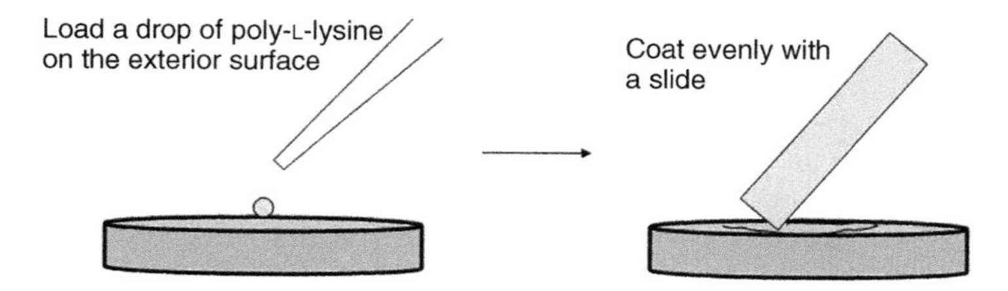

Fig. 3 Coating of the exterior surface of a Petri dish cover with poly-L-lysine.

- Remove the culture medium with a Pasteur pipette attached to vacuum line. *Note:* Pasteur pipette should be positioned at the edges of the coverslip.
- Put a piece of filter paper on the top of the poly-L-lysine-coated coverslip and apply light pressure on the top of the coverslip by placing a rubber cork (20 mm in diameter) or by slightly pushing down with a finger for 20 s.
- Lift off the poly-L-lysine-coated coverslip from the cell monolayer with a pair of curved forceps.
- The samples are either fixed for electron microscopy (see Section III.G) or scraped off, centrifuged, concentrated in sucrose density gradients as previous described (Raychowdhury et al., 2005), and then frozen for biochemical analysis.
- To asses the efficiency of cilia isolation, immunofluorescence microscopy and SEM are used to examine the samples from the same preparation. Both samples (i.e., poly-L-lysine-coated coverslip with attached isolated cilia and remaining after isolation cells) are stained with acetylated α-tubulin and processed for SEM (see Section III.G).

D. Procedure of Primary Cilia Isolation by Slide-Pulling Technique

- Wash the cells with culture medium without serum three times and add 15 ml of fresh medium plus protease inhibitors (0.01% soybean trypsin inhibitor and 0.1 mM PMSF). Put a poly-L-lysine-coated (6 cm in diameter) Petri dish cover upside down and place on top of the medium (Fig. 2).
- Shake using an AROS 160^{TM} adjustable reciprocating orbital shaker at 70–100 rpm at room temperature for 5 min.
- Collect the culture medium in a 50-ml polypropylene centrifuge tube and keep on ice bath.
- Vortex supernatants in polypropylene centrifuge tube for 1 min and place on ice for 20–30 min to allow the debris to settle.
- Pipet off supernatant in a 50-ml polypropylene centrifuge tube and centrifuge at 500 × *g* at 4°C for 10 min to further remove whole cells and cellular debris.
- Pipet off supernatant and transfer it to polyallomer tubes for the ultracentrifugation.
- The remaining pellets are resuspended with PBS, centrifuged, and resuspended two more times.
- For enriching cilia, collected supernatant is centrifuged at 56,000 rpm (308,000 × g) for 30 min and supernatant is removed. Pellets are resuspended with PBS for biochemical analysis (if needed).

E. Purification of Isolated Cilia

- For purification of isolated cilia follow the method described by Raychowdhury et al. (2005).
- Combine partially purified cilia supernatant.
- Load the supernatant on top of a prechilled solution containing 45% sucrose/PBS/ 10 mM $CaCl_2$ and centrifuge for 1 h at 100,000 × *g*.
- Collect the interface of sucrose-supernatant band and dilute in 1:2–1:4 PBS/10 mM $CaCl_2$ solution and centrifuge again for 1 h at 100,000 × *g*.
- For TEM, pellets are resuspended in 2% glutaraldehyde in 0.1 M phosphate buffer (pH 7.4). For protein analysis, pellets are resuspended in PBS and stored in (−80°C). Samples for western blotting are resuspended in equal volume 2× PAGE sample buffer.

F. Immunofluorescence Analysis of Isolated Cilia

- Isolate primary cilia as described (Section II.B or II.C).
- Fix the coverslips containing primary cilia isolated by the peel-off technique, as well as the remaining cells, in 4% PFA for 30 min at room temperature.
- Primary cilia isolated by the slide-pulling approach are loaded on poly-L-lysine coverslips, air dried, and fixed in 4% PFA for 30 min at room temperature.
- Permeabilize with 0.2% Triton X-100 in PBS for 5 min.
- Rinse five times with PBS.
- Quench in 0.01 M glycine, pH 7.4, for 10 min.
- Block with 10% FCS/PBS solution for 20 min.
- Incubate in first antibody diluted with 10% FCS/PBS for 1 h at room temperature.
- Rinse five times with PBS.
- Incubate in secondary antibody diluted with 10% FCS/PBS for 1 h at room temperature.
- Wash with five times with PBS.
- Mount with antifading solution.

G. Electron Microscope Analysis of Isolated Cilia

1. Scanning Electron Microscopy

Note: All the fixation process should be performed in a fume hood.

- Fix coverslips with 2% glutaraldehyde in 0.1 M phosphate buffer (pH 7.4) for 1 h at room temperature or at 4°C overnight.
- Rinse specimens with 0.1 M phosphate buffer five times (each wash 2–5 min).
- Immerse the specimens in phosphate-buffered 1% osmium tetroxide for 1 h on ice.
- Wash with distilled water five times (each wash is 2–5 min).
- Dehydrate in serial ethanol (50–70–95–100–100–100%), each step for 5 min. *Note:* Specimens should be always completely immersed in ethanol.

- Dry in critical point dryer. *Note:* It is essential to keep the specimens with a little 100% ethanol to avoid the drying artifact when transferring specimens into the chamber of critical dryer.
- Sputter coating with gold–palladium for 90 s.
- Observe under the scanning electron microscope at 3 kV.

2. Transmission Electron Microscopy

- Fix coverslips or ciliary pellets with 2% glutaraldehyde in 0.1 M phosphate buffer (pH 7.4) for 1 h at room temperature or at 4°C overnight.
- Wash two times with 0.1 M phosphate buffer for 6 min, then five times in water.
- Stain with 4% uranyl acetate in water for 2 h or overnight at 4°C.
- Rinse five times in water for 15 min.
- Dehydrate in serial ethanol (50–70–95–100–100–100%), each step for 5 min. *Note:* Specimens should be always immersed in ethanol to avoid dry artifacts.
- Infiltrate in serial resin (Spurr) at the ratio of 1:2, 1:1, and 2:1 with 100% ethanol, 100% resin for two times (each step for 30 min), then further infiltrate in fresh made resin overnight.
- Embed in the plastic flat embedding mold or beam capsules.
- Polymerize at 70°C for 24 h.
- Section the resin blocks at 90-nm thickness.
- Stain with lead citrate for 5 min.
- Observe the grids under TEM at 80 kV.

IV. Results and Discussion

A. Isolation and Characterization of Primary Cilia by Pee-Off Technique

Details of primary cilia isolation by peel-off technique are described in Sections II and III and depicted in Fig. 1. Figure 4 shows the morphological features of cilia isolated by peel-off technique. After the procedure, isolated cilia attached to poly-L-lysine coverslips and those remaining on cultured cells after isolation are stained with the ciliary marker, acetylated α-tubulin to ensure successful isolation. Representative immunofluorescent confocal (Fig. 4A) images demonstrate the presence of primary cilia on poly-L-lysine-coated coverslip after isolation. Importantly, only a limited number of cells retain primary cilia after isolation (Fig. 4B). Up to 70% of primary cilia could be detached from the apical membrane of cultured cells by this technique. Application of SEM is important to confirm the successful isolation and to check the integrity of isolated cilia. Figure 4C shows the scanning electron microphotograph of isolated primary cilia with well preserved morphological structure. The cultured cells after isolation (Fig. 4D) are largely devoid of primary cilia (asterisks depict the site of individual cell from which primary cilium was ripped off). Further confirmation of successful isolation of primary cilia should be done using TEM (not shown).

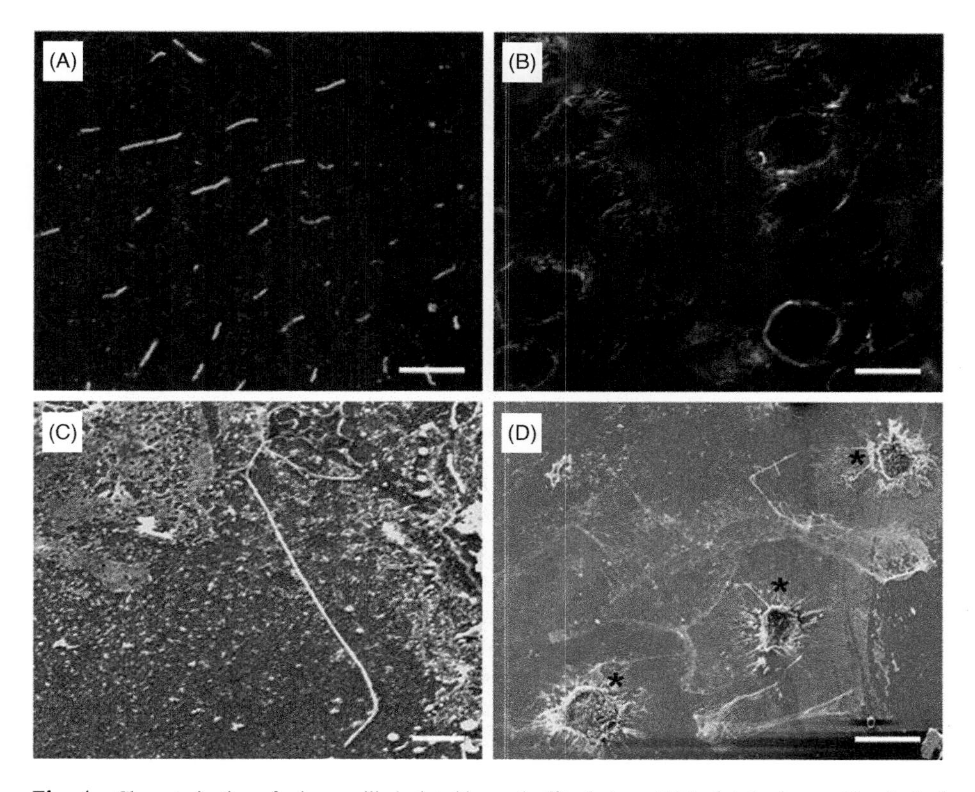

Fig. 4 Characterization of primary cilia isolated by peel-off technique. (A) Isolated primary cilia attached to poly-L-lysine coverslip and (B) cells remaining after isolation are stained with the ciliary marker, acetylated α-tubulin. SEM microphotographs show isolated individual cilium (C) and cells remaining after isolation (D). Asterisks depicts area of the cell from which primary cilium was detached. Bars: 10 μm (A and B), 1 μm (C), and 25 μm (D).

B. Isolation and Characterization of Primary Cilia by Slide-Pulling Technique

Details of primary cilia isolation by the slide-pulling technique are described in Sections II and III and depicted in Fig. 2. Cilia are detached from the apical surface of the cell monolayer by generating a pulling force and then collected by centrifugation. As for peel-off isolation approach, primary cilia isolated by slide-pulling technique are characterized by immunofluorescent confocal (Fig. 5A) and transmission electron (Fig. 5B) microscopy.

C. Characterization of Isolated Primary Cilia by Biochemical and Molecular Approaches

Biochemical analysis (i.e., western blot) is needed to confirm the purity of isolated cilia. Acetylated α-tubulin, a well-characterized ciliary marker, is used to establish that

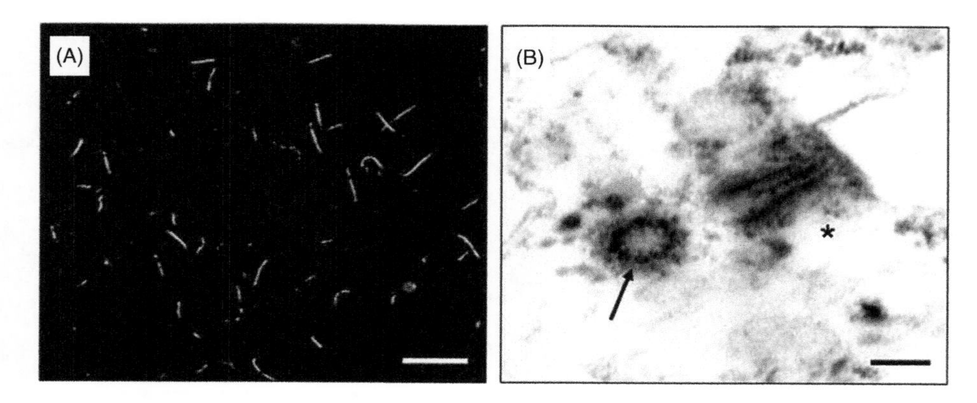

Fig. 5 Characterization of primary cilia isolated by slide-pulling technique. (A) Primary cilia collected after isolation by centrifugation are stained with the ciliary marker, acetylated α-tubulin. (B) TEM microphotographs demonstrate the cross section (arrow) and longitudinal (asterisk) fragments of isolated cilia. Bars: 10 m (A), 200 nm (B).

isolated pellets are indeed primary cilia. Moreover, a marker of the apical membrane (expressed only at the apical membrane but not on cilia) is also run on western blot in order to verify that the isolated ciliary fractions are not contaminated with fragments of cell membranes. Markers of the apical membrane of the particular cells used for isolation should be determined in advance. Figure 6 shows a verification of the purity of isolated cilia. We previously determined that the purinergic receptor, P2Y2, is exclusively localized to the apical membrane of cholangiocytes but not in cholangiocyte cilia. Thus, the P2Y2 receptor served as a marker of apical membrane of cholangiocytes. Western blots demonstrate the purity of primary cilia isolated by both approaches. After verification of the purity of isolated cilia, these fractions could be used for detection of any ciliary-associated proteins of interest by western blotting or by immunofluorescent confocal microscope. Figure 7 shows that isolated primary cilia of cholangiocytes are positively stained for ciliary-associated proteins, polycystin-1 and polycystin-2.

D. Evaluation of the Approaches

The success of primary cilia isolation by any approach depends on several factors. Whatever method of isolation is used, it is essential to ensure that cells are grown in healthy cultured conditions (without any contamination). Before isolation, check the presence of cilia at the maximal length (e.g., 7–10 μm for normal cholangiocytes) using immunofluorescent microscopy by staining of samples from each preparation with the ciliary marker, acetylated α-tubulin (see above).

Primary cilia, isolated by the peel-off technique, are directly attached to the poly-L-lysine-coated coverslips. As demonstrated in Fig. 4, isolated cilia are well preserved. This approach, however, has some disadvantages. First, besides cilia and

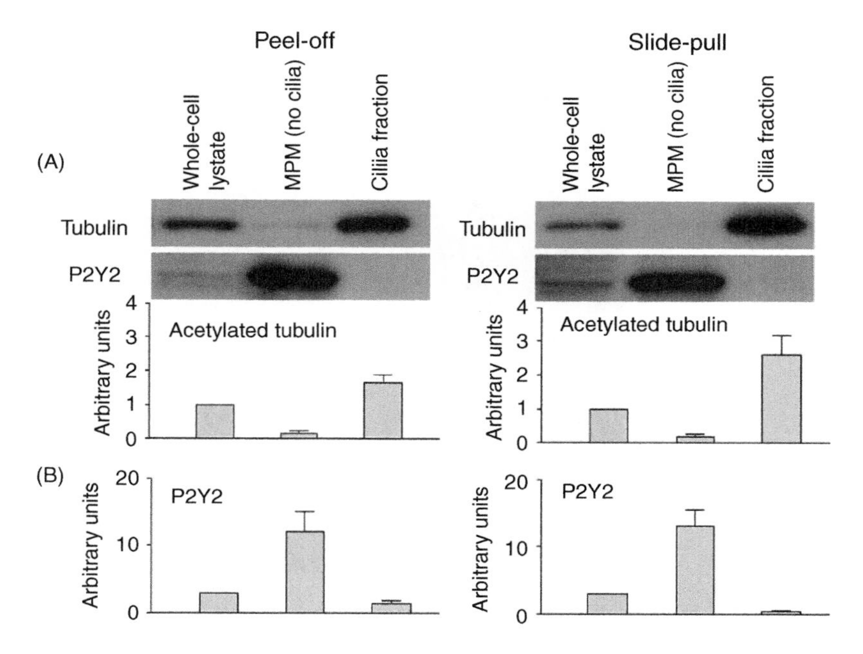

Fig. 6 Western blots of the whole-cell lysate fraction, mixed plasma membrane (MPM) fraction and ciliary fraction isolated by peel-off (left panels) and slide-pull (right panels) techniques. Acetylated α-tubulin is a ciliary marker and P2Y2 is a MPM marker. Bar graphs are densitometric analysis showing fractions of isolated cilia are enriched with ciliary marker, acetylated α-tubulin, while P2Y2 is mainly present in the MPM fraction ($p < 0.05$). Published with permission.

ciliary fragments, apical membranes of adherent cells or even some cells may also attach to poly-L-lysine-coated coverslip. To avoid contamination of ciliary fractions with cell membranes and to increase the yield of isolated cilia, conditions of the experiment (especially the force strength applied) should be optimized. The yield of isolated cilia may vary (35–150 μg) and might be relatively low for biochemical analysis (especially, for western blot). We recommend using this approach mostly for confocal immunofluorescent microscopy to examine the expression of ciliary-associated proteins.

The second approach (i.e., the slide-pull technique) also requires several precautions: (1) an appropriate volume of medium (10–15 ml) should be used to allow the cover of small Petri dish to float and move around the edge of the big Petri dish (in which cells were grown) generating pulling force to detach the primary cilia and also avoid lifting the cells from bottom of the Petri dish and (2) to avoid cell detachment from the bottom of the Petri dish, do not exceed speed (should be not more than 100 rpm) and shaking time (not more than 5 min). This approach yields a large quantity (150–270 μg) of total protein and is suitable for biochemical analyses; however, it also can be used for confocal microscopy.

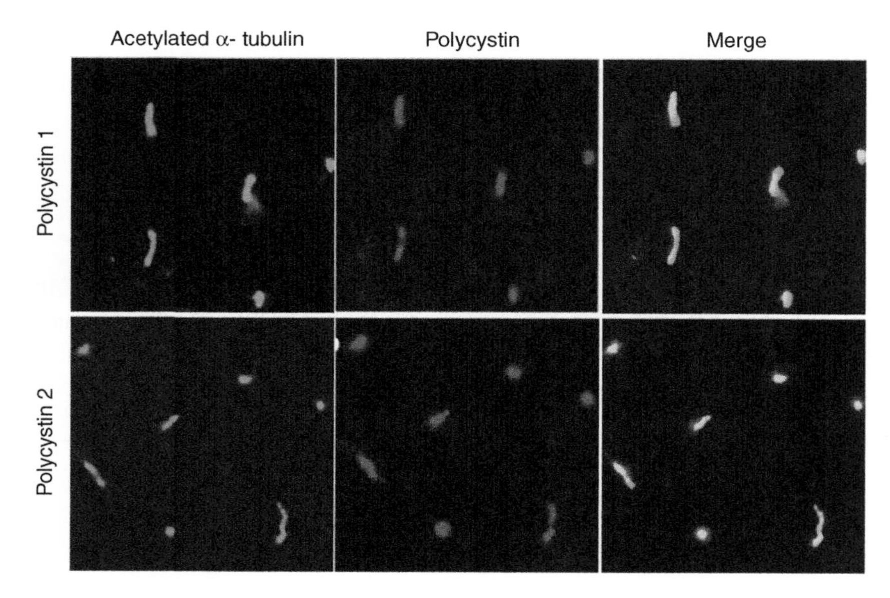

Fig. 7 Polycystin 1 and polycystin 2 (shown in red) are localized to isolated primary cilia (shown in green). Polycystin 1 is detected in cilia isolated by peel-off approach, polycystin 2 is detected in cilia isolated by slide-pulling technique. Merged images are shown in yellow. Magnification, ×100. (See Plate no. 4 in the Color Plate Section.)

V. Summary

Two approaches for isolation of primary cilia described here are for epithelial cells that develop primary cilia when grown as a polarized monolayer. They could be used to obtain fractions of isolated primary cilia of large quantity and high purity. These fractions are suitable for biochemical and molecular analysis of primary cilia in order to reveal the structure, content, and functions of these organelles.

Acknowledgments

This work was supported by the National Institutes of Health (grant DK 24031), by the PKD Foundation, and by the Mayo Foundation.

References

Christensen, S.T., Pedersen, L.B., Schneider, L., Satir, P. (2007). Sensory cilia and integration of signal transduction in human health and disease. *Traffic* **8**(2), 97–109.

D'Angelo, A., and Franco, B. (2009). The dynamic cilium in human diseases. *Pathogenetics* **2**(1), 3.

Dentler, W.L. (1995). Isolation and fractionation of ciliary membranes from Tetrahymena. *Methods Cell Biol.* **47**, 397–400.

Gerdes, J.M., Davis, E.E., Katsanis, N. (2009). The vertebrate primary cilium in development, homeostasis, and disease. *Cell* **137**(1), 32–45.

Huang, B.Q., Masyuk, T.V., Muff, M.A., Tietz, P.S., Masyuk, A.I., Larusso, N.F. (2006). Isolation and characterization of cholangiocyte primary cilia. *Am. J. Physiol. Gastrointest. Liver Physiol.* **291**(3), G500–509.

Masyuk, A.I., Masyuk, T.V., LaRusso, N.F. (2008). Cholangiocyte primary cilia in liver health and disease. *Dev. Dyn.* **237**(8), 2007–2012.

Mitchell, K.A., Gallagher, B.C., Szabo, G., Otero Ade, S. (2004). NDP kinase moves into developing primary cilia. *Cell Motil. Cytoskeleton* **59**(1), 62–73.

Muff, M.A., Masyuk, T.V. Stroope, A.J., Huang, B.Q., Splinter, P.L., Lee, S.O., LaRusso, N.F. (2006). Development and characterization of a cholangiocyte cell line from the PCK rat, an animal model of Autosomal Recessive Polycystic Kidney Disease. *Lab. Invest.* **86**(9), 940–950.

Nelson, D.L. (1995). Preparation of cilia and subciliary fractions from Paramecium. *Methods Cell Biol.* **47**, 17–24.

Raychowdhury, M.K., McLaughlin, M., Ramos, A.J., Montalbetti, N., Bouley, R., Ausiello, D. A., Cantiello, H.F. (2005). Characterization of single channel currents from primary cilia of renal epithelial cells. *J. Biol. Chem.* **280**(41), 34718–34722.

Veland, I.R., Awan, A., Pedersen, L.B., Yoder, B.K., Christensen, S.T. (2009). Primary cilia and signaling pathways in mammalian development, health and disease. *Nephron Physiol.* **111**(3), 39–53.

CHAPTER 6

Analyzing Primary Cilia by Multiphoton Microscopy

Cornelia E. Farnum*, **Rebecca M. Williams**†, *and* **Eve Donnelly**‡

*Department of Biomedical Sciences, Cornell University, Ithaca, New York 14865

†Department of Biomedical Engineering, Cornell University, Ithaca, New York 14853

‡Mineralized Tissues Laboratory, Hospital for Special Surgery, New York, New York 10021

Abstract

In this chapter, a technique is outlined for the use of immunohistochemistry (IHC) followed by multiphoton microscopy (MPM) for the analysis of incidence, length, and 3D orientation of the axoneme of the primary cilium. Although the application presented specifically emphasizes localizations in tenocytes and chondrocytes, the technique is

978-0-12-375024-2
DOI:10.1016/S0091-679X(08)94006-1

applicable to cells in a wide range of connective tissues. The primary advantages of utilizing MPM as opposed to TEM for these kinds of ciliary analyses are the rapidity of the technique for preparation of the samples and the ability to collect data from multiple cells simultaneously. Using MPM, the axoneme, basal body, and associated centriole can be visualized by specific IHC with localizing antibodies. However, the resolution achieved through TEM analyses allows the complex morphology of the primary cilium to be visualized, and this remains the primary advantage of TEM versus MPM. SHG, which occurs only with MPM, allows visualization of collagen fibrils and is particularly advantageous for localizing primary cilia associated with cells in connective tissues. This, and the deep penetration with less photobleaching, are the primary advantages of MPM compared to confocal microscopy. As with any microscopical technique, the protocol needs to be optimized for any given tissue. In particular, additional antigen retrieval techniques to enhance the unmasking of specific epitopes for antibody binding may be required for adaptation of this approach to other dense connective tissues with complex spatial organizations such as intervertebral disc or meniscus.

I. Introduction

In the last 15 years there has been a renewed interest in the primary cilium, a cellular organelle first observed by transmission electron microscopy (TEM) almost half a century ago in neurons (Barnes, 1961). The morphological complexity of the primary cilium has been well documented based on thorough morphological analyses by TEM throughout the 1960s and 1970s in multiple cellular types—neurons, muscle cells, epithelial cells, and connective tissue cells (early examples include Grillo and Palay, 1963; Sorokin, 1962; Wilsman, 1978). Indeed, in a recent review Wheatley commented that, while there has been extensive research in the last decade concerning the potential role of the primary cilium as a sensory organelle involved in multiple signal transduction pathways, the knowledge of the basic morphology of the primary cilium still is best documented by TEM studies starting a half century ago (Wheatley, 2008).

The structure of the primary cilium of articular chondrocytes was described in detail by Wilsman, based on TEM serial section reconstruction (Wilsman, 1978). The ciliary axoneme, containing nine radially symmetric microtubular doublets, nucleates from a centriolar basal body and projects into the extracellular space, either directly or by a tunnel-like invagination of the plasma membrane (Fig. 1A and B). The axoneme of the primary cilium, unlike that of actively motile cilia such as on tracheal epithelium, lacks centrally positioned microtubular singlets and dynein arms necessary for force generation. Numerous recent studies in several different cell types have demonstrated that the outer doublets are involved with both anterograde and retrograde intraflagellar transport (IFT), which is essential for the sensory signal transduction pathways hypothesized to be central to the primary cilium's function (Davenport and Yoder, 2005; Gerdes *et al.*, 2009; Ocbina and Anderson, 2008; Pazour and Rosenbaum, 2002; Pedersen and Rosenbaum, 2008; Pedersen *et al.*, 2006; Satir and Christensen, 2007; Song *et al.*, 2007).

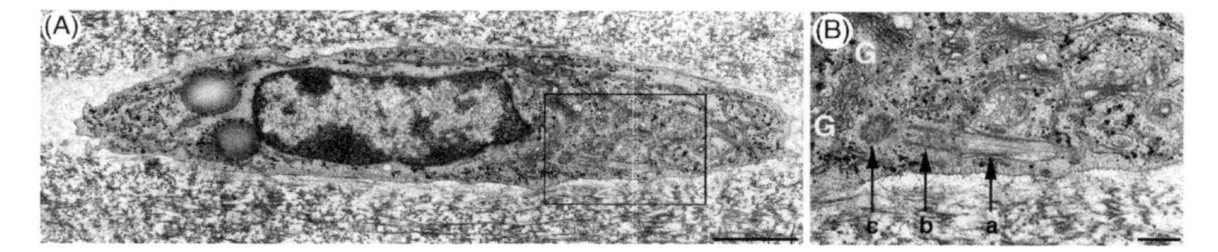

Fig. 1 Articular chondrocyte with a primary cilium. In this TEM image of a superficial zone articular chondrocyte, a profile of the axoneme of a primary cilium can be seen at low magnification in the whole cell in A, with the inset at higher magnification in B. The basal body (b) is connected to the ciliary axoneme (a), which emerges from the chondrocyte through a tunnel-like invagination of the plasma membrane. The extent to which the axoneme projects into the extracellular matrix could only be analyzed if adjacent serial sections were available. Note the close proximity of the centriole (c) to the Golgi apparatus of the cell (G). In TEM images the complex morphology of the primary cilium is maintained. However, it is rare to see images of this kind where a significant extent of the axoneme is visible on one section. (TEM sections fixed in presence of osmium ferrocyanide (Farnum and Wilsman, 1983; Wilsman *et al.*, 1980). Scale bar = 2 μm (A) and 1 μm (B).

The finding that the pathogenesis of polycystic kidney disease involves disruption of the function of primary cilia in renal tubule epithelial cells placed interest in the primary cilium squarely on center stage, and numerous studies have since focused on the complexity of functions of this organelle in epithelial cells (Davenport *et al.*, 2007; Lehman *et al.*, 2008; Low *et al.*, 2006; Yoder, 2007). Several of the diseases associated with primary ciliary dysfunction, and the mouse models used to study these diseases, also are characterized by an abnormal skeletal phenotype, leading to increased interest in the function of the primary cilium in skeletal tissue cells (Bisgrove and Yost, 2006; Haycraft and Serra, 2008; Kaushik *et al.*, 2009; Malone *et al.*, 2007; Serra, 2008; Sharma *et al.*, 2008). Primary cilia have been shown to be present in essentially all cells of connective tissue origin—from adipose cells to odontoblasts to osteocytes, as examples (reviewed in Donnelly *et al.*, 2008). The most extensive analyses for connective tissues have been done for chondrocytes in articular cartilage (Poole *et al.*, 1985; Wilsman, 1978; Wilsman and Fletcher, 1978).

Imaging of primary cilia in connective cells for functional studies is particularly challenging. The primary cilium projects into an extracellular matrix (ECM) and the ciliary axoneme makes direct attachments to ECM molecules (McGlashan *et al.*, 2006). The dense ECM and the irregular shape of many connective tissue cells hinder identification of cellular boundaries using conventional imaging modalities. Since the time required for ultrastructural studies makes TEM a poor choice for experimental analyses involving multiple variables, confocal microscopy has been the most commonly used imaging modality for functional studies of primary cilia in skeletal tissues, particularly cartilage (Haycraft *et al.*, 2007; Jensen *et al.*, 2004; McGlashan *et al.*, 2006, 2007; Poole *et al.*, 1997; Poole *et al.*, 2001; Serra, 2008). Confocal microscopy allows rapid optical sectioning for three-dimensional (3D) analyses and versatile immunohistochemical approaches using multiple labels. Multiphoton microscopy (MPM) has these same advantages but also enables visualization of the unstained ECM, which is critical for connective tissues, as well as deeper penetration than confocal microscopy for creation of z-stacks. Our goal was to develop approaches using MPM to meet the challenges of studying the presence, incidence and 3D orientation of primary cilia in connective tissues *in situ*.

Our protocols initially were developed using growth plate cartilage, where chondrocytes are isolated from each other and where the pericellular matrix of glycoproteins, hyaluronic acid, and type VI collagen enhances the ability to visualize individual cells (Farnum and Wilsman, 1983, 1986). We then extended our studies to primary cilia associated with cells of extensor tendons, where highly irregularly shaped tenocytes form a cellular network surrounded by a dense type I collagenous ECM (McNeilly *et al.*, 1996). In combination, this allowed us to characterize primary cilia *in situ* in cells of connective tissues using MPM.

II. Materials and Solutions

A. Materials

1. Bromodeoxyuridine (BrdU)
2. Methanol
3. 0.01 M phosphate-buffered saline (PBS)
4. Ethanol
5. Chloroform
6. Paraffin (Paraplast X-tra, Fisher, Pittsburgh, PA, USA)
7. Poly-L-lysine-coated slides
8. Xylene
9. Citric acid
10. Sodium citrate
11. Sodium chloride
12. Hydrochloric acid
13. Sodium borate
14. Triton X-100
15. Normal goat serum
16. Normal donkey serum
17. Primary antibodies:
    a. Ciliary antibody: monoclonal mouse antiacetylated α-tubulin [6-11B-1] (Abcam, Cambridge, MA, USA, ab24610)
    b. Centrosomal antibody: polyclonal rabbit anti-γ tubulin (Abcam ab16504)
    c. BrdU antibody: polyclonal sheep anti-BrdU (Abcam ab1983)
18. Secondary antibodies:
    a. Goat antimouse Alexa 488 (Invitrogen A11029)
    b. Goat antirabbit Alexa 568 (Invitrogen A11031)
    c. Donkey antisheep Alexa 568 (Invitrogen, Carlsbad, CA, USA, A21099)
19. Hoechst 33258 Nuclear dye (Invitrogen)
20. VectaShield Hard-Set with 4′,6-diamidino-2-phenylindole (DAPI) (Vector Labs, Burlingame, CA, USA)
21. VectaShield (Vector Labs, Burlingame, CA, USA)

B. Solutions

1. BrdU injection solution: BrdU 2.5 mg/ml in PBS
2. 10% neutral-buffered formalin
3. Graded ethanols (100, 90, 95, and 70%) diluted in dH_2O
4. Antigen retrieval solution: 0.2 M sodium citrate buffer, pH 3.5 (25 g sodium citrate-$2H_2O$ + 40 g citric acid + 7.2 g NaCl, fill to 1 L (for clarity-1 appears as a 1) dH_2O and adjust pH)
5. PBST: PBS–Triton permeabilizer (0.1% Triton X-100 in PBS)
6. Blocking buffer 1: 5% normal goat serum in PBST
7. Blocking buffer 2: 5% normal donkey serum in PBST
8. 0.1 M sodium borate buffer (pH 8.5)
9. Primary antibody solutions (diluted in blocking buffer):
   a. monoclonal mouse antiacetylated α-tubulin (1:200 for whole mount, 1:1000 for sections)
   b. polyclonal sheep anti-BrdU (1:50 for whole mount, 1:50 for sections)
   c. polyclonal rabbit anti-γ-tubulin (1:500 for whole mount)
10. Secondary antibody solutions (diluted in blocking buffer):
   a. Goat antimouse Alexa Fluor 488 (1:300 for sections, 1:200 for whole mount)
   b. Goat antirabbit Alexa Fluor 488 (1:300 for sections, 1:200 for whole mount)
   c. Donkey antisheep Alexa Fluor 568 (1:300 for sections, 1:200 for whole mount)

III. Methods

Immunofluorescent labeling of whole-mount and paraffin sections combined with MPM imaging allows 3D visualization of cells and organelles (nuclei, primary cilia, centrosomes) in their native ECM. Whole-mount specimens are used for analysis of ciliary orientation to preserve the 3D tissue structure, while paraffin sections are used for analysis of ciliary incidence to maximize antibody penetration. The experimental approach is summarized in Fig. 2.

A. Tissue Collection

If cell-cycle labeling is desired, rats or mice 3- to 5-weeks-old are injected with BrdU solution (25 mg/kg body weight, or 1 ml of solution for a 100-g rat) 1 h prior to euthanasia. Following euthanasia, the limbs are disarticulated, the skin is removed, and the limbs are immediately placed in fixative (10% neutral-buffered formalin for paraffin sections or cold methanol for whole-mount specimens). Tissues of interest are isolated in fixative under a dissecting microscope. For whole-mount preparation of common digital extensor tendons, the epitenon is carefully removed, and the specimen is separated into individual fascicles. For growth plate collections, ~0.1 mm of epiphyseal and metaphyseal bone are left adjacent to the cartilage to aid in orientation of the specimen under the microscope. Individual full-thickness slabs about 0.2 mm thick are cut by hand with an alcohol-washed razor blade.

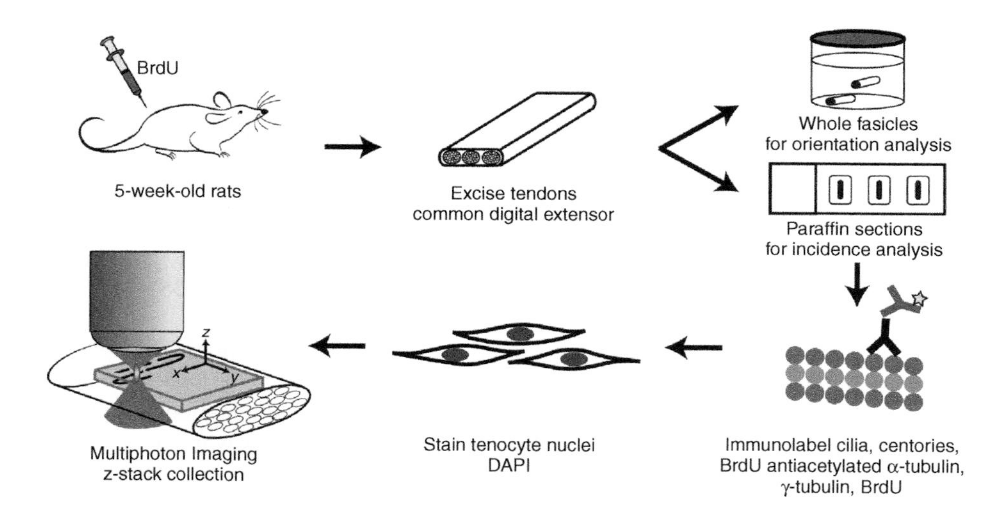

Fig. 2 Schematic summarizing collection, processing, immunolabeling, and imaging of extensor tendons.

B. Tissue Preparation

Specimens for whole-mount preparations are fixed in microcentrifuge tubes filled with methanol at 4°C for 2 h. After fixation, whole-mount specimens are washed in PBS (3x 5 min) and are ready for immunohistochemistry (IHC). At the time of collection, specimens for paraffin sections are placed in histological cassettes and fixed overnight in 10% neutral-buffered formalin. Following three 20-min rinses, the cassettes are placed in the paraffin tissue processor (Leica, Bannockburn, IL, USA, TP1020) and are processed through the following baths: 30 min each in a series of graded ethanols (70, 90, 95, 100, an 100%), chloroform (two sequential 30-min baths), and paraffin (Paraplast X-tra) at 58°C (two sequential baths, one 30 min and one overnight). The specimens are then embedded in paraffin using a histological embedding station (Leica EG1160), allowed to harden completely, and placed in the –20°C freezer prior to sectioning. The 5-μm-thick sections are microtomed, placed onto slides, and allowed to dry for 2 h on the slide warmer. The sections are then deparaffinized in xylene (3x 5 min) and rehydrated for 5 min each in graded ethanols (100, 100, 95, and 70%) and dH_2O (2x). After antigen retrieval in sodium citrate buffer (60 min, 48°C), the slides are washed in PBS (2x 3 min) and are ready for IHC.

C. Immunohistochemistry

Generally, the IHC procedures are similar for whole-mount specimens and paraffin sections, but with shorter incubation times and reduced antibody concentrations for the sections. IHC procedures are performed on whole tissues in microcentrifuge tubes on a rocker and on sections on slides in a humidified chamber. All specimens (whole mount and sections) are allocated into three groups: (1) single label for acetylated α-tubulin to label primary cilia, (2) single label for BrdU to label S-phase nuclei, and (3) double label for acetylated α-tubulin and γ-tubulin to examine colocalization of primary cilia and centrioles.

Single label for acetylated α-tubulin: Following incubation in blocking buffer 1 for 1 h at room temperature (RT), specimens are incubated in the mouse antiacetylated α-tubulin primary antibody (1:200 for whole mounts, 1:1000 for sections) for 2 h at RT (whole mounts only) and overnight at 4°C (whole mounts and sections). After washing 3x in PBS, tissues are incubated in Alexa Fluor 488 goat antimouse secondary antibody (1:200 for whole mounts, 1:300 for sections) overnight at 4°C and washed 3x in PBS. At this point the specimens are ready for MPM imaging.

Single label for BrdU: Following incubation in blocking buffer 2 for 1 h at RT, specimens are denatured in 2 N HCl for 20 min, neutralized in 0.1 M sodium borate buffer (pH 8.5) for 15 min at RT, and incubated in the sheep anti-BrdU primary antibody (1:50 for whole mounts and sections) overnight at 4°C. After washing 3x in PBS, the specimens are incubated in Alexa Fluor 568 donkey antisheep secondary antibody (1:200 overnight at 4°C for whole mounts, 1:300 for 60 min at RT for sections) and washed 3x in PBS.

Double label for acetylated α-tubulin and γ-tubulin: For the double-label specimens, following labeling for acetylated α-tubulin as described above, tissues are incubated in blocking buffer 1 for 1 h at RT. Then, specimens are incubated in rabbit anti-γ-tubulin primary antibody (1:500 for whole mounts and sections) for 2 h at RT (whole mounts only) and overnight at 4°C (whole mounts and sections) and washed 3x in PBS. Specimens are incubated in Alexa Fluor 568 goat antirabbit secondary antibody (1:200 overnight at 4°C for whole mounts, 1:300 for 60 min at RT for sections) and washed 3x in PBS.

Finally, for nuclear visualization whole-mount specimens are stained with Hoechst 33258 nuclear stain (0.01 μg/ml) for 30 min at 37°C, washed 3x in PBS, and stored in PBS until imaging. Sections are coverslipped in a mounting medium containing DAPI (VectaShield Hard-Set with DAPI). All specimens are stored <24 h in opaque containers at 4°C prior to imaging.

D. Multiphoton Microscopy

MPM and second harmonic generation (SHG) microscopy enable simultaneous visualization of cells, immunolabeled organelles, and ECM. In these techniques, nonlinear optical processes are restricted to the focal spot of a tightly focused beam (Denk *et al.*, 1990). Raster scanning the focal spot enables a detailed image to be acquired from specimens that are optically thick (hundreds of microns). In MPM, fluorophores are excited by a two-photon process in which the excitation source is set to half the excitation energy (twice the wavelength) of the fluorophores' excitation bands. A (generally femtosecond) pulsed source is necessary to reduce the effects of linear processes, such as heating and nonspecific molecular absorptions. After absorption, fluorophores relax slightly due to molecular vibrations and return to their ground state by emitting fluorescence. In contrast, SHG is the nonlinear equivalent of scattering; there is no absorption of energy by the specimen.

SHG is emitted at exactly twice the energy (half the wavelength) of the illuminating source emitting from macromolecular structures with a large hyperpolarizeabiltity (those that lack inversion symmetry). Collagen fibrils are extremely bright examples of such structures and make up the bulk of SHG from biological specimens (Williams *et al.*, 2005; Zipfel *et al.*, 2003a). No exogenous labels are required for SHG microscopy of collagen.

When illuminating with a raster-scanned beam, both two-photon-excited fluorescence (TPEF) and SHG are emitted simultaneously (Fig. 3A) and can be spectrally separated and collected to form 3D-resolved images of fluorophore location with respect to a collagenous matrix. The SHG wavelength depends only on the illumination wavelength, whereas the fluorescence emission wavelength depends on the electronic energy levels of the fluorophores used (Fig. 3B). Though collagen can be imaged with reflection confocal microscopy for certain applications, SHG provides collagen images that are significantly more detailed and less subject to nonspecific reflections (See confocal and SHG images of the same transverse tendon section in Figs. 3C and 3D, respectively.)

For these experiments, whole-mount tendons and growth plate cartilage are imaged in PBS or VectaShield, respectively, and sections on slides are imaged in their coverslipping medium. For visualization of fibrillar structure in a typical epi-detection mode, physiological salt content in the mounting medium is essential (Williams *et al.*, 2005).

Two similar MPM systems are used for imaging—one for growth plate and one for tendon. The details of both microscopes have been described previously (Williams *et al.*, 2005; Zipfel *et al.*, 2003a). Both systems are equipped with Ti:sapphire lasers (Tsunami, SpectraPhysics, Mountain View, CA) that generated ~100-fs pulses. Growth plate specimens are imaged with a Bio-Rad MRC-600 scanner interfaced with an upright, custom modified Olympus (Center Valley, PA) AX-70 with 780-nm excitation and a 20X/0.75 NA Flaur objective (Carl Zeiss MicroImaging, Thornwood, NY). Emission filters (EFs) are chosen for a blue/green separation [BGG22 and 580/150 EF with a 500 long-pass dichroic filter (LPD, Chroma Technology, Rockingham, VT)], allowing endogenous SHG at 390 nm to be separated from TPEF of the fluorescent immunolabels. Tendons are imaged with 800-nm excitation using a Bio-Rad 1024 scanner interfaced with an inverted Olympus IX-70 microscope and a 40X/1.15 NA water immersion objective (Olympus UApo/340). The MPM system is equipped with filters corresponding to blue, green, and red emission: 490 LPD reflecting to a BGG22 EF, a 550 LPD to a 525/50 EF, and a 670 LPD to a 605/90 EF. These filters enable separation of three signals: SHG, blue–green fluorescence from the Alexa 488-labeled primary cilia, and yellow fluorescence from the Alexa 568-labeled BrdU or centrioles. Additionally, blue fluorescence from the DAPI-stained nuclei is collected in the first two channels and visualized as a merged signal. The beam is scanned within the specimen and stepped at 1- to 2-μm increments to generate image stacks.

E. Image Analysis

Three-dimensional image restoration: Due to optical diffraction at the focus, there is an asymmetry in the in-plane (x–y) and out-of-plane (z) resolutions of the MPM system [as is characteristic of virtually all optical microscopies (Zipfel *et al.*, 2003b)]. Structures in the raw MPM image stacks appear "stretched" in the z-direction and must be

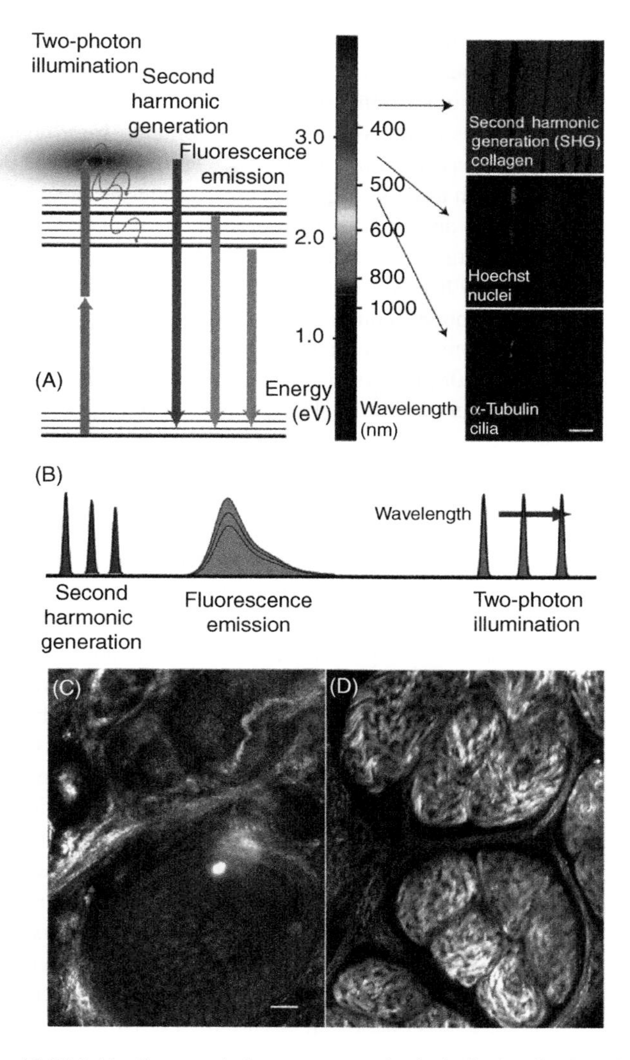

Fig. 3 Principles of MPM. Nonlinear optical processes are intrinsically localized to the focal plane, providing intrinsic optical sectioning for image acquisition in optically thick tissues. Excitation requires "simultaneous" interaction of several photons with the specimen energy levels (black lines in A). SHG from collagen emits at one half the illumination wavelength (representing the total energy input with no gain or loss), whereas fluorescence emission wavelengths are dependent on the particular fluorophores used. All emissions are collected and spectrally filtered to form a pseudocolored image at a particular plane in the specimen. Figure B demonstrates that changing the color of the illumination beam results in a color change of the SHG, but merely an intensity change of the fluorescence. This property enables researchers to tune the SHG wavelength for a clean spectral separation from the fluorescence wavelength. (C, D) Reflection confocal images of tissue collagen such as this cross section of rat tail tendon are possible (C) but show less specificity than those collected using SHG (D). Scale bar = 10 μm. (See Plate no. 5 in the Color Plate Section).

numerically corrected. Image deconvolution assigns diffracted photons back to their proper locations and is accomplished in this case with software for restoration of MPM stacks using a point spread function calculated from the incident wavelength and the properties of the system optics (Volocity, Improvision, Waltham, MA, USA).

Analysis of ciliary incidence: The incidence of primary cilia in fully differentiated cells such as adult articular chondrocytes, when analyzed by serial section TEM *in situ*, has been documented to approach one per cell (Wilsman *et al.*, 1980). By comparison, cells in tissue culture generally have a lower incidence of primary cilia (Wheatley *et al.*, 1996). Because the mother centriole, which typically forms one of the mitotic spindle poles, comprises the ciliary basal body, primary cilia are resorbed prior to mitosis but reassembled and present throughout the majority of interphase (Christensen *et al.*, 2008; Pedersen *et al.*, 2008; Plotnikova *et al.*, 2008; Quarmby and Parker, 2005; Sharma *et al.*, 2008). Thus, actively dividing cells, whether *in situ* or in tissue culture, are expected to have a ciliary incidence of less than one per cell. One goal for using MPM analysis is to develop a relatively rapid way to assess ciliary incidence in populations of connective tissue cells *in situ*.

For the incidence analysis on tendon, one full-thickness stack is collected from each of two sections for each animal. Nuclei and their associated cilia are tracked throughout each image stack in ImageJ [National Institutes of Health (NIH)] to determine the ratios of: (1) nuclei with cilia to total nuclei and (2) BrdU-labeled nuclei to total nuclei. Nuclei on the edges of the stack are excluded from the analysis.

Analysis of 3D ciliary orientation: For primary cilia in connective tissues, the significance of understanding the projection of the ciliary axoneme in 3D space relates to hypotheses of the cilium as a mechanosensory organelle. The directionality of projection of the axoneme is hypothesized to be related either to the directionality of sensory input or to the directionality of secretion of molecules associated with the ECM (Anderson *et al.*, 2008; Poole *et al.*, 1985; Quarmby and Parker, 2005; Whitfield, 2008). Growth plate cartilage is utilized for the initial analysis of ciliary axonemal orientation because: (1) the organization of chondrocytes in columns is highly anisotropic (with a change in direction of the long axis of the chondrocyte relative to the long axis of the bone in different zones) (Fig. 4A) and (2) our previous TEM studies of articular cartilage had demonstrated that the direction the axoneme projects relative to the long axis of the cell is variable (Wilsman *et al.*, 1980). Figure 4B–D demonstrates three orientations of the ciliary axoneme in superficial zone articular chondrocytes as seen by TEM.

Ciliary length and 3D orientation can be determined by following individual axonemal profiles through z-series spanning their entire length. For the ciliary orientation analysis, deconvolved image stacks from whole-mount preparations of growth plate cartilage and tendon fascicles are analyzed with a custom 3D morphometric program in which nuclei and cilia are modeled as ellipsoids and lines, respectively (Ascenzi *et al.*, 2007). The following parameters are calculated for each ciliated cell: the ciliary length, the in-plane angle θ with respect to the x-axis, and the elevation angle ϕ with respect to the z-axis (Fig. 5A). If a cilium is observed on only one frame, the axoneme is assumed to lie entirely in the imaging plane, which is assigned an elevation angle of 90°. For tendon, the

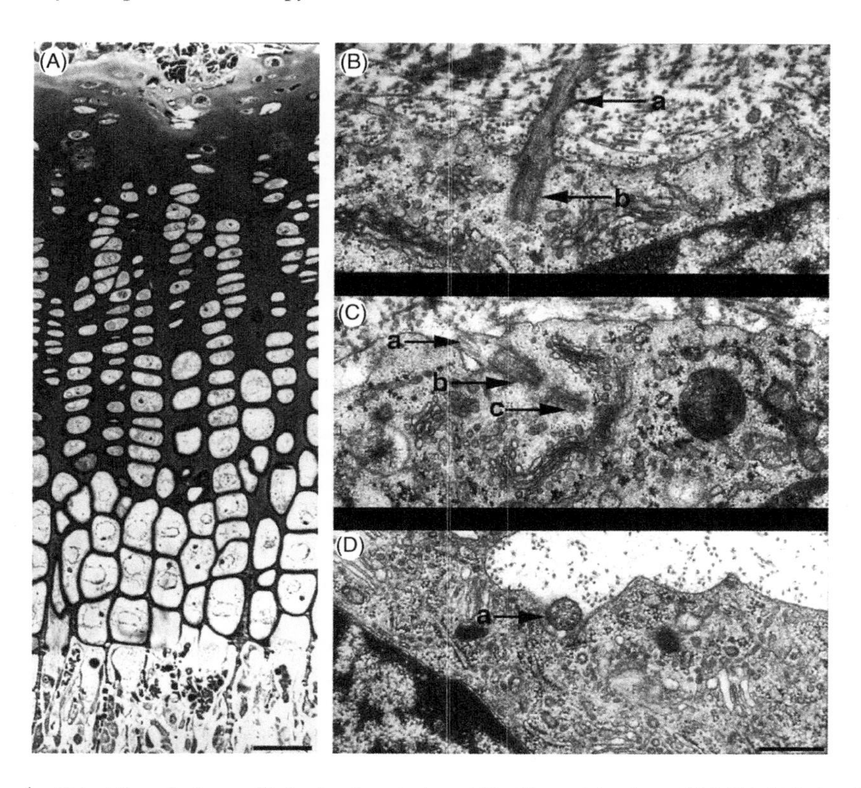

Fig. 4 Orientation of primary cilia in chondrocytes is variable. Figure A is a 1-μm-thick histological section of the proximal tibial growth plate from a 4-week-old mouse, fixed in 0.2% glutaraldehyde/2% paraformaldehyde with 0.7% ruthenium hexamine trichloride (RHT) (Hunziker *et al.*, 1983), and embedded in Epon araldite. Staining is with basic fuchsin, toluidine blue, and methylene blue. Note the vertical columns of growth plate chondrocytes, with a change in the direction of the long axis of the cell relative to the long axis of the bone in different zones. Scale bar = 40 μm. Figures B–D are a sequence of TEM images demonstrating three different orientations of the ciliary axoneme in articular chondrocytes. In B, the basal body (b) of the cilium can be seen with attachments to the plasma membrane, with the axoneme (a) projecting at an angle of almost 90° relative to the cell. In C, a grazing section of the ciliary axoneme (a) is seen projecting diagonally into the extracellular matrix. The ciliary basal body (b) and its attachment to the plasma membrane also are shown in this image, as is a segment of the associated centriole (c). In D, a cross-sectional profile of the ciliary axoneme (a) is seen adjacent to the cellular plasma membrane. TEM sections fixed in presence of osmium ferrocyanide (Farnum and Wilsman, 1983; Wilsman *et al.*, 1980). Scale bar = 0.5 μm (B–D).

x- and *z*-axes are the proximal–distal and cranial–caudal axes of the limb, respectively, chosen because we thought it most biologically meaningful to consider the *x*-axis as paralleling the major collagen fiber direction of the tendon. By contrast, for the growth plate, we thought it most biologically meaningful to consider the chondrocytic columnar arrangement as the *y*-axis, paralleling the long axis of the limb. Therefore, for the growth plate, the *x*- and *z*-axes are the medial–lateral and the cranial–caudal axes, respectively. Thus the selection of axes can be chosen empirically, depending upon what is considered to be most biologically meaningful for a given tissue.

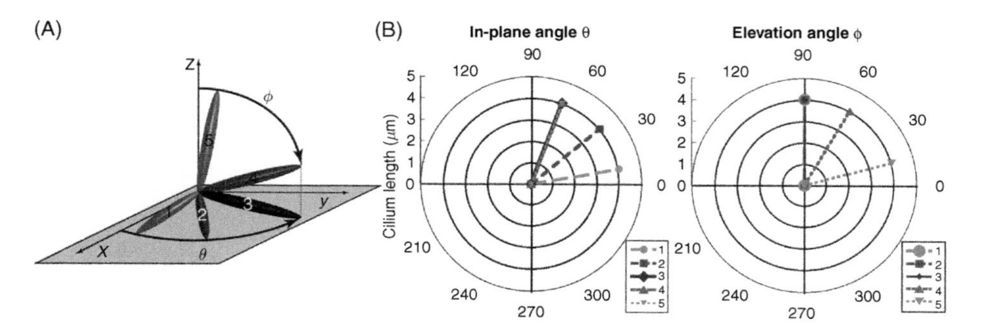

Fig. 5 Schematic showing ciliary orientation angles. Figure A is a schematic showing five cilia of equal length and varying orientations. Ciliary angle definitions are depicted for cilium 4: the in-plane angle θ between the cilium and the x-axis and elevation angle ϕ between the cilium and the z-axis. Cilia 1–3 have the same elevation angle (90°) but different in-plane angles, and cilia 3–5 have the same in-plane angle (70°) but different elevation angles. Figure B demonstrates polar plots summarizing the angular orientations of the five cilia depicted in A.

For the growth plate, the different morphology of the epiphyseal and metaphyseal bone allows us to discriminate proximal from distal during imaging, and thus the elevation angle ϕ is defined over the range $0° \leq \phi < 180°$. During imaging of tendon, although the sample could be aligned along the proximal/distal axis of the limb, there were no distinguishing marks to allow discrimination of proximal from distal. Therefore, the elevation angle ϕ for tendon is reported over the range $0° \leq \phi < 90°$. If it were deemed important for either tendon or growth plate to assess the value of ϕ through the entire 360° range that exists in a living animal, collection for imaging would need to be made marking each sample in a way that clearly distinguished proximal versus distal and cranial versus caudal so that the specimen could be placed on the imaging stage consistent with its limb position *in vivo*.

The orientation data can be summarized in polar plots in which the orientation of each cilium is expressed in coordinates of (r, angle), where r is the ciliary length and *angle* is the in-plane angle θ or the elevation angle ϕ (Fig. 5B). Thus, the length and angular orientation of a "population" of cilia can be expressed on two plots for visualization of trends in their orientation distributions.

IV. Results and Discussion

Imaging by MPM proved to be an excellent way to visualize ciliary axonemes in both growth plate chondrocytes and extensor tenocytes to a depth of 100 μm from the exposed surface of whole mounts of the tissue, and throughout 5-μm sections. Collection of z-stacks allowed rapid collection of images of large numbers of cells and associated cilia for image analysis without significant loss of signal due to photobleaching. SHG demonstrated collagen orientation and could be separated from the fluorescent axonemal signal. This was helpful for specimen orientation in both tissues, but critical for visualizing primary cilia projecting into the dense type I collagen of the tendon ECM. Nonspecific background staining was minimal in both tissues; a positive reaction for acetylated α-tubulin was frequently seen in the Golgi, but by form and locale it was easily distinguished from that of the ciliary axoneme.

Our results to date demonstrate that ciliary length has a wide variation in a given field of cells, from 2 μm to almost 13 μm in growth plate chondrocytes (unpublished data). For tenocytes, ciliary lengths ranged from 1.3 to 11.3 μm (Donnelly *et al.*, in press). This is an interesting finding based on MPM imaging data, but its biological significance currently is unknown.

A. Primary Cilia Associated with Growth Plate Chondrocytes

IHC localizing acetylated α-tubulin on methanol-fixed whole sections of growth plate allowed visualization of primary cilia in columns of chondrocytes at low magnifications. Imaging through z-stacks using MPM significantly shortens the time required to analyze features of a field of cells and their primary cilia, compared to doing an analysis by serial sectioning TEM. By MPM, columns of growth plate chondrocytes could be followed in z-steps at 1 or 2-μm intervals (Fig. 6A and B),

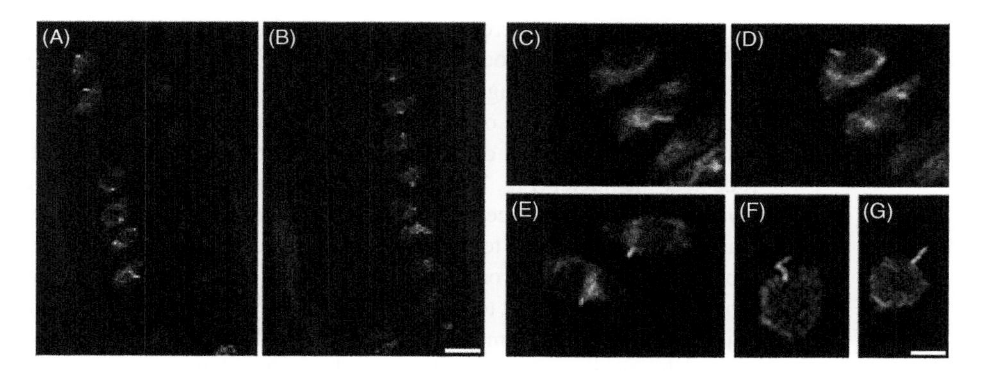

Fig. 6 MPM images of growth plate chondrocytes immunolabeled for acetylated α-tubulin. Figures A and B are low magnification images of neighboring columns of growth plate chondrocytes from the same series. Figure B is 4 μm deeper into the growth plate than A. Note the intensity of axonemal staining relative to overall cellular staining. Although final determination of the presence of a ciliary axoneme could not be determined without analyzing through a series of adjacent 2-μm steps, these images demonstrate that cilia for cells of a given column often appear in the same plane. Scale bar = 20 μm. (C–G) Note that in C the uppermost cell lacks an axonemal profile, while one is clearly seen in the second cell of the column. Four microns deeper, in D, the situation is reversed. In D bright juxtanuclear punctate staining is of acetylated α-tubulin in the Golgi. Figure E demonstrates adjacent chondrocytes with a difference in the position of the cilium relative to the nucleus (dark circle within the chondrocyte), but with similar (but not identical) values of θ as seen in this x-y projection. Figures F and G demonstrate that both curved (F) and straight (G) profiles of the ciliary axoneme could be identified. The significance of this observation is not known. Scale bar = 10 μm (C–G).

allowing one to analyze the characteristics of a given cell and its cilium and compare that directly to the characteristics of the cilium of neighboring cells. This is of considerable value compared to TEM analysis, where cilia on individual cells can only be assessed at high magnification one cell at a time.

Figure 6C–G shows MPM images of primary cilia associated with growth plate chondrocytes. Our observations indicate that, although a cilium can project from any surface along the contour of the chondrocyte, commonly the cilium projects from a position near the cell's center (Fig. 6C and D). The position that the cilium projects from the cell is independent of its axonemal orientation as defined by the angles ϕ and θ, as shown in Fig. 6E of two cells where the *position* that the axoneme projects from the cell relative to the nucleus differs, but θ is essentially the same. Ciliary axonemes almost always were straight; the significance of bending of the ciliary axonene is unknown (Fig. 6F and G).

B. Primary Cilia Associated with Tenocytes

Cilia in fields of tenocytes were visualized *in situ* throughout sections and whole fascicles of extensor tendon in 1-μm z-steps. The SHG and Two photon excited fluorescence (TPEF) signals could be detected to a depth of ~100 μm, allowing collection of z-stacks encompassing the entire fascicle thickness. By comparison to chondrocytes in growth plates, where nuclear and cellular boundaries could be identified with no additional staining, nuclear staining was essential for identification of tenocyte nuclei and discrete cellular boundaries could not be seen.

MPM images of tenocytes showed ellipsoidal Hoechst-stained nuclei aligned in rows between the collagen fibers. The collagen appeared bright due to its strong endogenous generation of SHG, while the elongated cytoplasmic spaces appeared dark due to the absence of collagen (Fig. 7). Primary cilia appeared as bright dots or rods near the nuclei (Fig. 7A), distinguishable by their high intensity and focal staining from the more diffuse acetylated α-tubulin staining associated with the Golgi that also was observed in many tenocytes. Double labeling with γ-tubulin and acetylated α-tubulin to localize centrosomes and primary cilia, respectively, showed punctuate γ-tubulin staining adjacent to dot- or rod-like acetylated α-tubulin staining (Fig. 7B). The adjacent staining confirmed that the acetylated α-tubulin immunolabel localized primary cilia associated with a basal body, rather than a nonciliary component of the cytoskeleton. BrdU labeling, colocalized with Hoechst nuclear stain, was visible in a relatively small fraction of tenocyte nuclei (Fig. 7C), indicating the presence of some cycling cells undergoing DNA replication.

Because ciliary incidence depends on the cell cycle duration, both incidence and BrdU labeling index were quantified in immunolabeled tendon sections. In our experience, sequential double labeling for acetylated α-tubulin and BrdU results in unreliable and generally poor ciliary labeling due to the acid denaturation step required for BrdU labeling. Thus, ciliary incidence and the BrdU labeling index were separately assessed by analyzing MPM stacks for the percentages of ciliated nuclei and BrdU-labeled nuclei, respectively. Ciliary incidence was 64 ± 6%, and the BrdU labeling index was 6

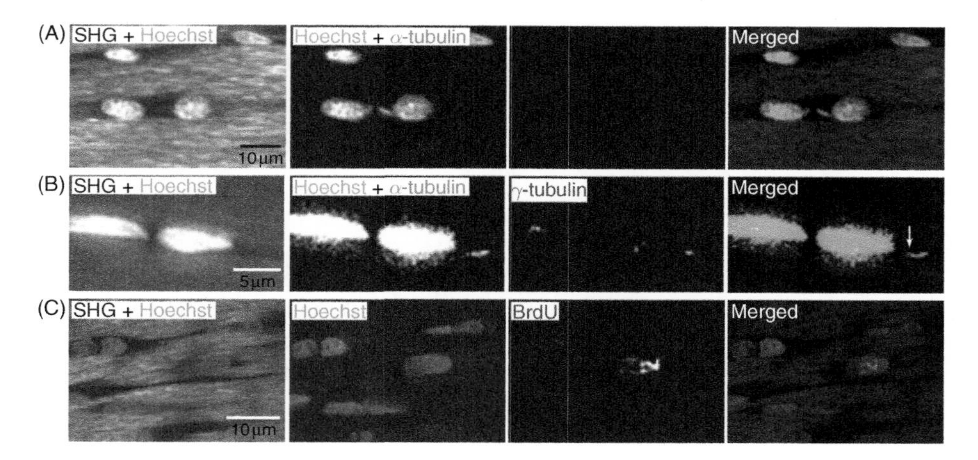

Fig. 7 Multiphoton images of tendon fascicles. Tendon fascicles are immunolabeled for: (A) acetylated α-tubulin (primary cilia), (B) acetylated α-tubulin (primary cilia) and γ-tubulin (centrioles), or (C) BrdU (S-phase nuclei). The three leftmost columns show the signals from the blue, green, and red detector channels, which arise, respectively, from endogenous collagen SHG, Alexa 488-labeled acetylated α-tubulin, and Alexa 568-labeled γ-tubulin or BrdU. The Hoechst signal appears in both the blue and green channels. In the merged images in the fourth column, the collagen, tenocyte nuclei, primary cilia, and γ-tubulin/S-phase nuclei appear blue, cyan, green, and red, respectively. The merged image in B demonstrates labeling of the centriolar basal body (highlighted with arrow) adjacent to the primary cilium. (See Plate no. 6 in the Color Plate Section).

± 2% in a prior study of primary cilia in tenocytes (Donnelly *et al.*, in press). No systematic spatial variation in ciliated or unciliated cells within the tendon was noted.

Experimental design for analysis of ciliary incidence involves balancing two factors: (1) maximizing antibody penetration, which favors thin specimens and (2) minimizing undercounting of ciliated cells in the case of nuclei whose associated cilia were located in a subsequent section, which favors thick specimens (whole mounts). In pilot studies of MPM image stacks of whole tendon fascicles, immunolabeling of cell membranes did not distinguish the boundaries of individual tenocytes because tenocyte processes form a large, interconnected network (McNeilly *et al.*, 1996). Nevertheless, primary cilia were localized close to the nuclei, typically in the same imaging plane (<10 μm from the nucleus in the x–y plane and <1 μm from the nucleus in the z direction). Therefore, 5-μm-thick sections were used for the analysis of incidence in tenocytes. Nuclei that intersected any of the boundaries of the imaging volume were excluded to minimize undercounting of ciliated nuclei. Even with these precautions, incidence values assessed from sections rather than whole tissues may represent a lower bound on the actual incidence. Selection of section thickness depends critically on the cell morphology and ciliary orientation and must be tailored to each cell type.

In the analysis of ciliary orientation, the tenocyte primary cilia were highly oriented with respect to the nuclei and the surrounding ECM. Three-dimensional reconstruction and morphometric analysis of image stacks enabled quantification of ciliary orientation. The cilia lay primarily in the x–y (frontal) plane, parallel to the y-axis (proximal–distal axis) and the collagen fibers (Donnelly *et al.*, in press). Although tenocyte ciliary orientation had not before been quantified prior to our studies, our orientation data are consistent with a TEM study of embryonic chick flexor tendon, which showed the tenocyte ciliary axoneme oriented parallel to the collagen fibers surrounding the cell (Poole *et al.*, 1985).

C. Evaluation of the Technique

The major limitation of using whole mounts for IHC is that penetration of the antibodies (primary, secondary, or both) into the tissue becomes limited with depth, and so the TPEF signal diminishes with depth into the section. These effects can be mitigated with tissue choice: the digital extensor tendons are relatively flat (thickness <100 μm in the rat), and the TPEF and SHG signals show minimal systematic variation with imaging depth in these tissues. Nevertheless, whole mounts are not considered suitable for studies of ciliary incidence, where it is critical for the localizing antibodies to reach all of the cilia, unless additional steps are taken to enhance tissue penetrability. We consider that paraffin embedded 5-μm-thick sections using postembedment immunohistochemical localization are the method of choice for analyzing ciliary incidence.

Unlike TEM, MPM cannot resolve ultrastructural features. As demonstrated in Fig. 1, the primary cilium and its associated basal body have a complex morphology whose ultrastructural details can be resolved with TEM but not with MPM. Since the centrioles are short (~1 μm), fluorescence associated with localizing antibodies allows centriolar visualization, but not discrimination of length of the centriole or its orientation relative to the ciliary basal body and axoneme, using our MPM system. However, in terms of time and ease of preparation, MPM offers a way to analyze presence, length, and orientation of the ciliary axoneme under experimental conditions where z-stacks taken at 1-μm intervals containing fields of ~50 cells can be collected in ~5 min.

When recording results we present axonemal orientation (θ and ϕ) for populations of cells (Donnelly *et al.*, in press), although θ and ϕ can also be tracked for individual cells. The presentation of axonemal orientation data for either populations of cells or for individual cells mirroring the orientation in the live animal requires that excised tissue be labeled or marked at the time of collection such that it can be positioned during imaging exactly as it was *in vivo* along all three axes of interest. For the tendon and growth plate we have not found this level of detail necessary for our studies to date.

Our experimental approach combining IHC and MPM provides a versatile means for *in situ* characterization of the primary cilium in skeletal tissues. Several organelles can be simultaneously visualized in their native ECM by collection of endogenous SHG and exogenous immunofluorescent signals. Reconstruction of z-stacks collected from whole tissues enabled quantification of 3D ciliary length and orientation. Relative to confocal microscopy, MPM allows visualization of unstained collagen, deeper imaging (up to 3× for tendon), and substantially reduced specimen photobleaching because SHG and TPEF excitation processes are limited to the focal volume. The reduced specimen preparation time and rapidity of image collection of MPM compare favorably to TEM and allow analysis of relatively large sample sizes. Our methods can be extended to other connective tissues to characterize relationships between primary cilia, cells, and the surrounding ECM.

Acknowledgments

The authors thank Maria-Grazia Ascenzi for creating the algorithm for analysis of 3D orientation; Maria Serrat, Barbara Linnehan, and Linell Bigelow for assistance with tissue preparation; and Serena Chang both for helpful discussions and for her continuing work on ciliary incidence in growth plate chondrocytes. Special thanks are also given to Jen Patterson for help with the figures. This work was funded by NIH grant 5R21 AR053849 to Cornelia Farnum.

References

Anderson, C., Castillo, A., Brugmann, S., Helms, J., Jacob, C., and Stearns, T. (2008). Primary cilia: Cellular sensors for the skeleton. *Anat. Rec.* **291**, 1074–1078.

Ascenzi, M-G., Tinsley, M., and Farnum, C.E. (2007). Analysis of the orientation of primary cilia in growth plate cartilage: A mathematical method based on multiphoton microscopical images. *J. Struct. Biol.* **158**, 293–306.

Barnes, B.G. (1961). Ciliated secretory cells in the pars distalis of the mouse hypophysis. *J. Ultrastruct. Res.* **5**, 453–367.

Bisgrove, B.W. and Yost, H.J. (2006). The roles of cilia in developmental disorders and disease. *Development* **133**, 4131–4143.

Christensen, S.T., Pedersen, S.F., Satir, P., Veland, I.R., and Schneider, L. (2008). The primary cilium coordinates signaling pathways in cell cycle control and migration during development and tissue repair. *Curr. Top. Dev. Biol.* **85**, 261–301.

Davenport, J.R., Watts, A., Roper, V.C., Croyle, M.J., van Groen, T., Wyss, J.M., Nagy, T.R., Kesterson, R.A., and Yoder, B.K. (2007). Disruption of intraflagellar transport in adult mice leads to obesity and slow-onset cystic kidney disease. *Curr. Biol.* **17**, 1586–1594.

Davenport, J.R., and Yoder, B.K. (2005). An incredible decade for the primary cilium: A look at a one-forgotten organelle. *Am. J. Physiol. Renal Physiol.* **289**, F1159–F1169.

Denk, W., Strickler, J.H., and Webb, W.W. (1990). Two-photon laser scanning fluorescence microscopy. *Science* **248**, 73–76.

Donnelly, E., Ascenzi, M-G., and Farnum, C.E. (in press). Primary cilia are highly oriented with respect to collagen direction and long axis of extensor tendon. *J. Orthop. Res.*, DOI: 10.1002/jor.20946, URL: <http://dx.doi.org/10.1002/jor.20946>http://dx.doi.org/10.1002/jor.20946.

Donnelly, E., Williams, R., and Farnum, C.E. (2008). The primary cilium of connective tissue cells: Imaging by multiphoton microscopy. *Anat. Rec.* **291**, 1062–1073.

Farnum, C.E., and Wilsman, N.J. (1983). The pericellular matrix of growth plate chondrocytes: A study using postfixation with osmium-ferrocyanide. *J. Histochem. Cytochem.* **31**, 765–775.

Farnum, C.E., and Wilsman, N.J. (1986). *In situ* localization of lectin-binding glycoconjugates in the matrix of growth-plate cartilage. *Am. J. Anat.* **76**, 65–82.

Gerdes, J., Davis, E., and Katsanis, N. (2009). The vertebrate primary cilium in development, homeostasis, and disease. *Cell* **137**, 32–45.

Grillo, M.A., and Palay, S.L. (1963). Ciliated Schwann cells in the autonomic nervous system of the adult rat. *J. Cell Biol.* **16**, 430–436.

Haycraft, C.J., and Serra, R. (2008). Cilia involvement in patterning and maintenance of the skeleton. *Curr. Top. Dev. Biol.* **85**, 303–332.

Haycraft, C.J., Zhang, Q., Song, B., Jackson, W.S., Detloff, P.J., Serra, R., and Yoder, B.K. (2007). Intraflagellar transport is essential for endochondral bone formation. *Development* **134**, 307–316.

Jensen, C.G., Poole, C.A., McGlashan, S.R., Marko, M., Issa, Z.I., Vujcich, K.V., and Bowser, S.S. (2004). Ultrastructural, tomographic and confocal imaging of the chondrocyte primary cilium in situ. *Cell Biol. Int.* **28**, 101–110.

Kaushik, A.P., Martin, J.A., Zhang, Q., Sheffield, V.C., and Morcuende, J.A. (2009). Cartilage abnormalities associated with defects of chondrocytic primary cilia in Bardet-Biedl syndrome in mutant mice. *J. Orthop. Res.* **27,** 1093–1099.

Lehman, J.M., Michaud, E.J., Schoeb, T.R., Aydin-Son, Y., Miller, M., and Yoder, B.K. (2008). The Oak Ridge polycystic kidney mouse: Modeling ciliopathies of mice and men. *Dev. Dyn.* **237**, 1960–1971.

Low, S.H., Vasanth, S., Larson, C.H., Mukherjee, S., Sharma, N., Kinter, M., Kane, M., Obara, T., and Weimbs, T. (2006). Polycystin-1, STAT6, and P100 function in a pathway that transduces ciliary mechanosensation and is activated in polycystic kidney disease. *Dev. Cell* **10**, 57–69.

Malone, A., Anderson, C., Tummala, P., Kwon, R., Johnston, T., Stearns, T., and Jacobs, C. (2007). Primary cilia mediate mechanosensing in bone cells by a calcium-independent mechanism. *Proc. Natl. Acad. Sci. USA* **104**, 13325–13330.

McGlashan, S.R., Haycraft, C.J., Jensen, C.G., Yoder, B.K., Poole, C.A. 2007. Articular cartilage and growth plate defects are associated with chondrocyte cytoskeletal abnormalities in Tg737^{orpk} mice lacking the primary cilia protein polaris. *Matrix Biol.* **26**, 234–246.

McGlashan, S.R., Jensen, C.C., and Poole, C.A. (2006). Localisation of extracellular matrix receptors on the chondrocyte primary cilium. *J. Histochem. Cytochem.* **54**, 1005–1014.

McNeilly, C.M., Banes, A.J., Benjamin, M., and Ralphs, J.R. (1996). Tendon cells in vivo form a three-dimensional network of cell processes linked by gap junctions. *J. Anat.* **189**, 593–600.

Ocbina, P., and Anderson, K.V. (2008). Intraflagellar transport, cilia, and mammalian hedgehog signaling: Analysis in mouse embryonic fibroblasts. *Dev. Dyn.* **237**, 2030–2038.

Pazour, G.J. and Rosenbaum, J.L. (2002). Intraflagellar transport and cilia-dependent diseases. *Trends Cell Biol.* **12**, 551–555.

Pedersen, L.B., and Rosenbaum, J.L. (2008). Intraflagellar transport (IFT): Role in ciliary assembly, resorption and signaling. *Curr. Top. Dev. Biol.* **85**, 23–61.

Pedersen, L.B., Veland, I.R., Schroder, J.M., and Christensen, S.T. (2008). Assembly of primary cilia. *Dev. Dyn.* **237**, 1993–2006.

Plotnikova, O.V., Golemis, E.A., and Pugacheva, E.N. (2008). Cell cycle-dependent ciliogenesis and cancer. *Cancer Res.* **68**, 2058–2061.

Poole, C.A., Flint, M.H., and Beaumont, B.W. (1985). Analysis of the morphology and function of primary cilia in connective tissues: A cellular cybernetic probe? *Cell Motil.* **5**, 175–193.

Poole, C.A., Jensen, C.G., Snyder, J.A., Gray, C.G., Hermanutz, V.L., and Wheatley, D.N. (1997). Confocal analysis of primary cilia structure and colocalization with the golgi apparatus in chondrocytes and aortic smooth muscle cells. *Cell Biol. Int.* **21**, 483–494.

Poole, C.A., Zhang, Z., and Ross, J.M. (2001). The differential distribution of acetylated and detyrosinated alpha-tubulin in the microtubular cytoskeleton and primary cilia of hyaline cartilage chondrocytes. *J. Anat.* **199**, 393–405.

Quarmby, L.M., and Parker, J.D.K. (2005). Cilia and the cell cycle? *J. Cell Biol.* **169**, 707–710.

Santos, N., and Reiter, J.F. (2008). Building it up and taking it down: The regulation of vertebrate ciliogenesis. *Dev. Dyn.* **237**, 1972–1981.

Satir, P., and Christensen, S.T. (2007). Overview of structure and function of mammalian cilia. *Annu. Rev. Physiol.* **69**, 377–400.

Serra, R. (2008). Role of intraflagellar transport and primary cilia in skeletal development. *Anat. Rec.* **291**, 1049–1061.

Sharma, N., Berbari, N.F., and Yoder, B.K. (2008). Ciliary dysfunction in developmental abnormalities and diseases. *Curr. Top.Dev. Biol.* **85**, 371–427.

Song, B., Haycraft, C.J., Seo, H., Yoder, B.K., and Serra, R. (2007). Development of the post-natal growth plate requires intraflagellar transport proteins. *Dev. Biol.* **305**, 202–216.

Sorokin, S. (1962). Centrioles and the formation of rudimentary cilia by fibroblasts and smooth muscle cells. *J. Cell Biol.* **15**, 363–377.

Wheatley, D.N. (2008). Nanobiology of the primary cilium—Paradigm of a multifunctional nanomachine complex. *Methods Cell Biol.* **90**, 139–156.

Wheatley, D.N., Wang, A.M., and Strugnell, G.E. (1996). Expression of primary cilia in mammalian cells. *Cell Biol. Int.* **20**, 73–81.

Whitfield, J.F. (2008). The solitary (primary) cilium—A mechanosensory toggle switch in bone and cartilage cells. *Cell. Signal.* **20**, 1019–1024.

Williams, R.M., Zipfel, W.R., and Webb, W.W. (2005). Interpreting second harmonic generation images of collagen I fibrils. *Biophys. J.* **88**, 1377–1386.

Wilsman, N. (1978). Cilia of adult canine articular chondrocytes. *J. Ultrastruct. Res.* **64**, 270–281.

Wilsman, N.J., and Fletcher, T.F. (1978). Cilia of neonatal articular chondrocytes: Incidence and morphology. *Anat. Rec.* **190**, 871–889.

Wilsman, N.J., Farnum, C.E., and Reed-Aksamit, D.K. (1980). Incidence and morphology of equine and murine chondrocytic cilia. *Anat. Rec.* **197**, 355–361.

Yoder, B.K. (2007). Role of primary cilia in the pathogenesis of polycystic kidney disease. *J. Am. Soc. Nephrol.* **18**, 1381–1388.

Zipfel, W.R., Williams, R.M., Christie, R., Nikitin, A.Y., Hyman, B.T., and Webb, W.W. (2003a). Live tissue intrinsic emission microscopy using multiphoton-excited native fluorescence and second harmonic generation. *Proc. Natl. Acad. Sci. USA* **100**, 7075–7080.

Zipfel, W.R., Williams, R.M., and Webb, W.W. (2003b). Nonlinear magic: Multiphoton microscopy in the biosciences. *Nat. Biotechnol.* **21**, 1369–1377.

CHAPTER 7

Primary Cilia and the Cell Cycle

Olga V. Plotnikova*[,†], Elena N. Pugacheva[‡], *and* Erica A. Golemis*

[*]Program in Molecular and Translational Medicine, Fox Chase Cancer Center, Philadelphia, Pennsylvania 19111

[†]Department of Molecular Biology and Medical Biotechnology, Russian State Medical University, Moscow, Russia

[‡]Mary Babb Randolph Cancer Center, West Virginia University, Morgantown, West Virginia 26506

Abstract

Cilia are microtubule-based structures that protrude from the cell surface and function as sensors for mechanical and chemical environmental cues that regulate cellular differentiation or division. In metazoans, ciliary signaling is important during organismal development and in the homeostasis controls of adult tissues,

978-0-12-375024-2
DOI: 10.1016/S0091-679X(08)94007-3

with receptors for the Hedgehog, platelet derived growth factor (PDGF), Wnt, and other signaling cascades arrayed and active along the ciliary membrane. In normal cells, cilia are dynamically regulated during cell cycle progression: present in G0 and G1 cells, and usually in S/G2 cells, but almost invariably resorbed before mitotic entry, to reappear post-cytokinesis. This periodic resorption and reassembly of cilia, specified by the intrinsic cell cycle the intrinsic cell cycle machinery, influences the susceptibility of cells to the influence of extrinsic signals with cilia-associated receptors. Pathogenic conditions of mammals associated with loss of or defects in ciliary integrity include a number of developmental disorders, cystic syndromes in adults, and some cancers. With the continuing expansion of the list of human diseases associated with ciliary abnormalities, the identification of the cellular mechanisms regulating ciliary growth and disassembly has become a topic of intense research interest. Although these mechanisms are far from being understood, a number of recent studies have begun to identify key regulatory factors that may begin to offer insight into disease pathogenesis and treatment. In this chapter we will discuss the current state of knowledge regarding cell cycle control of ciliary dynamics, and provide general methods that can be applied to investigate cell cycle-dependent ciliary growth and disassembly.

I. Introduction

A. Physical Components of Cilia

Cilia (also known as flagella) are found throughout the evolutionary tree, in organisms spanning the green algae *Chlamydomonas reinhardtii*, the protist *Tetrahymena thermophila, Drosophila melanogaster, Caenorhabditis elegans*, and all vertebrates so far examined. Detailed studies of cilia date back over 50 years (Gilula and Satir, 1972; Lewin, 1952b; Wheatley, 1969, 1995; Wheatley *et al.*, 1994). (For a more complete historical perspective, readers should consult Chapter 1 by Bloodgood, this volume.) The structure of a cilium is highly conserved. The cytoskeletal scaffold of a cilium is termed the axoneme, and is composed of microtubule triplets arrayed in a $9 + 0$ configuration in nonmotile cilia (9 triplets in a hollow ring), or a $9 + 2$ configuration (with 2 additional central triplets) in motile cilia. The ciliary axoneme is anchored at the cell body-proximal surface of the ciliary membrane by the basal body, which provides a point of connection to the cell surface. As discussed in detail below, the basal body is derived from the mother centriole in a cell during specific phases of the cell cycle, requiring its release from cilia and utilization as a microtubule-organizing center (MTOC) in a different capacity during mitosis. In vertebrates, this oscillation between basal body and MTOC identity is typically accompanied by extension of the cilium in postmitotic (G1 or G0) cells, and resorption of the cilium later in the cell cycle, prior to mitotic entry (Archer and Wheatley, 1971; Ho and Tucker, 1989; Quarmby, 2004; Sorokin, 1968a,b; Tucker *et al.*, 1979a,b, 1983). The exact point of ciliary resorption during the cell cycle depends on the cell type, with some cells resorbing cilia in S-phase, and

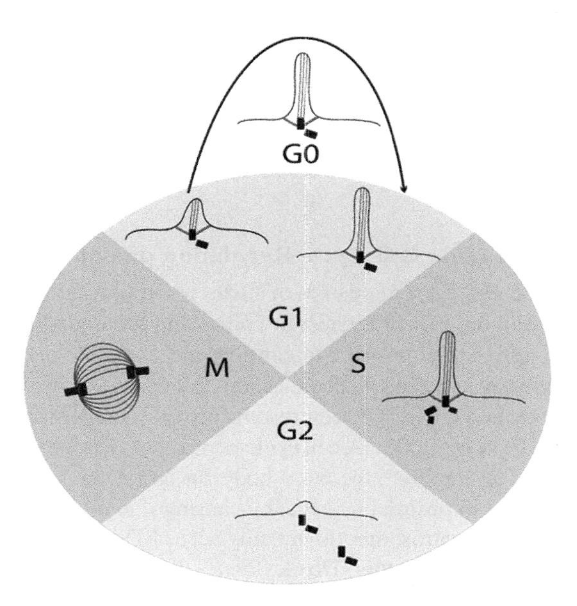

Fig. 1 Ciliary protrusion, resorption, and the cell cycle in metazoans. In most cells, primary cilium formation first occurs post-cytokinesis, during G1. Upon entry into S phase, the centrioles and the DNA initiate replication. Ciliary disassembly typically occurs at the G2–M transition. Once cell division is complete, the cilia reassemble in G1. Typically, the length of cilia and the proportion of a cell population with cilia are increased in quiescent (G0) rather than in cycling cells.

others at the G2–M transition (Jensen *et al.*, 1987; Rash *et al.*, 1969; Rieder *et al.*, 1979), with the latter pattern more typical (Fig. 1).

The basic physical process of ciliary assembly and disassembly requires contributions of the cell secretory machinery and of specialized motors that transport cargo toward and away from the cilia "tip." The cilium typically emerges from the cell surface in close conjunction with the Golgi apparatus, with association of a vesicle with the distal surface of the basal body preceding extension (Sorokin, 1962). Recent studies have implicated the Arl and Rab GTPases of the Ras superfamily and their regulators in cilium formation and function (Grayson *et al.*, 2002; Qin *et al.*, 2007; Yoshimura *et al.*, 2007). Rab family GTPases are known to promote interactions between membrane and the cytoskeleton, and to control specific interactions between membranes (Zerial and McBride, 2001). In the context of cilium formation, Rabs might therefore provide a means to control the interaction of the basal body with the cell surface, and integrate microtubule growth and membrane elongation. Intraflagellar transport (IFT), which delivers ciliary proteins to and from the ciliary tip, is mediated by IFT complexes A and B: IFT complex B associates with the motor kinesin 2 to transport vesicles from the cell body toward the tip, while complex A associates with the motor dynein to transport packets from the tip toward the cell body. Cell cycle-regulated ciliary assembly and resorption relies in part on highly coordinated anterograde and retrograde cargo delivery systems managed through the IFT proteins (Rosenbaum, 2002; Rosenbaum and Witman, 2002), and there are close connections between proteins involved in IFT and in core cell cycle controls. As we will discuss herein, a number of discrete signaling systems have now been identified as contributing to coordination of cell cycle with ciliogenesis.

B. Normal and Pathogenic Cell Cycle Regulation of Cilia

As cell cycle regulation of ciliary resorption and extension is strongly conserved through evolution, it is of interest to understand the underlying basis (if any) for selection of this coupling. It is possible to consider the needs of the cell cycle, or the needs of ciliation, as primary in driving the connection. For example, the dual function of a centriole in the basal body and the MTOC/centrosome provides one obvious point of connection (Doxsey *et al.*, 2005; Nigg, 2006). A cell cycle-centric perspective might ask if the selection is based on the need to "recover" the basal body for use as an MTOC in mitosis, or to "incapacitate" an MTOC in postmitotic cells, by causing it to undergo differentiation to a basal body status. Further, centrosomes do not only act as MTOCs, but also serve as major signaling hubs for cell cycle regulators (Doxsey *et al.*, 2005). The process of differentiation between centrosome and basal body may cause sequential displacement of factors that contribute to an earlier cell cycle stage, allowing them to be replaced by groups of proteins that control a later cell cycle stage. Ciliary resorption/shortening during the progression from G2 to M phase might be important for timing of the cell cycle, altering accessibility of cells to growth factors that use cilia-localized receptors.

In this context, it is striking that the majority of tumor cells arising from ciliated normal precursor cells do not possess cilia at any stage of the cell cycle (Seeley *et al.*, 2009; Wheatley, 1995, 1996). The loss of ciliation in many tumor cells may indicate either that the deregulated cell cycle, which is a hallmark of cancer cells, requires loss of a cell cycle-restrictive signal mediated through cilia, or that the loss of tumor cells' ability to readily enter quiescence causes defects in ciliogenesis. Besides tumor cells, it has also been found that the cystogenesis found in polycystic kidney disease (PKD) or due to other mutations is accompanied by both defects in cell cycle progression and shortened or absent cilia (Deane and Ricardo, 2007; Veland *et al.*, 2009). The increasingly close connections apparent between inappropriate ciliation and disease state make understanding of the mechanisms involved of potentially great therapeutic value.

Although it is not unreasonable to think of cell cycle requirements as a primary driver of the ciliation cycle, some studies argue against easy generalizations or development of absolute rules. For example, a number of vertebrate cell lineages, such as lymphocytes, lack cilia entirely, arguing that it is not necessary for a centriole to pass through a phase as a basal body for a cell to undergo a normal cycle. In exciting recent work, Wong and coworkers demonstrated that cilia could positively or negatively regulate tumorigenesis, based on the role of the Hedgehog pathway in distinct cell types (Wong *et al.*, 2009). As noted above, different lineages of cells resorb cilia at different phases of the cell cycle, again arguing against a strict requirement for a centrosome versus a basal body at a specific stage. In many cell types, cells with ablated centrosomes are able to undergo mitosis (Mahoney *et al.*, 2006), while ablation of centrosomes is accompanied by severe defects in primary cilia assembly (Mikule *et al.*, 2007). With few exceptions in highly specialized conditions, such as the acentriolar pathway used for the formation of hundreds of cilia from deuterosomes in multiciliated cells (Dirksen, 1991; Duensing *et al.*, 2007; Habedanck *et al.*, 2005; Hagiwara *et al.*, 2004; Kleylein-Sohn *et al.*, 2007; Rodrigues-Martins *et al.*, 2007a; Sorokin, 1968b; Vladar and Stearns, 2007), a centriole is absolutely required for cilia formation. These latter observations suggest that centrioles may be more important for formation of cilia than the cell cycle, an idea developed at length by Marshall (2008). Finally, in adult organisms, most cells exist in a quiescent state, with cell cycle considerations secondary to the need to respond to transient environmental signals sensed through the cilia. Hence, in a cilia-centric view, brief "borrowing" of the centriole to act as an MTOC during mitosis may be tolerated as an efficient way to optimize cell division.

II. Signaling Systems Regulating Ciliary Protusion and Resorption

A. Sources of Mechanistic Insight: The Leading Role of *Chlamydomonas*

Although insights into the regulation of cell cycle-coordinated ciliary extension and disassembly have emerged from many model systems, most paradigms for these processes have emerged from analysis of the biflagellate, unicellular alga *C. reinhardtii* (Dutcher, 1995; Haimo and Rosenbaum, 1981; Mitchell *et al.*, 2004; Mitchell and Rosenbaum, 1985; Rosenbaum and Carlson, 1969). Many structural characteristics of the cilia and basal body in *Chlamydomonas* are identical to those of higher eukaryotes. The short life cycle of *Chlamydomonas*, the genetic potential of the organism, and the ability to manipulate environmental stimuli combine to create a robust capacity to analyze signals controlling ciliary dynamics. Of the regulatory signals discussed below as relevant to ciliary control in vertebrates, some were identified first based on experiments performed in *Chlamydomonas*, and it is likely that further "mining" of this source will be fruitful.

In comparing the regulatory controls of *Chlamydomonas* and vertebrates, it is important to remain aware that the stimuli leading to loss of cilia may not be identical, given the different life cycles of unicellular organisms and metazoa. *Chlamydomonas* can reproduce asexually (either by vegetative cell division, or more often by zoospore formation) or sexually. During the vegetative cell cycle, the flagella regress before cell division begins, gradually becoming shorter over a period of approximately 30 min (Lewin, 1952a). After cell division occurs, flagella grow again from the basal body. In non-vegetative cells, the related process of flagella resorption is directly coupled to the cell cycle, with shortening occurring in preprophase and extension at the beginning of G1. In the sexual cycle, flagellar resorption begins a few hours after biflagellate + and − gametes have fused to form a quadriflagellate zygote, and proceeds gradually (Cavalier-Smith, 1974). This disassembly and resorption occurs during preprophase, in parallel with basal body replication: the duplicated basal bodies remain associated with the plasma membrane, and serve as poles for the mitotic spindle (Quarmby, 2004). Genes involved in regulating this mating-associated resorption process contain many parallels with the cell cycle-regulated loss of cilia in mammals, as discussed below.

Flagellar loss also occurs in response to a diverse set of environmental and stress-associated signals that are not always paralleled by stimuli occurring physiologically in a complex metazoan (Quarmby, 2004). For example, increasing intracellular concentration of Ca^{2+} (Quarmby and Hartzell, 1994) or addition of 1 mM Ca^{2+} in the culture medium (Lohret *et al.*, 1998) results in rapid deflagellation. Additional triggers of flagellar disassembly include exposure to low pH (Lewin *et al.*, 1982), increase in temperature above 40°C (Lewin *et al.*, 1982), or treatment with the local anesthetic dibucaine (Butterworth and Strichartz, 1990). For many of these stimuli, the mechanism of flagellar removal is not controlled resorption, but instead a rapid process termed variously deflagellation, excision, shedding, or autotomy (meaning "self-severing") (Quarmby, 2004). This process involves direct severing of the axoneme and rapid changes in IFT (Parker and Quarmby, 2003) and vesicular transport (Overgaard *et al.*, 2009). This process is much less well documented in vertebrates, although it clearly exists. The ciliated epithelia of the oviduct and lung transiently shed cilia in response to triggers including smoke and infection (Quarmby, 2004). A recent fascinating study documents chemical stress-induced ciliary shedding in cultured epithelial cells, and documents that the process of shedding is linked to enhanced tight junction association and transepithelial barrier function (Overgaard *et al.*, 2009). The near-complete dearth of literature on this topic makes it a fertile ground for future studies; however, at present, it is not clear how directly genes involved in *Chlamydomonas* deflagellation, rather than flagellar resorption, will be relevant to cell cycle regulation of the ciliary cycle in vertebrates.

B. IFT Proteins

IFT is the cellular process essential for the formation and maintenance of eukaryotic cilia and flagella (Fig. 2). The IFT protein complexes are nonmembrane-bound particles moving along the axonemal doublet microtubules from the base to the tip of the organelle, and subsequently returning. The relative rate of transport of building materials to and from the tip governs the length of the primary cilium. During resorption, disassembly of the axoneme counterintuitively is associated with an increased rate of anterograde IFT particles entering the cilium, but with those particles having a higher rate of empty cargo sites (Pan and Snell, 2005; Rosenbaum and Witman, 2002). Defects in IFT proteins commonly result in failure to assemble cilia, with this phenotype the primary outcome of mutants affecting IFT-associated genes.

Interestingly, several studies have suggested that a limited number of specific IFT proteins may play dual roles in controlling ciliary protrusion and resorption, and in regulating the cell cycle. Depletion of IFT27, a Rab-like small G-protein, results in the loss of flagella as well as the inhibition of cytokinesis (Qin *et al.*, 2007). Rosenbaum and colleagues suggested that IFT27 may be involved in regulation of membrane dynamics during cytokinesis, limiting the vesicular trafficking necessary for

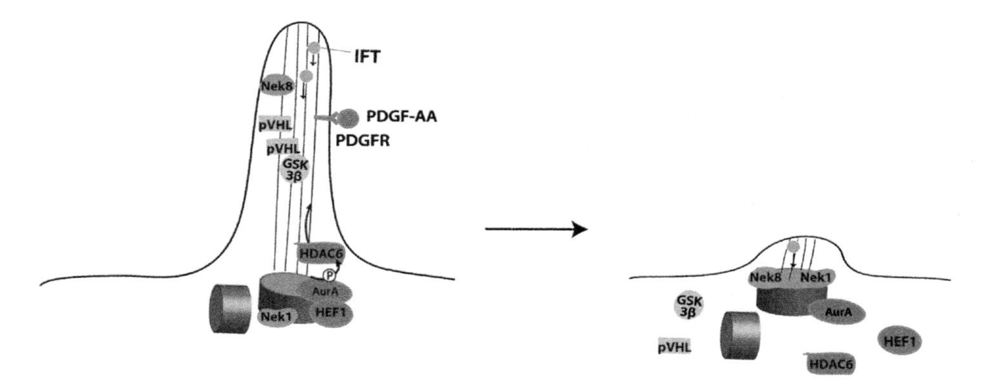

Fig. 2 Cell cycle and signaling proteins associated with control of ciliogenesis and disassembly. See text for details: PDGF-AA binding to the PDGFRαα receptor located at the ciliary membrane induces phosphorylation and activation of the number of different signaling pathways that have a role in ciliary resorption and cell cycle regulation. Among these, HEF1 binds to Aurora-A kinase, promoting its activation. Aurora-A in turn phosphorylates and activates cilia-associated HDAC6, resulting in HDAC6-mediated deacetylation of substrates in the ciliary axomene, causing ciliary resorption. IFT proteins (IFT27 and IFT88) directly coordinate cell cycle machinery and trafficking of particles within the axonome, contributing to timed disassembly. Nek1 and Nek8 act at the basal body and axoneme during cell cycle-regulated disassembly. pVHL and GSK-3β localize to primary cilia, act together to maintain the primary cilium, and may be targeted during the disassembly process.

abscission. Potentially, IFT27 might act as a cell cycle-repressive checkpoint protein, limiting cell cycle progression until the cilia are resorbed and the centrioles can be used in cell division (Qin *et al.*, 2007). As another example, mice with mutant IFT88/polaris/Tg737 (the Tg737/*orpk* mouse model) die soon after birth with cells containing abnormally short cilia, suggesting a defect in ciliogenesis or ciliary maintenance (Pazour *et al.*, 2000; Taulman *et al.*, 2001), In mammalian cells, IFT88 is not only required for the formation of primary cilia, but also serves as a centrosomal protein that regulates G1–S transition in nonciliated HeLa cells (Robert *et al.*, 2007). The G1–S transition control may involve the interactions documented between IFT88 and Che-1, a regulator for S phase entry that binds and inactivates the tumor suppressor Rb: IFT88 overexpression causes failure to enter S, and stimulates apoptosis, while IFT88 depletion drives cell cycle progression through S and G2 (Robert *et al.*, 2007). No cell cycle-specific functions have yet been described for other IFTs, nor have IFT mutations been specifically associated with resorption phenotypes.

C. NEK Kinases

Although Rieder first predicted that ciliary resorption might be controlled by factors regulating mitotic progression in 1979 (Jensen *et al.*, 1987; Rieder *et al.*, 1979), more than two decades elapsed before this insight was validated (Fig. 2). The "never-in-mitosis" or NIMA-related kinase (NRK) family of proteins are cell cycle kinases that are essential for mitosis regulation, G2–M transition, and centrosome separation (Quarmby and Parker, 2005). In 2004, the Quarmby group first identified an NRK family member, Nek2/FA2, as important for both cell cycle progression and deflagellation in *Chlamydomonas* (Mahjoub *et al.*, 2004; Quarmby and Parker, 2005) The Fa2p protein localizes to a specific region of the proximal cilium in interphase cells, the site of flagellar autotomy (SOFA). Fa2p is essential for Ca^{2+}-mediated axonemal microtubule severing during deflagellation. Cells carrying a complete deletion of the Fa2 gene also have a significant delay at the G2–M transition, slowing their transit of the cell cycle (Mahjoub *et al.*, 2004); whether the severing and cell cycle defects are directly related or partially independent needs further study. Nek2 kinases control centrosomal maturation in nonciliated cycling cells (Fry *et al.*, 2000), suggesting a possible independent function; further, subsequent studies by the Quarmby group indicated the ciliary function is kinase-dependent, whereas the cell cycle function is kinase-independent (Mahjoub *et al.*, 2004). During resorption, Fa2p relocates from the SOFA to the proximal end of the basal bodies; during mitosis, it is associated with the polar region of the mitotic spindle. As the cells exit mitosis, Fa2p accumulates at the proximal end of the basal bodies and moves out to the SOFA as soon as ciliogenesis is initiated. These localizations position Fa2p appropriately to directly phosphorylate proteins that regulate severing and resorption, although at present critical substrates have not been identified.

A growing number of NRK proteins have been implicated in the control of ciliary resorption and/or deflagellation coordinated by the cell cycle. The Quarmby group has more recently shown that the NRK Cnk2p is found along entire length of the axoneme of cilia/flagella, and regulates flagellar length by promoting flagellar disassembly. Cnk2p also regulates cell size, by affecting the assessment of cell size prior to mitosis (Bradley and Quarmby, 2005). An extensive phylogenetic analysis of NRKs in the highly ciliated organism *Tetrahymena* has nominated multiple NRKs from three different subfamilies as regulating ciliary resorption (Wloga *et al.*, 2006): in this case, specific NRKs induced the resorption of specific sets of cilia on the organismal surface, suggesting specialized control mechanisms. Although analysis of the NRKs in vertebrates is not as advanced as in lower eukaryotes, Fa2p localization and likely activity are equivalent in murine kidney cells and *Chlamydomonas* (Mahjoub *et al.*, 2004). In both mice and humans, mutations in the genes Nek1 and Nek8 cause polycystic kidney disease (Liu *et al.*, 2002; Upadhya *et al.*, 2000), and these proteins have been shown to localize to cilia and regulate both centrosomal and ciliary integrity (Otto *et al.*, 2008; Trapp *et al.*, 2008; White and Quarmby, 2008). More work to establish the downstream targets of the NRKs is clearly required.

D. Aurora-A and HEF1/NEDD9

Another source of insight into ciliary resorption in humans came from studies of the *Chlamydomonas* kinase CALK, which was identified by the Snell group as an important trigger for flagellar resorption (Pan and Snell, 2003, 2005; Pan *et al.*, 2004, 2005). Kinesin II translocates CALK into the flagella in response to fertilization, where its action promotes ciliary shortening. Beyond fertilization, the CALK kinase becomes activated by a number of different stimuli inducing loss or shortening of flagella, and depletion of CALK by RNA interference (RNAi) inhibits both resorption and deflagellation.

Although CALK is as much an environmental sensor as a cell cycle kinase in *Chlamydomonas*, it is evolutionarily related to the cell cycle regulatory Aurora-A kinase in vertebrates (Fig. 2). Aurora-A is a centrosomally associated kinase that becomes activated at mitotic entry to phosphorylate substrates including Cdc25B, TPX2, Eg5, histone 3 and others that promote progression through the stages of mitosis (Vader and Lens, 2008). Aurora-A activity is reduced and much of the kinase pool is degraded as cells reenter G1. Given the known tendency of ciliary resorption to coincide with the G2–M transition, the suggestive timing of this activity profile and the centrosomal localization of Aurora-A together led our group to investigate Aurora-A as a regulator of ciliary resorption (Pugacheva *et al.*, 2007). We found that microinjection of Aurora-A into ciliated cells caused rapid (<2 min) ciliary disassembly, while treatment with a specific small molecule inhibitor of Aurora-A (PHA-680632) or siRNA depleting Aurora-A prevented serum-induced disassembly of primary cilia, indicating activated Aurora-A is necessary and sufficient for ciliary shortening. In addition, stimulation of cells with serum to initiate a disassembly program induced Aurora-A activation and autophosphorylation at the basal body of the cilium immediately preceding ciliary resorption. Unexpectedly, this study demonstrated that cilia were resorbed in two waves: a first at 1–2 h postserum stimulation, when cells were still in G0/G1 phase, and a second at 18–24 h postserum stimulation, when cells were entering mitosis. Aurora-A became active before disassembly even in the earlier, nonmitotic wave, providing the first evidence for a nonmitotic activation of this kinase in vertebrates, but paralleling the environmental activation of CALK seen in *Chlamydomonas*.

The mechanism of Aurora-A action in promoting disassembly is likely to involve Aurora-A dependent activation of the tubulin deacetylase HDAC6, which deacetylates α-tubulin *in vitro* and *in vivo* leading to destabilization of microtubules. Supporting this idea, pretreatment of cells with the HDAC6 inhibitor tubacin (Haggarty *et al.*, 2003) limited Aurora-A induced ciliary resorption, while Aurora-A phosphorylation of HDAC6 stimulated its activity *in vitro* (Pugacheva *et al.*, 2007). However, it is uncertain whether HDAC6 is the only substrate of Aurora-A involved in ciliary resorption, while studies of *Chlamydomonas* with nonacetylatable mutants of α-tubulin do not have notable defects in ciliary dynamics (Kozminski *et al.*, 1993); this topic requires more investigation.

The activation of Aurora-A for ciliary disassembly following serum stimulation is mediated in part by interaction with the adaptor protein HEF1/NEDD9 (Singh *et al.*, 2007), which was previously shown to directly bind and activate Aurora-A during the G2–M transition (Pugacheva and Golemis, 2005). Serum stimulates HEF1 expression and hyper-phosphorylation preceding each wave of Aurora-A activation, while siRNA depletion of HEF1 reduces ciliary resorption (Pugacheva *et al.*, 2007). Upstream of HEF1, constituents of serum previously associated with HEF1 activation that might be relevant to ciliary disassembly include ligands of integrin receptors (Law *et al.*, 1996), GPCR (Zhang *et al.*, 1999), Ca^{2+} (Zhang *et al.*, 2002), and growth factors such as TGFβ (Liu *et al.*, 2000). We have assessed a number of these and other components individually for their ability to induce ciliary disassembly in starved ciliated cells [Pugacheva *et al.* (2007), and unpublished results]. In tests of PDGF, TGFβ, EGF, estrogen, hydrocortisone, progesterone, and ionic calcium, zinc, and magnesium, only PDGF elicited a partial response. The role of PDGFαα at cilia has been discussed in detail (Christensen *et al.*, 2008).

E. GSK3 and VHL

In contrast to the preceding examples, in which the identified proteins have clear roles in ciliary resorption in the cell cycle, GSK3 and von Hippel-Lindau (VHL) have been described as associated with the process of ciliary maintenance (Fig. 2), and their cell cycle function is less clear (Thoma *et al.*, 2007). VHL disease is marked by renal cysts and high incidence of renal cell carcinomas: the VHL gene mutated in this syndrome encodes a protein that regulates oxygen sensing, but also microtubule dynamics. Although the VHL protein localizes to the ciliary axoneme, loss of VHL as a single event does not affect formation of cilium in serum-starved cells, but does increase the rate of ciliary loss in cells stimulated with serum (Thoma *et al.*, 2007). VHL is a substrate of the kinase GSK3β, which regulates its functions at microtubules (Hergovich *et al.*, 2006). Moreover, while *Chlamydomonas* has no ortholog for VHL, the GSK3β ortholog in this organism regulates flagellar assembly and length control (Zhou and Hung, 2005). Thoma and colleagues determined that simultaneous loss of GSK3β- and VHL-intensified serum-induced ciliary loss, and in some contexts (GSK3β-deficient MEFs) reduced the overall formation of ciliation following serum starvation. These activities were specifically linked to the microtubule-stabilizing rather than oxygen-sensing activities of VHL. Finally, renal cells from the cysts of patients with VHL disease tended to become deciliated more readily when treated with agents that inhibited GSK3β activity.

It is interesting to note that Aurora-A and GSK3β have been shown to directly interact in a number of studies (Dar *et al.*, 2009; Fumoto *et al.*, 2008), with phosphorylation by Aurora-A altering some functional activities of GSK3β (e.g., toward β-catenin). At present, potential interactions between these proteins, or between NEK kinases, GSK3β, and VHL have not yet been examined in regard to cilium or cell cycle: such studies would be of considerable interest, particularly given the observation that inactivation of VHL contributes to ciliary loss. In future experiments, it may also be interesting to consider that GSK3β is regulated by Lkb1, a tumor suppressor that influences the cell cycle and cell polarity (Green, 2004), and regulates expression of HEF1/NEDD9 (Ji *et al.*, 2007), the Aurora-A activator (Pugacheva *et al.*, 2007).

F. CEP97, CPP110, CEP290, CEP164

In the past 2 years, a series of studies have identified an interacting cluster of proteins that specifically suppress ciliary protrusion (Fig. 3) (Graser *et al.*, 2007;

Spektor *et al.*, 2007; Tsang *et al.*, 2008). CP110, a centrosomal protein required for centrosomal duplication and cytokinesis, was identified as binding the additional centrosomal proteins Cep97 (Spektor *et al.*, 2007), and subsequently CEP290 (Tsang *et al.*, 2008): mutations in CEP290 have been implicated in a set of "ciliary diseases," including Bardet–Biedl syndrome (BBS), nephronophthisis, and others. Importantly, depletion of Cep97 or CP110 caused protrusion of cilia in quiescent cells, while overexpression of CP110 repressed ciliary protrusion in quiescent cells without affecting their cell cycle (Spektor *et al.*, 2007). Mutants of CP110 that were unable to bind CEP290 were unable to repress ciliary protrusion (Tsang *et al.*, 2008). CEP290 binds to the Rab8a GTPase that controls vesicular trafficking into the centrosome; Tsang and coworkers showed that Rab8a binding to CEP290 was necessary for ciliogenesis. Taken together, these results supported a model in which ciliogenesis is actively inhibited in cycling cells, based on the action of CP110 + CEP97 and CP110 + CEP290 + Rab8a complexes acting at the centrosome. Serum starvation led to the removal of CP110 from these complexes, allowing a CEP290 + Rab8a complex to contribute to ciliary protrusion (Tsang *et al.*, 2008). Importantly, these results indicated that it is possible to at least partially decouple ciliogenesis from the cell cycle in vertebrate systems, although some abnormalities associated with the cilia-like structures emerging from CP110-depleted cells have been noted (Keller and Marshall, 2008).

An independent study based on the siRNA depletion of proteins associated with the basal body (Graser *et al.*, 2007) also identified CEP290 as required for ciliogenesis. This thorough study showed that a considerable number of basal body and centrosomal functions influenced ciliogenesis, with depletion of CEP290, pericentrin, and CEP164 yielding particularly strong phenotypes. Interestingly, CEP164 localizes specifically to mature centrioles, and associates with distal appendages (Graser *et al.*, 2007). In the normal centrosome/ciliary cycle, as cells emerge from quiescence and resorb cilia, the CDK2–cyclinE complex progressively associates with the centrosome and triggers the G1–S transition. Centrosome duplication and DNA synthesis are tightly connected through cyclin E/Cdk2 activation (Hinchcliffe and Sluder, 2001; Schnackenberg *et al.*, 2008). Both centrioles disengage at this point as a new "daughter" is formed from the "mother," or older, centriole. Cilia invariably protrude from mothers rather than daughters; significantly, in normally cycling cells, no centriole is capable of forming a cilium until it has gone through one phase of mitosis (Rodrigues-Martins *et al.*, 2007b; 2008). Of relevance to CEP164, between S phase and mitosis, the new mother centriole undergoes a maturation process during which it gains a set of mother centriole-specific proteins and appendages (Pearson *et al.*, 2009; Pearson and Winey, 2009). The appendages, which include subdistal proteins such as ninein and CEP170, and distal proteins including ODF2, are necessary for anchoring the cilium to the apical side of the cell, and nucleation of the axoneme (Bornens, 2002; Bornens and Piel, 2002; Palazzo *et al.*, 2000; Pedersen and Rosenbaum, 2008; Piehl *et al.*, 2004). The identification of CEP164 as a novel distal appendage protein suggests that it may serve as a mark allowing protrusion of the ciliary axonome in postmitotic cells, connecting the basal body and plasma membrane.

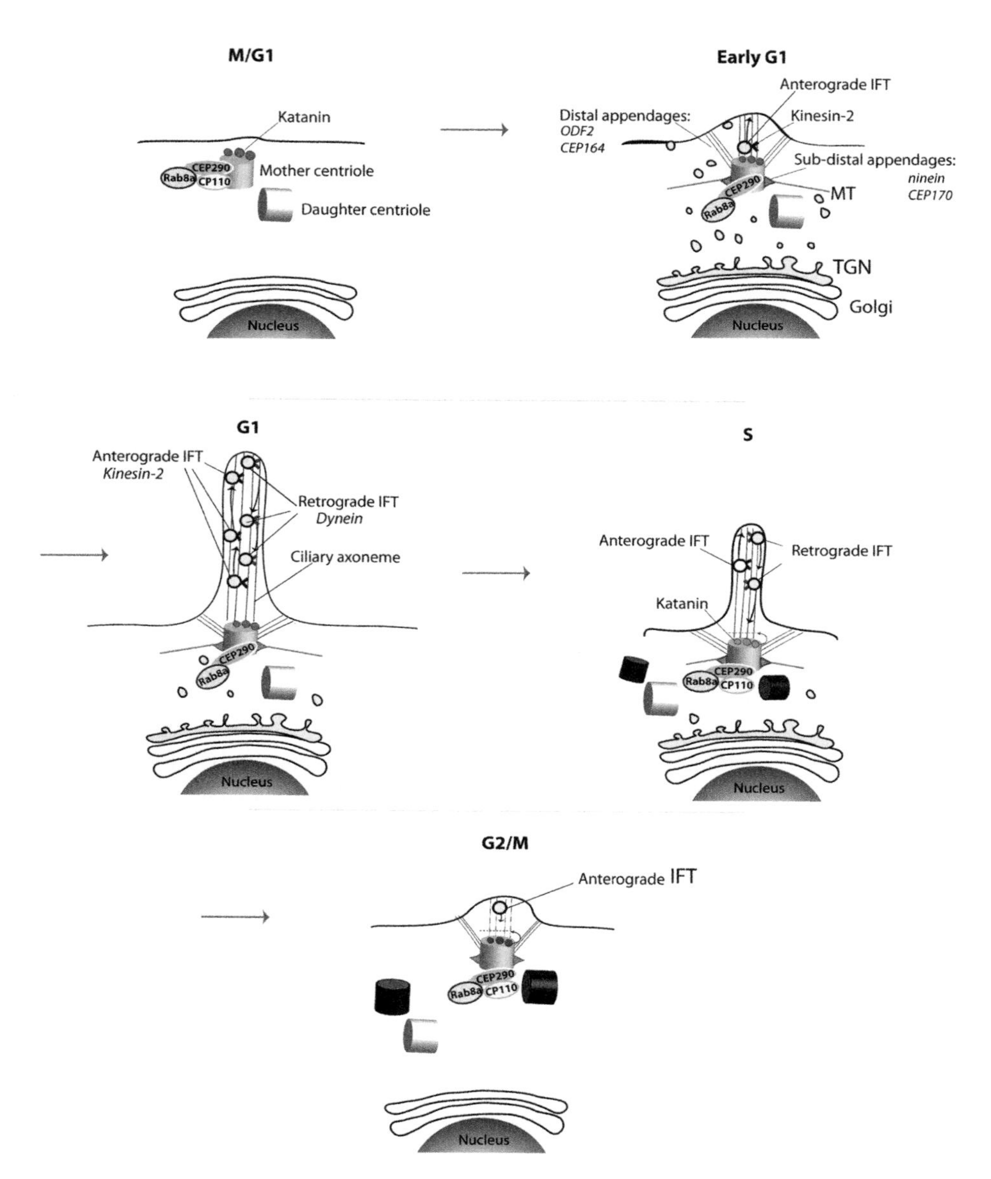

Fig. 3 (*Continued*)

G. New Candidates: Katanin, Kinesin, and Screens

In addition to the regulatory pathways discussed above, a number of other proteins have been implicated in the gain and loss of cilia in *Clamydomonas* and/or in metazoans. Some interesting recent findings are summarized here.

As noted above, resorption likely involves activity at the base of cilia, in addition to the well-established disassembly that occurs at the ciliary tip (Marshall and Rosenbaum, 2001; Parker and Quarmby, 2003), while deflagellation is caused by precise severing of the axoneme at a specific site between the axoneme and the flagellar transition zone, known as the SOFA (Mahjoub *et al.*, 2004). The microtubule-severing ATPase katanin induces breakage of the outer doublet microtubules during deflagellation in *Chlamydomonas* (Lohret *et al.*, 1998). In vertebrates, katanin is best known for its role in cell cycle progression (Buster *et al.*, 2002; Toyo-Oka *et al.*, 2005; Zhang *et al.*, 2007). Now, using an RNAi approach to reduce levels of katanin expression in *Chlamydomonas,* Rasi *et al.* (2009) have recently demonstrated that there is a second site of axonemal-severing proximal to the transition zone, and that severing occurs at this site before complete premitotic resorption of flagella (Fig. 3). Cells deficient in katanin fail to release resorbing flagella from the basal bodies. These findings suggest that there are at least two parallel programs active during resorption: (1) reducing the axoneme from the tip by modulation of IFT, resulting in a net transport of disassembled ciliary components from the tip back to the cell body; and (2) severing of the basal body from ciliary remnants before functional reassignment of the basal bodies to the spindle poles/centrosome. It will be interesting to establish if katanin-dependent loss of cilia occurs in mammalian cells.

Kinesins are important regulators of microtubule dynamics in mitosis (Fig. 3), and one member of the family, kinesin-2, is well established for a role of anterograde IFT in the flagella of *Chlamydomonas*. In a recent study by Piao *et al.* (2009), a new

Fig. 3 The centrosome-basal body transition in cell cycle. Following mitotic exit, primary cilium formation occurs as cells move into G0/G1 following centrosomal docking to the membrane. Distal appendages that include ODF2 and CEP164 help to anchor of mother centrosome (basal body) to the cell membrane. Subdistal appendages serve as a major site for microtubule (MT) anchoring, an activity requiring component proteins ninein and CEP170. CEP290 cooperates with the vesicular transport regulatory GTPase Rab8a to promote cilium formation: in phases of the cell cycle at which active cilium assembly is not occurring, CEP290, CP110, and Rab8a form an inactive centrosomal complex. Ciliary assembly is accomplished by action of the intraflagellar transport (IFT) proteins, which relies on the microtubule motor proteins kinesin 2 for anterograde transport of vesicles from the trans-Golgi network (TGN), and dynein for retrograde transport from the ciliary tip. Anterograde transport builds the ciliary axoneme, which extends triplet microtubules directly from the mother centriole. Ciliary resorption (which invariably occurs at G2/M, and in some cases occurs in cells in G1/S) involves decreased IFT-dependent transport of cargo into the cilia, activation of retrograde transport of IFT complexes, and katanin-dependent separation of the basal body via severing of axonemal microtubules. With the maturation and separation of the two centrosomes at the G2–M transition, the cilium is disassembled and the centrioles function in mitotic spindle formation.

Chlamydomonas kinesin, Crkinesin-13, was identified as phosphorylated upon the initiation of flagellar protrusion, and dephosphorylated when flagella reached maximal length. Depletion of Crkinesin-13 delayed flagellar shortening after physiological stimuli such as recovery from pH shock. Crkinesin-13 physically is moved into the flagella by an IFT process during flagellar shortening, while depletion of Crkinesin-13 also inhibited the shortening process. Taken together, these data indicate a central role for this protein in regulating microtubule dynamics during all phases of flagellar remodeling. No studies of Crkinesin orthologs have been undertaken relative to their role in the regulation of metazoan cilia.

Finally, data from a number of high throughput profiling and screening studies have begun to identify a diverse set of proteins that influence ciliary resorption, maintenance, and protrusion. Examples of these approaches include a differential gel electrophoretic (DIGE) analysis of the flagellar tip complex of *Chlamydomonas* with long versus short flagella, which identified a novel protein methylation pathway specifically activated during flagellar resorption (Schneider *et al.*, 2008). Proteins in the MetE complex are associated with the axoneme, and appear to be cell cycle regulated; a role for methylation in regulation of metazoan cilia remains to be investigated. Stolc and coworkers performed a genome-wide transcriptional analysis of *Chlamydomonas* cells undergoing flagellar regeneration (Stolc *et al.*, 2005), following an earlier more limited profiling screen (Li *et al.*, 2004). This screen identified 220 strongly induced genes, additional repressed genes, and a number of orthologs of genes previously identified in zebrafish studies as associated with polycystic kidney disease, almost none of which have been studied in detail.

An siRNA-based kinome screen to identify modulators of Hedgehog (Hh) signaling (Evangelista *et al.*, 2008) predictably identified a number of known and new regulators of ciliogenesis, given the known dependence of the Hh signaling cascade on cilia-localized receptors. Implicated kinases included the cell cycle regulator Nek1, but also Prkra. The latter kinase is best known as a regulator of Pkr, which regulates interferon response, and also controls microRNA processing. As Evangelista and colleagues note, some microRNAs have been localized to the basal body of cilia (Deo *et al.*, 2006)—what functional role they might play in resorption or ciliogenesis remains obscure. Finally, repeated analyses of the *Chlamydomonas* centriolar proteome (Keller *et al.*, 2005, 2009) have led to the identification of many proteins orthologous with ciliary disease genes. The rich results of these various studies in *Chlamydomonas* suggest that the next several years will be extremely rich in disease-relevant discoveries, as findings from *Chlamydomonas* are used to inform studies in humans.

III. Methods

A number of immortalized cell lines can readily be made to form cilia. Features that make cell lines particularly useful for studying ciliary function and structure include a homogeneous growth profile, a high frequency of ciliation under induced conditions, and availability of reagents (e.g., antibodies) suitable for use in the species of derivation. Cell lines we have evaluated and which are suitable for studies include MDCK (canine polarized epithelial kidney cells) (ATCC, CCL-34) (Praetorius and Spring, 2003), IMCD3 (mouse kidney collecting duct cells) (ATCC, CRL-212) (Mai *et al.*, 2005), Caki-1 (human renal cells) (ATCC, HTB-46) (Glube and Langguth, 2008), LLCPK1 (porcine kidney cells) (ATCC, CL-101) (Bendayan *et al.*, 1994), hTERT-RPE1 (human retinal pigmented epithelial cells) (ATCC, CRL-4000) (Rambhatla *et al.*, 2002), MEFs (primary mouse embryo fibroblasts) (Wheatley, 1972), and HK-2 (normal human proximal tubule cells) (ATCC, CRL-2190) (van Rooijen *et al.*, 2008). Typically, only 10–30% of the overall population of these cells is ciliated while they are exponentially growing in medium with 10% fetal bovine serum, while >90% of the population can become ciliated under starvation conditions.

In 1979 Tucker showed that 80–90% of 3T3 fibroblasts form cilia after growing for 48 h in low serum medium, at a confluence of 80–100% (Tucker *et al.*, 1979a). For the cell lines discussed here, the time of incubation in serum-free medium required for the induction of cilia varies, and should be adjusted for each cell type individually. For example, for hTERT-RPE1 cells, full ciliation is typically achieved in 48 h, while for renal cell carcinoma cell lines containing reexpressed VHL, primary cilium formation required maintenance of confluent cultures for 7 days in serum-free medium (Lutz and Burk, 2006). Some cell lines, such as LLCPK1 cells, need a prolonged starvation period of up to 15–20 days to form cilia.

For most of the cell lines we have examined, there is a separate contribution of high cell density to the ciliation process: many cell lines require maintenance at >70% confluence for efficient ciliation. However, particularly when studying disassembly of cilia, it is useful to spend some time investigating minimum and maximum densities contributing to efficient deciliation, as plating at too high a density may in some cases limit reentry into the cell cycle and ultimate degree of ciliary resorption.

Working with fibroblasts, Tucker described three phases following stimulation with serum: "first, an initial but transient deciliation within 1–2 h; second, a return of the cilium by 6–8 h; and third, a subsequent final deciliation of the centriole coincident with the initiation of DNA synthesis at 12–24 h." In our work with other cell lines, we have more typically observed two phases: an initial loss at 1–2 h, and a more complete loss at 18–24 h, which in a number of cases we have determined coincides with mitotic entry rather than initiation of DNA synthesis. We strongly suggest parallel measurement of cell cycle status when using live cell imaging or fluorescence microscopy of fixed cells to measure ciliary resorption in cell lines that have not previously been assessed. The protocol presented below was initially optimized in the hTERT-RPE1 cell line, but also works well for IMCD3, Caki-1, MDCK, HK-2 cell lines and MEFs.

Compared to the well-studied list of chemical and physical stimuli that induce flagella resorption in *Chlamydomonas*, relatively little is known about the specific chemical factors that may play a role in induction of disassembly of primary cilium in mammals. As alternative to full serum, PDGF causes partial ciliary disassembly for at least some cell types (Christensen *et al.*, 2008; Pugacheva *et al.*, 2007). Some cell lines, such as human umbilical vein endothelial cells (HUVECs) disassemble under such mechanical stimulation as laminar shear stress (Iomini *et al.*, 2004). Flow chambers suitable for generating defined amounts of flow forces can be purchased from Bioptechs, Butler, PA (the FC2 system).

A. Cilium Disassembly Protocol

1. Plate cells on cover slips in medium containing full serum (typically 10% fetal bovine serum), and grow for 24–48 h, or until they have achieved 50–70% confluency. For cell lines that adhere poorly or unevenly, coating slides with collagen, fibronectin, or poly-L-lysine before plating can assist in the subsequent visualization of cilia using immunofluorescence.
2. To induce formation of cilia, replace the medium to serum free (Opti-MEM medium, Invitrogen) and culture for 48 h. With hTERT-RPE1 cells, >85% of cells normally have clearly visible cilia, and little further increase is seen with additional days of culture. For a new cell line, it would be prudent to initially compare cultures maintained for 48 and 96 h in Opti-MEM. If a low degree of ciliation is observed (<50%), or more cilia are seen at 96 h than at 48 h, it would be useful to also explore longer incubation periods. Fresh Opti-MEM should be added to cells every 2 days for longer culture periods.
3. To initiate ciliary disassembly, replace Opti-MEM with medium containing 10% fetal bovine serum. Ciliary resorption should commence within 1–2 h, and extend over the following 24 h (see Fig. 4). Usually addition of serum leads to resorption of the primary cilium in hTERT-RPE1, Caki, MDCK, or IMCD3 cells in two waves: at 2 h (G0/G1 phase) and at 18–24 h (most cells entering mitosis). Standard techniques for FACS analysis, BrDU staining, and direct observation of condensed DNA and mitotic figures (following staining with DAPI and antibodies to α-tubulin) are useful approaches to establish the cell cycle phase during the resorption process.
4. Cilia can be visualized by immunofluorescence microscopy, using primary antibodies to mark the axoneme and adjacent basal body. For example, acetylation of tubulin results in stabilization of tubulin polymers during mitosis in cilia (Kannarkat *et al.*, 2006; Matsuyama *et al.*, 2002), and in quiescent interphase cells acetylated tubulin is strongly enriched in microtubules within cilia. Tubulin within the ciliary axoneme also accumulates additional posttranslational modifications, such as glutamylation (Wheatley *et al.*, 1994; Wloga *et al.*, 2008). Hence, antiacetylated α-tubulin (clone 6–11B-1, Sigma, or clone K(Ac)40, Biomol) or alternatively, antiglutamylated α-tubulin (T9822, Sigma) (Million *et al.*, 1999) is useful to mark the axoneme. Antibodies against the ciliary proteins katanin, IFT polypeptides, and tektin can also be used to visualize the axoneme. Antibodies to γ-tubulin, centrin, and other proteins associated with the centriole and/or pericentriolar mass (PCM) are useful to visualize the basal body and (in cells with disassembled cilia) the centrosome.
5. Antibodies to T^{288}-phosphorylated Aurora-A (Cell Signaling Technology, Beverly, MA) allow measurement of Aurora-A activation during the ciliary resorption process (Pugacheva *et al.*, 2007). These antibodies can be used as a control for

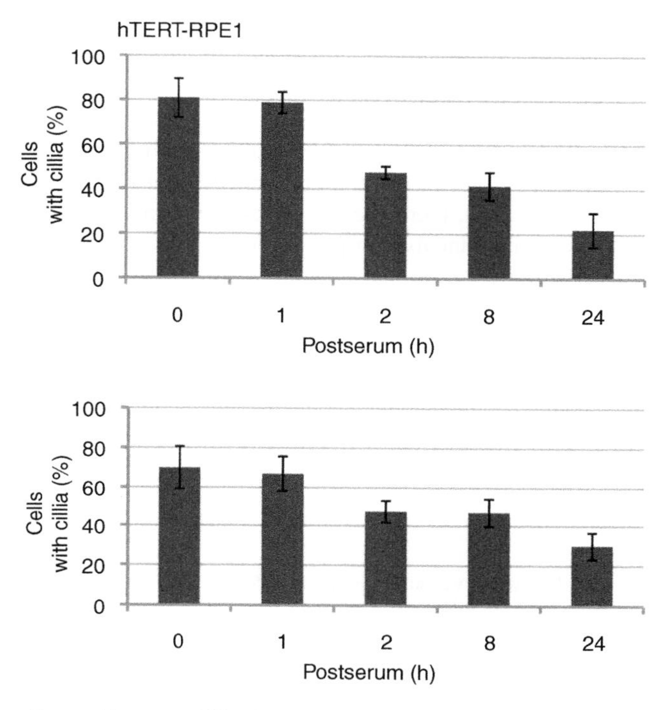

Fig. 4 Ciliary disassembly curve. Ciliated cells were induced by serum stimulation after 48 h of serum starvation in hTERT-RPE1 and HK2 cell lines. An average of 150 cells were counted in each of three experiments; error bars show the standard deviation.

use with immunofluorescence to mark the timing of peak ciliary disassembly. The antibodies currently available work well in human cell lines, but are less effective in the mouse cell lines we have tested. Unfortunately, there are currently no commercially available antibodies to endogenous phosphorylated Aurora-A that work well on Western blots.

B. Blocking Ciliary Disassembly

Small molecule inhibitors targeting specific molecules such as U0126 (an MEK1/2 inhibitor, EMD Chemicals Inc., Cat. No. 662005), roscovitine (a Cdk1/2 inhibitor, EMD Chemicals Inc Cat. No. 557360), TSA [a broad spectrum histone deacetylase (HDAC) inhibitor, Sigma, T1952], PHA-680632 [an Aurora-A inhibitor, Soncini *et al.* (2006)] and tubacin [an HDAC6 inhibitor, Haggarty *et al.* (2003)] have been used for studies of the contribution of growth factors or specific cell cycle regulators to ciliary resorption. It needs to be taken into account that some of these inhibitors will block cell cycle progression at certain stages. Depending upon the length of preincubation with the drug, and the stage at which characteristic blockage occurs, it is important to minimize these potentially complicating effects for the interpretation of studies of resorption (i.e., in establishing whether resorption effects are secondary to, or independent of, effects on the cell cycle). Controls for cell cycle stage are particularly important when using inhibitors.

For studies in hTERT-RPE1 cells, we first established IC50 curves for inhibitors of Aurora-A and HDAC6 inhibitors (0.3 μM for PHA-680632 and 2.5 μM for tubacin, respectively), then selected doses at approximately IC75 values, followed by a relatively short incubation period. To achieve effective inhibition and to minimize both drug toxicity and secondary effects, drugs are added in Opti-MEM typically 1–3 h before addition of serum to starved cells ensure inactivation of their target of interest prior to initiation of ciliary disassembly. Depending upon the stability of a particular inhibitor, it is sometimes necessary to repeat treatment during the course of study, since a full cell cycle in some cells might take 30–40 h. Effective inhibition should be confirmed using an assay for a well-defined substrate for the targeted protein, at the beginning and end of the time of assay.

Acknowledgments

This work was supported by NIH R01 CA-63366 and R01 CA-113342, DOD W81XWH-07-1-0676 from the Army Materiel Command, and Tobacco Settlement funding from the State of Pennsylvania (to EAG); by NCI core grant CA-06927 and support from the Pew Charitable Fund to Fox Chase Cancer Center. Additional funds were provided by Fox Chase Cancer Center via institutional support of the Kidney Cancer Keystone Program. ENP was supported by a Strategic Research Foundation Grant from West Virginia University, and by an MBRCC Pilot Grant.

References

Archer, F.L., and Wheatley, D.N. (1971). Cilia in cell-cultured fibroblasts. II. Incidence in mitotic and post-mitotic BHK 21-C13 fibroblasts. *J. Anat.* **109**, 277–292.

Bendayan, R., Lo, B., and Silverman, M. (1994). Characterization of cimetidine transport in LLCPK1 cells. *J. Am. Soc. Nephrol.* **5**, 75–84.

Bornens, M. (2002). Centrosome composition and microtubule anchoring mechanisms. *Curr. Opin. Cell Biol.* **14**, 25–34.

Bornens, M., and Piel, M. (2002). Centrosome inheritance: Birthright or the privilege of maturity? *Curr. Biol.* **12**, R71–R73.

Bradley, B.A., and Quarmby, L.M. (2005). A NIMA-related kinase, Cnk2p, regulates both flagellar length and cell size in Chlamydomonas. *J. Cell Sci.* **118**, 3317–3326.

Buster, D., McNally, K., and McNally, F.J. (2002). Katanin inhibition prevents the redistribution of gamma-tubulin at mitosis. *J. Cell Sci.* **115**, 1083–1092.

Butterworth, J.F., and Strichartz, G.R. (1990). Molecular mechanisms of local anesthesia: a review. *Anesthesiology.* **72**, 711–734.

Cavalier-Smith, T. (1974). Basal body and flagellar development during the vegetative cell cycle and the sexual cycle of *Chlamydomonas reinhardtii*. *J. Cell. Sci.* **16**, 529–556.

Christensen, S.T., Pedersen, S.F., Satir, P., Veland, I.R., and Schneider, L. (2008). The primary cilium coordinates signaling pathways in cell cycle control and migration during development and tissue repair. *Curr. Top. Dev. Biol.* **85**, 261–301.

Dar, A.A., Belkhiri, A., and El-Rifai, W. (2009). The aurora kinase A regulates GSK-3beta in gastric cancer cells. *Oncogene* **28**, 866–875.

Deane, J.A., and Ricardo, S.D. (2007). Polycystic kidney disease and the renal cilium. *Nephrology (Carlton)*. **12**, 559–564.

Deo, M., Yu, J.Y., Chung, K.H., Tippens, M., and Turner, D.L. (2006). Detection of mammalian microRNA expression by in situ hybridization with RNA oligonucleotides. *Dev. Dyn.* **235**, 2538–2548.

Dirksen, E.R. (1991). Centriole and basal body formation during ciliogenesis revisited. *Biol. Cell* **72**, 31–38.

Doxsey, S., Zimmerman, W., and Mikule, K. (2005). Centrosome control of the cell cycle. *Trends Cell Biol.* **15**, 303–311.

Duensing, A., Liu, Y., Perdreau, S.A., Kleylein-Sohn, J., Nigg, E.A., and Duensing, S. (2007). Centriole overduplication through the concurrent formation of multiple daughter centrioles at single maternal templates. *Oncogene* **26**, 6280–6288.

Dutcher, S.K. (1995). Purification of basal bodies and basal body complexes from *Chlamydomonas reinhardtii*. *Methods Cell Biol.* **47**, 323–334.

Evangelista, M., Lim, T.Y., Lee, J., Parker, L., Ashique, A., Peterson, A.S., Ye, W., Davis, D.P., and de Sauvage, F.J. (2008). Kinome siRNA screen identifies regulators of ciliogenesis and hedgehog signal transduction. *Sci. Signal* **1**, ra7.

Fry, A.M., Descombes, P., Twomey, C., Bacchieri, R., and Nigg, E.A. (2000). The NIMA-related kinase X-Nek2B is required for efficient assembly of the zygotic centrosome in *Xenopus laevis*. *J. Cell Sci.* **113**(Pt. 11), 1973–1984.

Fumoto, K., Lee, P.C., Saya, H., and Kikuchi, A. (2008). AIP regulates stability of Aurora-A at early mitotic phase coordinately with GSK-3beta. *Oncogene* **27**, 4478–4487.

Gilula, N.B., and Satir, P. (1972). The ciliary necklace. A ciliary membrane specialization. *J. Cell Biol.* **53**, 494–509.

Glube, N., and Langguth, P. (2008). Caki-1 cells as a model system for the interaction of renally secreted drugs with OCT3. *Nephron Physiol.* **108**, 18–28.

Graser, S., Stierhof, Y.D., Lavoie, S.B., Gassner, O.S., Lamla, S., Le Clech, M., and Nigg, E.A. (2007). Cep164, a novel centriole appendage protein required for primary cilium formation. *J. Cell Biol.* **179**, 321–330.

Grayson, C., Bartolini, F., Chapple, J.P., Willison, K.R., Bhamidipati, A., Lewis, S.A., Luthert, P.J., Hardcastle, A.J., Cowan, N.J., and Cheetham, M.E. (2002). Localization in the human retina of the X-linked retinitis pigmentosa protein RP2, its homologue cofactor C and the RP2 interacting protein Arl3. *Hum. Mol. Genet.* **11**, 3065–3074.

Green, J.B. (2004). Lkb1 and GSK3-beta: Kinases at the center and poles of the action. *Cell Cycle*. **3**, 12–14.

Habedanck, R., Stierhof, Y.D., Wilkinson, C.J., and Nigg, E.A. (2005). The Polo kinase Plk4 functions in centriole duplication. *Nat. Cell Biol.* **7**, 1140–1146.

Haggarty, S.J., Koeller, K.M., Wong, J.C., Grozinger, C.M., and Schreiber, S.L. (2003). Domain-selective small-molecule inhibitor of histone deacetylase 6 (HDAC6)-mediated tubulin deacetylation. *Proc. Natl. Acad. Sci. USA* **100**, 4389–4394.

Hagiwara, H., Ohwada, N., and Takata, K. (2004). Cell biology of normal and abnormal ciliogenesis in the ciliated epithelium. *Int. Rev. Cytol.* **234**, 101–141.

Haimo, L.T., and Rosenbaum, J.L. (1981). Cilia, flagella, and microtubules. *J. Cell Biol.* **91**, 125s–130s.

Hergovich, A., Lisztwan, J., Thoma, C.R., Wirbelauer, C., Barry, R.E., and Krek, W. (2006). Priming-dependent phosphorylation and regulation of the tumor suppressor pVHL by glycogen synthase kinase 3. *Mol. Cell Biol.* **26**, 5784–5796.

Hinchcliffe, E.H., and Sluder, G. (2001). "It takes two to tango": Understanding how centrosome duplication is regulated throughout the cell cycle. *Genes Dev.* **15**, 1167–1181.

Ho, P.T., and Tucker, R.W. (1989). Centriole ciliation and cell cycle variability during G1 phase of BALB/c 3T3 cells. *J. Cell Physiol.* **139**, 398–406.

Iomini, C., Tejada, K., Mo, W., Vaananen, H., and Piperno, G. (2004). Primary cilia of human endothelial cells disassemble under laminar shear stress. *J. Cell Biol.* **164**, 811–817.

Jensen, C.G., Davison, E.A., Bowser, S.S., and Rieder, C.L. (1987). Primary cilia cycle in PtK1 cells: effects of colcemid and taxol on cilia formation and resorption. *Cell Motil. Cytoskeleton* **7**, 187–197.

Ji, H., Ramsey, M.R., Hayes, D.N., Fan, C., McNamara, K., Kozlowski, P., Torrice, C., Wu, M.C., Shimamura, T., Perera, S.A., Liang, M.C., Cai, D., *et al.*, (2007). LKB1 modulates lung cancer differentiation and metastasis. *Nature* **448**, 807–810.

Kannarkat, G.T., Tuma, D.J., and Tuma, P.L. (2006). Microtubules are more stable and more highly acetylated in ethanol-treated hepatic cells. *J. Hepatol.* **44**, 963–970.

Keller, L.C., Geimer, S., Romijn, E., Yates, J., 3rd, Zamora, I., and Marshall, W.F. (2009). Molecular architecture of the centriole proteome: The conserved WD40 domain protein POC1 is required for centriole duplication and length control. *Mol. Biol. Cell* **20**, 1150–1166.

Keller, L.C., and Marshall, W.F. (2008). Isolation and proteomic analysis of *Chlamydomonas* centrioles. *Methods Mol. Biol.* **432**, 289–300.

Keller, L.C., Romijn, E.P., Zamora, I., Yates, J.R., 3rd, and Marshall, W.F. (2005). Proteomic analysis of isolated *Chlamydomonas* centrioles reveals orthologs of ciliary-disease genes. *Curr. Biol.* **15**, 1090–1098.

Kleylein-Sohn, J., Westendorf, J., Le Clech, M., Habedanck, R., Stierhof, Y.D., and Nigg, E.A. (2007). Plk4-induced centriole biogenesis in human cells. *Dev. Cell* **13**, 190–202.

Kozminski, K.G., Diener, D.R., and Rosenbaum, J.L. (1993). High level expression of nonacetylatable alpha-tubulin in *Chlamydomonas reinhardtii*. *Cell Motil. Cytoskeleton* **25**, 158–170.

Law, S.F., Estojak, J., Wang, B., Mysliwiec, T., Kruh, G., and Golemis, E.A. (1996). Human enhancer of filamentation 1, a novel p130cas-like docking protein, associates with focal adhesion kinase and induces pseudohyphal growth in *Saccharomyces cerevisiae*. *Mol. Cell Biol.* **16**, 3327–3337.

Lewin, R.A. (1952a). The primary zygote membrane in *Chlamydomonas moewusii*. *J. Gen. Microbiol.* **6**, 249–250.

Lewin, R.A. (1952b). Ultraviolet induced mutations in *Chlamydomonas moewusii* Gerloff. *J. Gen. Microbiol.* **6**, 233–248.

Lewin, R.A., Lee, T.H., and Fang, L.S. (1982). Effects of various agents on flagellar activity, flagellar autotomy and cell viability in four species of *Chlamydomonas* (chlorophyta: volvocales). *Symp. Soc. Exp. Biol.* **35**, 421–437.

Li, J.B., Gerdes, J.M., Haycraft, C.J., Fan, Y., Teslovich, T.M., May-Simera, H., Li, H., Blacque, O.E., Li, L., Leitch, C.C., Lewis, R.A., Green, J.S., *et al.*, (2004). Comparative genomics identifies a flagellar and basal body proteome that includes the BBS5 human disease gene. *Cell* **117**, 541–552.

Liu, S., Lu, W., Obara, T., Kuida, S., Lehoczky, J., Dewar, K., Drummond, I.A., and Beier, D.R. (2002). A defect in a novel Nek-family kinase causes cystic kidney disease in the mouse and in zebrafish. *Development* **129**, 5839–5846.

Liu, X., Elia, A.E., Law, S.F., Golemis, E.A., Farley, J., and Wang, T. (2000). A novel ability of Smad3 to regulate proteasomal degradation of a Cas family member HEF1. *EMBO J.* **19**, 6759–6769.

Lohret, T.A., McNally, F.J., and Quarmby, L.M. (1998). A role for katanin-mediated axonemal severing during *Chlamydomonas* deflagellation. *Mol. Biol. Cell* **9**, 1195–1207.

Lutz, M.S., and Burk, R.D. (2006). Primary cilium formation requires von Hippel-Lindau gene function in renal-derived cells. *Cancer Res.* **66**, 6903–6907.

Mahjoub, M.R., Qasim Rasi, M., and Quarmby, L.M. (2004). A NIMA-related kinase, Fa2p, localizes to a novel site in the proximal cilia of *Chlamydomonas* and mouse kidney cells. *Mol. Biol. Cell* **15**, 5172–5186.

Mahoney, N.M., Goshima, G., Douglass, A.D., and Vale, R.D. (2006). Making microtubules and mitotic spindles in cells without functional centrosomes. *Curr. Biol.* **16**, 564–569.

Mai, W., Chen, D., Ding, T., Kim, I., Park, S., Cho, S.Y., Chu, J.S., Liang, D., Wang, N., Wu, D., Li, S., Zhao, P., *et al.*, (2005). Inhibition of Pkhd1 impairs tubulomorphogenesis of cultured IMCD cells. *Mol. Biol. Cell* **16**, 4398–4409.

Marshall, W.F. (2008). Basal bodies platforms for building cilia. *Curr. Top. Dev. Biol.* **85**, 1–22.

Marshall, W.F., and Rosenbaum, J.L. (2001). Intraflagellar transport balances continuous turnover of outer doublet microtubules: Implications for flagellar length control. *J. Cell Biol.* **155**, 405–414.

Matsuyama, A., Shimazu, T., Sumida, Y., Saito, A., Yoshimatsu, Y., Seigneurin-Berny, D., Osada, H., Komatsu, Y., Nishino, N., Khochbin, S., Horinouchi, S., and Yoshida, M. (2002). In vivo destabilization of dynamic microtubules by HDAC6-mediated deacetylation. *EMBO J.* **21**, 6820–6831.

Mikule, K., Delaval, B., Kaldis, P., Jurcyzk, A., Hergert, P., and Doxsey, S. (2007). Loss of centrosome integrity induces p38-p53-p21-dependent G1-S arrest. *Nat. Cell Biol.* **9**, 160–170.

Million, K., Larcher, J., Laoukili, J., Bourguignon, D., Marano, F., and Tournier, F. (1999). Polyglutamylation and polyglycylation of alpha- and beta-tubulins during in vitro ciliated cell differentiation of human respiratory epithelial cells. *J. Cell. Sci.* **112**(Pt. 23), 4357–4366.

Mitchell, B.F., Grulich, L.E., and Mader, M.M. (2004). Flagellar quiescence in *Chlamydomonas*: Characterization and defective quiescence in cells carrying sup-pf-1 and sup-pf-2 outer dynein arm mutations. *Cell Motil. Cytoskeleton* **57**, 186–196.

Mitchell, D.R., and Rosenbaum, J.L. (1985). A motile *Chlamydomonas* flagellar mutant that lacks outer dynein arms. *J. Cell. Biol.* **100**, 1228–1234.

Nigg, E.A. (2006). Cell biology: A licence for duplication. *Nature* **442**, 874–875.

Otto, E.A., Trapp, M.L., Schultheiss, U.T., Helou, J., Quarmby, L.M., and Hildebrandt, F. (2008). NEK8 mutations affect ciliary and centrosomal localization and may cause nephronophthisis. *J. Am. Soc. Nephrol.* **19**, 587–592.

Overgaard, C.E., Sanzone, K.M., Spiczka, K.S., Sheff, D.R., Sandra, A., and Yeaman, C. (2009). Deciliation is associated with dramatic remodeling of epithelial cell junctions and surface domains. *Mol. Biol. Cell* **20**, 102–113.

Palazzo, R.E., Vogel, J.M., Schnackenberg, B.J., Hull, D.R., and Wu, X. (2000). Centrosome maturation. *Curr. Top. Dev. Biol.* **49**, 449–470.

Pan, J., and Snell, W.J. (2003). Kinesin II and regulated intraflagellar transport of *Chlamydomonas aurora* protein kinase. *J. Cell. Sci.* **116**, 2179–2186.

Pan, J., and Snell, W.J. (2005). *Chlamydomonas* shortens its flagella by activating axonemal disassembly, stimulating IFT particle trafficking, and blocking anterograde cargo loading. *Dev. Cell* **9**, 431–438.

Pan, J., Wang, Q., and Snell, W.J. (2004). An aurora kinase is essential for flagellar disassembly in *Chlamydomonas. Dev. Cell.* **6**, 445–451.

Pan, J., Wang, Q., and Snell, W.J. (2005). Cilium-generated signaling and cilia-related disorders. *Lab. Invest.* **85**, 452–463.

Parker, J.D., and Quarmby, L.M. (2003). *Chlamydomonas* fla mutants reveal a link between deflagellation and intraflagellar transport. *BMC. Cell Biol.* **4**, 11.

Pazour, G.J., Dickert, B.L., Vucica, Y., Seeley, E.S., Rosenbaum, J.L., Witman, G.B., and Cole, D.G. (2000). *Chlamydomonas* IFT88 and its mouse homologue, polycystic kidney disease gene tg737, are required for assembly of cilia and flagella. *J. Cell Biol.* **151**, 709–718.

Pearson, C.G., and Winey, M. (2009). Basal body assembly in ciliates: The power of numbers. *Traffic* **10**, 461–471.

Pearson, C.G., Giddings, T.H., Jr., and Winey, M. (2009). Basal body components exhibit differential protein dynamics during nascent basal body assembly. *Mol. Biol. Cell* **20**, 904–914.

Pedersen, L.B., and Rosenbaum, J.L. (2008). Intraflagellar transport (IFT) role in ciliary assembly, resorption and signalling. *Curr. Top. Dev. Biol.* **85**, 23–61.

Piao, T., Luo, M., Wang, L., Guo, Y., Li, D., Li, P., Snell, W.J., and Pan, J. (2009). A microtubule depolymerizing kinesin functions during both flagellar disassembly and flagellar assembly in *Chlamydomonas. Proc. Natl. Acad. Sci. USA* **106**, 4713–4718.

Piehl, M., Tulu, U.S., Wadsworth, P., and Cassimeris, L. (2004). Centrosome maturation: measurement of microtubule nucleation throughout the cell cycle by using GFP-tagged EB1. *Proc. Natl. Acad. Sci. USA* **101**, 1584–1588.

Praetorius, H.A., and Spring, K.R. (2003). Removal of the MDCK cell primary cilium abolishes flow sensing. *J. Membr. Biol.* **191**, 69–76.

Pugacheva, E.N., and Golemis, E.A. (2005). The focal adhesion scaffolding protein HEF1 regulates activation of the Aurora-A and Nek2 kinases at the centrosome. *Nat. Cell Biol.* **7**, 937–946.

Pugacheva, E.N., Jablonski, S.A., Hartman, T.R., Henske, E.P., and Golemis, E.A. (2007). HEF1-dependent Aurora A activation induces disassembly of the primary cilium. *Cell* ***129***, 1351–1363.

Qin, H., Wang, Z., Diener, D., and Rosenbaum, J. (2007). Intraflagellar transport protein 27 is a small G protein involved in cell-cycle control. *Curr. Biol.* **17**, 193–202.

Quarmby, L.M. (2004). Cellular deflagellation. *Int. Rev. Cytol.* **233**, 47–91.
Quarmby, L.M., and Hartzell, H.C. (1994). Dissection of eukaryotic transmembrane signalling using *Chlamydomonas*. *Trends Pharmacol. Sci.* **15**, 343–349.
Quarmby, L.M., and Parker, J.D. (2005). Cilia and the cell cycle? *J. Cell Biol.* **169**, 707–710.
Rambhatla, L., Chiu, C.P., Glickman, R.D., and Rowe-Rendleman, C. (2002). In vitro differentiation capacity of telomerase immortalized human RPE cells. *Invest. Ophthalmol. Vis. Sci.* **43**, 1622–1630.
Rash, J.E., Shay, J.W., and Biesele, J.J. (1969). Cilia in cardiac differentiation. *J. Ultrastruct. Res.* **29**, 470–484.
Rasi, M.Q., Parker, J.D., Feldman, J.L., Marshall, W.F., and Quarmby, L.M. (2009). Katanin knockdown supports a role for microtubule severing in release of basal bodies before mitosis in *Chlamydomonas*. *Mol. Biol. Cell* **20**, 379–388.
Rieder, C.L., Jensen, C.G., and Jensen, L.C. (1979). The resorption of primary cilia during mitosis in a vertebrate (PtK1) cell line. *J. Ultrastruct. Res.* **68**, 173–185.
Robert, A., Margall-Ducos, G., Guidotti, J.E., Bregerie, O., Celati, C., Brechot, C., and Desdouets, C. (2007). The intraflagellar transport component IFT88/polaris is a centrosomal protein regulating G1-S transition in non-ciliated cells. *J. Cell Sci.* **120**, 628–637.
Rodrigues-Martins, A., Bettencourt-Dias, M., Riparbelli, M., Ferreira, C., Ferreira, I., Callaini, G., and Glover, D.M. (2007a). DSAS-6 organizes a tube-like centriole precursor, and its absence suggests modularity in centriole assembly. *Curr. Biol.* **17**, 1465–1472.
Rodrigues-Martins, A., Riparbelli, M., Callaini, G., Glover, D.M., and Bettencourt-Dias, M. (2007b). Revisiting the role of the mother centriole in centriole biogenesis. *Science.* **316**, 1046–1050.
Rodrigues-Martins, A., Riparbelli, M., Callaini, G., Glover, D.M., and Bettencourt-Dias, M. (2008). From centriole biogenesis to cellular function: Centrioles are essential for cell division at critical developmental stages. *Cell Cycle* **7**, 11–16.
Rosenbaum, J. (2002). Intraflagellar transport. *Curr. Biol.* **12**, R125.
Rosenbaum, J.L., and Carlson, K. (1969). Cilia regeneration in *Tetrahymena* and its inhibition by colchicine. *J. Cell Biol.* **40**, 415–425.
Rosenbaum, J.L., and Witman, G.B. (2002). Intraflagellar transport. *Nat. Rev. Mol. Cell Biol.* **3**, 813–825.
Schnackenberg, B.J., Marzluff, W.F., and Sluder, G. (2008). Cyclin E in centrosome duplication and reduplication in sea urchin zygotes. *J. Cell Physiol.* **217**, 626–631.
Schneider, M.J., Ulland, M., and Sloboda, R.D. (2008). A protein methylation pathway in *Chlamydomonas* flagella is active during flagellar resorption. *Mol. Biol. Cell* **19**, 4319–4327.
Seeley, E.S., Carriere, C., Goetze, T., Longnecker, D.S., and Korc, M. (2009). Pancreatic cancer and precursor pancreatic intraepithelial neoplasia lesions are devoid of primary cilia. *Cancer Res.* **69**, 422–430.
Singh, M., Cowell, L., Seo, S., O'Neill, G., and Golemis, E. (2007). Molecular basis for HEF1/NEDD9/Cas-L action as a multifunctional co-ordinator of invasion, apoptosis and cell cycle. *Cell Biochem. Biophys.* **48**, 54–72.
Soncini, C., Carpinelli, P., Gianellini, L., Fancelli, D., Vianello, P., Rusconi, L., Storici, P., Zugnoni, P., Pesenti, E., Croci, V., Ceruti, R., Giorgini, M.L., *et al.* (2006). PHA-680632, a novel Aurora kinase inhibitor with potent antitumoral activity. *Clin. Cancer Res.* **12**, 4080–4089.
Sorokin, S. (1962). Centrioles and the formation of rudimentary cilia by fibroblasts and smooth muscle cells. *J. Cell Biol.* **15**, 363–377.
Sorokin, S.P. (1968a). Centriole formation and ciliogenesis. *Aspen Emphysema Conf.* **11**, 213–216.
Sorokin, S.P. (1968b). Reconstructions of centriole formation and ciliogenesis in mammalian lungs. *J. Cell Sci.* **3**, 207–230.
Spektor, A., Tsang, W.Y., Khoo, D., and Dynlacht, B.D. (2007). Cep97 and CP110 suppress a cilia assembly program. *Cell* **130**, 678–690.
Stolc, V., Samanta, M.P., Tongprasit, W., and Marshall, W.F. (2005). Genome-wide transcriptional analysis of flagellar regeneration in *Chlamydomonas reinhardtii* identifies orthologs of ciliary disease genes. *Proc. Natl. Acad. Sci. USA* **102**, 3703–3707.
Taulman, P.D., Haycraft, C.J., Balkovetz, D.F., and Yoder, B.K. (2001). Polaris, a protein involved in left-right axis patterning, localizes to basal bodies and cilia. *Mol. Biol. Cell* **12**, 589–599.

Thoma, C.R., Frew, I.J., Hoerner, C.R., Montani, M., Moch, H., and Krek, W. (2007). pVHL and GSK3beta are components of a primary cilium-maintenance signalling network. *Nat. Cell Biol.* **9**, 588–595.

Toyo-Oka, K., Sasaki, S., Yano, Y., Mori, D., Kobayashi, T., Toyoshima, Y.Y., Tokuoka, S.M., Ishii, S., Shimizu, T., Muramatsu, M., Hiraiwa, N., Yoshiki, A., *et al.* (2005). Recruitment of katanin p60 by phosphorylated NDEL1, an LIS1 interacting protein, is essential for mitotic cell division and neuronal migration. *Hum. Mol. Genet.* **14**, 3113–3128.

Trapp, M.L., Galtseva, A., Manning, D.K., Beier, D.R., Rosenblum, N.D., and Quarmby, L.M. (2008). Defects in ciliary localization of Nek8 is associated with cystogenesis. *Pediatr. Nephrol.* **23**, 377–387.

Tsang, W.Y., Bossard, C., Khanna, H., Peranen, J., Swaroop, A., Malhotra, V., and Dynlacht, B.D. (2008). CP110 suppresses primary cilia formation through its interaction with CEP290, a protein deficient in human ciliary disease. *Dev. Cell* **15**, 187–197.

Tucker, R.W., Pardee, A.B., and Fujiwara, K. (1979a). Centriole ciliation is related to quiescence and DNA synthesis in 3T3 cells. *Cell.* **17**, 527–535.

Tucker, R.W., Scher, C.D., and Stiles, C.D. (1979b). Centriole deciliation associated with the early response of 3T3 cells to growth factors but not to SV40. *Cell* **18**, 1065–1072.

Tucker, R.W., Meade-Cobun, K.S., Jayaraman, S., and More, N.S. (1983). Centrioles, primary cilia and calcium in the growth of Balb/c 3T3 cells. *J. Submicrosc. Cytol.* **15**, 139–143.

Upadhya, P., Birkenmeier, E.H., Birkenmeier, C.S., and Barker, J.E. (2000). Mutations in a NIMA-related kinase gene, Nek1, cause pleiotropic effects including a progressive polycystic kidney disease in mice. *Proc. Natl. Acad. Sci. USA* **97**, 217–221.

Vader, G., and Lens, S.M. (2008). The Aurora kinase family in cell division and cancer. *Biochim. Biophys. Acta* **1786**, 60–72.

van Rooijen, E., Giles, R.H., Voest, E.E., van Rooijen, C., Schulte-Merker, S., and van Eeden, F.J. (2008). LRRC50, a conserved ciliary protein implicated in polycystic kidney disease. *J. Am. Soc. Nephrol.* **19**, 1128–1138.

Veland, I.R., Awan, A., Pedersen, L.B., Yoder, B.K., and Christensen, S.T. (2009). Primary cilia and signaling pathways in mammalian development, health and disease. *Nephron Physiol.* **111**, 39–53.

Vladar, E.K., and Stearns, T. (2007). Molecular characterization of centriole assembly in ciliated epithelial cells. *J. Cell Biol.* **178**, 31–42.

Wheatley, D.N. (1969). Cilia in cell-cultured fibroblasts. I. On their occurrence and relative frequencies in primary cultures and established cell lines. *J. Anat.* **105**, 351–362.

Wheatley, D.N. (1972). Cilia in cell-cultured fibroblasts. IV. Variation within the mouse 3T6 fibroblastic cell line. *J. Anat.* **113**, 83–93.

Wheatley, D.N. (1995). Primary cilia in normal and pathological tissues. *Pathobiology* **63**, 222–238.

Wheatley, D.N., Feilen, E.M., Yin, Z., and Wheatley, S.P. (1994). Primary cilia in cultured mammalian cells: Detection with an antibody against detyrosinated alpha-tubulin (ID5) and by electron microscopy. *J. Submicrosc. Cytol. Pathol.* **26**, 91–102.

Wheatley, D.N., Wang, A.M., and Strugnell, G.E. (1996). Expression of primary cilia in mammalian cells. *Cell Biol. Int.* **20**, 73–81.

White, M.C., and Quarmby, L.M. (2008). The NIMA-family kinase, Nek1 affects the stability of centrosomes and ciliogenesis. *BMC Cell Biol.* **9**, 29.

Wloga, D., Camba, A., Rogowski, K., Manning, G., Jerka-Dziadosz, M., and Gaertig, J. (2006). Members of the NIMA-related kinase family promote disassembly of cilia by multiple mechanisms. *Mol. Biol. Cell* **17**, 2799–2810.

Wloga, D., Rogowski, K., Sharma, N., Van Dijk, J., Janke, C., Edde, B., Bre, M.H., Levilliers, N., Redeker, V., Duan, J., Gorovsky, M.A., Jerka-Dziadosz, M., *et al.* (2008). Glutamylation on alpha-tubulin is not essential but affects the assembly and functions of a subset of microtubules in *Tetrahymena thermophila*. *Eukaryot. Cell.* **7**, 1362–1372.

Wong, S.Y., Seol, A.D., So, P.L., Ermilov, A.N., Bichakjian, C.K., Epstein, E.H., Jr., Dlugosz, A.A., and Reiter, J.F. (2009). Primary cilia can both mediate and suppress Hedgehog pathway-dependent tumorigenesis. *Nat. Med.* **15**, 1055–1061.

Yoshimura, S., Egerer, J., Fuchs, E., Haas, A.K., and Barr, F.A. (2007). Functional dissection of Rab GTPases involved in primary cilium formation. *J. Cell Biol.* **178**, 363–369.

Zerial, M., and McBride, H. (2001). Rab proteins as membrane organizers. *Nat. Rev. Mol. Cell Biol.* **2**, 107–117.

Zhang, D., Rogers, G.C., Buster, D.W., and Sharp, D.J. (2007). Three microtubule severing enzymes contribute to the "Pacman-flux" machinery that moves chromosomes. *J. Cell. Biol* **177**, 231–242.

Zhang, Z., Hernandez-Lagunas, L., Horne, W.C., and Baron, R. (1999). Cytoskeleton-dependent tyrosine phosphorylation of the p130(Cas) family member HEF1 downstream of the G protein-coupled calcitonin receptor. Calcitonin induces the association of HEF1, paxillin, and focal adhesion kinase. *J. Biol. Chem.* **274**, 25093–25098.

Zhang, Z., Neff, L., Bothwell, A.L., Baron, R., and Horne, W.C. (2002). Calcitonin induces dephosphorylation of Pyk2 and phosphorylation of focal adhesion kinase in osteoclasts. *Bone* **31**, 359–365.

Zhou, B.P., and Hung, M.C. (2005). Wnt, hedgehog and snail: sister pathways that control by GSK-3beta and beta-Trcp in the regulation of metastasis. *Cell Cycle* **4**, 772–116.

SECTION III

Function

CHAPTER 8

Utilization of Conditional Alleles to Study the Role of the Primary Cilium in Obesity

Robert A. Kesterson[*], **Nicolas F. Berbari**[†], **Raymond C. Pasek**[†], *and* **Bradley K. Yoder**[†]

[*]Department of Genetics, University of Alabama at Birmingham Medical School, Birmingham, Alabama 35294

[†]Department of Cell Biology, University of Alabama at Birmingham Medical School, Birmingham, Alabama 35294

Abstract

Ciliopathies are a group of human diseases that involve dysfunction of the cilium. Human patients with mutations in ciliary proteins can exhibit a wide range of phenotypes, one of which is obesity. This is seen in patients with Bardet–Biedl syndrome

978-0-12-375024-2
DOI: 10.1016/S0091-679X(08)94008-5

(BBS) and Alström syndrome (ALMS). Both of these disorders are caused by mutations in proteins that localize to the cilium or the basal body at the base of the cilium. These rare human disorders and their corresponding mouse models together with genetic approaches to disrupt cilia on specific cell types are beginning to uncover the connection between the cilium and energy homeostasis. Here we will review the current data on how cilia are thought to be involved in energy homeostatic pathways and discuss several key factors to consider when utilizing conditional approaches to evaluate ciliary function and their link to obesity.

I. Introduction

Obesity is a complex disorder associated with diminished quality of life due to increased risk for many other disease states such as diabetes, stroke, heart disease, and hypertension. In simplest terms, obesity is a disturbance of energy balance that occurs when energy intake exceeds energy expenditure due to either genetic or environmental influences, or a combination of the two. While the recognition of the worldwide obesity epidemic has promoted a great deal of attention to this debilitating disorder, much of our understanding of the underlying molecular mechanisms stems from research on unique animal models of obesity and leanness, and from patients with rare human obesity syndromes such as Bardet–Biedl syndrome (BBS, OMIM #209900) or Alström syndrome (ALMS, OMIM #203800). In particular, mice harboring spontaneous, as well as genetically engineered, mutations (resulting in loss, or more rarely, gain of function) have expanded our knowledge of energy homeostatic pathways (Bolze and Klingenspor, 2009). Importantly, mouse models were instrumental in establishing a connection between the primary cilium and obesity by demonstrating that cilia have effects on both central nervous system (CNS) mechanisms controlling feeding behavior and energy balance, as well as in the periphery through regulation of adipogenesis.

A comprehensive discussion and review of all animal models of disturbed energy balance, as well as potential mechanisms of action associated with cilia implications, is not feasible in the current venue. Therefore, we focus on hypothalamic pathways uncovered using mouse models of disturbed energy balance and the recently defined roles of cilia in energy homeostasis, before finally detailing strategies needed for analyzing conditional alleles for the study of cilia and obesity.

II. CNS Mechanisms of Energy Balance

Obesity is one of the most common nutritional disorders that now affects more than a third of the population in the United States. The regulation of energy balance is a complex process with numerous signaling circuits involving peripherally derived hormones that must be coordinated with neuroendocrine factors and neuronal activity in the CNS to elicit a response (Flier, 2004). One of the most studied and best characterized of these pathways is the leptin–melanocortin signaling axis (Cone, 2005). Leptin is produced by adipocytes and is released into the circulation where levels of leptin correlate with the extent of adiposity. When leptin crosses the blood–brain barrier it exerts an anorexigenic effect. Although leptin acts on several brain regions, it appears to primarily influence feeding activity through actions on the pro-opiomelanocortin (POMC) and neuropeptide Y (NPY)/agouti-related peptide (AgRP) neurons in the arcuate nucleus (ARC) of the hypothalamus (Fig. 1). Both NPY/AgRP and POMC neurons express the leptin receptor but are differentially influenced by the hormone. Whereas leptin increases expression of the anorexigenic hormone alpha-melanocyte stimulating hormone (α-MSH) in POMC neurons, it reduces NPY expression and inhibits AgRP activity by the NPY/AgRP neurons. In the presence of leptin, α-MSH is released from POMC neurons and activates the G-protein-coupled melanocortin 4 receptor (MC4R) on higher order CNS neurons. This in turn regulates endocrine outputs (such as growth and fertility through control of pituitary function), behavioral outputs (including feeding and arousal state), and autonomic outputs (energy expenditure, insulin secretion, and glucose homeostasis). AgRP, whose activity is suppressed by leptin, functions as a selective antagonist of "brain" melanocortin receptors blocking α-MSH interaction with MC4R. Defects in this CNS pathway cause morbid obesity in mammals (Huszar *et al.*, 1997), with mutations in the MC4R gene being the most common monogenetic cause of obesity known in humans (Farooqi and O'Rahilly, 2006).

In addition to leptin, there is another neuroendocrine signaling pathway known to regulate feeding behavior through hypothalamic neurons that warrants further mention due to its recent association with the cilium; the G-protein-coupled receptor melanin-concentrating hormone receptor 1 (MCHR1) (Berbari *et al.*, 2008a). Administration of melanin-concentrating hormone (MCH) into the brain induces a rapid increase in feeding behavior while chronic administration of MCHR1 antagonists elicits an anorexigenic response (Pissios *et al.*, 2006). Furthermore, transgenic mice overexpressing MCH are obese, while mice lacking expression of either MCH (Shimada *et al.*, 1998) or MCHR1 (Marsh *et al.*, 2002) are lean. The recent demonstration that MCHR1 localizes to cilia in regions of the brain known to regulate feeding behavior suggests that localization of MCHR1 to cilia may be important for signaling through the receptor and proper regulation of these feeding processes (Berbari *et al.*, 2008a). Moreover, the MCHR1 fails to localize correctly in neuronal cilia of BBS mutant mice (BBS) further supporting a role for the MCH pathway in cilia-mediated regulation of feeding behavior (Berbari *et al.*, 2008b).

III. Human Ciliopathies and Obesity

The connection between cilia and obesity was initially established in 2003 when genes responsible for BBS (OMIM #209900) were identified and found to localize in or at the base of the cilium (Ansley *et al.*, 2003). The ciliary connection was further

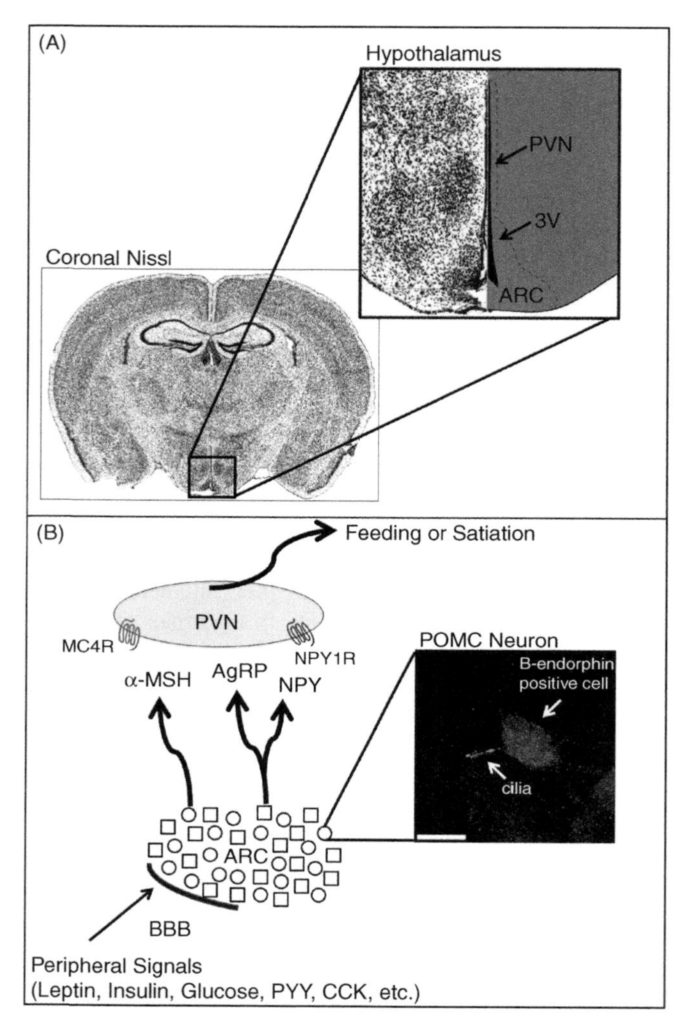

Fig. 1 Hypothalamus and feeding. The ARC of the hypothalamus is located near the third ventricle on the ventral side of the brain. Panel A shows a Nissl-stained coronal section of the mouse brain. The inset shows the ARC of the hypothalamus, the third ventricle (3V), and the paraventricular nucleus (PVN) of the hypothalamus. Images modified from the Allen Brain Atlas http://www.brain-map.org/. Panel B shows two populations of neurons (POMC in red and AgRP/NPY in green) involved in feeding behavior and are located in the ARC. The peptides released by these neurons act on receptors located on 'higher' second-order neurons of the PVN. This signaling ultimately results in changes in neuronal activity that led to either feeding behavior or satiation. The inset shows cultured hypothalamic neurons labeled for NPY in green and the cilia receptor MCHR1 in green. Cilia are indicated by arrows. BBB corresponds to the blood–brain barrier. (See Plate no. 7 in the Color Plate Section.)

strengthened in 2005 with the discovery that the gene ALMS1, whose function is disrupted in ALMS (OMIM #203800), is also present at the base of the cilium (Hearn *et al.*, 2005).

BBS patients typically exhibit early-onset truncal obesity, retinopathy, renal abnormalities, and cognitive impairments (Beales *et al.*, 1999). BBS is a heterogeneous disorder with at least 12 causative genes. The proteins associated with BBS localize to the cilia or the basal bodies at the base of the cilium. Many of the BBS proteins have been shown to function as part of large complex called the BBSome that has a role in regulating receptor entry into the cilium and cilia membrane biogenesis (Nachury *et al.*, 2007).

ALMS is a recessive disorder caused by mutations in the gene ALMS1 (Collin *et al.*, 2002). Similar to BBS, ALMS is a multisystemic disease with cone–rod retinal dystrophy, hearing defects, insulin resistance, and childhood obesity. The ALMS1 protein is present at the basal body at the base of the cilium and at the centrioles and analysis of mutant cell lines from ALMS patients indicate that cilia assembly is not disrupted (Hearn *et al.*, 2005). The exact function of ALMS1 is still unknown.

IV. Cilia Mutant Mouse Models of Obesity

A. Mouse Models of BBS

While the list of ciliopathies continues to expand, it is interesting to note that only BBS and ALMS are associated with obesity (Sen Gupta *et al.*, 2009). Mouse models of these genetic disorders have been developed and largely mimic the phenotypes observed in the human disorders. While the molecular etiology behind the obesity phenotype remains unclear, the current literature offers several clues and potential candidate pathways that may be behind this phenotype (Berbari *et al.*, 2008b; Davenport *et al.*, 2007; Seo *et al.*, 2009). These range from the implication of the BBSome in leptin receptor or melenocyte concentrating hormone (MCHR1) receptor trafficking to a potential role for primary cilia in adipocyte differentiation.

The genetic heterogeneity observed in BBS combined with the broad range of clinical features allows for the opportunity to examine one feature such as obesity in the context of several mutant mouse models. To date, there have been several congenital BBS and ALMS mouse models reported, and many of these have hyperphagia-induced obesity as is observed in patients (Arsov *et al.*, 2006; Collin *et al.*, 2005; Li *et al.*, 2007; Mykytyn *et al.*, 2004; Nishimura *et al.*, 2004). In fact, one study demonstrated a knock-in of a humanized point mutation that displayed the same phenotypes observed in knockout mouse models and human patients (Davis *et al.*, 2007).

Recent work in BBS models has begun to identify some potential candidate pathways that may play a role in the hyperphagia-induced obesity phenotype. In one example, BBS mutations have been found to affect the cilia localization of G-protein-coupled receptors, one of which is the MCHR1 (Berbari *et al.*, 2008b). In neurons of BBS mutants, MCHR1 cilia localization is absent. This observation is intriguing because MCHR1 plays an important role in feeding behavior and genetic and pharmacological data suggest that the ligand MCH has orexigenic effects (Borowsky *et al.*, 2002; Chen *et al.*, 2002; Ludwig *et al.*, 2001; Qu *et al.*, 1996; Shimada *et al.*, 1998). However, how loss of cilia localization of MCHR1 in BBS mutants may contribute to excess energy intake is unknown and remains an active area of research.

Other studies of BBS mutant mice have indicated a connection between cilia and the leptin signaling axis (Rahmouni *et al.*, 2008). Leptin normally suppresses feeding through binding of its receptor on CNS neurons. In the case of BBS mutant mice, intracerebroventricular leptin administration was found to be ineffective in reducing food intake (Seo *et al.*, 2009). This was the case even when mice are maintained in a lean state and endogenous leptin levels are suppressed. These data raise the possibility that leptin resistance occurs through mistrafficking of the leptin receptor in the BBS mutants. How cilia may function to coordinate signaling of the leptin and MCH pathways to regulate energy homeostasis is not yet understood.

Another possible connection between cilia and obesity is data showing that signaling through the cilia may regulate adipocyte differentiation (Forti *et al.*, 2007; Marion *et al.*, 2009). BBS10 and BBS12 knockdown studies have shown that the levels of peroxisome proliferator-activated receptor gamma (PPAR-γ) and glycogen synthase kinase3β (GSK3β) are elevated, suggesting promotion of the adipogenic lineage (Marion *et al.*, 2009). Both PPARγ and GSK3β act downstream of hedgehog-and Wnt-signaling pathways, and the cilium has been implicated as a key regulator of each of these pathways (reviewed in Berbari *et al.*, 2009). Thus, the data suggest that loss of cilia may release repression of adipogenesis, possibly through hedgehog and Wnt signaling; however, these interpretations are based solely on knockdown approaches and there are limited data indicating that adipocytes are directly linked to the development of obesity in cilia models.

One thing is clear, both leptin and MCH pathways are major regulators of human energy homeostasis; thus, establishing a connection between the cilium and these pathways in mice represents a significant advance in understanding the molecular and cellular mechanisms driving the human obesity epidemic. One potential way to further tease out the role of cilia in energy homeostasis is through the use of conditional mutation approaches.

B. Analysis of the Obesity Phenotype in Conditional Cilia Mutant Mice

Although once thought to be of minimal importance, primary cilia actually have critical roles during early development and are necessary for viability. This is a complication when wanting to directly analyze ciliary function in postnatal life. Further, hypomorphic mutations in ciliogenic genes, such as seen in the *$IFT88^{TgN737RPW}$* mutants (Oak Ridge Polycystic Kidney mouse; hereafter called ORPK), result in hydrocephalus and cystic kidneys, livers, and pancreas that likely mask other potential cilia relevant phenotypes (Moyer *et al.*, 1994). To overcome this issue, Davenport *et al.* (2007) recently used conditional alleles of two ciliogenic genes (*$ift88^{tm1Bky}$*, and *$kif3a^{tm2Gsn}$*) and a systemically expressed Cre (CAGG-CreER, *Tg (CAG-cre/Esr1)5Amc* (Hayashi and McMahon, 2002)) line whose activity is regulated

Table I
Cre-Mediated Loss of Cilium and Obesity

Cre line	Site of expression	Obesity in cilia mutants (relative)	Regulation/ induction	Source	References
CAGG-CreERTM	Ubiquitous	+++	With tamoxifen	Jackson Laboratory	Hayashi and McMahon (2002)
Synapsin1-Cre	Most CNS neurons	+++	Congenital	Jackson Laboratory	Zhu *et al.* (2001)
POMC-Cre	ARC (hypothalamus), nucleus tract solitarius (brain stem), pituitary, skin	++	Congenital		Xu *et al.* (2007)

by tamoxifen to induce cilia loss in the context of an adult mouse (Table I). Interestingly, when cilia loss was induced at 8 weeks of age, the mice did not develop phenotypes predicted by the ORPK mutants, but rather became obese. This phenotype was due to a marked increase in energy intake with no obvious changes in physical activity or basal metabolic rate. In fact, when the adult conditional mutants had their energy intake restricted by being fed an equivalent amount of food as unrestricted control mice with intact cilia (pair-fed study), the mutant animals did not become obese. The mutant mice also developed other characteristics associated with human obesity including: hyperinsulinemia, hyperleptinemia, and hyperglycemia. However, these were secondary to the onset of obesity as the pair-fed lean mice do not show this type-II diabetes like phenotype. A similar result was obtained with the BBS mouse models as well as in human ALMS (Lee *et al.*, 2009; Seo *et al.*, 2009).

To further explore the connection between the cilium and satiation, cilia were disrupted specifically on neurons in the CNS using the synapsin-1-Cre (Syn1-Cre, B6.Cg-Tg(Syn1-Cre)671Jxm, Zhu *et al.*, 2001). As seen with the CAGG-CreER line, these mice also became markedly obese with many reaching upwards of 80 g compared to 40 g for littermate controls (Fig. 2). However, in contrast to the CAGG-CreER, Syn1-Cre is expressed congenitally and onset of the obesity phenotype was more delayed. There are several potential explanations for this delay and apparent differential phenotype. First, the differences could be due to Syn1-Cre being less efficient in deleting the floxed alleles than the CAGG-CreER, particularly in the hypothalamus. Unfortunately, detailed analyses of complete neuronal patterns of expression and relative levels of Cre activity have yet to be done for the Syn1-Cre and CAGG-CreER lines, although it is generally accepted that Syn1-Cre targets most CNS neurons, whereas the CAGG promoter is thought to be expressed ubiquitously. Second, the data may be an indication that cilia have functions outside of the CNS to regulate satiation. Therefore, peripheral loss of cilia function in tamoxifen-treated CAGG-CreER mutants may stimulate pathways and hormonal signals outside of the CNS, to subsequently augment perceived negative energy

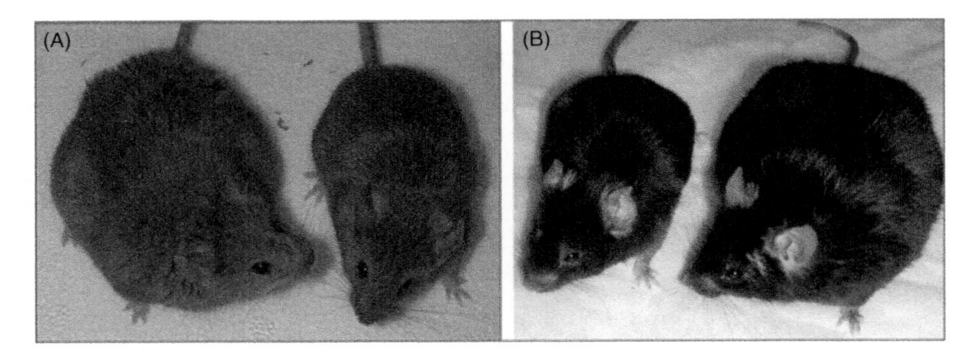

Fig. 2 Obese cilia mutant mouse models. Panel A shows an obese *bbs4* mutant mouse next to his normal weight wild-type littermate (Image courtesy of Val Sheffield and Kirk Myktyn). Panel B shows an obese conditional *ift88* mutant that expresses the pan neuronal Cre transgene synapsin1-Cre. (Image reproduced from Sharma *et al.* (2008). *Curr. Top. Dev. Biol.* **85**, 371–427.)

balance that leads to enhanced feeding. Finally, the congenital mutants may be able to compensate for changes that occur early in development that may not be feasible in the adult-induced cilia mutants due to the CNS feeding circuits in the brain being "hard-wired" prior to tamoxifen-induced activation of Cre activity.

Importantly, the development of obesity in the conditional cilia mutants generated with Syn1-Cre indicates that cilia on CNS neurons function in a pathway to regulate satiation responses. The region of the brain involved in this cilia-mediated response was further defined using a POMC-Cre (*Tg.PomcCre*) (Xu *et al.*, 2007) line to disrupt cilia on a small group of POMC neurons in the hypothalamus and brain stem. These mice also became obese due to hyperphagia, albeit to a lesser extent than that seen in the tamoxifen-treated CAGG-CreER mutants. This differential degree of obesity can again be due to several possibilities. While these data establish a role for cilia on POMC neurons to control energy balance, it may be the case that cilia on other neuronal populations similarly exert control over feeding behavior pathways. While less likely, peripheral POMC expression could be playing a role, although the lack of disruption of the hypothalamic–pituitary–adrenal (HPA) axis in POMC-Cre cilia mutants suggests otherwise (Davenport *et al.*, 2007). As above, "hard-wired" circuits may be more susceptible to cilia loss in an adult animal; therefore, animals with congenital POMC-Cre loss of cilia may be more malleable to compensatory pathways being established developmentally to control feeding behavior. Interestingly, the projection of hypothalamic arcuate POMC neurons to distal innervation sites is developmentally determined by a perinatal surge in leptin signaling, which if disrupted (as in the case of leptin-deficient ob/ob mice) leads to permanent neuroanatomical loss of this key inhibitory feeding circuit (Bouret *et al.*, 2004). Overall, these data with POMC-Cre cilia mutants along with the results from the Syn1-Cre line establish that cilia, most likely on the hypothalamic POMC neurons, are critical for normal energy homeostasis; however, a role on other POMC expressing cells of the body cannot be completely eliminated.

V. Neuronal Cilia

Primary cilia are present on nearly every cell type in the mammalian body including neurons, a fact that often comes as a surprise to many neurobiologists (Fig. 3) (Fuchs and Schwark, 2004). Interestingly, glia and neural precursor cells also possess primary cilia (Bishop *et al.*, 2007; Doetsch *et al.*, 1999) and recent findings suggest that cilia on the neuronal precursors are important for adult neurogenesis (Breunig *et al.*, 2008). The cilia of CNS cells generally extend off the cell body and are of varying length ranging from <2 μm to nearly 10 μm (Fuchs and Schwark, 2004). In addition to cilia displaying basic morphological differences, they also have differences in the complement of signaling components that they contain suggesting distinct signaling properties (Berbari *et al.*, 2007, 2008b).

While it has been known for decades that cilia are present in the CNS (Cohen and Meininger, 1987; Dahl, 1963), they have been grossly understudied. In part, this was due to the pervading misconception that primary cilia are vestigial organelles; however, there are several additional challenges that have contributed to the delay in analyzing the functional importance of neuronal cilia.

One of the major challenges has been the inability to readily visualize the organelle. The visualization of neuronal cilia utilizing the traditional cilia antibody markers of stabilized forms of tubulin (i.e., acetylated-α tubulin) have proved difficult. Neuronal processes and soma have high levels of these forms of tubulin making the use of such markers for cilia visualization comparable to the proverbial needle in the haystack. As such, it was not until the discoveries that specific receptors and signaling components

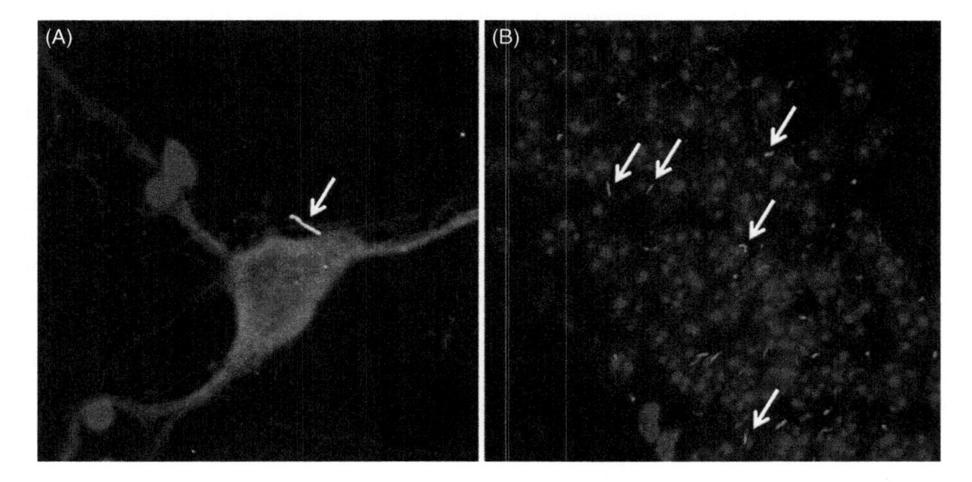

Fig. 3 Neuronal cilia. Neurons form primary cilia both *in vitro* and *in vivo*. Panel A shows an ACIII positive cilia (green) on a cultured neuron. (Image courtesy of Kirk Mykytyn.) A neuron-specific marker (neuronal filament) antibody labeling in red and cytox nuclear labeling in blue. Panel B shows neuronal cilia *in vivo* labeled with ACIII within the dentate gyrus. Hoechst nuclear labeling in blue. (See Plate no. 8 in the Color Plate Section.)

such as somatostatin receptor 3 (Sstr3), serotonin receptor 6 (5HT6), and adenylate cyclase III (ACIII), preferentially localized to the cilia membrane in certain regions of the brain, that the study of neuronal cilia signaling could begin in earnest (Berbari *et al.*, 2007; Brailov *et al.*, 2000; Fuchs and Schwark, 2004; Handel *et al.*, 1999). While a growing number of signaling proteins have now been found in the neuronal cilium, the functional relevance of their localization to this organelle remains unknown (for a review on neuronal cilia, see Domire and Mykytyn, 2009).

A second complication is that the fixation technique employed can dramatically affect the ability to detect cilia in the brain. In general, fixation through perfusion is highly recommended, while the type of fixation can vary depending on the requirements of the antibody being used. In the methods section below, we describe approaches to visualize neuronal cilia and include a list of antibodies, sources, and fixation conditions for labeling neuronal cilia.

VI. Strategies and Methods of Analyzing Obesity in Cilia Mouse Models

A. Detection of Primary Cilia on Neurons

Retention of the primary cilium is dependent on rapid and thorough fixation. For optimal immunofluorescent detection of the cilium, animals must be perfused. To accomplish this, mice are anesthetized using a 0.1 ml/10 g intraperitoneal injection of 2.5% tribromoethanol (Sigma-Aldrich, St. Louis, Missouri, USA), killed by cardiac puncture and perfusion with phosphate-buffered saline (PBS) followed by a mixture of 2% paraformaldehyde (PFA) and HistoChoice (Amresco, Solon, OH, USA). The brains of the mice are then isolated and postfixed for an additional 24 h in 2% PFA/HistoChoice at 4°C, followed by cryoprotection in 30% sucrose in PBS for 16–24 h at 4°C. The brain is trimmed and embedded in optimal cutting temperature (OCT) compound, and snap frozen using a dry ice ethanol bath. Sections of the brain are cut with a cryostat at a thickness of 10 μm. Tissue sections are washed with PBS to remove OCT and are then permeabilized and blocked with 0.1% Triton X-100 in PBS with 2% normal secondary antibody host serum, 0.02% sodium azide, and 10 mg/ml bovine serum albumin. Sections are incubated for 16–24 h at 4°C in primary antibodies (Table II) followed by at least three 5-min washes in PBS. The sections are incubated with secondary antibody for 1 h at room temperature. Nuclei are visualized using Hoechst (1:1000, Sigma, #33258, St. Louis, Missouri, USA) and slides are mounted using Immu-Mount (Themo Scientific, Pittsburgh, PA, USA).

B. Inducing Cilia Loss with Tamoxifen Regulated Cre lines

To analyze ciliary function after neural connections and circuits have been established in an adult mouse, it may be necessary to utilize a line where Cre activity can be regulated. This can be accomplished using a CreER or CreERT2 line where Cre is

Table II
List of Primary Antibodies Verified for Detection of Neuronal Cilia

Protein	Source	Catalog #	Recommended fixation	Recommended dilution
Adenylyl cyclase III (ACIII)	Santa Cruz Biotech	SC-588	4% PFA	1:500
Somatostatin receptor 3 (SSTR3)	Santa Cruz Biotech	SC-11617	4% PFA	1:500
Somatostatin receptor 3 (SSTR3)	Gramsch	SS-830	2% PFA:Histochoice	1:250
Melanin-concentrating hormone receptor I (MCHR1)	Santa Cruz Biotech	SC-5534	2% PFA:Histochoice	1:500

fused with a modified form of the estrogen receptor. This modified form no longer binds to estrogen but does to the estrogen derivative tamoxifen that can be added exogenously. In the absence of tamoxifen, Cre is excluded from the nucleus; however, once tamoxifen is injected, Cre translocates to the nucleus and mediates the excision of DNA between the loxP sites to generate a deletion and mutant alleles.

To induce Cre activity in tamxoifen-regulated CreER lines, a 10 mg/ml solution of tamoxifen (Sigma, St. Louis, MO, USA) is dissolved in corn oil (Sigma). Tamoxifen can be difficult to resuspend and may required intense agitation for several hours. Once in suspension, tamoxifen is administered for five consecutive days through IP (interperitoneal) injection in mice at the desired age. Injections are typically given at a dose of 6.0 mg per 40.0 g body weight (BW) as described (Hayashi and McMahon, 2002).

The next step will be to evaluate whether Cre activity is efficient in generating the deletion, and how long after induction it takes for cilia loss to occur. The efficiency of the Cre can be monitored using one of several transgenic reporter lines such as *Rosa26* (*Gt(ROSA)26Sor*^{tm1Sho}) (Soriano, 1999), *Z/AP(Tg(CAG-βgeo/ALPP)1Lbe)* (Lobe *et al.*, 1999), or *mT/mG* (*Gt(ROSA)26Sor*$^{tm4(ACTB\text{-}tdTomato,\text{-}EGFP)Luo}$) (Muzumdar *et al.*, 2007); but for detection of the cilium, one of the most useful has been the Z/EG (*Tg(CAG-βgeo/GFP)21Lbe*) (Novak *et al.*, 2000). This is a dual reporter line generated in the laboratory of Dr. Lobe. In the absence of Cre activity (i.e., no tamoxifen), cells express the β-galactosidase (β-gal) reporter gene; however, once Cre is activated, the β-gal gene is deleted by virtue of floxed sequences that flank the gene and cells indelibly express the enhanced green fluorescent protein (Z/EG line). By analyzing enhanced green fluorescent protein (EGFP) expression along with immunofluorescence using a neuronal cilia marker (Fig. 3, Table II), it is possible to assess whether cilia loss is occurring in an efficient manner on the target cells

C. Monitoring Feeding Behavior and Weight Gain

If a disturbance in energy balance is anticipated in a given cilia mouse model, a series of experiments to begin assessing the phenotype should be followed. First, accurate records of body weight (BW) and growth rates must be maintained wherein it is recommended that the BW of each animal in a similarly housed cohort is recorded weekly, and at a consistent time of day. Standard housing conditions include a 12-h light cycle and free access to food (chow diet) and water. Initial experiments can be done with group housed animals; however, singly housed mice are ideal in order to minimize dominant and subordinate behaviors that can influence food intake. A simple plot of BW versus time will reveal an overt obese phenotype often within 4–6 months of age when statistically compared using repeated measures analysis of variance. An important consideration is that the animals may be obese (with increased body mass index), yet not display a difference in overall BW. This is seen in the melanocortin-3 receptor knockout (MC3R KO) mice (Butler *et al.*, 2000; Chen *et al.*, 2000), which show an increase in adipose mass accompanied by a reduction in lean mass. If no apparent BW differences are seen, more detailed analyses of body composition may be needed. This may be accomplished a number of ways using complex instrumentation including: Dual-energy X-ray absorptiometry that can determine total body fat, soft-lean tissue, and bone mineral content of mice using a PIXImus densitometer (GE-Lunar PIXImus, Madison, WI, USA) that has been previously validated (Nagy and Clair, 2000), or via magnetic resonance technologies that can determine lean and fat content in unanesthetized mice (Echo Medical or Bruker Optics). However, if not available, body composition can be determined by chemical carcass analysis, or individual fat pads can be isolated and weighed upon necropsy.

Once determined to be obese, the driving force behind the increased fat mass needs to be addressed in mutant animals. The most common culprits leading to a positive energy balance are increased food intake, reduced physical activity, or reduced metabolic rate. To determine if mutant mice have increased feeding behavior compared to controls, animals must be singly housed under carefully regulated conditions that are held as invariant as possible for the unobtrusive monitoring of food intake. Prior to the onset of obesity, mice of similar age, weight, and sex should be given free access to a defined diet that has been pre-measured to determine accurately the amount consumed during each period of study (daily is most common). Containers of food should be weighed daily at a consistent time of day, with food replaced as needed. Average daily food intake should be determined from several independent measures (3 or more) for accuracy. While the use of high caloric rodent chow can be used to more quickly exacerbate the development of obesity, standard rodent chow is normally used in feeding studies prior to the onset of the obese phenotype.

If hyperphagia is detected, a common experiment to perform in determining if increased feeding is solely influencing weight gain is a pair-feeding study. In this case, a third group of animals is added to a standard growth curve experiment (BW $\times$ time) containing control and mutant mice with unrestricted feeding. The third group of "pair-fed" mice only receive the amount of food consumed by the control mice that are freely feeding. Using singly housed mice, the average amount of food eaten on a given day by mice in the control group is then provided to the individual pair-fed mice the next day. For instance, if the average food intake by the wild-type control mice on a given day is 3.5 g of rodent chow (an approximate amount for a 25 g C57BL/6 mouse), then each mouse in the mutant pair-fed group only receives 3.5 g of chow that day. If this restriction in food intake completely blocks the development of obesity, then the primary cause of the phenotype is overeating or hyperphagia. However, if the pair-fed mice still become obese while consuming a "normal" amount of food, then other mechanisms are driving the phenotype (most likely either a disturbance in locomotor activity, or basal metabolic rate). An interesting phenomenon in conducting these experiments is that animals in the pair-fed group will alter their feeding behavior to consume the food provided almost immediately after presentation as the experiment goes on.

Obesity without a detectable increase in energy intake suggests that animals have decreased energy expenditure. A general measure of rodent physical activity can be made by monitoring locomotor activity, which is determined by movement around the home cage, as well as "wheel running" behavior if provided. Spontaneous locomotor activity can be followed several ways via video tracking, implantable transmitters, or infrared beam monitors. Infrared beam monitors are most commonly available in institutional behavior core facilities. These are preferable to the use of implantable transmitters that may have surgical complications when using young mice such as those needed for analyzing the primary defects prior to the onset of obesity. More difficult to detect are changes in basal metabolic rate that depend upon the use of expensive equipment to measure energy expenditure via indirect calorimetry. From measures of CO_2 produced and O_2 consumed by an animal, calculations of energy output, metabolic rate, and respiratory quotient (relative estimation of carbohydrate and lipid oxidation) can be made. It is especially important to use appropriate controls in these kinds of studies since there is some debate about appropriate calculations that need to be made in order to normalize between different sized animals, which can be problematic if trying to use age-matched controls that deviate significantly in body size. Moreover, caution should be taken when making cross model comparisons, especially if mutations were made in differing genetic background strains (e.g., 129 and B6), where basal phenotypes can deviate significantly between strains.

D. Additional Factors to Consider When Analyzing Obesity Using Inducible Cilia Mutant Mouse Models

In addition to analyzing deletion of your gene of interest along with where and when Cre activity occurs, other factors that should be considered are the stability of your protein and, if you are generating ciliogenic mutants, how long it may require for the cilium to regress. Cre activity may be detected early, but the targeted protein or the cilium may remain for several days after deletion of the DNA occurs. This is likely to be influenced by the rate of proliferation of the target cell type, as the cilium must be reabsorbed prior to entering mitosis and rebuilt again after division. Thus, potential phenotypes associated with the loss of the protein or cilium may not manifest for several days to weeks after the DNA deletion occurs. This should be evaluated using a combination of approaches including western blot analysis, and if possible, by immunofluorescence. Western analysis may be complicated if the target gene is expressed broadly, and the study is being conducted with a cell-type-specific Cre line (e.g., POMC-Cre or AGRP-Cre), or by the availability of specific antibodies.

Another area that should be considered, especially in the case of analyzing neuronal cilia, is whether potential phenotypes may be caused by secondary factors that are not directly relevant to the cilium, or whether your protein may have roles outside of the cilium. For example, it is not known whether loss of the cilium influences proliferation or apoptosis of neuronal populations, or whether loss of signaling activity regulated through the cilium could alter neural networks and projections. In fact, in the BBS mutants, data suggest that dysfunction of the cilium in these mice may cause a reduction in the number of POMC neurons (Seo *et al.*, 2009), and thus it is feasible that this could contribute to the obesity phenotype secondarily to cilia dysfunction. Assessing these possibilities would require comparing neuronal populations between mutant and control mice, evaluating proliferation and cell death with Bromodeoxyuracil (BrdU) and Terminal Deoxynucleotidyl Transferase dUTP Nick End Labeling (TUNEL), and conducting tracing studies along with assessing changes in neuronal activity.

A final factor to consider with regard to assessing possible signaling pathways associated with the cilium is whether the state of being obese may complicate interpretation of data. For instance, data from the BBS mice suggest that the leptin pathway may be impaired due to loss of ciliary function. However, obese mice and humans develop leptin resistance regardless of whether a cilium is present or not. Thus, in the BBS studies it was important that the analysis be done on mutants maintained on a lean state through pair-feeding to keep endogenous leptin levels low (Seo *et al.*, 2009). It is important to evaluate candidate pathways under both obese and lean conditions to make sure that any effects are not caused by secondary events of becoming obese.

Acknowledgments

We would like to thank C. Williams for critical comments on the manuscript and helpful discussions. This work was supported by grant R01DK075996 to B. K. Y., P30 DK074038 (UAB Recessive Polycystic Kidney Disease Core Center, Dr L. Guay-Woodford), and P30 CA-13148 UAB Comprehensive Cancer Center (PI: Partridge) and P60 DK079626 UAB Diabetes Research and Training Center (PI: Garvey) to R. A. Kesterson.

References

Ansley, S. J., Badano, J. L., Blacque, O. E., Hill, J., Hoskins, B. E., Leitch, C. C., Kim, J. C., Ross, A. J., Eichers, E. R., Teslovich, T. M., Mah, A. K., Johnsen, R. C., *et al.* (2003). Basal body dysfunction is a likely cause of pleiotropic Bardet-Biedl syndrome. *Nature* **425**, 628–633.

Arsov, T., Silva, D. G., O'Bryan, M. K., Sainsbury, A., Lee, N. J., Kennedy, C., Manji, S. S., Nelms, K., Liu, C., Vinuesa, C. G., de Kretser, D. M., Goodnow, C. C., *et al.* (2006). Fat aussie—a new Alstrom syndrome mouse showing a critical role for ALMS1 in obesity, diabetes, and spermatogenesis. *Mol. Endocrinol.* **20**, 1610–1622.

Beales, P. L., Elcioglu, N., Woolf, A. S., Parker, D., and Flinter, F. A. (1999). New criteria for improved diagnosis of Bardet-Biedl syndrome: results of a population survey. *J. Med. Genet.* **36**, 437–446.

Berbari, N. F., Bishop, G. A., Askwith, C. C., Lewis, J. S., and Mykytyn, K. (2007). Hippocampal neurons possess primary cilia in culture. *J. Neurosci. Res.* **85**, 1095–1100.

Berbari, N. F., Johnson, A. D., Lewis, J. S., Askwith, C. C., and Mykytyn, K. (2008a). Identification of ciliary localization sequences within the third intracellular loop of G protein-coupled receptors. *Mol. Biol. Cell* **19**, 1540–1547.

Berbari, N. F., Lewis, J. S., Bishop, G. A., Askwith, C. C., and Mykytyn, K. (2008b). Bardet-Biedl syndrome proteins are required for the localization of G protein-coupled receptors to primary cilia. *Proc. Natl. Acad. Sci. USA* **105**, 4242–4246.

Berbari, N. F., O'Connor, A. K., Haycraft, C. J., and Yoder, B. K. (2009). The primary cilium as a complex signaling center. *Curr. Biol.* **19**, R526–R535.

Bishop, G. A., Berbari, N. F., Lewis, J. S., and Mykytyn, K. (2007). Type III adenylyl cyclase localizes to primary cilia throughout the adult mouse brain. *J. Comp. Neurol.* **505**, 562–571.

Bolze, F., and Klingenspor, M. (2009). Mouse models for the central melanocortin system. *Genes Nutr* **4**, 129–134.

Borowsky, B., Durkin, M. M., Ogozalek, K., Marzabadi, M. R., DeLeon, J., Lagu, B., Heurich, R., Lichtblau, H., Shaposhnik, Z., Daniewska, I., Blackburn, T. P., Branchek, T. A., *et al.* (2002). Antidepressant, anxiolytic and anorectic effects of a melanin-concentrating hormone-1 receptor antagonist. *Nat. Med.* **8**, 825–830.

Bouret, S. G., Draper, S. J., and Simerly, R. B. (2004). Trophic action of leptin on hypothalamic neurons that regulate feeding. *Science* **304**, 108–110.

Brailov, I., Bancila, M., Brisorgueil, M. J., Miquel, M. C., Hamon, M., and Verge, D. (2000). Localization of 5-HT(6) receptors at the plasma membrane of neuronal cilia in the rat brain. *Brain Res.* **872**, 271–275.

Breunig, J. J., Sarkisian, M. R., Arellano, J. I., Morozov, Y. M., Ayoub, A. E., Sojitra, S., Wang, B., Flavell, R. A., Rakic, P., and Town, T. (2008). Primary cilia regulate hippocampal neurogenesis by mediating sonic hedgehog signaling. *Proc Natl. Acad. Sci. USA* **105**, 13127–13132.

Butler, A. A., Kesterson, R. A., Khong, K., Cullen, M. J., Pelleymounter, M. A., Dekoning, J., Baetscher, M., and Cone, R. D. (2000). A unique metabolic syndrome causes obesity in the melanocortin-3 receptor-deficient mouse. *Endocrinology* **141**, 3518–3521.

Chen, A. S., Marsh, D. J., Trumbauer, M. E., Frazier, E. G., Guan, X. M., Yu, H., Rosenblum, C. I., Vongs, A., Feng, Y., Cao, L., Metzger, J. M., Strack, A. M., *et al.* (2000). Inactivation of the mouse melanocortin-3 receptor results in increased fat mass and reduced lean body mass. *Nat. Genet.* **26**, 97–102.

Chen, Y., Hu, C., Hsu, C. K., Zhang, Q., Bi, C., Asnicar, M., Hsiung, H. M., Fox, N., Slieker, L. J., Yang, D. D., Heiman, M. L., and Shi, Y. (2002). Targeted disruption of the melanin-concentrating hormone receptor-1 results in hyperphagia and resistance to diet-induced obesity. *Endocrinology* **143**, 2469–2477.

Cohen, E., and Meininger, V. (1987). Ultrastructural analysis of primary cilium in the embryonic nervous tissue of mouse. *Int. J. Dev. Neurosci.* **5**, 43–51.

Collin, G. B., Cyr, E., Bronson, R., Marshall, J. D., Gifford, E. J., Hicks, W., Murray, S. A., Zheng, Q. Y., Smith, R. S., Nishina, P. M., and Naggert, J. K. (2005). Alms1-disrupted mice recapitulate human Alstrom syndrome. *Hum. Mol. Genet.* **14**, 2323–2333.

Collin, G. B., Marshall, J. D., Ikeda, A., So, W. V., Russell-Eggitt, I., Maffei, P., Beck, S., Boerkoel, C. F., Sicolo, N., Martin, M., Nishina, P. M., and Naggert, J. K. (2002). Mutations in ALMS1 cause obesity, type 2 diabetes and neurosensory degeneration in Alstrom syndrome. *Nat. Genet.* **31**, 74–78.

Cone, R. D. (2005). Anatomy and regulation of the central melanocortin system. *Nat. Neurosci.* **8**, 571–578.

Dahl, H. A. (1963). Fine structure of cilia in rat cerebral cortex. *Z. Zellforsch. Mikrosk. Anat.* **60**, 369–386.

Davenport, J. R., Watts, A. J., Roper, V. C., Croyle, M. J., van Groen, T., Wyss, J. M., Nagy, T. R., Kesterson, R. A., and Yoder, B. K. (2007). Disruption of intraflagellar transport in adult mice leads to obesity and slow-onset cystic kidney disease. *Curr. Biol.* **17**, 1586–1594.

Davis, R. E., Swiderski, R. E., Rahmouni, K., Nishimura, D. Y., Mullins, R. F., Agassandian, K., Philp, A. R., Searby, C. C., Andrews, M. P., Thompson, S., Berry, C. J., Thedens, D. R., *et al.* (2007). A knockin mouse model of the Bardet-Biedl syndrome 1 M390R mutation has cilia defects, ventriculomegaly, retinopathy, and obesity. *Proc. Natl. Acad. Sci. USA* **104**, 19422–19427.

Doetsch, F., Garcia-Verdugo, J. M., and Alvarez-Buylla, A. (1999). Regeneration of a germinal layer in the adult mammalian brain. *Proc. Natl. Acad. Sci. USA* **96**, 11619–11624.

Domire, J. S., and Mykytyn, K. (2009). Markers for neuronal cilia. *In* "Methods in Cell Biology" (S. M. King and G. J. Pazour, eds.), Vol. 91. Elsevier, 112–121.

Farooqi, S., and O'Rahilly, S. (2006). Genetics of obesity in humans. *Endocr. Rev.* **27**, 710–718.

Flier, J. S. (2004). Obesity wars: molecular progress confronts an expanding epidemic. *Cell* **116**, 337–350.

Forti, E., Aksanov, O., and Birk, R. Z. (2007). Temporal expression pattern of Bardet-Biedl syndrome genes in adipogenesis. *Int. J. Biochem. Cell Biol.* **39**, 1055–1062.

Fuchs, J. L., and Schwark, H. D. (2004). Neuronal primary cilia: A review. *Cell Biol. Int.* **28**, 111–118.

Handel, M., Schulz, S., Stanarius, A., Schreff, M., Erdtmann-Vourliotis, M., Schmidt, H., Wolf, G., and Hollt, V. (1999). Selective targeting of somatostatin receptor 3 to neuronal cilia. *Neuroscience* **89**, 909–926.

Hayashi, S., and McMahon, A. P. (2002). Efficient recombination in diverse tissues by a tamoxifen-inducible form of Cre: A tool for temporally regulated gene activation/inactivation in the mouse. *Dev. Biol.* **244**, 305–318.

Hearn, T., Spalluto, C., Phillips, V. J., Renforth, G. L., Copin, N., Hanley, N. A., and Wilson, D. I. (2005). Subcellular localization of ALMS1 supports involvement of centrosome and basal body dysfunction in the pathogenesis of obesity, insulin resistance, and type 2 diabetes. *Diabetes* **54**, 1581–1587.

Huszar, D., Lynch, C. A., Fairchild-Huntress, V., Dunmore, J. H., Fang, Q., Berkemeier, L. R., Gu, W., Kesterson, R. A., Boston, B. A., Cone, R. D., Smith, F. J., Campfield, L. A., *et al.* (1997). Targeted disruption of the melanocortin-4 receptor results in obesity in mice. *Cell* **88**, 131–141.

Lee, N. C., Marshall, J. D., Collin, G. B., Naggert, J. K., Chien, Y. H., Tsai, W. Y., and Hwu, W. L. (2009). Caloric restriction in Alstrom syndrome prevents hyperinsulinemia. *Am. J. Med. Genet. A.* **149A**, 666–668.

Li, G., Vega, R., Nelms, K., Gekakis, N., Goodnow, C., McNamara, P., Wu, H., Hong, N. A., and Glynne, R. (2007). A role for Alstrom syndrome protein, alms1, in kidney ciliogenesis and cellular quiescence. *PLoS Genet.* **3**, e8.

Lobe, C. G., Koop, K. E., Kreppner, W., Lomeli, H., Gertsenstein, M., and Nagy, A. (1999). Z/AP, a double reporter for cre-mediated recombination. *Dev. Biol.* **208**, 281–292.

Ludwig, D. S., Tritos, N. A., Mastaitis, J. W., Kulkarni, R., Kokkotou, E., Elmquist, J., Lowell, B., Flier, J. S., and Maratos-Flier, E. (2001). Melanin-concentrating hormone overexpression in transgenic mice leads to obesity and insulin resistance. *J. Clin. Invest.* **107**, 379–386.

Marion, V., Stoetzel, C., Schlicht, D., Messaddeq, N., Koch, M., Flori, E., Danse, J. M., Mandel, J. L., and Dollfus, H. (2009). Transient ciliogenesis involving Bardet-Biedl syndrome proteins is a fundamental characteristic of adipogenic differentiation. *Proc. Natl. Acad. Sci. USA* **106**, 1820–1825.

Marsh, D. J., Weingarth, D. T., Novi, D. E., Chen, H. Y., Trumbauer, M. E., Chen, A. S., Guan, X. M., Jiang, M. M., Feng, Y., Camacho, R. E., Shen, Z., Frazier, E. G., *et al.* (2002). Melanin-concentrating hormone 1 receptor-deficient mice are lean, hyperactive, and hyperphagic and have altered metabolism. *Proc. Natl. Acad. Sci. USA* **99**, 3240–3245.

Moyer, J. H., Lee-Tischler, M. J., Kwon, H. Y., Schrick, J. J., Avner, E. D., Sweeney, W. E., Godfrey, V. L., Cacheiro, N. L., Wilkinson, J. E., and Woychik, R. P. (1994). Candidate gene associated with a mutation causing recessive polycystic kidney disease in mice. *Science* **264**, 1329–1333.

Muzumdar, M. D., Tasic, B., Miyamichi, K., Li, L., and Luo, L. (2007). A global double-fluorescent Cre reporter mouse. *Genesis* **45**, 593–605.

Mykytyn, K., Mullins, R. F., Andrews, M., Chiang, A. P., Swiderski, R. E., Yang, B., Braun, T., Casavant, T., Stone, E. M., and Sheffield, V. C. (2004). Bardet-Biedl syndrome type 4 (BBS4)-null mice implicate Bbs4 in flagella formation but not global cilia assembly. *Proc. Natl. Acad. Sci. USA* **101**, 8664–8669.

Nachury, M. V., Loktev, A. V., Zhang, Q., Westlake, C. J., Peranen, J., Merdes, A., Slusarski, D. C., Scheller, R. H., Bazan, J. F., Sheffield, V. C., and Jackson, P. K. (2007). A core complex of BBS proteins cooperates with the GTPase Rab8 to promote ciliary membrane biogenesis. *Cell* **129**, 1201–1213.

Nagy, T. R., and Clair, A. L. (2000). Precision and accuracy of dual-energy X-ray absorptiometry for determining in vivo body composition of mice. *Obes. Res.* **8**, 392–398.

Nishimura, D. Y., Fath, M., Mullins, R. F., Searby, C., Andrews, M., Davis, R., Andorf, J. L., Mykytyn, K., Swiderski, R. E., Yang, B., Carmi, R., Stone, E. M., *et al.* (2004). Bbs2-null mice have neurosensory deficits, a defect in social dominance, and retinopathy associated with mislocalization of rhodopsin. *Proc. Natl. Acad. Sci. USA* **101**, 16588–16593.

Novak, A., Guo, C., Yang, W., Nagy, A., and Lobe, C. G. (2000). Z/EG, a double reporter mouse line that expresses enhanced green fluorescent protein upon Cre-mediated excision. *Genesis* **28**, 147–155.

Pissios, P., Bradley, R. L., and Maratos-Flier, E. (2006). Expanding the scales: The multiple roles of MCH in regulating energy balance and other biological functions. *Endocr. Rev.* **27**, 606–620.

Qu, D., Ludwig, D. S., Gammeltoft, S., Piper, M., Pelleymounter, M. A., Cullen, M. J., Mathes, W. F., Przypek, R., Kanarek, R., and Maratos-Flier, E. (1996). A role for melanin-concentrating hormone in the central regulation of feeding behaviour. *Nature* **380**, 243–247.

Fath, M. A., Seo, S., Thedens, D. R., Berry, C. J., Weiss, R., Nishimura, D. Y., and Sheffield, V. C. (2008). Leptin resistance contributes to obesity and hypertension in mouse models of Bardet-Biedl syndrome. *J. Clin. Invest.* **118**, 1458–1467.

Sen Gupta, P., Prodromou, N., and Chapple, J. (2009). Can faulty antennae increase adiposity? The link between cilia proteins and obesity. *J Endocrinol.* May 21, Epub ahead of print.

Seo, S., Guo, D. F., Bugge, K., Morgan, D. A., Rahmouni, K., and Sheffield, V. C. (2009). Requirement of Bardet-Biedl syndrome proteins for leptin receptor signaling. *Hum. Mol. Genet.* **18**(7), 1323–1331.

Sharma, N., Berbari, N. F., Yoder, B. K. (2008). Ciliary dysfunction in developmental abnormalities and diseases. *Curr. Top. Dev. Biol.*, **85**, 371–427.

Shimada, M., Tritos, N. A., Lowell, B. B., Flier, J. S., and Maratos-Flier, E. (1998). Mice lacking melanin-concentrating hormone are hypophagic and lean. *Nature* **396**, 670–674.

Soriano, P. (1999). Generalized lacZ expression with the ROSA26 Cre reporter strain. *Nat. Genet.* **21**, 70–71.

Xu, A. W., Ste-Marie, L., Kaelin, C. B., and Barsh, G. S. (2007). Inactivation of signal transducer and activator of transcription 3 in proopiomelanocortin (Pomc) neurons causes decreased pomc expression, mild obesity, and defects in compensatory refeeding. *Endocrinology* **148**, 72–80.

Zhu, Y., Romero, M. I., Ghosh, P., Ye, Z., Charnay, P., Rushing, E. J., Marth, J. D., and Parada, L. F. (2001). Ablation of NF1 function in neurons induces abnormal development of cerebral cortex and reactive gliosis in the brain. *Genes. Dev.* **15**, 859–876.

CHAPTER 9

Using Nucleofection of siRNA Constructs for Knockdown of Primary Cilia in P19.CL6 Cancer Stem Cell Differentiation into Cardiomyocytes

Christian A. Clement[*], **Lars A. Larsen**[†], *and* **Søren T. Christensen**[*]

[*]Department of Biology, Section of Cell and Developmental Biology, University of Copenhagen, DK-2100 Copenhagen OE, Denmark

[†]Wilhelm Johannsen Centre for Functional Genome Research, Department of Cellular and Molecular Medicine, University of Copenhagen, Copenhagen, Denmark

METHODS IN CELL BIOLOGY, VOL. 94

978-0-12-375024-2
DOI: 10.1016/S0091-679X(08)94009-7

Abstract

Primary cilia assemble as solitary organelles in most mammalian cells during growth arrest and are thought to coordinate a series of signal transduction pathways required for cell cycle control, cell migration, and cell differentiation during development and in tissue homeostasis. Recently, primary cilia were suggested to control pluripotency, proliferation, and/or differentiation of stem cells, which may comprise an important source in regenerative biology. We here provide a method using a P19.CL6 embryonic carcinoma (EC) stem cell line to study the function of the primary cilium in early cardiogenesis. By knocking down the formation of the primary cilium by nucleofection of plasmid DNA with siRNA sequences against genes essential in ciliogenesis (IFT88 and IFT20) we block hedgehog (Hh) signaling in P19.CL6 cells as well as the differentiation of the cells into beating cardiomyocytes (Clement *et al.*, 2009). Immunofluorescence microscopy, western blotting, and quantitative PCR analysis were employed to delineate the molecular and cellular events in cilia-dependent cardiogenesis. We optimized the nucleofection procedure to generate strong reduction in the frequency of ciliated cells in the P19.CL6 culture.

I. Introduction

Primary cilia are organelles that emanate from the surface of most growth-arrested mammalian cells. They consist of a microtubule (MT)-based axoneme organized in a $9+0$ axonemal ultrastructure ensheathed by a bilayer lipid membrane continuous with the plasma membrane, but which contains a distinct subset of receptors and other proteins engaged in signaling pathways in developmental processes and tissue homeostasis. Primary cilia are formed via a process termed intraflagellar transport (IFT), which is essential for the assembly and maintenance of almost all eukaryotic cilia and flagella (Cole and Snell, 2009; Pedersen *et al.*, 2008). Separating the two membrane compartments at the ciliary base is a region known as the "ciliary necklace" (Gilula and Satir, 1972), which is connected by fibers to the transition zone of the basal body, which may function as a pore where ciliary precursors and IFT proteins accumulate prior to entering the ciliary compartment via IFT, a process essential for assembly of virtually all cilia and flagella. IFT is a bidirectional transport system that tracks along the polarized MTs of the ciliary axoneme. IFT is composed of large protein complexes, known as IFT particles, and the motor proteins heterotrimeric kinesin-2 (kinesin-2) for anterograde (base to tip) transport of ciliary building blocks, and cytoplasmic dynein 2 for retrograde (tip to base) transport of ciliary turnover products (Pedersen *et al.*, 2008). The signaling pathways being coordinated by the developed primary cilium include Hh, Wingless (Wnt), platelet-derived growth factor receptor (PDGFR)α, Ca^{2+}, neuronal and purinergic receptor signaling, and communication with the ECM (Satir *et al.*, 2010). Accordingly, defects in assembly or function of the primary cilium are a major cause of human diseases and developmental abnormalities and disorders now commonly referred to as ciliopathies (reviewed in Lehman *et al.*, 2008).

Recent observations indicate that primary cilia in stem cells coordinate signaling pathways, including Hh signaling, in cell differentiation during embryonic development and potentially in regulation of stem cell maintenance and/or pluripotency (reviewed in Veland *et al.*, 2009). Stem cells hold great promises for their possible therapeutical abilities since they can give rise to all three germinal layers and differentiate to form specific cell types dependent on the environment and specific factors present. Stem cells may also be important targets against cancer. Hh regulates cell proliferation and differentiation in numerous embryonic tissues and Hh ligands are expressed in the notochord, the floorplate of the neural tube, the brain, the limb bud zone of polarizing activity, and the gut (Odent *et al.*, 1999). Hh signaling is further required in homeostasis of mature tissues and is implicated in human cancers (Beachy *et al.*, 2004) and neurodegenerative disorders (Bak *et al.*, 2003). A screen for embryonic patterning mutations characteristic of defective Hh signaling first indicated a link between IFT proteins, Hh signaling, and nervous system development (Huangfu *et al.*, 2003). Subsequent studies confirmed that Hh signaling is coordinated by the primary cilium to control targets of the Hh pathway by Gli transcription factors (Corbit *et al.*, 2005; Huangfu and Anderson, 2005; Liu *et al.*, 2005; Rohatgi *et al.*, 2007; reviewed in Wong and Reiter, 2008). Functioning Hh components, including Ptc-1, Smo, and Gli transcription factors are localized in primary cilia of human embryonic stem cells (Kiprilov *et al.*, 2008) and neuronal development proceeds ciliary Hh signaling in adult neural stem cell formation, specification of neural cell fate, hippocampal neurogenesis and development of cerebellum and neocortex (Breunig *et al.*, 2008; Han *et al.*, 2008; Komada *et al.*, 2008; Spassky *et al.*, 2008). Similarly, primary cilia are involved in the coordination of Hh signaling, for example, in limb bud formation (Haycraft *et al.*, 2007), skeletogenesis (Gouttenoire *et al.*, 2007), mammary gland development and ovarian function (Johnson *et al.*, 2009), molar tooth number (Ohazama *et al.*, 2009), and development of the pancreas (Nielsen *et al.*, 2008).

Primary cilia and Hh signaling are also implicated in early cardiogenesis as evidenced by defective heart development in knockout mice with defects in ciliary assembly, including decreased trabeculation, increased pericardial space, and malformations of the cardiac outflow tract (Clement *et al.*, 2009). Further, knock down of the primary cilium in the pluripotent P19.CL6 EC stem cell line blocked Hh signaling and differentiation of cells into beating cardiomyocytes *in vitro* (Clement *et al.*, 2009). The P19.CL6 cell line is a subclone from the P19 cell line that spontaneously differentiate into clusters of beating cardiomyocytes in the presence of dimethyl sulfoxide (DMSO) (Habara-Ohkubo, 1996). Further, P19.CL6 cells have no requirement for being cultured in suspension and form embryoid bodies before carrying out the analysis on cardiac differentiation (Uchida *et al.*, 2007). This allows the investigator to follow the function of the primary cilium in the initial phases of differentiation from day 1 through a 2-week period until the formation of beating cardiomyocytes.

II. Rationale

Here we provide a detailed and optimized method for nucleofecting P19.CL6 EC cells with IFT88 and IFT20 siRNA plasmid DNA to produce a high transfection percentage to knockdown primary cilia in cultures of P19.CL6 cells during their differentiation into cardiomyocytes. IFT88 is a subunit of the IFT particle complex B required for functional IFT and assembly of the primary cilium (Pedersen and Rosenbaum, 2008). IFT20 is associated with the Golgi apparatus, and knockdown of this IFT particle reduces ciliary assembly without affecting Golgi structure (Follit *et al.*, 2006). Following knockdown of the cilium cell differentiation can be assessed by light microscopy (LM), immunofluorescence microscopy (IFM), SDS-PAGE, western blotting (WB), and quantitative PCR (qPCR) analysis in order to follow changes in DMSO-induced formation of beating cardiomyocytes, expression and localization of stem cell and cardiomyocyte markers, and activation of Hh signaling.

III. Materials

A. Cell Line and Cell Culture Reagents

The P19.CL6 cell line is of mouse origin isolated from embryonal carcinoma tissue. The originator is Habara, Akemi and registered with Murofushi, Kimiko, Japan (ref nr. 2406 3467).

MEM Alpha medium (Invitrogen, Taastrup, Denmark, Cat#22561-021)
Penicillin/streptomycin (Invitrogen, Taastrup, Denmark, pen/strep, Cat#15140-148)
Phosphate-Buffered Saline (PBS)
Fetal bovine serum (FBS, Sigma Aldrich, Copenhagen, Denmark, Cat#F9665)
Trypsin (Trypsin-EDTA, Invitrogen, Taastrup, Denmark, Cat#15 400-054)
T75 cell culture flasks (Greiner Bio-One GmbH, Frickenhausen, Germany, Cat#658 170)
T25 cell culture flasks (Greiner Bio-One GmbH, Frickenhausen, Germany, Cat#690 175)
6-well trays (Sigma Aldrich, Copenhagen, Denmark, Cat#Z707759-126EA)
Petri dishes (Greiner Bio-One GmbH, Frickenhausen, Germany, 60 × 15 mm, Cat# 628 160)
DMSO (MERCK, Darmstadt, Germany, Cat#1.02952.1000)

B. Reagents and Solutions for Nucleofection

Nucleofector device II (Amaxa Biosystems Lonza, Switzerland)
Nucleofector Kit V (Amaxa Biosystems Lonza, Switzerland)
2 μg of IFT88 or IFT20 siRNA plasmid DNA (high grade, high concentration)
MEM Alpha medium (Gibco, Cat#22561-021)

C. Reagents and Solutions for IFM Analysis

Microscope slides and glass coverslips (12 mm diameter)
Concentrated HCl
Humidity chamber
Ethanol (96% v/v and 70% v/v)
Paraformaldehyde (PFA), 4% w/v solution
Blocking buffer: 2% w/v Bovine Serum Albumin (BSA) in PBS
Permeabilization buffer: 0.2% v/v Triton X-100, 1% w/v BSA in PBS
DAPI (4′,6-diamidino-2-phenylindole, dihydrochloride)
Mounting medium (PBS, 2% w/v *N*-propylgallate, 85% v/v glycerol in PBS)
Nail polish
Antibodies and fluorescent reagents (Table I)

D. Western Blot

NOVEX system (XcellSure Lock, Invitrogen, Taastrup, Denmark)
Precast NuPAGE 10% and 12% BIS-TRIS 12-well gels
Nucleospin kit (Macherey-Nagel, Germany, Cat#740 933.50) for RNA/protein isolation
BSA protein standard (Pierce Biotechnology, Rockford, IL, USA)
Protein assay (BioRAD, Copenhagen, Denmark, DC based on Lowry's method)
Running buffer (Invitrogen, Taastrup, Denmark, Cat#NP0001)
Transferbuffer (Invitrogen, Taastrup, Denmark, Cat#NP0006-1)
Non-fat dry milk blocking buffer
NuPAGE Antioxidant (Invitrogen, Taastrup, Denmark, Cat#NP0005)
Ethanol (96% v/v)
TBST (10 mM Tris, 150 mM NaCl, 0.1% v/v Tween-20, pH 7.6)
Antibodies (Table I)

Table I
Used Antibodies and Fluorescent Reagents

Primary Antibodies	*Dilution*
Mouse anti-α-actinin (Sarcomeric), (Sigma Aldrich, Copenhagen, Denmark, Cat#A-7811)	1:100
Goat anti-Nkx2 (N-19), (Santa Cruz, Cat#SC-8697)	1:100
Mouse anti-β-actin (Sigma Aldrich, Cat#A-5441)	1:5000
Mouse anti-acetylated Tubulin, (Sigma Aldrich, Copenhagen, Denmark, Cat#T7451)	1:1500
Rabbit anti-detyrosinated α-Tubulin (Glu-Tubulin, Chemicon, Millipore, MA, USA, Cat#AB3201)	1:600
Rabbit anti-IFT20 (see Follit *et al.* 2006)	1:1000
Secondary Antibodies and Fluorescent Agents	
Donkey anti-goat (DAG), Alexa Flour® 488 (Molecular Probes, Eugene, OR, USA, Cat#A11055)	1:600
Donkey anti-mouse (DAM), Alexa Flour® 568 (Molecular Probes, Cat#A10037)	1:600
Goat anti-rabbit (GAR), F(ab′)2-specific Alkaline phosphatase-conjugated (Sigma Aldrich, Cat#A3937)	1:5000
Goat anti-mouse (GAM), F(ab′)2-specific Alkaline phosphatase-conjugated (Sigma Aldrich, Cat#A1293)	1:5000
DAPI: 4′,6-diamidino-2-phenylindole, dihydrochloride (Molecular Probes, Cat#D1306)	1:1000

Table II
PCR Primers Used in this Study

Gene	Forward primer (5′–3′)	Reverse primer (5′–3′)	Annealing temp. (°C)
Mef2c	TGCCATCAGTGAATCAAAGG	ATGTTATGTAGGTGCTGCTGC	58
Myh6	CAAAGGAGGCAAGAAGAAAGG	GTCCCCATAGAGAATGCGG	56
Myh7	TGAGGGAACAGTATGAGGAGG	TCGATCTCATTCTGCAGCC	60
Gene	**Forward primer (5′–3′)**	**Reverse primer (5′–3′)**	**Annealing temp. (°C)**
Ptc-1	CCTGCCCACCAAGTGATTGT	CGTTGGGTTCCGAGGGTT	52
Gli1	GCGGAAGGAATTCGTGTGCC	CGACCGAAGGTGCGTCTTGA	62
Gene	**Forward primer (5′–3′)**	**Reverse primer (5′–3′)**	**Annealing temp. (°C)**
Gapdh	AACAGCAACTCCCACTCTTC	TGGTCCAGGGTTTCTTACTC	58
B2m	ATTTTCAGTGGCTGCTACTCG	ATTTTTTTCCCGTTCTTCAGC	58
Hprt	CAAAATGGTTAAGGTGCAAGC	TTTTACTGGCAACATCAACAGG	57
Psmd4	GCAAGATGGTGTTGGAGAGC	TTTGGGTTGGACAGTGTGG	59
Rp13a	GGAGAAACGGAAGGAAAAGG	CTCTATCCACAGGAGCAGTGC	57
Alas1	GGATCGGTGATCGGGATGGCGTCA	AGGTGGTGAAGATGAAGCCCGCAGCGTA	60
Pbgd2	CGTTTGCAGATGGCTCCAATAGTAAAG	TGGCATACAGTTTGAAATCATTGCTATGT	60

E. Quantitative RT-PCR

Nucleospin kit (Macherey-Nagel, Cat#740 933.50) for RNA/protein isolation
SuperScript™ II reverse transcriptase (Invitrogen, Cat#18064-014)
DNase I Amplification grade (Invitrogen, Cat#18068-015)
PCR-grade nuclease-free water
dNTP mix, 10 mM of each dNTP
RNAse inhibitor (rRNasin from Promega, VI, USA)
qPCR plate (Applied Biosystems, Foster City, CA, USA Cat#4346906)
Lightcycler® FastStart DNA Master^{PLUS} SYBR Green I (Roche Hvidovre, Denmark Cat#030515885001)
RNA purifying kit (Macherey-Nagel, Cat#740 933.50) for RNA/protein Ammonium Buffer with 15 mM $MgCl_2$ (Ampliqon III, Cat#AMP300305)
TAQ-polymerase (5 units/μl, Ampliqon III)
Agarose, ethidium bromide, TAE buffer (4.84 g Tris base, 1.14 ml glacial acetic acid, 2 ml 0.5 M Na_2 EDTA pH 8.0, ddH_2O up to 1 l), nucleic acid molecular weight marker
PCR primers (Table II)
Mouse universal reference total RNA (Stratagene, Cedar Creek, Texas, USA Cat#740100)
E.N.Z.A. gel extraction kit (Omega Bio-Tek inc., Norcross, GA, USA, Cat#D2500-00)

IV. Methods

A. Introductory Remarks and Experimental Outline

Growth arrest and formation of primary cilia in cultures of mammalian cells can be induced either by depletion of serum and/or by growing cells to confluency. In cultures of P19.CL6 EC stem cells primary cilia are formed in the presence of serum both in their pluripotent stage and during the differential steps induced by DMSO to form clusters of beating cardiomyocytes, most probably because physical contact between individual cells promotes the entrance into G_0 (Clement *et al.*, 2009). This allows the investigation of the function of primary cilia in stem cell differentiation and cardiomyogenesis *in vitro* by knocking down the cilium. Cultures of P19.CL6 cells in 9.6 cm^2 Petri dishes may form up to about 20 clusters of cardiomyocytes with a diameter of 0.2–0.6 mm that beat synchronously at a frequency of about 60 rhythmic contractions per minute around day 12 in the presence of DMSO (Clement *et al.*, 2009). Prior to differentiation cells are positive for stem cell markers Sox2 and Oct4, which are replaced by Gata4 positive cells at day 2, Nkx2–5 positive cells around day 8, and α-actinin positive cells marking the Z-line on α-cardiac muscle stress fibers around day 12 concomitantly with the onset of contractions of the mini hearts. During these steps of cell proliferation and differentiation, the expression of the transcriptional target genes for Hh signaling, *Gli1* and *Ptc-1* are highly upregulated: both cardiomyocytes differentiation and Hh signaling being blocked in the presence of the Smo antagonist and Hh-signaling inhibitor, cyclopamine (Clement *et al.*, 2009).

In this protocol we performed siRNA knockdown of the primary cilium in P19.CL6 cells by nucleofection, which is a transfection method that enables efficient and reproducible transfer of nucleic acids such as DNA, RNA, and siRNA into cells. Nucleofection is also known as Nucleofector Technology, which was invented by Amaxa (http://www.amaxa.com). Nucleofection uses a combination of optimized parameters generated by a device termed Nucleofector device II with cell-type specific reagents and buffers. The substrate is transferred directly into the cell nucleus and cytosol. Here we nucleofected P19.CL6 cells with IFT88 and IFT20 siRNA plasmid DNAs in order to knock down the cilium and follow its consequences in stem cell maintenance, cell differentiation, and Hh signaling. The IFT88 plasmid DNA is expressing GFP, which enables quantification of nucleofection rates and direct visualization of its consequences on ciliary formation and cell differentiation. Here we present a detailed protocol on the nucleofection procedure, followed by qualitative and quantitative analysis on cell differentiation and Hh signaling by IFM, SDS-PAGE, WB, and qPCR analysis. The protocol for qPCR analysis is presented comprehensively, while IFM, SDS-PAGE, and WB analyses are portrayed in less detail.

B. Cell Culturing and Passaging

The cells were grown for passaging in T25 cell culture flasks at 37°C, 5% CO_2, and 95% humidity. The cells were passaged every 2–3 days by trypsination, and grown in MEM Alpha cell culture media, containing 1% penicillin/streptomycin and 10% FBS. Prior to trypsination the cells were washed once in 37°C PBS and reseeded at 10–15% confluency (avoid growing cells into 100% confluency). To induce cardiomyocyte differentiation, the medium was supplemented with 1% DMSO and grown under normal incubator conditions. Experimental cells were seeded at a confluency of about 30% in T75 cell culture flasks, 6-well trays, or Petri dishes.

C. Nucleofection of P19.CL6 Cells with IFT80 and IFT20 siRNA (Plasmid DNA)

Sufficient cells were cultivated in T75 flasks, enough to provide each sample with 2×10^6 cells. Preferably, the cells should be passaged the day before nucleofection but if the experiments require the cells to enter a stage of differentiation (e.g., DMSO-induced differentiation that needs siRNA knockdown at day 2–3), it is still possible to carry out the nucleofection with a high transfection rate. This will disrupt cell culture morphology since the cells will need to be resuspended in nuclofection solution and reseeded.

Before the actual nucleofection step, prepare the following:

- 6-well trays/dishes with the appropriate working volume of differentiation media are put inside the incubator (37°C/5% CO_2) and prewarmed 20 min before nucleofection.
- Prewarm the supplemental Cell Line Nucleofector Solution V to room temperature.
- Thaw up your highly purified IFT88 and IFT20 siRNA plasmids. Amount of μg is determined by optimization (e.g., 2 μg DNA per sample, see Section IV.D).

The medium was removed from the T75 culture flasks and washed once in 37°C PBS. The cells were then trypsinated in 1 × Trypsin-EDTA for 5 min in 37°C/5% CO_2 and resuspended in an appropriate volume of 37°C growth medium (preferably freshly made). The volume was adjusted so that the cell density was high enough to collect 2×10^6 cells in a 1.5 ml Eppendorf tube. The cells were counted in a hemocytometer to determine the actual number of cells per ml cell suspension. The cells were then spun down at $90 \times g$ for 10 min at room temperature. The cell pellet was resuspended in 100 μl Cell Line Nucleofector Solution V per 2×10^6 cells per sample. Do not keep the cells in the Nucleofector Solution V for more than 15 min.

Each sample is nucleofected in the following steps:

- Add 2 μg of siRNA plasmid DNA into the 100 μl solution containing the cell pellet.
- Gently mix the DNA, cells, and Cell Line Nucleofector Solution V three times in a 1000 μl pipette and transfer the sample to an Amaxa-certified cuvette (make sure the sample covers the cuvette bottom with no air bubbles).
- Close the cuvette with the lid and insert into the Nucleofector device II and run program C-020.
- Transfer 500 μl media from the prewarmed 6-well tray/dish using the supplied plastic pipettes into the cuvette and transfer the whole sample back into the 6-well tray/dish. Avoid transferring white foam "popcorn" to the 6-well tray/dish since this contains only cell debris

The 6-well trays/dishes are put into the incubator as soon as possible in 37°C/5% CO_2 to provide as little stress as possible. Analysis of the samples and the effect of IFT88 and IFT20 siRNA knockdown can be performed 24 h after nucleofection. At this time point the maximal effect of the IFT88 and IFT20 knockdown can be found. The optimal time for siRNA plasmid expression can be determined by selecting different time points after nucleofection and verifying the level of mRNA knockdown with IF analysis of a GFP-reporter inserted in the siRNA plasmid.

D. Methods for Optimization with IFT88 siRNA Plasmid for High Transfection Rate

To find the optimal transfection rate, we used the pmaxGFP^{TM} positive control vector supplied by Amaxa biosystems. An experimental setup with variable amounts of pmaxGFP^{TM} (1 μg, 1.5 μg, 2 μg, 2.5 μg, 3 μg, and 4 μg) was nucleofected into 2×10^6 cells. The cells were grown on coverslips in 6-well trays at 37°C/5% CO_2 for 24 h. The cells were fixed in 4% paraformaldehyde for 15 min at RT, followed by two washes in PBS, and then permeabilized in 0.2% Triton X-100, 1% BSA for 12 min. Cells were then stained with DAPI to visualize total cell numbers on the coverslips and the total-cell/transfected-cell ratio could hereby be determined. Also variable time points after nucleofection (12, 24, 36, 48, and 72 h) were tested to determine at which time point the transfection rate was the highest. The recommended DNA/cell number ratio supplied by Amaxa biosystems was 2×10^6 cells with 2 μg DNA plasmid followed by analysis after 24 h. The optimal transfection rate for our experiments was found under the recommended conditions and was close to 80%. Cell viability was good—estimated to ~70%. Furthermore, it is highly recommendable to perform both positive and negative nucleofection controls (cells + solution + DNA – program) (cells + solution – DNA + program) to assess influences of nucleofection or purity of DNA on cell viability.

E. Nucleofection for Light and Immunofluorescence Microscopy Analysis

After nucleofection the cells were transferred to 6-well trays containing coverslips and grown in 37°C/5% CO_2 for at least 24 h to visualize the optimal effect of the transfection. After 1, 5, 8, and 12 days of culturing in differentiation medium the cells were subjected to IFM in order to analyze the frequency of ciliated cells. The protocol used for IFM and detection of primary cilia is identical to that described in Chapter 3 by Thorsteinsson *et al.* (this volume). For detection of primary cilia we used primary antibodies against acetylated α-tubulin (1:1500) and detyrosinated α-tubulin (1:600). Similarly, one can use antibodies directed against specific markers of cardiomyogenesis in order to follow the time course for differentiation, and how nucleofection with IFT88 and IFT20 siRNA constructs impinge on stem cell maintenance and cardiomyocte differentiation with antibodies directed against Oct4 and Sox2 (stem cell markers) and Gata4, Nkx2–5, and α-actinin (cardiomyocytes markers) (Clement *et al.*, 2009). Further, expression and localization of Hh signaling components such as Gli transcription factors are assessed by IFM and the formation of beating clusters of cardiomyocytes is easily detected by light microscopy.

F. Nucleofection for Western Blot Analysis

After nucleofection the cells were transferred to Petri dishes and grown in 37°C/5% CO_2 for at least 24 hours. Hereafter the cells were cultured for 1–12 days in differentiation medium. For WB analysis cells were washed in PBS, spun down at 500 × *g* for 5 min, and then added with 350 μl lysis buffer with 1% β-mercaptoethanol. Proteins were purified using Nucleospin kit protocol (Macherey-Nagel, Cat# 740 933.50) for

RNA/protein. After protein precipitation the cells were added with 2% SDS, 1% Glycerol lysis buffer and sonicated, centrifuged at 20,000 × *g* to precipitate nonsoluble material. The protein concentration was compared with a BCA protein standard (Pierce Biotechnology) and measured with a Protein assay so that an equal amount of protein could be loaded in each lane. For SDS-PAGE a NOVEX system was used with precast NuPAGE 10% and 12% BIS-TRIS 12-well gels (Schneider *et al.*, 2005).

G. Nucleofection for Quantitative Real-Time RT-PCR (qPCR) Analysis

After nucleofection the cells were treated as in the above for WB analysis in order to isolate total RNA accordingly to the Nucleospin kit protocol for RNA/protein and to assess transcriptional processes in Hh signaling and cell differentiation. RNA was treated with DNase I Amplification grade and cDNA was produced from 1 μg total RNA using SuperScript^{TM} II reverse transcriptase. PCR primers for amplification of housekeeping genes (*Gapdh, B2m, Hprt, Psmd4, Rp13a, Alas1, Pbgd2*), cardiomyocyte markers (*Mef2C, Myh6, Myh7)* and Hh signaling genes (*Smo, Ptc-1)* were designed using Oligo version 6.23 (Table II). PCR annealing temperature was optimized in a QuatroCycler temperature gradient thermocycler (VWR, West Chester, PA, USA) using universal mouse cDNA as template. For each primer pair PCR fragments were excised from an agarose gel and extracted using an E.N.Z.A. gel extraction kit. Standard curves were generated by 10-fold serial dilutions of the PCR fragments. The quantitative real-time RT-PCR (qPCR) reactions were performed in a 7500 fast Real-time PCR system (Applied Biosystems,) using a Lightcycler Fast Start DNA Master plus SYBR Green 1 kit (Roche, Copenhagen, Denmark).

For each gene the cycle threshold (C_t) value was converted to a relative expression (E_r) value using the standard curve. Due to the high sensitivity of qPCR, differences in the efficiency of the cDNA synthesis and differences in PCR reaction kinetics between samples may lead to incorrect quantitation of the expression. To correct for intersample variation, samples are normalized by dividing the E_r value of the gene of interest with the E_r value of an endogenous control. The ideal endogenous control for qPCR analysis is a gene that displays similar reaction kinetics and expression profiles in all samples (VanGuilder *et al.*, 2008). Usually, housekeeping genes are used as endogenous controls. However, not all housekeeping genes are resistant to experimental conditions; thus to ensure a robust expression profile of the endogenous control in all samples, we normalized using the average E_r value of at least three housekeeping genes with similar expression profiles across the samples. Only normalized E_r values were compared in the experiments.

V. Results and Discussion

A. Timetable and Markers of Differentiation in P19.CL6 Cells

In order to delineate the onset for DMSO-induced P19.CL6 stem cell differentiation we initially used light microscopy analysis to evaluate the time point for formation of beating clusters of cardiomyocytes. As depicted in Fig. 1A most cell cultures form

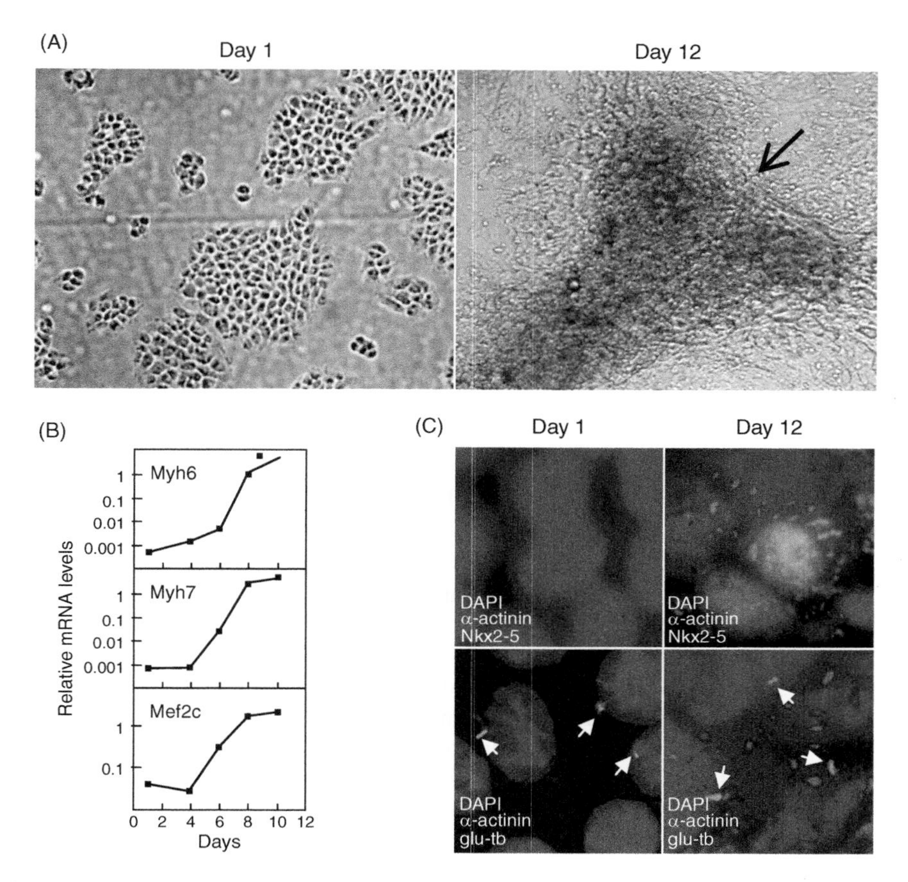

Fig. 1 Morphology and expression profile of cardiomyocyte markers in P19.CL6 cells during cardiomyocyte differentiation induced by 1% DMSO. (A) Light microscope images of P19.CL6 cell morphology at day 1 prior to differentiation and at day 12 where cells have formed beating clusters of cardiomyocytes (open arrow). (B) Quantitative RT-PCR analysis on Myh6, Myh7, and Mef2c mRNA levels relative to expression of housekeeping genes during days 1–10 of P19.CL6 differentiation. (C) Immunofluorescence microscopy analysis at days 1 and 12 of localization of α-actinin and primary cilia (lower panels) and α-actinin and Nkx2–5 (upper panels). Upper panels: Nkx2–5: green; α-actinin: red; DAPI: blue. Lower panels: primary cilia (arrows, detyrosinated α-tubulin, glu-tb): green; α-actinin: red; DAPI: blue. (See Plate no. 9 in the Color Plate Section.)

beating clusters around day 12, clusters appearing in small networks that beat at a frequency of about 60 rhythmic contractions per minute. These results are consistent with the findings that clusters of cells at this time point are positive for α-actinin that marks the Z-line on α-cardiac muscle stress fibers (Fig. 1C, upper panels) (Clement *et al.*, 2009). To follow differentiation of P19.CL6 cells in more detail we then measured the transcription of the cardiomyocyte markers *Gata4, Nkx2-5, Actc2,*

Mef2C, Myh6, and *Myh7* using qRT-PCR. As exemplified in Fig. 1B cardiomyocyte marker genes are transcriptionally upregulated around day 4 after DMSO addition, their expression levels peaking around day 10. These results are in agreement with previous data, showing that cardiomyocyte markers are upregulated at around day 2 and 6 in P19.CL6 cells stimulated with DMSO (Clement *et al.*, 2009). Finally, we performed IFM analysis to show that clusters of cells either at day 1 in undifferentiated cells or day 12 in differentiated cells (positive for Nkx2–5 and α-actinin; Fig. 1C, upper panels) formed primary cilia (Fig. 1C, lower panels) as evidenced by staining with antidetyrosinated α-tubulin (Glu-tb) that is highly enriched in primary cilia of vertebrate cells (Gundersen and Bulinski, 1986). This allows the investigator to examine the role of the primary cilium by RNAi methods in differentiation of P19. CL6 cells using qRT-PCR analysis on the expression of cardiomyocyte genes, *Myh6, Myh7,* and *Mef2c.*

B. Nucleofection with IFT88 and IFT20 siRNA and Its Consequences on Ciliary Formation in P19.CL6

Nucleofection was performed with IFT88 and IFT20 siRNA plasmid constructs in order to inhibit the formation of primary cilia in cultures of P19.CL6 cells and then subsequently analyze the effect of ciliary knockdown on cardiomyocyte differentiation. In this regard there are a number of analyses to perform to ensure optimal knockdown efficiency and minimize unwanted effects on cell behavior, such as cell viability caused by the transfection method. As described in Section IV.D, transfection rates can initially be determined with a pmaxGFP™ positive control vector, in which the rate of transfection can be monitored and estimated by green fluorescence in nucleofected cells. In this analysis we found that the optimal transfection rate of 80% was achieved by nucleofection of 2×10^6 P19.CL6 cells with 2 μg plasmid DNA 2 days after the addition of DMSO to the cell cultures, followed by analysis after 24 h. In this setup cell viability was estimated to about 70%. We then used these parameters to analyze cell viability and transfection rate after nucleofection with IFT88 siRNA GFP plasmid. The cells were brought into suspension and nucleofected with the plasmid, and after 24 h of subsequent cell culture the cells were fixed, stained with DAPI, and the transfection rate was calculated to be about 70–80% as judged by IFM (Fig. 2A). In total, the cells had 3 days of differentiation with 1% DMSO in the culture media, hence the day 3 in Fig. 2A.

In order to examine the effect of IFT88 knockdown on formation of primary cilia, we performed IFM analysis with antibodies directed against either detyrosinated α-tubulin or acetylated α-tubulin, the latter also marking primary cilia and other stable cellular MTs (Piperno and Fuller, 1985). As indicated in Fig. 2B, the number of primary cilia at day 6 was markedly reduced in cells nucleofected with IFT88 siRNA plasmid as compared to mock transfected cells. Further, primary cilia emerging from cells in IFT88 siRNA-nucleofected cells were often shorter than in the control cells, supporting the conclusion that IFT88 is required for ciliary assembly. The percentage of ciliated cells at day 3 was further enumerated by IFM analysis as

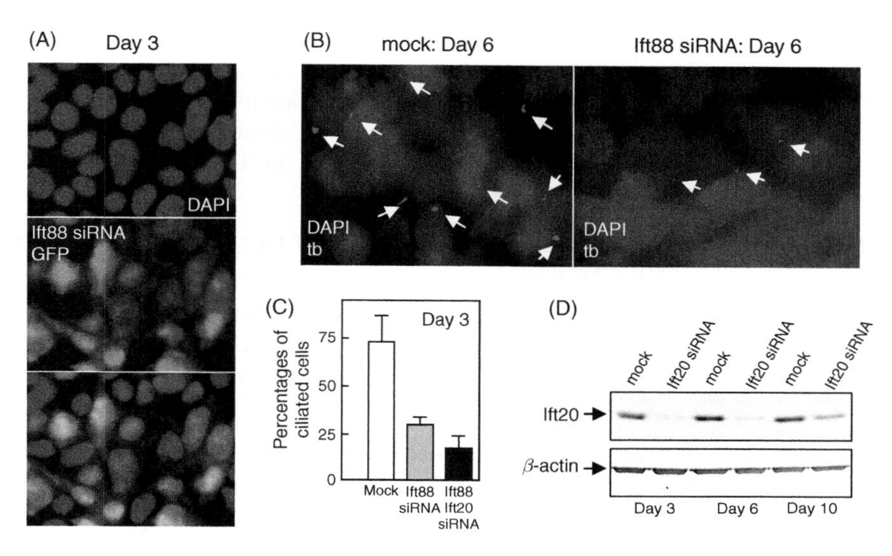

Fig. 2 Nucleofection of IFT88 and IFT20 plasmid siRNA in P19.CL6 cells. (A) Immunofluorescence microscopy analysis of cells nucleofected with GFP-expressing IFT88 siRNA plasmid at day 3. Transfection rate efficiency: ~70–80% (IFT88 siRNA GFP: green; DAPI: blue). (B) Immunofluorescence microscopy analysis of primary cilia at day 6 of differentiation after nucleofection with mock and IFT88 siRNA (primary cilia were localized with antiacetylated α-tubulin, tb: red; DAPI: blue). (C) Bar graph showing the percentage of ciliated cells in mock, IFT88, and IFT88 + IFT20 siRNA-nucleofected P19.CL6 cells at day 3. Reproduced with permission from Clement *et al.* (2009). (D) Western blot analysis of IFT20 (~20 kDa) and β-actin (~43 kDa) protein levels in P19.CL6 cells after 3, 6, and 10 days of IFT20 siRNA nucleofection versus mock transfected cells. (See Plate no. 10 in the Color Plate Section.)

shown in Fig. 2C. In controls, the percentage of ciliated cells was calculated to be about 75%, whereas this number was reduced to about 30% in IFT88 siRNA-nucleofected cells. As a further control, we found no differences in cell viability between mock and IFT88 siRNA-nucleofected cells, which was estimated to be about 70% in both cases (Clement *et al.*, 2009). We then went on to perform a double knockdown of the primary cilium in P19.CL6 cells using both IFT88 and IFT20 siRNA plasmid DNA. As indicated in Fig. 2C, the number of ciliated cells was further reduced to about 20%, showing that a reduction in both IFT88 and IFT20 has a stronger inhibitory effect on ciliary formation than IFT88 alone.

The level of IFT20 knockdown by nucleofection can also be verified on protein levels with WB analysis. We show here (Fig. 2D) an example of such an analysis of IFT20 protein levels in cells subjected to IFT20 siRNA knockdown compared to mock transfected cells on day 3, 6, and 10. Initially protein levels were greatly reduced on day 3 and 6 compared to control cells. At day 10, however, reduction in the protein level was less pronounced, indicating a loss in siRNA of IFT20. A similar phenomenon was observed with IFT88 siRNA (not shown). A plausible explanation for this is that a portion of the transfected cells may lose their siRNA plasmid over time as a consequence of continuous cell divisions such that the number of plasmid-containing cells in the culture is reduced. This also means that the percentage of ciliated cells in nucleofected cultures may increase over time around day 10. Since nucleofection requires that cells are kept in suspension, it was not an option to perform a second round of nucleofection of P19.CL6 cells at day 10 without disrupting cell morphology and clusters of cardiomyocytes, which begin to form at this time point after DMSO stimulation.

C. Nucleofection with IFT88 and IFT20 siRNA and Its Consequences on Differentiation and Hh Signaling in P19.CL6 Cells

To analyze the effects of IFT88 and IFT20 knockdown on cardiomyocte differentiation and Hh signaling we initially performed qRT-PCR analysis on the mRNA levels of myocyte enchancer factor 2c (Mef2c) and myosin heavy chain 7 (Myh7). As shown in Fig. 3A the

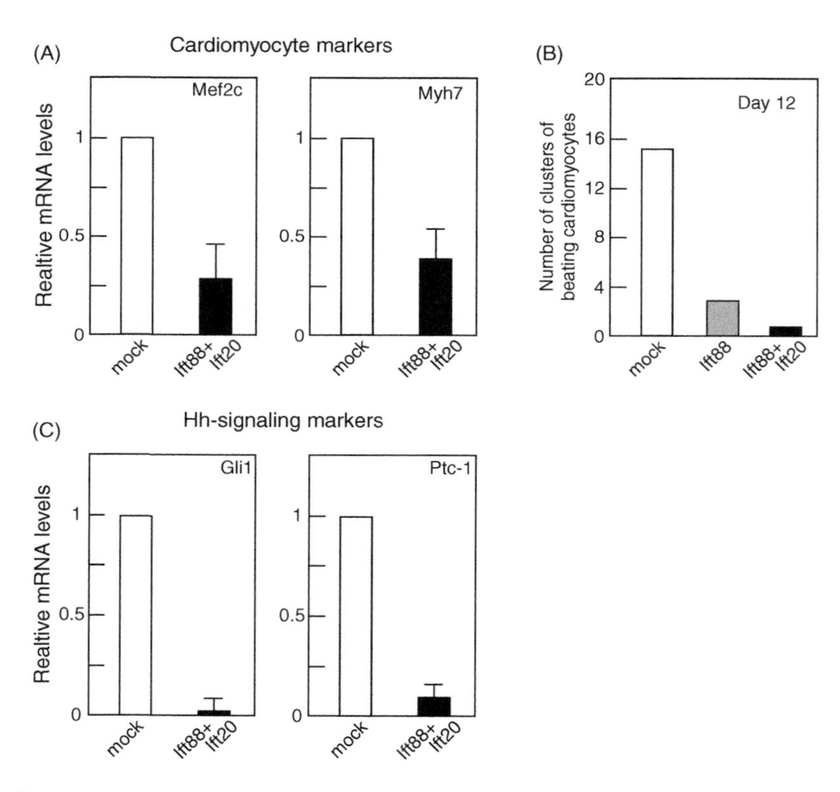

Fig. 3 Knockdown of IFT88 and IFT20 by siRNA nucleofection inhibits cardiomyocyte differentiation and Hh signaling in P19.CL6 cells. (A) Bar graph showing quantitative RT-PCR analysis on Mef2c and Myh7 mRNA levels relative to expression of housekeeping genes after siRNA nucleofection versus mock at day 5. (B) Number of beating cardiomyocyte clusters in mock, IFT88, and IFT88 + IFT20 siRNA-nucleofected P19.CL6 cells at day 12. (C) Bar graph showing quantitative RT-PCR analysis on Ptc1 and Gli1 mRNA levels relative to expression of housekeeping genes siRNA nucleofection versus mock at day 5. Reproduced with permission from Clement *et al.* (2009).

mRNA levels of both markers of cardiomyocytes differentiation were largely reduced by about 75 and 60%, respectively, compared to mock transfected cells. Similarly, it was shown that IFT88 and IFT20 knockdown reduced the mRNA and/or protein levels of Gata4, Nkx2–5, and α-actinin as judged by qRT-PCR, WB, and IFM analysis (Clement *et al.*, 2009), supporting the conclusion that the primary cilium is critical in the regulation of P19.CL6 cell differentiation into cardiomyocytes. This was further sustained by the observation that knock down of IFT proteins reduced the number of clusters of beating cardiomyocytes at day 12. Figure 3B presents data from a single and representative experiment in which 15 beating clusters in mock transfected cells were reduced to 3 and 1 clusters in IFT88 and IFT88 + 20 siRNA-nucleofected cells, respectively. Moreover, the clusters of cardiomyocytes in IFT siRNA-nucleofected cells were abnormally small with no or very little networks between the individual clusters. It is likely that these irregular clusters of cardiomyocytes are formed as a consequence of a partial loss of plasmid DNA in IFT88 and IFT88 + 20 siRNA-nucleofected cells.

Early cardiogenesis is regulated by a number of different signal transduction pathways, including Hh, Wnt, bone morphogenetic protein (BMP), and PDGFR signaling (Hirata *et al.*, 2007; Kwon *et al.*, 2008; van Wijk *et al.*, 2007; Washington Smoak *et al.*, 2005). In conjunction with defects in cardiomyogenesis qRT-PCR analysis showed that IFT88 20 siRNA inhibited Hh signaling in P19.CL6 cells. Under normal cardiomyocyte formation the key elements in Hh signaling, Gli1 and Ptc-1, are upregulated throughout differentiation from day 1 to 9 (Clement *et al.*, 2009). As shown in Fig. 3C, the relative mRNA levels of Gli1 and Ptc-1 at day 5 were reduced to about 5 and 10%, respectively, of the level in mock transfected cells (Clement *et al.*, 2009). These results indicate that Hh signaling in P19.CL6 cell differentiation is coordinated by the primary cilium. Additional experiments will be required to investigate whether IFT88 + 20 siRNA and ciliary knockdown in P19.CL6 cells also affects other signaling pathways, including Wnt, PDGFR, and BMP signaling, which may provide further insight into the function of the primary cilium in cardiomyogenesis and heart development.

VI. Summary

We have described a detailed siRNA-based nucleofection protocol for examining the role of the primary cilium in differentiation of P19.CL6 cancer stem cells into cardiomyocytes. Knockdown of IFT88 and IFT20 with their corresponding siRNA plasmid DNA inhibits ciliary formation, Hh signaling, and differentiation of cells into cardiomyocytes as judged by qRT-PCR, IFM, SDS-PAGE, and WB analysis. In addition to identifying cilia-related signaling pathways, the nucleofection method can be extended to identify other genes that are involved in P19.CL6 stem cell maintenance, proliferation, and differentiation.

Acknowledgments

This work was supported by the Lundbeck Foundation, the Danish Science Research Council (STC), the Danish Heart Association (LAL), and funds from the Department of Biology, University of Copenhagen,

Denmark (CAC). Wilhelm Johannsen Centre for Functional Genome Research is established by the Danish National Research Foundation. The authors would like to thank Stine Gry Kristensen for excellent help on qRT-PCR analysis. The authors would also like to thank Lillian Rasmussen, Kirsten Winther, and Laura Smedegaard Kruuse for technical assistance.

References

Bak, M., Hansen, C., Tommerup, N., and Larsen, L. A. (2003). The Hedgehog signaling pathway—Implications for drug targets in cancer and neurodegenerative disorders. *Pharmacogenomics* **4**, 411–429.

Beachy, P. A., Karhadkar, S. S., and Berman, D. M. (2004). Tissue repair and stem cell renewal in carcinogenesis. *Nature* **432**, 324–331.

Breunig, J. J., Sarkisian, M. R., Arellano, J. I., Morozov, Y. M., Ayoub, A. E., Sojitra, S., Wang, B., Flavell, R. A., Rakic, P., and Town, T. (2008). Primary cilia regulate hippocampal neurogenesis by mediating sonic hedgehog signaling. *Proc. Natl. Acad. Sci. USA* **105**, 13127–13132.

Clement, C. A., Kristensen, S. G., Møllgård, K., Pazour, G. J., Yoder, B. K., Larsen, L. A., and Christensen, S. T. (2009). The primary cilium coordinates early cardiogenesis and hedgehog signaling in cardiomyocyte differentiation. *J. Cell Sci.* **122**, 3070–3082.

Cole, D. G., and Snell, W. J. (2009). SnapShot: Intraflagellar transport. *Cell* **137**, 784–784.

Corbit, K. C., Aanstad, P., Singla, V., Norman, A. R., Stainier, D. Y., and Reiter, J. F. (2005). Vertebrate Smoothened functions at the primary cilium. *Nature* **437**, 1018–1021.

Gilula, N. B., and Satir, P. (1972). The ciliary necklace: A ciliary membrane specialization. *J. Cell Biol.* **53**, 494–509.

Gouttenoire, J., Valcourt, U., Bougault, C., Aubert-Foucher, E., Arnaud, E., Giraud, L., and Mallein-Gerin, F. (2007). Knockdown of the intraflagellar transport protein IFT46 stimulates selective gene expression in mouse chondrocytes and affects early development in zebrafish. *J. Biol. Chem.* **282**, 30960–30973.

Gundersen, G. G., and Bulinski, J. C. (1986). Microtubule arrays in differentiated cells contain elevated levels of a post-translationally modified form of tubulin. *Eur. J. Cell Biol.* **42**, 288–294.

Follit, J. A., Tuft, R. A., Fogarty, K. E., and Pazour, G. J. (2006). The intraflagellar transport protein IFT20 is associated with the Golgi complex and is required for cilia assembly. *Mol. Biol. Cell* **17**, 3781–3792.

Habara-Ohkubo, A. (1996). Differentiation of beating cardiac muscle cells from a derivative of P19 embryonal carcinoma cells. *Cell Struct. Funct.* **21**, 101–110.

Han, Y. G., Spassky, N., Romaguera-Ros, M., Garcia-Verdugo, J. M., Aguilar, A., Schneider-Maunoury, S., and Alvarez-Buylla, A. (2008). Hedgehog signaling and primary cilia are required for the formation of adult neural stem cells. *Nat. Neurosci.* **11**, 277–284.

Haycraft, C. J., Zhang, Q., Song, B., Jackson, W. S., Detloff, P. J., Serra, R., and Yoder, B. K. (2007). Intraflagellar transport is essential for endochondral bone formation. *Development* **134**, 307–316.

Hirata, H., Kawamata, S., Murakami, Y., Inoue, K., Nagahashi, A., Tosaka, M., Yoshimura, N., Miyamoto, Y., Iwasaki, H., Asahara, T., and Sawa, Y. (2007). Coexpression of platelet-derived growth factor receptor alpha and fetal liver kinase 1 enhances cardiogenic potential in embryonic stem cell differentiation in vitro. *J. Biosci. Bioeng.* **103**, 412–419.

Huangfu, D., and Anderson, K. V. (2005). Cilia and hedgehog responsiveness in the mouse. *Proc. Natl. Acad. Sci. USA* **102**, 11325–11330.

Huangfu, D., Liu, A., Rakeman, A. S., Murcia, N. S., Niswander, L., and Anderson, K. V. (2003). Hedgehog signalling in the mouse requires intraflagellar transport proteins. *Nature* **426**, 83–87.

Johnson, A. L., and Woods, D. C. (2009). Dynamics of avian ovarian follicle development: Cellular mechanisms of granulosa cell differentiation. *Gen Comp Endocrinol.* **163**, 12–17.

Kiprilov, E. N., Awan, A., Desprat, R., Velho, M., Clement, C. A., Byskov, A. G., Andersen, C. Y., Satir, P., Bouhassira, E. E., Christensen, S. T., and Hirsch, R. E. (2008). Human embryonic stem cells in culture possess primary cilia with hedgehog signaling machinery. *J. Cell Biol.* **180**, 897–904.

Komada, M., Saitsu, H., Kinboshi, M., Miura, T., Shiota, K., and Ishibashi, M. (2008). Hedgehog signaling is involved in development of the neocortex. *Development* **135**, 2717–2727.

Kwon, C., Cordes, K. R., and Srivastava, D. (2008). Wnt/beta-catenin signaling acts at multiple developmental stages to promote mammalian cardiogenesis. *Cell Cycle* **7**, 3815–3818.

Lehman, J. M., Michaud, E. J., Schoeb, T. R., Aydin-Son, Y., Miller, M., and Yoder, B. K. (2008). The Oak Ridge Polycystic Kidney mouse: Modeling ciliopathies of mice and men. *Dev. Dyn.* **237**, 1960–1971.

Liu, A., Wang, B., and Niswander, L. A. (2005). Mouse intraflagellar transport proteins regulate both the activator and repressor functions of Gli transcription factors. *Development* **132**, 3103–3111.

Nielsen, S. K., Møllgård, K., Clement, C. A., Veland, I. R., Awan, A., Yoder, B. K., Novak, I., and Christensen, S. T. (2008). Characterization of primary cilia and hedgehog signaling during development of the human pancreas and in human pancreatic duct cancer cell lines. *Dev. Dyn.* **237**, 2039–2052.

Odent, S., Atti-Bitach, T., Blayau, M., Mathieu, M., Aug, J., Delezo de A. L., Gall, J. Y., Le Marec, B., Munnich, A., David, V., and Vekemans, M. (1999). Expression of the Sonic hedgehog (SHH) gene during early human development and phenotypic expression of new mutations causing holoprosencephaly. *Hum. Mol. Genet.* **8**, 1683–1689.

Ohazama, A., Haycraft, C. J., Seppala, M., Blackburn, J., Ghafoor, S., Cobourne, M., Martinelli, D. C., Fan, C. M., Peterkova, R., Lesot, H., Yoder, B. K., and Sharpe, P. T. (2009). Primary cilia regulate Shh activity in the control of molar tooth number. *Development* **136**, 897–903.

Pedersen, L. B., and Rosenbaum, J. (2008). Intraflagellar transport (IFT) role in ciliary assembly, resorption and signalling. *Curr. Top. Dev. Biol.* **85**, 23–61.

Pedersen, L. B., Veland, I. R., Schrøder, J. M., and Christensen, S. T. (2008). Assembly of primary cilia. *Dev. Dyn.* **237**, 1993–2006.

Piperno, G., and Fuller, M. T. (1985). Monoclonal antibodies specific for an acetylated form of alpha-tubulin recognize the antigen in cilia and flagella from a variety of organisms. *J. Cell Biol.* **101**, 2085–2094.

Rohatgi, R., Milenkovic, L., and Scott, M. P. (2007). Patched1 regulates hedgehog signaling at the primary cilium. *Science* **317**, 372–376.

Satir, P., Pedersen, L. B., and Christensen, S. T. (2010). Primary Cilia at a Glance. *J. Cell Sci.* In press.

Schneider, L., Clement, C. A., Teilmann, S. C., Pazour, G. J., Hoffmann, E. K., Satir, P., and Christensen, S. T. (2005). PDGFRαα signaling is regulated through the primary cilium in fibroblasts. *Curr. Biol.* **15**, 1861–1866.

Spassky, N., Han, Y. G., Aguilar, A., Strehl, L., Besse, L., Laclef, C., Ros, M. R., Garcia-Verdugo, J. M., and Alvarez-Buylla, A. (2008). Primary cilia are required for cerebellar development and Shh-dependent expansion of progenitor pool. *Dev. Biol.* **317**, 246–259.

Uchida, S., Fuke, S., and Tsukahara, T. (2007). Upregulations of Gata4 and oxytocin receptor are important in cardiomyocyte differentiation processes of P19CL6 cells. *J. Cell Biochem.* **100**, 629–641.

VanGuilder, H. D., Vrana, K. E., and Freeman, W. M. (2008). Twenty-five years of quantitative PCR for gene expression analysis. *Biotechniques* **44**, 619–626.

van Wijk, B., Moorman, A. F., and van den Hoff, M. J. (2007). Role of bone morphogenetic proteins in cardiac differentiation. *Cardiovasc. Res.* **74**, 244–255.

Veland, I. R., Awan, A., Pedersen, L. B., Yoder, B. K., and Christensen, S. T. (2009). Primary cilia and signaling pathways in mammalian development, health and disease. *Nephron. Physiol.* **111**, 39–53.

Washington Smoak, I., Byrd, N. A., Abu-Issa, R., Goddeeris, M. M., Anderson, R., Morris, J., Yamamura, K., Klingensmith, J., and Meyers, E. N. (2005). Sonic hedgehog is required for cardiac outflow tract and neural crest cell development. *Dev. Biol.* **283**, 357–372.

Wong, S. Y., and Reiter, J. F. (2008). The primary cilium at the crossroads of mammalian hedgehog signaling. *Curr. Top. Dev. Biol.* **85**, 225–260.

CHAPTER 10

The Primary Cilium as a Hedgehog Signal Transduction Machine

Sarah C. Goetz[*], **Polloneal J.R. Ocbina**[*,†], *and* **Kathryn V. Anderson**[*]

[*]Developmental Biology Program, Sloan-Kettering Institute, 1275 York Avenue, New York, New York 10065

[†]Neuroscience Program, Weill Graduate School of Medical Sciences, Cornell University, New York, New York 10065

Abstract

The Hedgehog (Hh) signal transduction pathway is essential for the development and patterning of numerous organ systems, and has important roles in a variety of human cancers. Genetic screens for mouse embryonic patterning mutants first showed a connection between mammalian Hh signaling and intraflagellar transport (IFT), a process required for construction of the primary cilium, a small cellular projection

METHODS IN CELL BIOLOGY, VOL. 94

978-0-12-375024-2
DOI: 10.1016/S0091-679X(08)94010-3

found on most vertebrate cells. Additional genetic and cell biological studies have provided very strong evidence that mammalian Hh signaling depends on the primary cilium. Here, we review the evidence that defines the integral roles that IFT proteins and cilia play in the regulation of the Hh signal transduction pathway in vertebrates. We discuss the mechanisms that control localization of Hh pathway proteins to the cilium, focusing on the transmembrane protein Smoothened (Smo), which moves into the cilium in response to Hh ligand. The phenotypes caused by loss of cilia-associated proteins are complex, which suggests that cilia and IFT play active roles in mediating Hh signaling rather than serving simply as a compartment in which pathway components are concentrated. Hh signaling in *Drosophila* does not depend on cilia, but there appear to be ancient links between cilia and components of the Hh pathway that may reveal how this fundamental difference between the *Drosophila* and mammalian Hh pathways arose in evolution.

I. Introduction

The primary cilium is a single, small, hair-shaped projection that emanates from the surface of nearly all nondividing mammalian cells. Genetic studies of mouse mutants that affect the formation of primary cilia have shown that these structures are essential for developmental signaling through the Hh pathway (Huangfu and Anderson, 2005; Huangfu *et al.*, 2003; May *et al.*, 2005). This discovery was surprising, as the Hh-responsive cells in *Drosophila*, where the pathway was first characterized, are not ciliated. Nevertheless, the connection between cilia and mammalian Hh signaling is now clear, as all the key components of the Hh signal transduction pathway are enriched in the cilium (Corbit *et al.*, 2005; Haycraft *et al.*, 2005; Ocbina and Anderson, 2008; Rohatgi *et al.*, 2007). As the Hh pathway is required for normal development of every organ system in mammals (McMahon *et al.*, 2003) and Hh signaling depends on cilia in all tissues tested (Wong and Reiter, 2008), the primary cilium plays a central role in mammalian development. In addition, inappropriate Hh signaling causes tumors such as basal cell carcinoma and medulloblastoma (Jiang and Hui, 2008), and Hh appears to support the growth of a variety of other tumors (Ruiz i Altaba *et al.*, 2002; Yauch *et al.*, 2008).

As we describe here, mutations in different proteins required for cilia formation have distinct effects on Hh signaling: some block responses to Hh ligands, others cause ligand-independent pathway activation, and yet others change the spatial organization of pathway activation. Paralleling the complexity in the mouse embryo, mutations in a number of human genes encoding proteins that localize to primary cilia or basal bodies, which nucleate cilia, are associated with a set of pleiotropic human syndromes, including Bardet–Biedl (Tobin and Beales, 2007), Meckel–Gruber, Joubert, Oral–Facial Digital, Kartagener, and Alstrom syndromes (Badano *et al.*, 2006). These syndromes are associated with seemingly unrelated abnormalities as renal cysts, retinal degeneration, brain malformations, situs reversal, obesity, and polydactyly

(Badano *et al.*, 2006); it is not yet known how many of these abnormalities are due to the disruption of Hh signaling.

Because the primary cilium projects into the extracellular space, it has been proposed that the cilium acts as an antenna to facilitate the reception of Hh signals. This is an interesting perspective, but is only part of the story as it is now clear that the IFT machinery within the cilium provides an additional function that is essential for mammalian Hh signal transduction. Here, we focus on findings that define the cellular and molecular mechanisms by which IFT, a trafficking process required for the construction and maintenance of cilia, regulates Hh signaling. Based on the analysis of mutations that affect IFT and other cilia-associated components, we propose that, in addition to providing a compartment where signaling pathway components are enriched, ciliary proteins participate actively in the regulation of the Hh pathway.

A. Genetic Screens Reveal a Requirement for Cilia in Mammalian Hh Signaling

The connection between Hh signaling and primary cilia was first uncovered in forward genetic screens undertaken to identify genes required for patterning in the midgestation mouse embryo (Caspary and Anderson, 2006; Huangfu *et al.*, 2003). Because forward genetic screens are based on phenotypes, this approach can define unexpected functions of any gene, or group of genes, in the genome. As the central nervous system is prominent at midgestation, many mutations that affect patterning of the neural tube were identified (Garcia-Garcia *et al.*, 2005).

The specification of distinct types of neural progenitors along the dorsal–ventral axis of the neural tube requires highly regulated signaling events of several different pathways, including Wnt (Patapoutian and Reichardt, 2000), BMP (Liem *et al.*, 1997), retinioic acid (Appel and Eisen, 2003; Pierani *et al.*, 1999), and Sonic hedgehog (Shh) (Briscoe and Ericson, 1999; Dessaud *et al.*, 2008; Ho and Scott, 2002; Wilson and Maden, 2005). The initial source of the Shh signal in neural patterning is the notochord that lies below the ventral midline of the neural plate, and specification of cell types within the ventral neural tube depends on the levels of Shh (Fig. 1A). The ventral-most cell fates, floor plate and V3 interneuron progenitors, depend on high levels of Shh for their identities, motor neuron progenitors require intermediate levels of Shh, and V2, V1, and V0 interneurons require progressively lower levels of Hh signals for their specification. In the absence of any Hh signaling, as in *Smo* mutant embryos, all these ventral cell fates are missing and the neural tube is entirely dorsalized (Wijgerde *et al.*, 2002). Because the identity of cells within the ventral neural tube is highly sensitive to the levels and duration of Shh signal (Dessaud *et al.*, 2008), neural patterning provides a readout of Hh pathway activity.

A number of the mutations identified in the genetic screen based on altered neural patterning were found to disrupt components of the evolutionarily conserved Hh pathway (Caspary *et al.*, 2002). In addition, two mutants, *wimple* (*wim*) and *flexo* (*fxo*), exhibited exencephaly (the failure to close the anterior neural tube) as well as loss of Shh-dependent cell types in the neural tube, including floor plate, V3 interneurons, and motor neurons (Huangfu *et al.*, 2003) (Fig. 1B). The *wim* and *fxo* mutations were shown

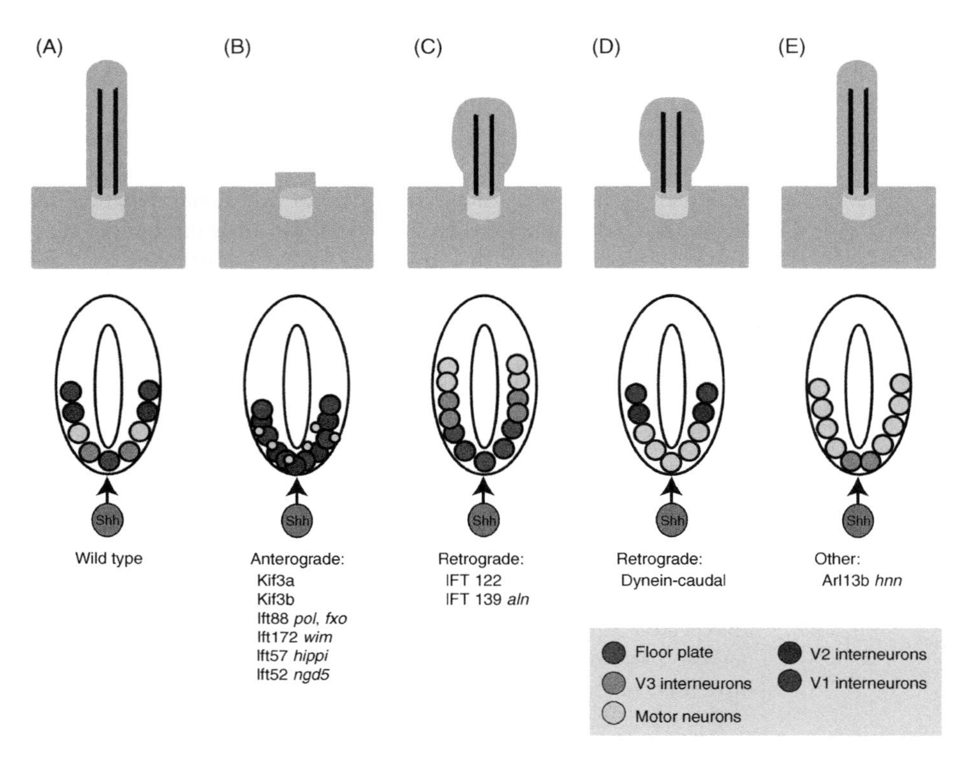

Fig. 1 Summary of IFT mutants and their associated neural patterning defects. Upper row: schematic of cilia morphology. Lower row: schematic of neural tube patterning in the same mutants. (A) In wild type, cilia are required for cells to respond to Shh, which is essential for the specification of a set of cell types in the ventral neural tube. The notochord (dark green) located ventral to the neural tube acts as the initial source of the Shh signal (arrow). In the absence of Shh, none of the indicated ventral neural cell types are specified. (B) In mutants that lack the anterograde IFT motor kinesin-2 or components of the IFT-B complex, cilia are severely shortened or absent; the most Shh-dependent ventral neural cell fates, floor plate and V3 interneurons, are absent; and the number of motor neurons is severely reduced (Houde *et al.*, 2006; Huangfu *et al.*, 2003; Liu, 2005). (C) IFT-A complex mutants have abnormal cilia with bulges and tend to be short. The neural tubes in these embryos show expansion of Hh-dependent ventral cell fates (Cortellino *et al.*, 2009; Tran *et al.*, 2008). (D) Dynein mutants, which are deficient in retrograde trafficking, have bulged cilia similar to those of IFT-A mutants, but their neural patterning phenotype is consistent with a loss of Hh signaling. Rostrally, the phenotype resembles that depicted in (B), while caudally motor neurons are specified (Huangfu and Anderson, 2005; May *et al.*, 2005). (E) *Arl13b^{hnn}* cilia have structural defects in the axoneme. Neural patterning in these mutants is unusual, with a loss of ventral-most cell fates, but an expansion of motor neurons, which depend on intermediate levels of Shh for their specification (Caspary *et al.*, 2007). Cilia in *Kif7* mutants appear normal, but motor neurons are expanded dorsally (not shown). (See Plate no. 11 in the Color Plate Section.)

to affect two IFT proteins: *Ift172 (wim)* and *Ift88 (fxo)*. In addition to a loss of Hh signaling, null mutants of these genes lack cilia (Huangfu *et al.*, 2003; Murcia *et al.*, 2000), consistent with phenotypes observed when homologues of these genes are mutated in *Chlamydomonas reinhardtii* (Pazour *et al.*, 2000; Pedersen *et al.*, 2005), the single-celled alga in which IFT was first described (Rosenbaum and Witman, 2002). A null mutation in *Kif3a*, which encodes a subunit of the kinesin-2 motor required for anterograde trafficking within the cilium, was also shown to disrupt Shh signaling and block the specification of Shh-dependent neural fates (Huangfu *et al.*, 2003), supporting a requirement for IFT in mammalian Hh signaling (Fig. 1B).

B. IFT Proteins Are Required at the Heart of the Hh Cytoplasmic Signal Transduction Pathway

The Hh signaling pathway plays critical roles in regulating embryonic development in metazoan animals from cnidarians (Matus *et al.*, 2008) to mammals (Jiang and Hui, 2008). Hh signaling was discovered and first characterized in *Drosophila*, where it is required for embryonic segment polarity (Mohler and Vani, 1992; Nüsslein-Volhard and Wieschaus, 1980) as well as patterning of the imaginal discs that give rise to external structures of the adult (Ma *et al.*, 1993; Tabata and Kornberg, 1994).

The core components and regulatory relationships are conserved between *Drosophila* and vertebrates (Fig. 2). In both flies and mammals, signaling is initiated upon

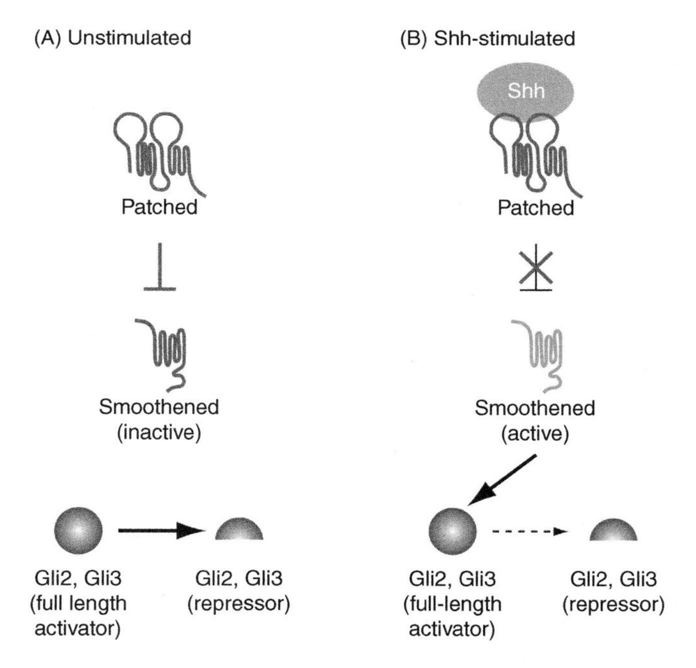

Fig. 2 The core Hh signaling pathway. The vertebrate Hedgehog (Hh) pathway in the absence (A) or presence (B) of Sonic hedgehog (Shh) ligand. (A) In the absence of ligand, the Shh receptor Ptch1 prevents the activation of another transmembrane domain protein, Smo. In this state, full-length Gli2 and Gli3 are proteolytically processed into a smaller repressor form. (B) Upon Shh ligand binding to Ptch1, the inhibition of Smo is relieved and Smo activates full-length Gli proteins and blocks production of Gli repressors.

binding of the secreted Hh ligand to the transmembrane receptor, Patched (Ptch) (Chen and Struhl, 1996; Marigo and Tabin, 1996; Marigo *et al.*, 1996a; Stone *et al.*, 1996). There is a single Hh ligand in *Drosophila*, whereas mammals have three Hh ligands—Sonic hedgehog (Shh), Indian hedgehog (Ihh), and Desert hedgehog (Dhh). Unlike traditional receptors, the Ptch receptor is a negative regulator of the pathway: in the absence of ligand, Ptch blocks the activation of another transmembrane protein Smoothened (Smo) (Marigo and Tabin, 1996). In this state, Cubitus interruptus (Ci), the transcription factor that acts as the effector of the *Drosophila* pathway, is proteolytically processed into a repressor form and blocks transcription of target genes. Upon Hh binding to Ptch, the repression of Smo is relieved, processing of the Ci repressor form is blocked, and the Ci transcriptional activator is produced, allowing the expression of target genes (Aza-Blanc *et al.*, 1997). There are three vertebrate homologues of Ci: Gli1, Gli2, and Gli3. Like Ci, Gli2 and Gli3 can both act as a proteolytically processed repressor or a full-length activator (Marigo *et al.*, 1996b; Pan *et al.*, 2006; Wang *et al.*, 2000). Gli3 functions primarily as a repressor and Gli2 as a transcriptional activator. Gli1 cannot be processed into a repressor and is a transcriptional target of the Hh pathway (Matise and Joyner, 1999) (Fig. 2).

Although mutations in *Drosophila* IFT genes do not affect Hh signaling, genetic and biochemical experiments showed that mouse IFT proteins are required for transmission of information from Shh, Ptch, and Smo to the Gli transcription factors that mediate the activity of the Hh pathway. For example, double mutant embryos that lack both an IFT complex B protein (IFT172 or IFT88) and an upstream signaling component (Shh, Ptch, or Smo) have phenotypes very similar to those seen in the IFT mutants, with respect to both overall morphology and specification of ventral neural cell fates (Huangfu *et al.*, 2003). IFT172, IFT88, and Kif3a are required for processing of Gli3 repressor (Huangfu and Anderson, 2005; Liu, 2005) and IFT172 has been shown to be required for Gli activator function (Ocbina *et al.*, 2008). Thus intraflagellar transport (IFT) is required for transduction of Hh signals from the membrane proteins Ptch and Smo to the downstream Gli transcription factors in the mouse, but not in *Drosophila*.

C. Hh Pathway Proteins Are Localized to Cilia

The phenotypes of mouse IFT mutants demonstrate a tight correlation between the process of IFT and the ability to properly transduce Hh signals. Further evidence that Hh signaling in mammals requires the primary cilium as an organelle came from studies showing that core components of the Hh signaling pathway are localized to cilia (Fig. 3). Ptch1, the mouse Hh receptor, is enriched within the primary cilium in the absence of ligand and moves out of the axonemal shaft to surround the base of the cilium after exposure to ligand (Rohatgi *et al.*, 2007). In a parallel time course, Smo protein becomes enriched in primary cilia in response to ligand (Corbit *et al.*, 2005; Rohatgi *et al.*, 2007). The Gli transcription factors are found at the tip of the cilium in both the presence and the absence of ligand. This localization to the ciliary tip is thought to be required for the activation and/or processing of the Gli transcription

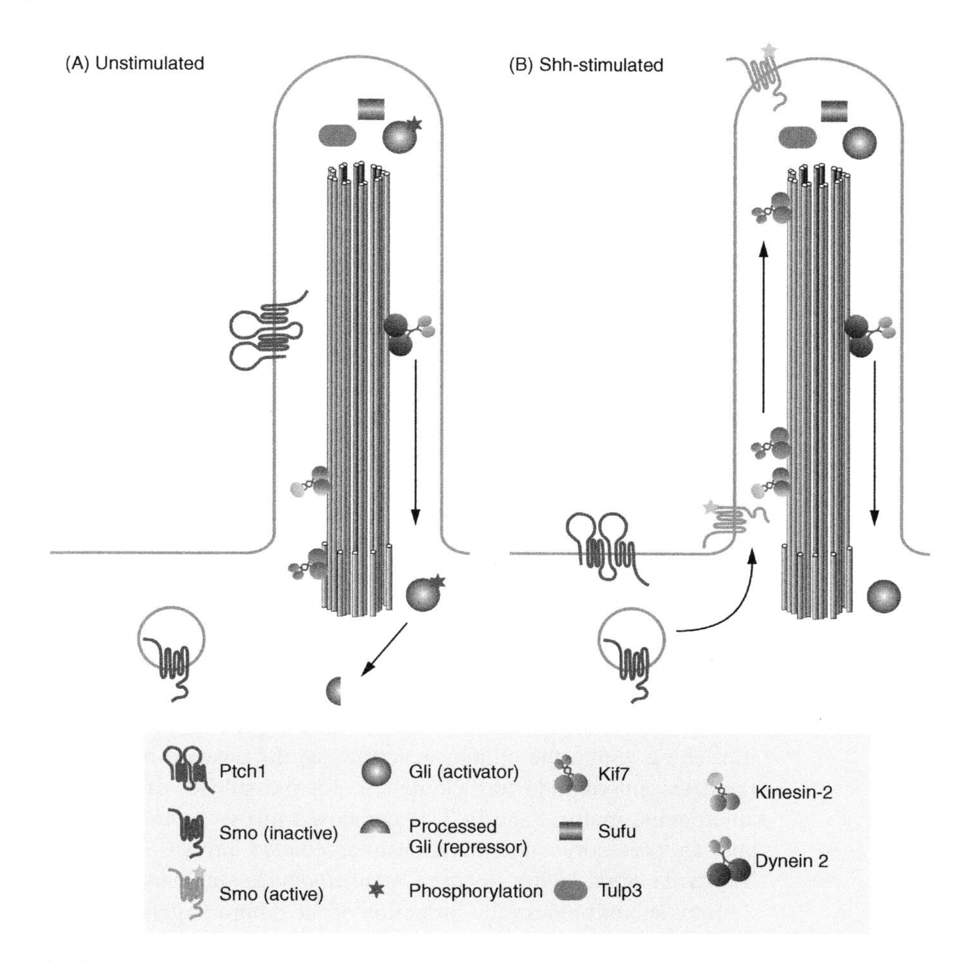

Fig. 3 Localization of Hh pathway components to the primary cilium in the presence and absence of Shh. (A) In the absence of Shh ligand, Ptch1 is enriched in primary cilia and Smo is not. Sufu and Gli are present at the cilia tip, and events that promote proteolytic processing of Gli3 to the transcriptional repressor form (red star) take place, causing repression of Hh target genes. Kif7, a kinesin homologous to *Drosophila* Cos2, is present at the base of the cilium and helps prevent activation of the pathway. (B) In the presence of Shh ligand, Shh binds to Ptch1, and Ptch1 moves out of the ciliary axoneme. In parallel, Smo is enriched in primary cilia and promotes activation rather than proteolytic processing of Gli proteins. Kif7 moves from the base to the tips of cilia and contributes to the activation of Gli proteins. Activated Gli proteins are transported out of the cilium to turn on Shh target gene expression in the nucleus. Sufu and Gli proteins are localized to the tip of the cilium in both the presence and the absence of ligand, as is Tulp3. (See Plate no. 12 in the Color Plate Section.)

factors (Haycraft *et al.*, 2005), although the mechanism by which localization to the cilium is required for activation of the Gli proteins in response to Hh signaling is not understood. Sufu, a negative regulator of the pathway, also localizes to the tip of the cilium in both the presence and the absence of ligand (Haycraft *et al.*, 2005). However, analysis of *Sufu/Ift88* double mutants suggests that Sufu is able to act as a repressor of Hh pathway even in the absence of cilia (Jia *et al.*, 2009).

D. Mutations in Different IFT Proteins Cause a Variety of Defects in Hh Signaling

Each primary cilium is formed of a ring of nine doublet microtubules that extends from a basal body, a modified form of one of the two centrioles of the cell—the mother centriole (Dutcher, 2003). Construction and maintenance of the ciliary axoneme depends on the bidirectional transport system of IFT (Fig. 4). The process of IFT, as well as its protein components, appears to be conserved among cilia and flagella across eukaryotes. Two protein complexes, IFT-A and IFT-B, form large protein platforms that deliver cargo to the tip of the cilium or flagellum (Cole, 2003). Genetic analysis has identified more than a dozen other cilia-associated proteins that are required for normal Hh signaling, and the roles of those proteins in cilia structure and Hh signaling are varied (Table I; Fig. 1).

As described above, the requirement for cilia in Hh signaling was initially discovered based upon the phenotypes of mutations in *Ift172* and *Ift88* (Huangfu *et al.*, 2003). In addition, mutations in *Ift52* and *Ift57* show the same loss of Shh-dependent ventral neural cell types (Houde *et al.*, 2006; Liu, 2005) (Fig. 1B). IFT172, IFT88, IFT52, and IFT57 are components of a single protein complex, IFT-B, and null alleles of these mouse genes lack cilia completely. The same phenotype, the failure to form cilia, is observed in *Chlamydomonas* mutants for components of the IFT-B complex (Pazour *et al.*, 2000; Scholey, 2008). The IFT-B complex is thought to mediate anterograde trafficking within the cilium; nevertheless, the detailed mechanisms by which IFT-B regulates anterograde trafficking are still the subject of investigation (Fig. 4). The anterograde motor, Kinesin-2, is composed of two motor subunits, Kif3a and Kif3b, and an accessory subunit, Kinesin-associated protein-3 (KAP-3) (Scholey, 1996; Yamazaki *et al.*, 1996). In both *Chlamydomonas* and mammals, loss of either Kinesin-2 motor subunit blocks the formation of the cilium/flagellum (Kozminski *et al.*, 1995; Marszalek *et al.*, 1999; Nonaka *et al.*, 1998), and mouse *Kif3a* mutants show the same loss of ventral neural cell types seen in IFT-B mutants (Huangfu *et al.*, 2003).

Once IFT complexes and motors have reached the tip of the cilium, they dissociate from the microtubules that make up the axoneme and release their cargo. In a process called tip turnaround, the IFT complexes are remodeled, Dynein is activated, Kinesin-2 is inactivated, and the complex takes on the cargo to be recycled to the base of the cilium (Iomini *et al.*, 2001; Pedersen *et al.*, 2005).

Retrograde transport from the tip of the ciliary axoneme to the base is mediated by cytoplasmic dynein 2 (Fig. 4). Mouse mutations in the Dynein heavy chain gene, *Dync2h1*, produce cilia that are shorter than wild type and have characteristic bulges along the axoneme (Huangfu and Anderson, 2005; May *et al.*, 2005; Ocbina and Anderson, 2008), similar to those seen in *Chlamydomonas* mutants lacking the Dynein heavy chain gene (*cDhc1b*) (Piperno *et al.*, 1998). Like *Kif3a* and IFT-B mutants, mouse *Dync2h1* mutants fail to specify Shh-dependent cell types in the ventral neural tube (Huangfu and Anderson, 2005; May *et al.*, 2005). As in IFT-B mutants, Gli3 is

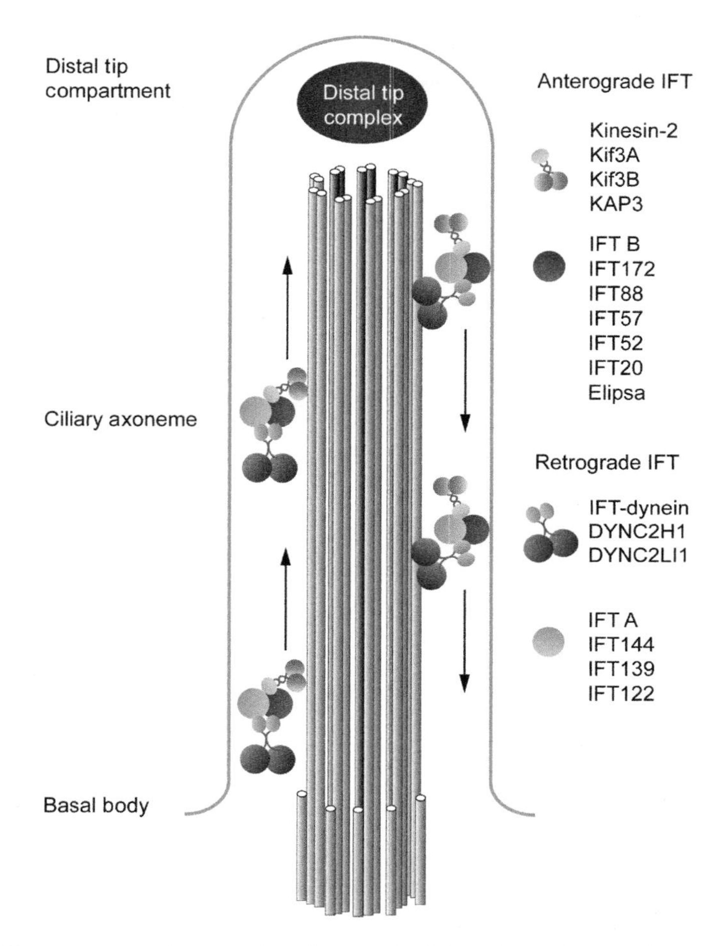

Fig. 4 The components of anterograde and retrograde IFT complexes. Anterograde transport from the basal body to the distal tip is driven by the heterotrimeric Kinesin-2 complex, composed of Kif3a, Kif3b (pink), and KAP3 (green), and the multiprotein IFT-B complex (red). At the tip of the cilium, IFT turnaround and remodeling of the IFT complexes occur in a poorly understood process (Pedersen *et al.*, 2006). Recycling of ciliary components to the base of the cilium is mediated by cytoplasmic Dynein 2 (blue), together with the IFT-A complex (orange). (See Plate no. 13 in the Color Plate Section.)

not properly proteolytically processed in *Dync2h1* mutants (Huangfu and Anderson, 2005; May *et al.*, 2005) and Gli activator function is also blocked (Ocbina and Anderson, 2008). Nevertheless, the *Dync2h1* mutants are not identical to the IFT-B complex mutants: the effect on neural patterning is stronger in the rostral region than in the caudal region of the *Dync2h1* mutant neural tube (Huangfu and Anderson, 2005) (Fig. 1D), in contrast to the uniformly strong effect of the IFT-B complex mutants. Mouse embryos lacking the light intermediate chain component of Dynein 2, *Dync2li1*

Table I
Cilia-Associated Proteins and Their Effects on Cilia Morphology and Mouse Hh Signaling

	Role in cilia	Vertebrate gene	Mutant cilia phenotype	Neural tube patterning phenotype	Reference
Kinesin-2	Kinesin-2 subunit	*Kif3a*	Absent	Loss of ventral cell types	Huangfu et al. (2003)
	Kinesin-2 subunit	*Kif3b*	Absent	Loss of ventral cell types (inferred)	Nonaka et al. (1998)
Kinesin-like	Kinesin-like	*Kif7*	Normal	Expansion of lateral cell types	Cheung et al. (2009); Endoh-Yamagami et al. (2009); Liem et al. (2009)
Dynein	Heavy chain	*Dync2h1*	Abnormal bulges	Loss of ventral cell types	Huangfu and Anderson (2005); May et al. (2005)
	Light intermediate chain	*Dync2li1*	n.d.	Loss of ventral cell types	Rana et al. (2004)
IFT complex B	Anterograde IFT	*Ift172*	Absent	Loss of ventral cell types	Huangfu et al. (2003)
		Ift88	Absent	Loss of ventral cell types	Huangfu et al. (2003); Liu (2005)
		Ift52	n.d.	Loss of ventral cell types	Liu (2005)
		Ift57	Absent	Loss of ventral cell types	Houde et al. (2006)
		Ift20	Short/absent	n.d	Follit et al. (2006)
		Elipsa/Ift54	Short	n.d	Omori et al. (2008)
IFT complex A	Retrograde IFT	*Ift139*	Short, with abnormal bulges	Expansion of ventral cell types	Tran et al. (2008)
		Ift122	Bulged	Expansion of ventral and lateral cell types	Cortellino et al. (2009)
Basal body	Basal body protein	*Ofd1*	Absent	Loss of ventral cell types	Ferrante et al. (2006); Romio et al. (2004);
	Basal body protein	*Rpgrip1l/Ftm*	Sparse and short	Loss of ventral cell types	Vierkotten et al. (2007)
	Basal body protein	*Evc*	Normal	n.d.	Ruiz-Perez et al. (2000, 2003)
	Basal body protein	*C2cd3 (hearty)*	Short/absent	Loss of ventral cell types	Hoover et al. (2008)
Arl13b	Localizes to axoneme	*Arl13b*	Short, abnormal structure	Expansion of lateral cell types	Caspary et al. (2007)
Tulp3	Localizes to cilia tip	*Tulp3*	Normal	Expansion of ventral cell types	Norman et al. (2009); Patterson et al. (2009)

n.d., not determined.

(also known as *mD2LIC*), show the same loss of floor plate markers seen in *Dync2h1* mutants, although neural patterning was not further characterized (Rana *et al.*, 2004). Thus loss of either anterograde or retrograde IFT blocks normal Hh signal transduction.

The IFT-A complex appears to cooperate with Dynein to mediate retrograde transport (Pedersen *et al.*, 2008) and may play a role in tip turnaround (Iomini *et al.*, 2001), as the IFT-A complex mutants in both *Chlamydomonas* and mammals caused bulged cilia/flagella, similar to those seen in the Dynein mutants (Cortellino *et al.*, 2009; Pedersen *et al.*, 2005; Piperno *et al.*, 1998; Tran *et al.*, 2008) (Fig. 1C).

In contrast to the loss of Hh signaling seen in the IFT-B and Dynein mutants, mouse mutations identified in two different genes that encode IFT complex A proteins have the opposite effect on Hh signaling: they cause ectopic activation of the Hh pathway in the dorsal neural tube (Cortellino *et al.*, 2009; Tran *et al.*, 2008) (Fig. 1C). The *alien (aln)* mutation disrupts the function of tetratricopeptide repeat-containing Hedgehog modulator-1 (THM-1) (Tran *et al.*, 2008), one of two mouse homologues of *Chlamydomonas* IFT139 (http://www.ncbi.nlm.nih.gov/protein/156766595). In *aln* mutant embryos, Shh-dependent cell fates including floor plate, V3 interneurons, and motor neurons are dorsally expanded (Tran *et al.*, 2008). A similar expansion of ventral cell fates is seen in mutants that lack IFT122, another component of the IFT complex A (Cortellino *et al.*, 2009).

The opposing effects of IFT-A and IFT-B mutants on Hh signaling are surprising and difficult to explain based on what is known about the two classes of proteins. IFT-A and IFT-B were identified as subcomplexes of a single large complex (Cole *et al.*, 1998). In both *Chlamydomonas* and *Caenorhabditis elegans*, the two subcomplexes move coordinately in the same particle (Ou *et al.*, 2005; Qin *et al.*, 2005). Studies in *Chlamydomonas* and *C. elegans* suggest that the IFT complex B is required for anterograde IFT and transports complex A proteins to the ciliary tip where IFT-A proteins are required for retrograde IFT (Pedersen *et al.*, 2005, 2006). However, studies on the phenotype of *C. elegans dyf-2* mutants suggest that the distinction between the functions of the two IFT subcomplexes may be an oversimplification. The *Caenorhabditis elegans dyf-2* gene encodes the homologue of *Chlamydomonas* IFT144 (called Wdr19 in the mouse), an IFT complex A protein (Cole, 2003). *dyf-2* mutants have very short sensory cilia, similar to those of complex B mutants, but they also accumulate the IFT complex B proteins in cilia, like complex A mutants (Efimenko *et al.*, 2006). Thus at least one IFT-A protein, DYF-2/IFT144, appears to play roles in both anterograde and retrograde IFT.

It is particularly surprising that mutations in the IFT Dynein and IFT-A proteins, which are thought to mediate retrograde IFT, cause similar defects in cilia morphology but exert the opposite effects on the Hh pathway. Both the IFT-A mutants analyzed (*THM-1/aln* and *Ift122*) have cilia defects similar to the phenotypes seen in *Dync2h1* mutants (Cortellino *et al.*, 2009; Huangfu and Anderson, 2005; May *et al.*, 2005; Tran *et al.*, 2008). The opposing phenotypes of Dynein 2 and IFT-A mutants suggest that these two classes of protein have distinct effects on the trafficking of Hh components.

E. Other Cilia-Associated Proteins and Hedgehog Signaling

Another mouse mutation, *hennin* (*hnn*), has a novel effect on neural patterning: cell types within the neural tube that depend on high levels of Hh signaling, such as floor plate, are lost, whereas cell types that depend on lower levels of Hh, such as motor neurons, are expanded dorsally (Caspary *et al.*, 2007) (Fig. 1E). Positional cloning showed that *hnn* inactivates Arl13b, a member of the Arl subfamily of Ras guanosine triphosphatases (GTPases) that is highly enriched in cilia. In addition to the defect in Hh signaling, loss of Arl13b causes a specific disruption in the microtubules of the ciliary axoneme, the failure to close the B tubule of the outer doublet microtubules of the ciliary axoneme. Loss of Arl13b does not affect Gli3 processing and instead causes Hh-independent, low-level activation of Gli proteins (Caspary *et al.*, 2007). Thus perturbation of the structure of the ciliary axoneme can have specific, unexpected effects on the activity of the Hh pathway.

Another cilia-associated protein that plays an important role in Hh signaling is tubby-like protein 3 (Tulp3). Tulp3 is one of four TULP members related to tubby (Tub), a regulator of late-onset obesity (Coleman and Eicher, 1990). All TULPs, including Tub, are expressed at highest levels in the retina (Ikeda *et al.*, 2002). Tulp3 is more widely expressed throughout the nervous system and *Tulp3* mutant embryos fail to close the anterior neural tube (Ikeda *et al.*, 2001), but its role in neural tube patterning was not investigated until recently. Like the IFT-A mutants, *Tulp3* mutants show an expansion of Shh-dependent ventral neural cell types, indicating that Tulp3 is a negative regulator of the pathway (Cameron *et al.*, 2009; Norman *et al.*, 2009; Patterson *et al.*, 2009). Tulp3 is required at the same step in the genetic hierarchy as the IFT proteins: downstream of Smo and upstream of the Gli proteins. Tulp3 protein localizes to the tips of primary cilia in the neural tube, as well as to both the plasma membrane and the nucleus (Norman *et al.*, 2009; Patterson *et al.*, 2009). *Tulp3* mutant cells do not show obvious defects in ciliogenesis (Norman *et al.*, 2009; Patterson *et al.*, 2009); however, the localization of Tulp3 to the tip of the cilium suggests that it could regulate the Gli transcription factors in that compartment (Fig. 3).

F. Hedgehog Pathway Proteins That Affect Cilia Structure and Trafficking

Thus far, we have focused on how mutations that disrupt cilia structure and trafficking also alter Hh signaling. Several recent studies suggest that there is a deep and ancient connection between Hh signaling and cilia: two proteins, Costal2 (Cos2) and Fused, first identified for their roles in *Drosophila* Hh signaling (Lum *et al.*, 2003; Robbins *et al.*, 1997; Sisson *et al.*, 1997), have roles in cilia function in vertebrates.

Drosophila Fused is a serine/threonine kinase required for phosphorylation of the atypical kinesin Cos2, Sufu, and perhaps other components of the Hh pathway to promote signal transduction (Aikin *et al.*, 2008). Although Fused is important for Hh signaling in zebrafish (Wolff *et al.*, 2003), the single mouse Fused homologue apparently does not participate in mammalian Shh signaling (Chen *et al.*, 2005; Merchant *et al.*, 2005). Instead, mouse Fused is required for the formation of the central pair of microtubules in motile cilia in both zebrafish and mouse (Wilson *et al.*, 2009b).

Cos2 is a kinesin-related protein that serves as an essential scaffold for *Drosophila* Hh signaling protein complexes. Recent work has shown that the mouse kinesin Kif7, a homologue of Cos2, acts as both a positive and a negative regulator of the Shh pathway (Cheung *et al.*, 2009; Endoh-Yamagami *et al.*, 2009; Liem *et al.*, 2009). Kif7 moves to the tip of the cilium in response to the activation of the pathway (Endoh-Yamagami *et al.*, 2009; Liem *et al.*, 2009) (Fig. 3). Movement of Kif7 to the cilia tip depends on its motor domain, suggesting that it acts as an anterograde motor within the cilium, in parallel with kinesin-2 (Liem *et al*, 2009). Perhaps, like the *C. elegans* homodimeric kinesin Osm-3/Kif17 (Ou *et al.*, 2005), Kif7 cooperates with Kinesin-2 to control the cargo and rate of trafficking within the ciliary axoneme.

G. Certain Basal Body Proteins Are Required for Cilia Formation and Hh Signaling

The ciliary basal body is a microtubule-based structure that forms from the mother centriole. A number of proteins that localize to the centriole/basal body have been shown to be important for both the assembly of primary cilia and for Hh signaling. The chicken *talpid3* mutation was first described over 40 years ago based on limb polydactyly (Ede and Kelly, 1964a,b) and causes a set of developmental defects that are consistent with disrupted Hh signaling (Lewis *et al.*, 1999). The *talpid3* is caused by a mutation in KIAA0586, a centrosomal protein. *talpid3* mutant embryos lack cilia, apparently due to a failure of the basal body to dock at the apical membrane (Yin *et al.*, 2009).

Several basal body proteins associated with human disorders have also been implicated in primary cilia formation and Hh signaling. Mutations in *Ofd1* cause the human disorder oral-facial-digital type 1 syndrome (Ferrante *et al.*, 2001), and mutations in *Ftm/Rpgrip1l* cause Joubert syndrome type B and Meckel syndrome (Delous *et al.*, 2007). Null mutations in mouse *Ftm* and *Ofd1* cause embryonic lethality and phenotypes characteristic of abnormal Hh signaling, including loss of ventral neural cell fates, polydactyly, and reduced expression of Hh target genes (Ferrante *et al.*, 2006; Vierkotten *et al.*, 2007). Both Ftm and Ofd1 proteins localize to the basal body (Romio *et al.*, 2004; Vierkotten *et al.*, 2007). Cilia in *Ftm* null embryos are reduced in number and the cilia that do form on fibroblasts derived from *Ftm* mutant embryos are short (Vierkotten *et al.*, 2007). In male mouse embryos hemizygous for a deletion of X-linked *Ofd1*, no cilia can be detected (Ferrante *et al.*, 2006). Thus, in addition to proteins associated with IFT, some proteins associated with the basal body are also required for primary cilia formation and therefore for Hh signaling.

Another basal body-associated protein, Ellis–van Creveld (EVC), is required for Hh signaling in a tissue-restricted manner (Ruiz-Perez *et al.*, 2007). The *Evc* gene is mutated in the human disorder EVC syndrome, a chondroectodermal dysplasia (Ruiz-Perez *et al.*, 2000, 2003). *Evc* is expressed specifically within the developing orofacial region and the cartilaginous components of the axial skeleton. In chondrocytes, EVC protein is present at the distal end of the mother centriole, which becomes the basal body. However, in contrast with *Ofd1* hemizygous and *Ftm* null embryos, *Evc* null mouse mutants can survive to adult stages, but show skeletal and craniofacial abnormalities. Cilia are present, even in chondrocytes, but the skeletal phenotypes appear to be due to reduced Ihh signaling: the mutants show premature differentiation in growth plates of the long bones and reduced expression of *Ptch* and *Gli1* in these tissues (Ruiz-Perez *et al.*, 2007). Thus this basal body protein apparently does not affect ciliogenesis, but is required for Hh signaling in specific cell types.

H. Vesicle Trafficking, Cilia Formation, and Hh Signaling

Vesicle trafficking from the Golgi to the cilium appears to be important in the initial stages of cilia formation from the mother centriole, as well as the subsequent growth and maintenance of the cilium (Pedersen *et al.*, 2008; Sorokin, 1962). IFT20, an IFT B protein, is localized to both cilia and the Golgi complex (Follit *et al.*, 2006) and may define one direct link between vesicle trafficking and cilia formation.

Further evidence for a role of vesicle trafficking in cilia formation comes from a study linking proteins associated with Bardet–Biedl syndrome (BBS) to the small GTPase Rab8, which is known to be important in vesicle trafficking and membrane fusion (Nachury *et al.*, 2007). The majority of the BBS proteins appear to form a stable complex that localizes to the base of the cilium and recruits factors important for membrane and vesicle trafficking, including Rab8 (Nachury *et al.*, 2007). Rab8 traffics within the cilium and can promote the fusion of membrane vesicles to the growing cilium. Overexpression of Rab8 or expression of a constitutively activated form of Rab8 in cells results in extension of the ciliary membrane uncoupled from extension of the axoneme. Conversely, a dominant-negative Rab8 blocks cilia formation in cells (Nachury *et al.*, 2007); however, mouse *Rab8* mutant mice are viable until about 5 weeks of age (Sato *et al.*, 2007), which is not consistent with an essential role in ciliogenesis.

A study in zebrafish identified a cilia-localized protein, Elipsa, that also appears to link IFT with vesicle trafficking via Rab8 (Omori *et al.*, 2008). Elipsa is the homologue of IFT54 (Follit *et al.*, 2009) and interacts with IFT20 and the Rab effector Rabaptin5. Rabaptin5 was shown to interact biochemically and genetically with Rab8, thus providing further evidence that the IFT complex B is linked with a pathway important for vesicle trafficking (Omori *et al.*, 2008).

There are also indications that regulation of trafficking of vesicles from the Golgi to the cilium can affect Hh signaling. The mouse *N*-ethyl-*N*-nitrosourea (ENU)-induced mutation *hearty* (*hty*), which disrupts C2cd3, a C2 domain-containing protein (Hoover *et al.*, 2008), exhibits a number of defects consistent with a loss of Shh signaling, including loss of ventral neural cell fates and polydactyly. These Shh pathway defects are apparently the result of cilia defects, as embryos homozygous for *hty*, which is a hypomorphic allele of *C2cd3*, show a reduced number of cilia. Embryos homozygous for a gene trap allele of *C2cd3*, which causes a more severe disruption of the protein, fail to form any detectable cilia (Hoover *et al.*, 2008). As C2 domain-containing proteins are involved with calcium-dependent lipid binding (Nalefski and Falke, 1996), it was suggested that C2cd3 may be involved with vesicle trafficking. Green fluorescent protein (GFP)-tagged C2cd3 localizes to the basal body (Hoover *et al.*, 2008), but a direct role in vesicle trafficking has not yet been established.

Furthermore, Rab23, a small GTPase disrupted in mouse *open brain* (*opb*) mutants, is a critical negative regulator of the Shh pathway in the mouse (Eggenschwiler *et al.*, 2001). Rab23 acts as downstream of Smo and upstream of the Gli receptors (Eggenschwiler *et al.*, 2006), the same step in the pathway as that affected by IFT. However, it is not known whether Rab23 regulates trafficking of vesicles to the cilium or affects a different kind of vesicle trafficking.

I. Trafficking of Smo to and Within the Cilium

Trafficking both to and within the cilium also appears to play a crucial role in the regulation of the seven-pass transmembrane protein Smo. A key step in the activation of the Hh pathway is the movement of Smo into cilia when cells are exposed to Shh (Fig. 3). Localization of Smo to the primary cilium is tightly correlated with Hh pathway activation. For example, mutations in Smo that cause ligand-independent activation of the pathway (e.g., SmoM2; Taipale *et al.*, 2000) also cause constitutive localization of the mutant Smo to primary cilia (Corbit *et al.*, 2005; Han *et al.*, 2008), and SmoM2 mutant protein cannot activate the pathway in the absence of cilia (Han *et al.*, 2008; Ocbina and Anderson, 2008). The transmembrane receptor Ptch1 is a negative regulator of pathway activity, and Smo is constitutively localized to cilia in *Ptch1* mutant cells (Ocbina and Anderson, 2008; Rohatgi *et al.*, 2007); however, the mechanism by which Ptch1 regulates Smo localization and activation is not known. Ptch is similar in structure to bacterial transporters, and it has been suggested that Ptch may modulate the concentration of a small molecule that controls Smo activation (Taipale *et al.*, 2002). Treatment of NIH3T3 fibroblasts with 20α-hydroxycholesterol (20α-OHC), an oxysterol derivative of cholesterol, is sufficient to promote Smo translocation to cilia and activate the pathway, which suggests that an oxysterol-like molecule might regulate Smo movement to the cilium (Rohatgi *et al.*, 2007), although the identity of that molecule is unknown.

The localization of Smo to cilia is necessary for its activation. For example, amino acid substitutions in Smo that prevent its localization to the cilium also block activation of the pathway (Aanstad *et al.*, 2009; Corbit *et al.*, 2005). However, a variety of experiments have shown that Smo localization in cilia is not sufficient for pathway activity. Overexpression of wild-type Smo in NIH3T3 cells leads to Smo enrichment in cilia without fully activating the pathway (Rohatgi *et al.*, 2007), and loss of the Dynein motor that directs retrograde IFT leads to constitutive ciliary Smo localization without pathway activation (Ocbina and Anderson, 2008). Furthermore, cyclopamine is a small molecule that inhibits the activity of the pathway, but it promotes Smo localization to cilia (Rohatgi *et al.*, 2009; Wang *et al.*, 2009; Wilson *et al.*, 2009a). These observations have suggested the hypothesis that Smo is activated in a two-step process: the first step being localization to the cilium and a second step being a conformational change of Smo within the cilium that triggers downstream signaling events (Rohatgi *et al.*, 2009) (Fig. 5).

The trafficking events that target Smo to the cilium are not known. The accumulation of Smo in the cilia of IFT dynein mutants, in which retrograde IFT is disrupted,

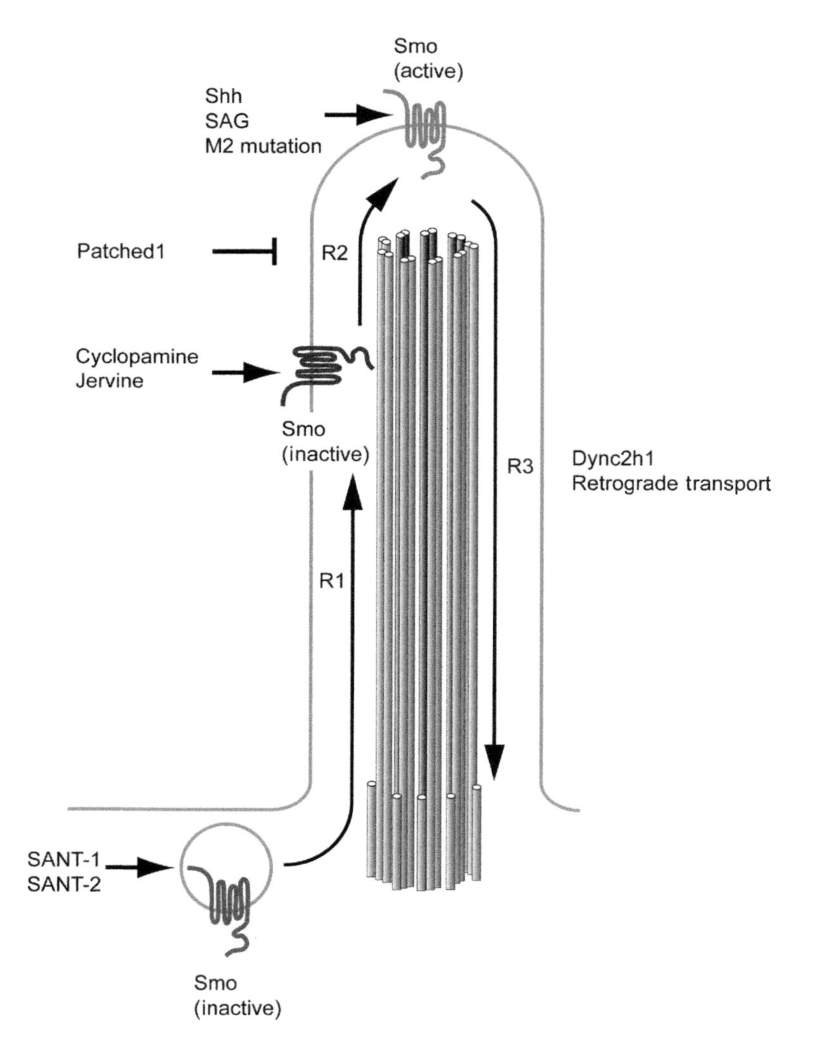

Fig. 5 Smo activation requires localization to the primary cilium. Smo is activated through a series of steps and conformations that can be stabilized by small molecule regulators of Hh signaling. Full activation of Smo requires the transport of inactive cytoplasmic Smo (red) from the cytoplasm into the cilium controlled by step R1. The Smo antagonists SANT-1 and SANT-2 stabilize Smo in this inactive state. Smo within the cilium can exist in an inactive state (blue) that is trafficked out of the cilium when no further activation is achieved (R3). This inactive form is stabilized by the antagonist cyclopamine. Smo can be activated (green) through a second step, R2, to promote activation of downstream pathway components such as Gli proteins. This state is promoted by Shh signal, and can be stabilized by the Smo agonist SAG as well as activating mutations such as SmoM2. Dynein 2 mediates retrograde transport of Smo (R3) in either an active or inactive state, as *Dync2h1* mutant cilia constitutively localize Smo without Shh pathway activation. The mechanisms that deliver Smo to the cilium are not known. Modified from Rohatgi *et al.* (2009). (See Plate no. 14 in the Color Plate Section.)

suggests that Smo is trafficked through the cilium at a low basal rate in the absence of ligand (Ocbina and Anderson, 2008). Several G-protein-coupled receptors that localize to the cilia of sensory neurons require the basal body-associated proteins BBS2 and BBS4 for their localization (Berbari *et al.*, 2008), but Hh signaling and cilia formation are normal in *Bbs2* and *Bbs4* mouse mutants (Kulaga *et al.*, 2004; Mykytyn *et al.*, 2004; Nishimura *et al.*, 2004), and there is no evidence of a role for these proteins in trafficking Smo. A recent study has suggested that β-arrestins mediate the interaction of Smo with Kif3a in NIH3T3 cells (Kovacs *et al.*, 2008). However, the exact function of ß-arrestins within the cilium and their relationship to Smo are controversial (Molla-Herman *et al.*, 2008), as mouse knockouts for either ß-arrestin-1 or ß-arrestin-2 do not affect early mouse development (Bohn *et al.*, 1999; Conner *et al.*, 1997). Thus additional genetic and cell biological experiments will be needed to define how Smo moves into the cilium.

II. Summary

Given the vital role of Shh signaling in regulating the patterning of the early embryo and the development of nearly every organ and the roles of the Shh pathway in cancer, it is critical to understand mammalian-specific regulation of Hh signaling by primary cilia. Studies conducted over the past several years have revealed the complexity of the mechanisms active in primary cilia to regulate the Hh pathway. If the cilium were simply a compartment where components of the Hh pathway are enriched, mutations that disrupt cilia formation would be predicted to have similar effects on pathway activity. However, as we have described, mutations in different cilia-associated proteins can increase, decrease, or change the spatial distribution of the activity of the pathway. A striking example of the unexplained complexity of the pathway is that different mutations in two kinds of components that affect retrograde IFT, Dynein and IFT-A proteins, have opposite effects on Shh signaling.

Proteins associated with the basal body also differ in terms of their effects on the Hh pathway. Some basal body proteins are clearly required for Hh signaling, whereas others, most notably BBS proteins, do not appear to be individually required to regulate Hh signaling. Of the basal body proteins that are required for Hh signaling, some are clearly required for cilia formation. Loss of function of other basal body proteins, however, including EVC, does not result in overt defects in cilia formation. Thus, the specific functions of proteins associated with the basal body with respect to cilia formation need to be defined.

Although defining the mechanisms that control Smo localization to cilia in response to Hh ligands is essential for understanding why and how the mammalian Hh pathway is tied to the primary cilium, much remains to be learned. *In vivo*, Shh acts as a morphogen: different concentrations of Shh direct different cell fates (Jiang and Hui, 2008; Ruiz i Altaba *et al.*, 2003). The question of whether the gradient of Shh activity depends on regulation of Smo localization to the cilium or activation within the cilium has yet to be tackled.

It remains a mystery why vertebrate, but not *Drosophila*, Hh signaling is coupled to cilia. The Hh pathway is ancient and must have been present in the common ancestor of both protostomes and deuterostomes (Matus *et al.*, 2008) and may have been coopted from more primitive eukaryotes before the advent of multicellular animals (Hausmann *et al.*, 2009). Cilia and IFT are also conserved across evolution from the single-celled alga *Chlamydomonas* to mammals. It will be important to determine whether Hh signaling in other invertebrate metazoans depends on cilia. It is interesting to note that *Chlamydomonas* IFT proteins play an integral role in a signaling cascade important during fertilization (Wang *et al.*, 2006), suggesting that the role of cilia in cell signaling goes far back in evolutionary history.

Acknowledgments

Work in our laboratory in this area was supported by NIH grant NS044385. SCG is an American Cancer Society postdoctoral fellow.

References

Aanstad, P., Santos, N., Corbit, K.C., Scherz, P.J., Trinh le, A., Salvenmoser, W., Huisken, J., Reiter, J.F., and Stainier, D.Y. (2009). The extracellular domain of Smoothened regulates ciliary localization and is required for high-level Hh signaling. *Curr. Biol.* **19**, 1034–1039.

Aikin, R.A., Ayers, K.L., and Therond, P.P. (2008). The role of kinases in the Hedgehog signalling pathway. *EMBO Rep.* **9**, 330–336.

Appel, B., and Eisen, J.S. (2003). Retinoids run rampant: Multiple roles during spinal cord and motor neuron development. *Neuron* **40**, 461–464.

Aza-Blanc, P., Ramirez-Weber, F.A., Laget, M.P., Schwartz, C., and Kornberg, T.B. (1997). Proteolysis that is inhibited by hedgehog targets Cubitus interruptus protein to the nucleus and converts it to a repressor. *Cell* **89**, 1043–1053.

Badano, J.L., Mitsuma, N., Beales, P.L., and Katsanis, N. (2006). The ciliopathies: an emerging class of human genetic disorders. *Annu. Rev. Genomics Hum. Genet.* **7**, 125–148.

Berbari, N.F., Lewis, J.S., Bishop, G.A., Askwith, C.C., and Mykytyn, K. (2008). Bardet-Biedl syndrome proteins are required for the localization of G protein-coupled receptors to primary cilia. *Proc. Natl. Acad. Sci. USA.* **105**, 4242–4246.

Bohn, L.M., Lefkowitz, R.J., Gainetdinov, R.R., Peppel, K., Caron, M.G., and Lin, F.T. (1999). Enhanced morphine analgesia in mice lacking beta-arrestin 2. *Science* **286**, 2495–2498.

Briscoe, J., and Ericson, J. (1999). The specification of neuronal identity by graded Sonic Hedgehog signalling. *Semin. Cell. Dev. Biol.* **10**, 353–362.

Cameron, D.A., Pennimpede, T., and Petkovich, M. (2009). Tulp3 is a critical repressor of mouse hedgehog signaling. *Dev. Dyn.* **238**, 1140–1149.

Caspary, T., and Anderson, K.V. (2006). Uncovering the uncharacterized and unexpected: unbiased phenotype-driven screens in the mouse. *Dev. Dyn.* **235**, 2412–2423.

Caspary, T., Garcia-Garcia, M.J., Huangfu, D., Eggenschwiler, J.T., Wyler, M.R., Rakeman, A.S., Alcorn, H. L., and Anderson, K.V. (2002). Mouse Dispatched homolog1 is required for long-range, but not juxtacrine, Hh signaling. *Curr. Biol.* **12**, 1628–1632.

Caspary, T., Larkins, C.E., and Anderson, K.V. (2007). The graded response to Sonic Hedgehog depends on cilia architecture. *Dev. Cell.* **12**, 767–778.

Chen, M.H., Gao, N., Kawakami, T., and Chuang, P.T. (2005). Mice deficient in the fused homolog do not exhibit phenotypes indicative of perturbed hedgehog signaling during embryonic development. *Mol. Cell Biol.* **25**, 7042–7053.

Chen, Y., and Struhl, G. (1996). Dual roles for patched in sequestering and transducing Hedgehog. *Cell* **87**, 553–563.

Cheung, H.O., Zhang, X., Ribeiro, A., Mo, R., Makino, S., Puviindran, V., Law, K.K., Briscoe, J., and Hui, C.C. (2009). The kinesin protein Kif7 is a critical regulator of Gli transcription factors in mammalian hedgehog signaling. *Sci. Signal.* **2**, ra29.

Cole, D.G., (2003). The intraflagellar transport machinery of Chlamydomonas reinhardtii. *Traffic* **4**, 435–442.

Cole, D.G., Diener, D.R., Himelblau, A.L., Beech, P.L., Fuster, J.C., and Rosenbaum, J.L. (1998). Chlamydomonas kinesin-II-dependent intraflagellar transport (IFT): IFT particles contain proteins required for ciliary assembly in Caenorhabditis elegans sensory neurons. *J. Cell Biol.* **141**, 993–1008.

Coleman, D.L., and Eicher, E.M. (1990). Fat (fat) and tubby (tub): two autosomal recessive mutations causing obesity syndromes in the mouse. *J. Hered.* **81**, 424–427.

Conner, D.A., Mathier, M.A., Mortensen, R.M., Christe, M., Vatner, S.F., Seidman, C.E., and Seidman, J.G. (1997). beta-Arrestin1 knockout mice appear normal but demonstrate altered cardiac responses to beta-adrenergic stimulation. *Circ. Res.* **81**, 1021–1026.

Corbit, K.C., Aanstad, P., Singla, V., Norman, A.R., Stainier, D.Y., and Reiter, J.F. (2005). Vertebrate Smoothened functions at the primary cilium. *Nature* **437**, 1018–1021.

Cortellino, S., Wang, C., Wang, B., Bassi, M.R., Caretti, E., Champeval, D., Calmont, A., Jarnik, M., Burch, J., Zaret, K.S., Larue, L., and Bellacosa, A. (2009). Defective ciliogenesis, embryonic lethality and severe impairment of the Sonic Hedgehog pathway caused by inactivation of the mouse complex A intraflagellar transport gene Ift122/Wdr10, partially overlapping with the DNA repair gene Med1/Mbd4. *Developmental Biology* **325**, 225–237.

Delous, M., Baala, L., Salomon, R., Laclef, C., Vierkotten, J., Tory, K., Golzio, C., Lacoste, T., Besse, L., Ozilou, C., Moutkine, I., and Hellman, N.E., *et al.* (2007). The ciliary gene RPGRIP1L is mutated in cerebello-oculo-renal syndrome (Joubert syndrome type B) and Meckel syndrome. *Nat. Genet.* **39**, 875–881.

Dessaud, E., Mcmahon, A., and Briscoe, J. (2008). Pattern formation in the vertebrate neural tube: a sonic hedgehog morphogen-regulated transcriptional network. *Development* **135**, 2489–2503.

Dutcher, S.K. (2003). Elucidation of basal body and centriole functions in *Chlamydomonas reinhardtii. Traffic* **4**, 443–451.

Ede, D.A., and Kelly, W.A. (1964a). Developmental Abnormalities in the Head Region of the Talpid Mutant of the Fowl. *J. Embryol. Exp. Morphol.* **12**, 161–182.

Ede, D.A., and Kelly, W.A. (1964b). Developmental Abnormalities in the Trunk and Limbs of the Talpid3 Mutant of the Fowl. *J. Embryol. Exp. Morphol.* **12**, 339–356.

Efimenko, E., Blacque, O.E., Ou, G., Haycraft, C.J., Yoder, B.K., Scholey, J.M., Leroux, M.R., and Swoboda, P. (2006). *Caenorhabditis elegans* DYF-2, an orthologue of human WDR19, is a component of the intraflagellar transport machinery in sensory cilia. *Mol. Biol. Cell* **17**, 4801–4811.

Eggenschwiler, J., Bulgakov, O., Qin, J., Li, T., and Anderson, K. (2006). Mouse Rab23 regulates Hedgehog signaling from Smoothened to Gli proteins. *Developmental Biology* **290**, 1–12.

Eggenschwiler, J.T., Espinoza, E., and Anderson, K.V. (2001). Rab23 is an essential negative regulator of the mouse Sonic hedgehog signalling pathway. *Nature* **412**, 194–198.

Endoh-Yamagami, S., Evangelista, M., Wilson, D., Wen, X., Theunissen, J.W., Phamluong, K., Davis, M., Scales, S.J., Solloway, M.J., de Sauvage, F.J., and Peterson, A.S. (2009). The mammalian Cos2 homolog Kif7 plays an essential role in modulating Hh signal transduction during development. *Curr. Biol.* **19**, 1320–1326.

Ferrante, M., Zullo, A., Barra, A., Bimonte, S., Messaddeq, N., Studer, M., Dollé, P., and Franco, B. (2006). Oral-facial-digital type I protein is required for primary cilia formation and left-right axis specification. *Nat. Genet.* **38**, 112–117.

Ferrante, M.I., Giorgio, G., Feather, S.A., Bulfone, A., Wright, V., Ghiani, M., Selicorni, A., Gammaro, L., Scolari, F., Woolf, A.S., Sylvie, O., and Bernard, L., *et al.* (2001). Identification of the gene for oral-facial-digital type I syndrome. *Am. J. Hum. Genet.* **68**, 569–576.

Follit, J.A., Tuft, R.A., Fogarty, K.E., and Pazour, G.J. (2006). The intraflagellar transport protein IFT20 is associated with the Golgi complex and is required for cilia assembly. *Mol Biol Cell.* **17**, 3781–3792.

Follit, J.A., Xu, F., Keady, B., and Pazour, G.J. (2009). Characterization of mouse IFT complex B. *Cell Motil. Cytoskeleton* **66**, 457–468.

Garcia-Garcia, M.J., Eggenschwiler, J.T., Caspary, T., Alcorn, H.L., Wyler, M.R., Huangfu, D., Rakeman, A.S., Lee, J.D., Feinberg, E.H., Timmer, J.R., and Anderson, K.V. (2005). Analysis of mouse embryonic patterning and morphogenesis by forward genetics. *Proc. Natl. Acad. Sci. U S A.* **102**, 5913–5919.

Han, Y., Spassky, N., Romaguera-Ros, M., Garcia-Verdugo, J., Aguilar, A., Schneider-Maunoury, S., and Alvarez-Buylla, A. (2008). Hedgehog signaling and primary cilia are required for the formation of adult neural stem cells. *Nat. Neurosci.* **11**, 277–284.

Hausmann, G., von Mering, C., and Basler, K. (2009). The hedgehog signaling pathway: where did it come from? *PLoS Biol.* **7**, e1000146.

Haycraft, C., Banizs, B., Aydin-Son, Y., Zhang, Q., Michaud, E.J., and Yoder, B. (2005). Gli2 and Gli3 localize to cilia and require the intraflagellar transport protein polaris for processing and function. *PLoS Genet.* **1**, e53.

Ho, K.S., and Scott, M.P. (2002). Sonic hedgehog in the nervous system: functions, modifications and mechanisms. *Curr. Opin. Neurobiol.* **12**, 57–63.

Hoover, A.N., Wynkoop, A., Zeng, H., Jia, J., Niswander, L.A., and Liu, A. (2008). C2cd3 is required for cilia formation and Hedgehog signaling in mouse. *Development* **135**, 4049–4058.

Houde, C., Dickinson, R.J., Houtzager, V.M., Cullum, R., Montpetit, R., Metzler, M., Simpson, E.M., Roy, S., Hayden, M.R., Hoodless, P.A., and Nicholson, D.W. (2006). Hippi is essential for node cilia assembly and Sonic hedgehog signaling. *Dev. Biol.* **300**, 523–533.

Huangfu, D., and Anderson, K.V. (2005). Cilia and Hedgehog responsiveness in the mouse. *Proc. Natl. Acad. Sci. USA.* **102**, 11325–11330.

Huangfu, D., Liu, A., Rakeman, A.S., Murcia, N.S., Niswander, L., and Anderson, K.V. (2003). Hedgehog signalling in the mouse requires intraflagellar transport proteins. *Nature* **426**, 83–87.

Ikeda, A., Ikeda, S., Gridley, T., Nishina, P.M., and Naggert, J.K. (2001). Neural tube defects and neuroepithelial cell death in Tulp3 knockout mice. *Hum. Mol. Genet.* **10**, 1325–1334.

Ikeda, A., Nishina, P.M., and Naggert, J.K. (2002). The tubby-like proteins, a family with roles in neuronal development and function. *J. Cell Sci.* **115**, 9–14.

Iomini, C., Babaev-Khaimov, V., Sassaroli, M., and Piperno, G. (2001). Protein particles in Chlamydomonas flagella undergo a transport cycle consisting of four phases. *J. Cell Biol.* **153**, 13–24.

Jia, J., Kolterud, Å., Zeng, H., Hoover, A., Teglund, S., Toftgård, R., and Liu, A. (2009). Suppressor of Fused inhibits mammalian Hedgehog signaling in the absence of cilia. *Dev. Biol.* **330**, 452–460.

Jiang, J., and Hui, C.C. (2008). Hedgehog signaling in development and cancer. *Dev. Cell.* **15**, 801–812.

Kovacs, J.J., Whalen, E.J., Liu, R., Xiao, K., Kim, J., Chen, M., Wang, J., Chen, W., and Lefkowitz, R.J. (2008). Beta-arrestin-mediated localization of smoothened to the primary cilium. *Science* **320**, 1777–1781.

Kozminski, K.G., Beech, P.L., and Rosenbaum, J.L. (1995). The Chlamydomonas kinesin-like protein FLA10 is involved in motility associated with the flagellar membrane. *J. Cell Biol.* **131**, 1517–1527.

Kulaga, H., Leitch, C., Eichers, E., Badano, J., Lesemann, A., Hoskins, B., Lupski, J., Beales, P., Reed, R., and Katsanis, N. (2004). Loss of BBS proteins causes anosmia in humans and defects in olfactory cilia structure and function in the mouse. *Nat. Genet.* **36**, 994–998.

Lewis, K.E., Drossopoulou, G., Paton, I.R., Morrice, D.R., Robertson, K.E., Burt, D.W., Ingham, P.W., and Tickle, C. (1999). Expression of ptc and gli genes in talpid3 suggests bifurcation in Shh pathway. *Development* **126**, 2397–2407.

Liem, K.F., Jr., He, M., Ocbina, P.J., and Anderson, K.V. (2009). Mouse Kif7/Costal2 is a cilia-associated protein that regulates Sonic hedgehog signaling. *Proc. Natl. Acad. Sci. U S A.* **106**, 13377–13382.

Liem, K.F., Jr., Tremml, G., and Jessell, T.M. (1997). A role for the roof plate and its resident TGFbeta-related proteins in neuronal patterning in the dorsal spinal cord. *Cell* **91**, 127–138.

Liu, A., Wang, B., and Niswander, L.A. (2005). Mouse intraflagellar transport proteins regulate both the activator and repressor functions of Gli transcription factors. *Development* **132**, 3103–3111.

Lum, L., Zhang, C., Oh, S., Mann, R.K., von Kessler, D.P., Taipale, J., Weis-Garcia, F., Gong, R., Wang, B., and Beachy, P.A. (2003). Hedgehog signal transduction via Smoothened association with a cytoplasmic complex scaffolded by the atypical kinesin, Costal-2. *Mol. Cell* **12**, 1261–1274.

Ma, C., Zhou, Y., Beachy, P.A., and Moses, K. (1993). The segment polarity gene hedgehog is required for progression of the morphogenetic furrow in the developing Drosophila eye. *Cell* **75**, 927–938.

Marigo, V., Davey, R.A., Zuo, Y., Cunningham, J.M., and Tabin, C.J. (1996). Biochemical evidence that patched is the Hedgehog receptor. *Nature* **384**, 176–179.

Marigo, V., Johnson, R.L., Vortkamp, A., and Tabin, C.J. (1996). Sonic hedgehog differentially regulates expression of GLI and GLI3 during limb development. *Dev. Biol.* **180**, 273–283.

Marigo, V., and Tabin, C.J. (1996). Regulation of patched by sonic hedgehog in the developing neural tube. *Proc. Natl. Acad. Sci. U S A.* **93**, 9346–9351.

Marszalek, J.R., Ruiz-Lozano, P., Roberts, E., Chien, K.R., and Goldstein, L.S. (1999). Situs inversus and embryonic ciliary morphogenesis defects in mouse mutants lacking the KIF3A subunit of kinesin-II. *Proc. Natl. Acad. Sci. U S A.* **96**, 5043–5048.

Matise, M.P., and Joyner, A.L. (1999). Gli genes in development and cancer. *Oncogene.* **18**, 7852–7859.

Matus, D.Q., Magie, C.R., Pang, K., Martindale, M.Q., and Thomsen, G.H. (2008). The Hedgehog gene family of the cnidarian, Nematostella vectensis, and implications for understanding metazoan Hedgehog pathway evolution. *Developmental Biology* **313**, 501–518.

May, S.R., Ashique, A.M., Karlen, M., Wang, B., Shen, Y., Zarbalis, K., Reiter, J., Ericson, J., and Peterson, A.S. (2005). Loss of the retrograde motor for IFT disrupts localization of Smo to cilia and prevents the expression of both activator and repressor functions of Gli. *Developmental Biology* **287**, 378–389.

McMahon, A.P., Ingham, P.W., and Tabin, C.J. (2003). Developmental roles and clinical significance of hedgehog signaling. *Curr. Top. Dev. Biol.* **53**, 1–114.

Merchant, M., Evangelista, M., Luoh, S.M., Frantz, G.D., Chalasani, S., Carano, R.A., van Hoy, M., Ramirez, J., Ogasawara, A.K., McFarland, L.M., Filvaroff, E.H., and French, D.M., *et al.* (2005). Loss of the serine/threonine kinase fused results in postnatal growth defects and lethality due to progressive hydrocephalus. *Mol. Cell Biol.* **25**, 7054–7068.

Mohler, J., and Vani, K. (1992). Molecular organization and embryonic expression of the hedgehog gene involved in cell-cell communication in segmental patterning of Drosophila. *Development* **115**, 957–971.

Molla-Herman, A., Boularan, C., Ghossoub, R., Scott, M.G., Burtey, A., Zarka, M., Saunier, S., Concordet, J.P., Marullo, S., and Benmerah, A. (2008). Targeting of beta-arrestin2 to the centrosome and primary cilium: role in cell proliferation control. *PLoS One* **3**, e3728.

Murcia, N.S., Richards, W.G., Yoder, B.K., Mucenski, M.L., Dunlap, J.R., and Woychik, R.P. (2000). The Oak Ridge Polycystic Kidney (orpk) disease gene is required for left-right axis determination. *Development* **127**, 2347–2355.

Mykytyn, K., Mullins, R.F., Andrews, M., Chiang, A.P., Swiderski, R.E., Yang, B., Braun, T., Casavant, T., Stone, E.M., and Sheffield, V.C. (2004). Bardet-Biedl syndrome type 4 (BBS4)-null mice implicate Bbs4 in flagella formation but not global cilia assembly. *Proc. Natl. Acad. Sci. U S A.* **101**, 8664–8669.

Nachury, M., Loktev, A., Zhang, Q., Westlake, C., Peranen, J., Merdes, A., Slusarski, D., Scheller, R., Bazan, J., and Sheffield, V. (2007). A Core Complex of BBS Proteins Cooperates with the GTPase Rab8 to Promote Ciliary Membrane Biogenesis. *Cell* **129**, 1201–1213.

Nalefski, E.A., and Falke, J.J. (1996). The C2 domain calcium-binding motif: structural and functional diversity. *Protein Sci.* **5**, 2375–2390.

Nishimura, D.Y., Fath, M., Mullins, R.F., Searby, C., Andrews, M., Davis, R., Andorf, J.L., Mykytyn, K., Swiderski, R.E., Yang, B., Carmi, R., and Stone, E.M., *et al.* (2004). Bbs2-null mice have neurosensory deficits, a defect in social dominance, and retinopathy associated with mislocalization of rhodopsin. *Proc. Natl. Acad. Sci. U S A.* **101**, 16588–16593.

Nonaka, S., Tanaka, Y., Okada, Y., Takeda, S., Harada, A., Kanai, Y., Kido, M., and Hirokawa, N. (1998). Randomization of left-right asymmetry due to loss of nodal cilia generating leftward flow of extraembryonic fluid in mice lacking KIF3B motor protein. *Cell* **95**, 829–837.

Norman, R.X., Ko, H.W., Huang, V., Eun, C.M., Abler, L.L., Zhang, Z., Sun, X., and Eggenschwiler, J.T. (2009). Tubby-like protein 3 (TULP3) regulates patterning in the mouse embryo through inhibition of Hedgehog signaling. *Hum. Mol. Genet.* **18**, 1740–1754.

Nusslein-Volhard, C., and Wieschaus, E. (1980). Mutations affecting segment number and polarity in Drosophila. *Nature* **287**, 795–801.

Ocbina, P.J., and Anderson, K.V. (2008). Intraflagellar transport, cilia, and mammalian Hedgehog signaling: analysis in mouse embryonic fibroblasts. *Dev. Dyn.* **237**, 2030–2038.

Omori, Y., Zhao, C., Saras, A., Mukhopadhyay, S., Kim, W., Furukawa, T., Sengupta, P., Veraksa, A., and Malicki, J. (2008). elipsa is an early determinant of ciliogenesis that links the IFT particle to membrane-associated small GTPase Rab8. *Nat. Cell. Biol.* **10**, 437–444.

Ou, G., Blacque, O.E., Snow, J.J., Leroux, M.R., and Scholey, J.M. (2005). Functional coordination of intraflagellar transport motors. *Nature* **436**, 583-587.

Pan, Y., Bai, C.B., Joyner, A.L., and Wang, B. (2006). Sonic hedgehog signaling regulates Gli2 transcriptional activity by suppressing its processing and degradation. *Mol. Cell. Biol.* **26**, 3365–3377.

Patapoutian, A., and Reichardt, L.F. (2000). Roles of Wnt proteins in neural development and maintenance. *Curr. Opin. Neurobiol.* **10**, 392–399.

Patterson, V.L., Damrau, C., Paudyal, A., Reeve, B., Grimes, D.T., Stewart, M.E., Williams, D.J., Siggers, P., Greenfield, A., and Murdoch, J.N. (2009). Mouse hitchhiker mutants have spina bifida, dorso-ventral patterning defects and polydactyly: identification of Tulp3 as a novel negative regulator of the Sonic hedgehog pathway. *Hum. Mol. Genet.* **18**, 1719–1739.

Pazour, G.J., Dickert, B.L., Vucica, Y., Seeley, E.S., Rosenbaum, J.L., Witman, G.B., and Cole, D.G. (2000). Chlamydomonas IFT88 and its mouse homologue, polycystic kidney disease gene tg737, are required for assembly of cilia and flagella. *J. Cell Biol.* **151**, 709–718.

Pedersen, L.B., Geimer, S., and Rosenbaum, J.L. (2006). Dissecting the molecular mechanisms of intraflagellar transport in chlamydomonas. *Curr. Biol.* **16**, 450–459.

Pedersen, L.B., Miller, M.S., Geimer, S., Leitch, J.M., Rosenbaum, J.L., and Cole, D.G. (2005). Chlamydomonas IFT172 is encoded by FLA11, interacts with CrEB1, and regulates IFT at the flagellar tip. *Curr. Biol.* **15**, 262–266.

Pedersen, L.B., Veland, I.R., Schroder, J.M., and Christensen, S.T. (2008). Assembly of primary cilia. *Dev. Dyn.* **237**, 1993–2006.

Pierani, A., Brenner-Morton, S., Chiang, C., and Jessell, T.M. (1999). A sonic hedgehog-independent, retinoid-activated pathway of neurogenesis in the ventral spinal cord. *Cell.* **97**, 903–915.

Piperno, G., Siuda, E., Henderson, S., Segil, M., Vaananen, H., and Sassaroli, M. (1998). Distinct mutants of retrograde intraflagellar transport (IFT) share similar morphological and molecular defects. *J. Cell Biol.* **143**, 1591–1601.

Qin, H., Burnette, D., Bae, Y., Forscher, P., Barr, M., and Rosenbaum, J. (2005). Intraflagellar Transport Is Required for the Vectorial Movement of TRPV Channels in the Ciliary Membrane. *Current Biology.* **15**, 1695–1699.

Rana, A.A., Barbera, J.P., Rodriguez, T.A., Lynch, D., Hirst, E., Smith, J.C., and Beddington, R.S. (2004). Targeted deletion of the novel cytoplasmic dynein mD2LIC disrupts the embryonic organiser, formation of the body axes and specification of ventral cell fates. *Development* **131**, 4999–5007.

Robbins, D.J., Nybakken, K.E., Kobayashi, R., Sisson, J.C., Bishop, J.M., and Therond, P.P. (1997). Hedgehog elicits signal transduction by means of a large complex containing the kinesin-related protein costal2. *Cell* **90**, 225–234.

Rohatgi, R., Milenkovic, L., Corcoran, R.B., and Scott, M.P. (2009). Hedgehog signal transduction by Smoothened: Pharmacologic evidence for a 2-step activation process. *Proc. Natl. Acad. Sci. USA.* **106**, 3196–3201.

Rohatgi, R., Milenkovic, L., and Scott, M. (2007). Patched1 Regulates Hedgehog Signaling at the Primary Cilium. *Science* **317**, 372–376.

Romio, L., Fry, A.M., Winyard, P.J., Malcolm, S., Woolf, A.S., and Feather, S.A. (2004). OFD1 is a centrosomal/basal body protein expressed during mesenchymal-epithelial transition in human nephrogenesis. *J. Am. Soc. Nephrol.* **15**, 2556–2568.

Rosenbaum, J.L., and Witman, G.B. (2002). Intraflagellar transport. *Nat. Rev. Mol. Cell. Biol.* **3**, 813–825.

Ruiz i Altaba, A., Nguyên, V., and Palma, V. (2003). The emergent design of the neural tube: prepattern, SHH morphogen and GLI code. *Curr. Opin. in Genet. Dev.* **13**, 513–521.

Ruiz i Altaba, A., Sanchez, P., and Dahmane, N. (2002). Gli and hedgehog in cancer: tumours, embryos and stem cells. *Nat. Rev. Cancer.* **2**, 361–372.

Ruiz-Perez, V.L., Blair, H.J., Rodriguez-Andres, M.E., Blanco, M.J., Wilson, A., Liu, Y.N., Miles, C., Peters, H., and Goodship, J.A. (2007). Evc is a positive mediator of Ihh-regulated bone growth that localises at the base of chondrocyte cilia. *Development.* **134**, 2903–2912.

Ruiz-Perez, V.L., Ide, S.E., Strom, T.M., Lorenz, B., Wilson, D., Woods, K., King, L., Francomano, C., Freisinger, P., Spranger, S., Marino, B., and Dallapiccola, B., *et al.* (2000). Mutations in a new gene in Ellis-van Creveld syndrome and Weyers acrodental dysostosis. *Nat. Genet.* **24**, 283–286.

Ruiz-Perez, V.L., Tompson, S.W., Blair, H.J., Espinoza-Valdez, C., Lapunzina, P., Silva, E.O., Hamel, B., Gibbs, J.L., Young, I.D., Wright, M.J., and Goodship, J.A. (2003). Mutations in two nonhomologous genes in a head-to-head configuration cause Ellis-van Creveld syndrome. *Am. J. Hum. Genet.* **72**, 728–732.

Sato, T., Mushiake, S., Kato, Y., Sato, K., Sato, M., Takeda, N., Ozono, K., Miki, K., Kubo, Y., Tsuji, A., Harada, R., and Harada, A. (2007). The Rab8 GTPase regulates apical protein localization in intestinal cells. *Nature* **448**, 366–369.

Scholey, J.M. (1996). Kinesin-II, a membrane traffic motor in axons, axonemes, and spindles. *J Cell Biol.* **133**, 1–4.

Scholey, J.M. (2008). Intraflagellar transport motors in cilia: moving along the cell's antenna. *J. Cell Biol.* **180**, 23–29.

Sisson, J.C., Ho, K.S., Suyama, K., and Scott, M.P. (1997). Costal2, a novel kinesin-related protein in the Hedgehog signaling pathway. *Cell* **90**, 235–245.

Sorokin, S. (1962). Centrioles and the formation of rudimentary cilia by fibroblasts and smooth muscle cells. *J. Cell Biol.* **15**, 363–377.

Stone, D.M., Hynes, M., Armanini, M., Swanson, T.A., Gu, Q., Johnson, R.L., Scott, M.P., Pennica, D., Goddard, A., Phillips, H., Noll, M., and Hooper, J.E., *et al.* (1996). The tumour-suppressor gene patched encodes a candidate receptor for Sonic hedgehog. *Nature* **384**, 129–134.

Tabata, T., and Kornberg, T.B. (1994). Hedgehog is a signaling protein with a key role in patterning Drosophila imaginal discs. *Cell* **76**, 89–102.

Taipale, J., Chen, J.K., Cooper, M.K., Wang, B., Mann, R.K., Milenkovic, L., Scott, M.P., and Beachy, P.A. (2000). Effects of oncogenic mutations in Smoothened and Patched can be reversed by cyclopamine. *Nature* **406**, 1005–1009.

Taipale, J., Cooper, M.K., Maiti, T., and Beachy, P.A. (2002). Patched acts catalytically to suppress the activity of Smoothened. *Nature* **418**, 892–897.

Tobin, J.L., and Beales, P.L. (2007). Bardet-Biedl syndrome: beyond the cilium. *Pediatr. Nephrol.* **22**, 926–936.

Tran, P., Haycraft, C., Besschetnova, T., Turbe-Doan, A., Stottmann, R., Herron, B., Chesebro, A., Qiu, H., Scherz, P., Shah, J., Yoder, B., and Beier, D. (2008). THM1 negatively modulates mouse sonic hedgehog signal transduction and affects retrograde intraflagellar transport in cilia. *Nat. Genet.* **40**, 403–410.

Vierkotten, J., Dildrop, R., Peters, T., Wang, B., and Ruther, U. (2007). Ftm is a novel basal body protein of cilia involved in Shh signalling. *Development* **134**, 2569–2577.

Wang, B., Fallon, J.F., and Beachy, P.A. (2000). Hedgehog-regulated processing of Gli3 produces an anterior/posterior repressor gradient in the developing vertebrate limb. *Cell* **100**, 423–434.

Wang, Q., Pan, J., and Snell, W.J. (2006). Intraflagellar transport particles participate directly in cilium-generated signaling in Chlamydomonas. *Cell* **125**, 549–562.

Wang, Y., Zhou, Z., Walsh, C.T., and Mcmahon, A. (2009). Selective translocation of intracellular Smoothened to the primary cilium in response to Hedgehog pathway modulation. *Proc. Natl. Acad. Sci. USA.* **106**, 2623–2628

Wijgerde, M., McMahon, J.A., Rule, M., and McMahon, A.P. (2002). A direct requirement for Hedgehog signaling for normal specification of all ventral progenitor domains in the presumptive mammalian spinal cord. *Genes. Dev.* **16**, 2849–2864.

Wilson, C.W., Chen, M.H., and Chuang, P.T. (2009a). Smoothened adopts multiple active and inactive conformations capable of trafficking to the primary cilium. *PLoS One* **4**, e5182.

Wilson, C.W., Nguyen, C.T., Chen, M.H., Yang, J.H., Gacayan, R., Huang, J., Chen, J.N., and Chuang, P.T., (2009b). Fused has evolved divergent roles in vertebrate Hedgehog signalling and motile ciliogenesis. *Nature* **459**, 98–102.

Wilson, L., and Maden, M. (2005). The mechanisms of dorsoventral patterning in the vertebrate neural tube. *Dev. Biol.* **282**, 1–13.

Wolff, C., Roy, S., and Ingham, P.W. (2003). Multiple muscle cell identities induced by distinct levels and timing of hedgehog activity in the zebrafish embryo. *Curr. Biol.* **13**, 1169–1181.

Wong, S.Y., and Reiter, J.F. (2008). The primary cilium at the crossroads of mammalian hedgehog signaling. *Curr. Top. Dev. Biol.* **85**, 225–260.

Yamazaki, H., Nakata, T., Okada, Y., and Hirokawa, N. (1996). Cloning and characterization of KAP3: a novel kinesin superfamily-associated protein of KIF3A/3B. *Proc. Natl. Acad. Sci. U S A.* **93**, 8443–8448.

Yauch, R.L., Gould, S.E., Scales, S.J., Tang, T., Tian, H., Ahn, C.P., Marshall, D., Fu, L., Januario, T., Kallop, D., Nannini-Pepe, M., and Kotkow, K., *et al.* (2008). A paracrine requirement for hedgehog signalling in cancer. *Nature* **455**, 406–410.

Yin, Y., Bangs, F., Paton, I.R., Prescott, A., James, J., Davey, M.G., Whitley, P., Genikhovich, G., Technau, U., Burt, D.W., and Tickle, C. (2009). The Talpid3 gene (KIAA0586) encodes a centrosomal protein that is essential for primary cilia formation. *Development* **136**, 655–664.

CHAPTER 11

Detecting the Surface Localization and Cytoplasmic Cleavage of Membrane-Bound Proteins

Hannah C. Chapin[*], **Vanathy Rajendran**[†], **Anna Capasso**[‡], *and* **Michael J. Caplan**[*,†]

[*]Department of Cell Biology, Yale University, New Haven, Connecticut, 06520

[†]Department of Cellular and Molecular Physiology, Yale University, New Haven, Connecticut, 06520

[‡]University of Naples, Yale University, New Haven, Connecticut, 06520

978-0-12-375024-2
DOI: 10.1016/S0091-679X(08)94011-5

Abstract

Polycystin-1 (PC1) is a large, membrane-bound protein that localizes to the cilia and is implicated in the common ciliopathy autosomal-dominant polycystic kidney disease. The physiological function of PC1 is dependent upon its subcellular localization as well as specific cleavages that release soluble fragments of its C-terminal tail. The techniques described here allow visualization and quantification of these aspects of the biology of the PC1 protein. To visualize PC1 at the plasma membrane, a live-cell surface labeling immunofluorescence protocol paired with the labeling of an internal antigen motif allows a robust detection of the surface population of this protein. This technique is modified to generate a surface enzyme-linked immunosorbent assay (ELISA), which quantitatively measures the amount of surface protein as a fraction of the total amount of the protein expressed in that cell population. These assays are powerful tools in the assessment of the small but biologically important pool of PC1 that reaches the cell surface. The C-terminal tail cleavage of PC1 constitutes an interesting modification that allows PC1 to extend its functional role into the nucleus. A reporter assay based on Gal4/VP16 luciferase can be used to quantitate the amount of PC1 C-terminal tail that reaches the nucleus. This assay can be paired with quantitative measurement of the protein expression in the cell, allowing a more complete understanding of the pattern of PC1 cleavage and the nuclear localization of the resultant.

I. Introduction

Mounting evidence illuminates the crucial role that cilia play in mechanosensation and signal transduction, linking extracellular conditions to changes in intracellular signaling pathways. Ciliary proteins such as polyductin and the polycystins localize to the cilia and plasma membrane where they can alter intracellular conditions directly, through changes in ion concentration, or indirectly, by releasing soluble cytoplasmic fragments that can partner with intracellular signaling molecules to affect processes such as gene transcription. Regulation of the correct physiological functioning of these ciliary proteins therefore involves their localization and cleavage; the immunofluorescence and reporter-driven assays described here provide ways to analyze these crucial aspects of ciliary protein biology.

Polycystin-1 (PC1) is the product of the polycystic kidney disease 1 (*PKD1*) gene that, along with *PKD2* (encoding PC2), harbors the mutations that cause autosomal-dominant polycystic kidney disease (ADPKD). This common genetic disease affects approximately 1 in 1000 individuals. A significant manifestation of the disease is the progressive appearance and growth of renal cysts. These cysts displace and destroy adjacent renal parenchyma, leading to end-stage renal disease in approximately 50% of cases. There are also cardiovascular, musculoskeletal, and gastrointestinal abnormalities associated with ADPKD (Gabow, 1993). The connection between the *PKD1* and

PKD2 genes and ADPKD was first shown by genetic linkage studies and later verified in animal models. Cysts form when both somatic copies of either polycystin gene are mutated or knocked out (Lu *et al.*, 1997; Qian *et al.*, 1996). Cysts can also arise when the level of PKD1 expression is significantly up- or downregulated (Lantinga-van Leeuwen *et al.*, 2004; Pritchard *et al.*, 2000).

The complex subcellular localization of PC1 reflects the broad range of this protein's cellular functions. There is an extensive literature documenting the localization of PC1 and PC2 to primary cilia. This localization is thought to permit the PC1/PC2 complex to play a role in sensing fluid flow (Chauvet *et al.*, 2004; Nauli *et al.*, 2003). PC1 may also play a role in establishing cell–cell connections; it is found at desmosomes and in the basolateral membrane of Madin–Darby canine kidney (MDCK) cells (Bukanov *et al.*, 2002) and can stimulate junction formation by binding E-cadherin (Streets *et al.*, 2009). It is also thought to contribute to ion channel activity when it localizes with PC2 to the plasma membrane (Hanaoka *et al.*, 2000). These studies have primarily used costaining with specific antibodies to identify PC1's presence in specific membrane domains through colocalization, but until now there has been no technique that has allowed a quantitative assessment of the overall delivery of PC1 to the plasma membrane.

PC1 participates in a variety of signaling pathways in the cell, and the cleavage of the PC1 protein's C-terminal tail may allow PC1 to affect a variety of diverse intracellular processes in response to stimuli such as extracellular fluid flow. One cleavage occurs within the cytoplasmic tail and releases a protein fragment that translocates to the nucleus and interacts with STAT6 and p100 (Low *et al.*, 2006). Another cleavage releases a larger soluble portion of the tail that activates the activator protein 1 pathway (Chauvet *et al.*, 2004) and that inhibits canonical Wnt signaling (Lal *et al.*, 2008). Rates of cleavage at both cleavage sites increase with the cessation of fluid flow, suggesting a link between PC1's roles in mechanosensation and the modulation of signaling pathways. To date, the principal approaches to studying the C-terminal PC1 cleavage and nuclear translocation have involved Western blotting and immunofluorescence, which can report on the occurrence of cleavage and translocation under some conditions but do not allow a very nuanced understanding of how environmental cues affect the extent of cleavage and the subcellular location of the soluble PC1 tail fragment. The Gal4/VP16 assay as described below permits the measurement of PC1 cleavage and nuclear translocation, allowing a more detailed analysis of these processes under varying physiological conditions.

II. Assay Rationale and History

A. Surface Immunofluorescence and Enzyme-Linked Immunosorbent Assay

PC1 has two distinct subcellular distributions when it is exogenously expressed by transfection in cell culture. A significant portion of the protein is found in the endoplasmic reticulum when heterologously expressed in cell culture, but it has also been shown to localize to the plasma membrane and the primary cilium and this localization is likely critical for the function of the PC1 protein as a channel or flow sensor. Since the location of PC1 may have an effect on its function, it is useful to know under what conditions the protein reaches the plasma membrane and whether this can be altered by coexpression of other proteins or the application of drugs to change the intra- or extracellular environment. While cell surface biotinylation is often used as a standard method for quantifying the amount of a given protein that reaches the surface (Hurley *et al.*, 1985), PC1's large size and relatively low level of detectable surface delivery led us to develop another method for identifying and quantifying surface PC1.

Visualizing the surface pool of PC1 protein is most effectively accomplished using an immunofluorescence protocol that yields a view of protein distribution in which the surface protein is tagged with one fluorescent marker and the internal pool is labeled with another color (Fig. 1). This provides a clear and relatively simple way to image the distribution of PC1 and provides an assay system that can then be perturbed with drugs or coexpressed proteins to reveal the effects of these manipulations on PC1's surface localization. The Alexa class of fluorescent dyes produce a bright signal that renders it easy to detect even small populations of surface-localized PC1, thus ensuring a high degree of sensitivity of the assay for surface localization.

Although immunofluorescence is an ideal method for qualitatively looking at protein distribution at the level of individual cells, effort is required to ensure that it is quantifiable. Quantitative immunofluorescence requires both a lengthy labeling protocol and a substantial investment of time to examine each coverslip on a microscope. We therefore developed a whole-cell enzyme-linked immunosorbent assay (ELISA) to provide a high-throughput assay system that quantifies the amount of surface PC1 relative to the total amount expressed in the cells. This assay can easily be performed in 96-well plates, allowing us to screen up to 16 conditions. Since the assay

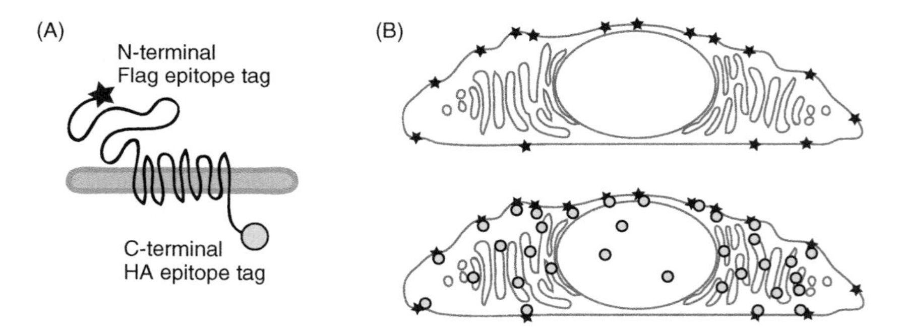

Fig. 1 Surface immunofluorescence detects the pool of polycystin-1 (PC1) at the plasma membrane. The N-terminal Flag epitope tag on PC1 is marked with one antibody (represented by a star), while the intracellular C-terminal HA epitope tag is detected with a second antibody (represented by a circle) (A). The protocol for surface immunofluorescence is optimized to mark only the PC1 that has reached the plasma membrane (B, upper panel), while adding the HA antibody after permeabilizing the cells allows a visualization of the total amount of PC1 expressed in the same cell (B, lower panel).

is sensitive to changes in cell number, it is best suited to assessing the consequences of manipulations that do not cause cell death.

B. Gal/VP Luciferase Assay

Regulated intramembrane proteolysis is a mechanism by which a membrane-bound protein, such as Notch, amyloid β-protein precursor, or EpCAM (Ebinu and Yankner, 2002; Maetzel *et al.*, 2009), is cleaved within a transmembrane domain to release a soluble cytoplasmic peptide that translocates to the nucleus and modifies gene expression or alters intracellular signaling pathways. The study of these types of cleavage and translocation events was facilitated through the generation of a complementary DNA (cDNA) reporter construct that could be transcriptionally activated by the nuclear translocation of a tagged protein. The DNA-binding domain of the yeast Gal4 transcription factor, combined with the transcriptional activation domain of the viral VP16 protein, drives strong expression of genes that are downstream of the Gal4-binding upstream activating sequence (UAS) (Sadowski *et al.*, 1988). This allows a membrane protein of interest, fused to both Gal4 and VP16, to induce the expression of a UAS-driven reporter protein upon the cleavage, release, and nuclear translocation of its cytoplasmic domain. Since the development of this system, various UAS-driven reporters have been used in a variety of assay systems to illuminate processes such as the *in vivo* location of Notch cleavage in the developing *Drosophila* embryo (Struhl and Greenwald, 1999) and to assay for small molecules that affect the cleavage and translocation of the amyloid precursor protein (Bakshi *et al.*, 2005).

The cleavage and trafficking of PC1 have many parallels with the life cycles of other proteins cleaved by regulated intramembrane proteolysis, in that the soluble C-terminal tail is cleaved and translocates to the nucleus (see review by Guay-Woodford, 2004). Studying PC1 cleavage has been complicated by the fact that there are at least two different C-terminal fragments that can be released by cleavage (Chauvet *et al.*, 2004; Low *et al.*, 2006). Expressing a soluble form of the C-terminal fragment could yield information about the effects of these peptides on intracellular signaling pathways, but this gives no insight into the processes that generate the soluble peptides and may actually produce results that are difficult to interpret in a physiological context (Basavanna *et al.*, 2007). The endogenous fragments produced by these cleavage events within the full-length protein are often short-lived and found in low abundance. Their relatively small size makes it difficult to use sodium dodecyl sulfate polyacrylamide gel electrophoresis (SDS-PAGE) gels to quantify their production in relation to the amount of the full-length protein or to determine the relative percent of the cleaved fragments that translocate to the nucleus under various physiological conditions. These limitations made PC1 an ideal candidate for a Gal4/VP16 reporter assay to quantify its cleavage and nuclear translocation.

To create PC1-Gal4/VP16, we inserted the binding and activation domains of Gal4/VP16 at the C-terminus of a cDNA encoding full-length PC1 in the mammalian expression vector pcDNA3.1, using a unique restriction site in a 3× hemagglutinin (HA) tag already present at the end of the PC1 sequence. The placement of the Gal4/VP16 sequence

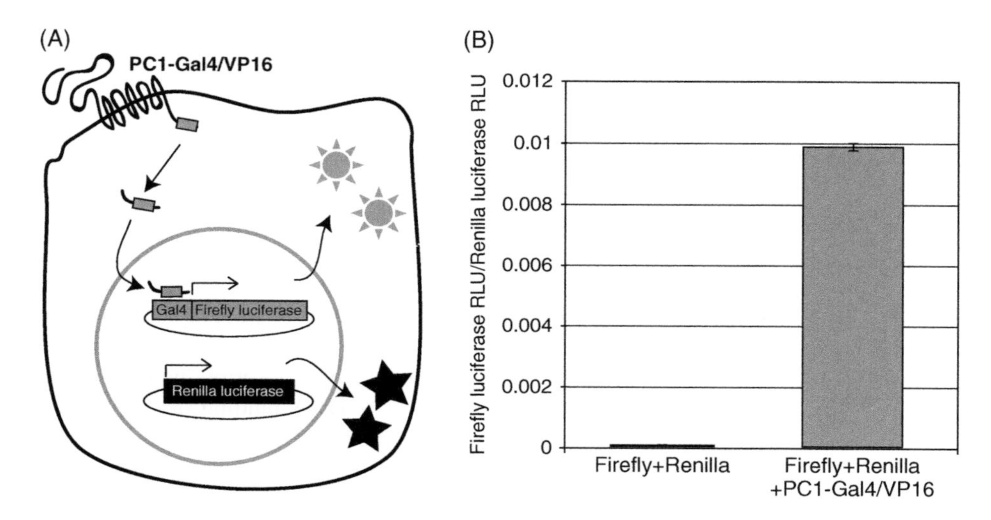

Fig. 2 The PC1-Gal4/VP16 reporter assay. A diagram illustrating the cleavage of PC1-Gal4/VP16 releasing a soluble C-terminal fragment that, upon nuclear translocation, binds to the Gal4 UAS DNA-binding domain and activates the transcription of the downstream firefly luciferase gene (A). Constitutive expression *Renilla* luciferase from a cotransfected plasmid allows normalization to account for variations in transfection efficiency. The assay has a very low background, as measured by the relative light units (RLUs) produced by cells expressing only the firefly and *Renilla* plasmids (B). Expression of PC1-Gal4/VP16 provides a specific increase in the firefly luciferase RLU, as normalized to the *Renilla* RLU count.

means that any fragment produced by a C-terminal cleavage contains the binding and activating domains required to drive the transcription of the chosen reporter construct. To study changes in PC1 cleavage over relatively short time courses in cell culture-based assays, we chose to use a UAS-luciferase from Promega containing a sequence of amino acids rich in proline, glutamic acid, serine and threonine (a PEST sequence) that promotes protein degradation. It is important to note that this reporter system requires robust cotransfection and expression of both the PC1-Gal4/VP16 and reporter proteins, so experimental conditions or treatments that alter transfection efficiency or protein synthesis could give false-positive results in assays evaluating manipulations that might perturb PC1 cleavage. To compensate for this, we cotransfected the *Renilla* luciferase gene under the constitutively active herpes simplex virus (HSV) promoter. We can then compensate for variations in transfection efficiency and protein synthesis by normalizing the PC1-induced firefly luciferase signal to the amount of *Renilla* luciferase activity (Fig. 2).

III. Materials

A. Enzyme-Linked Immunosorbent Assay

Blocking buffer: 5% FBS, 0.5% BSA in PBS^{++} (phosphate buffered saline with 100 μM $CaCl_2$ and 1 mM $MgCl_2$)

Permeabilization buffer: 5% FBS, 0.5% BSA, and 0.5% TritonX-100
Antibody against an epitope that is extracellular when the protein of interest reaches the plasma membrane (for illustrative purposes these instructions will refer to polyclonal anti-Flag, Sigma-Aldrich, St. Louis, MO)
Secondary antibody, conjugated to horseradish peroxidase (HRP), directed against the primary antibody's isotype
Ultra tetramethylbenzidine (TMB)-ELISA (Thermo Scientific, Waltham, MA)
Sulfuric acid, 1 N
Plate reader that can read absorbance at 450 nm

B. Surface Immunofluorescence

Primary antibody directed against an epitope that is exposed at the extracellular surface when the protein of interest reaches the plasma membrane (we use the same antibody as for the ELISA, which is polyclonal anti-Flag from Sigma-Aldrich)
Primary antibody against an intracellular epitope of the protein of interest. This antibody should not cross-react with the primary antibody against the extracellular epitope and should be produced in a different species (we use a monoclonal anti-HA from Covance, Princeton, NJ)
PBS^{++} with 100 μM $CaCl_2$ and 1 mM $MgCl_2$
Blocking Buffer: 0.1% BSA in PBS^{++}
Permeabilization buffer: 0.3% TritonX-100, 0.1% BSA in PBS^{++}
Goat serum dilution buffer (GSDB): 16% goat serum, 120 mM sodium phosphate, 0.3% Triton X-100, and 450 mM NaCl
Secondary antibodies, conjugated to fluorophores of choice, against immunoglobulin G (IgG) from the appropriate species used for the primary antibodies (we use the Alexa Fluor 594 anti-rabbit IgG and 488 anti-mouse IgG for the polyclonal anti-Flag and monoclonal anti-HA, respectively, both supplied by Invitrogen, Carlsbad, CA)

C. Gal/VP Luciferase Assay

UAS-promoted firefly luciferase cDNA plasmid [pGL4.31(*luc2p/GAL4*UAS/Hygro) from Promega, Madison, WI].
Constitutively expressed *Renilla* luciferase cDNA plasmid (pRL-TK from Promega)
PC1-Gal/VP: PC1 with the Gal4 DNA-binding domain and VP16-activation domains added at the C-terminus of the protein
Kit for reading the luciferase signals (Dual Luciferase kit from Promega).
Luminometer, either single-read or configured as a plate-reader. If using a plate reader, whether or not it has injection capabilities will influence your choice of assay kit, since some require addition of a buffer immediately before reading the light output

IV. Methods

A. Surface Immunofluorescence

This protocol is optimized for use on adherent cell lines grown on either glass coverslips or filters. The broad outline of the protocol requires that the surface antibody be applied in the cold prior to fixation, and then the internal antibody is added after fixation and permeabilization. As long as the surface and internal antibodies are from different source species, the secondary antibodies can be added together in the final labeling step. For PC1, immunofluorescence against the N-terminal Flag tag illuminates the pool of PC1 at the plasma membrane, while postpermeabilization immunofluorescence against the C-terminal HA tag detects the total population of PC1 (Fig. 3). This protocol is also applicable to polarized porcine kidney cells (LLC-PK) in which the ciliary localization of PC1 is demonstrated by the ciliary staining of both externally applied anti-Flag and internal anti-HA antibodies (Fig. 3).

Cells for surface immunofluorescence should be grown on either filters or glass coverslips, although cells that are weakly adherent should be plated on coverslips coated with poly-L-lysine to minimize cell loss. Cells should be transfected according to the method of choice for the particular cell type when the cells have reached an appropriate level of confluency. The cells should then be allowed to express the protein of interest for 24 h before beginning the immunofluorescence protocol. When the cells are ready to be assayed, they may be first incubated in the blocking buffer at 4°C for 30 min to prevent nonspecific antibody binding. We have found that signal-to-noise ratio of the polyclonal anti-Flag antibody we use does not improve dramatically with this surface-blocking step, but it has been a useful step for antibodies that have higher amounts of nonspecific binding. Cells not being blocked are rinsed once with blocking buffer, and then all coverslips are inverted onto a small volume of blocking buffer containing primary antibody against the extracellular epitope for 1 h at 4°C in a humidified chamber. If the cells are grown on filters, then the surface antibody may be selectively applied to the plasma membrane domain (apical or basolateral) in which the protein of interest resides. The primary antibody is applied at 4°C to prevent trafficking and redistribution of the surface population of the antigen during the course of the primary antibody incubation. After this incubation, the coverslips are returned to their wells and gently washed a minimum of one time in PBS^{++} and fixed in 4% paraformaldehyde for 20 min. All steps including and following the fixation are performed at room temperature. The protocol may be paused after the paraformaldehyde is washed out and the coverslips may be stored overnight at 4°C in PBS.

After fixation, the coverslips are washed three times in PBS^{++} and then incubated at room temperature with permeabilization buffer for 15 min. They are then blocked with GSDB for 30 min before an hour-long incubation with the antibody against the intracellular epitope diluted in GSDB. After this incubation, the cells are washed three times with permeabilization buffer and then incubated in a darkened, humidified chamber on a drop of the secondary antibodies diluted in GSDB for 45 min. We have found that the choice of secondary antibody makes a difference when detecting proteins, such as PC1,

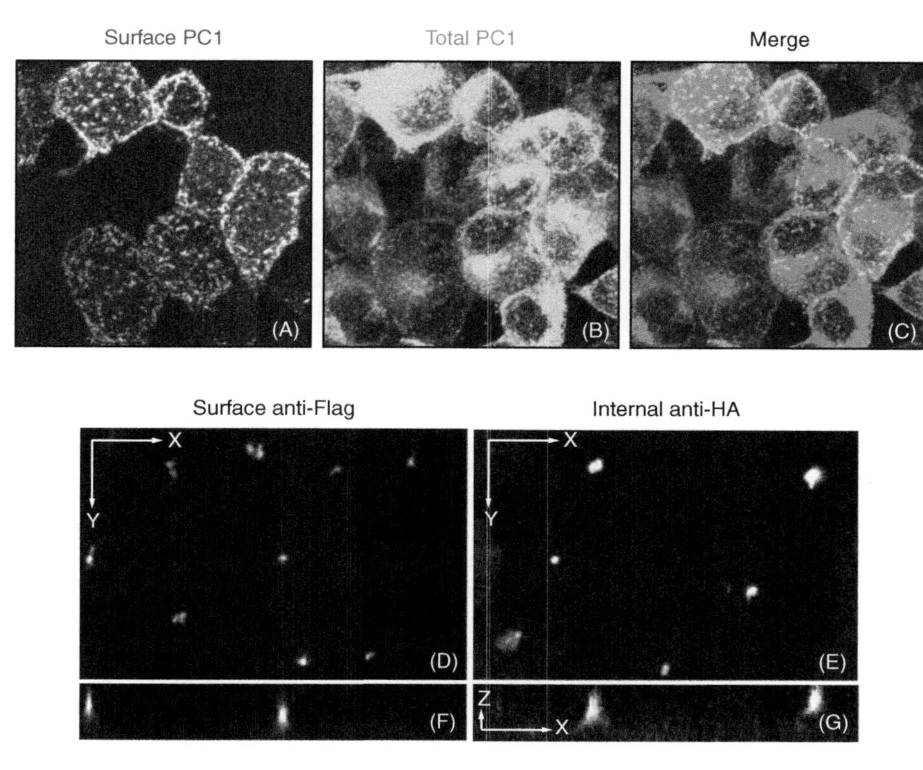

Fig. 3 Examples of surface immunofluorescence under conditions that promote PC1 delivery to the plasma membrane and cilia. In nonpolarized HEK293 stably expressing PC1, the anti-Flag antibody shows the surface pool of PC1 (A), while the internal anti-HA antibody shows the overall PC1 expression (B). The merged image reveals the colocalization between internal and external markers (C). Images (A–C) were generated by flattening a *z*-stack of images obtained using confocal microscopy. Polarized LLCPK cells stably expressing PC1 have a ciliary localization of PC1, as revealed with both the surface anti-Flag antibody (D, F) and the internal anti-HA antibody (E, G); images shown in *x*–*y* and *x*–*y* projections. A slice along the *z*-axis shows that the HA antibody labels both ciliary and intracellular protein pools (G), while the surface anti-Flag antibody labels only cilia (F). The scale bar for (D–G) is 10 μm. (See Plate no. 15 in the Color Plate Section.)

that have low levels of surface localization under some conditions. The high signal-to-noise obtained with bright fluorophores such as the Alexa dyes makes it much easier to detect the surface signal than does the use of rhodamine- or fluorescein isothiocyanate (FITC)-conjugated secondary antibodies.

Following incubation in secondary antibody, the coverslips are washed again in PBS^{++} and then the nuclei are stained. We routinely use Hoechst stain to mark the nuclei, although the nuclear stain should be chosen based on which secondary antibodies are being used and on the detection capabilities of the microscope. The stained cells are then mounted on a drop of anti-fade mounting media on glass slides, the edges are sealed with nail polish, and the slides are then ready to be viewed on the microscope of choice. We use an upright Zeiss Axiophot microscope equipped with a charge-coupled device (CCD) camera for viewing the slides and taking pictures for routine quantification, but we use a Zeiss LSM 510 Meta confocal microscope when we wish to generate images of higher quality.

In addition to providing a means to visualize the presence or absence of protein at the surface, this technique can also be adapted to quantify how the pool of surface protein changes in response to biological or biochemical treatments. To quantify the intensity of surface immunofluorescence signal, we take images of each experimental condition with identical settings for zoom and exposure. While taking images with confocal microscopy, we image the entire cell in the vertical dimension, generating a *z*-stack of images. The thickness of each slice and the number of slices should be optimized for the cell type being used in the experiment. We then flatten the resulting stack by exporting an image from the extended focus view in Volocity (Improvision, PerkinElmer). Once flattened, the confocal image file can be treated like that of an image obtained using an epifluorescence microscope.

To quantify the average pixel intensity per cell, an image's total pixel intensity is calculated and divided by the total number of cells in each image that are positive for surface immunofluorescence signal. We perform this calculation by exporting the image's pixel intensity histogram from Image J (National Institutes of Health), then multiplying each intensity value (0–255) by the number of pixels that have that intensity. Summing all values above background then gives the total pixel intensity for the image. The cutoff for the background should be calculated by evaluating the intensity histogram for an image taken of a microscopic field without surface signal and determining the pixel intensity value below which most of this background signal is found. The number of surface-positive cells can be counted from the raw images or, more accurately, it can be counted after subtracting the background cutoff value from each pixel using Image J, thereby showing exactly how many of the cells are being included in the final pixel intensity sum. Having calculated the average pixel intensity per cell, independent experiments can then be compared to determine the effects that specific treatments have on the surface pool of protein.

B. Enzyme-Linked Immunosorbent Assay

The ELISA, to compare amounts of surface and total protein, uses a spectrophotometric assay to measure the amount of protein detected using conditions in which labeling is performed with or without permeabilization. Given that this assay is amenable to being used in relatively high-throughput applications, we plate cells into 96-well plates and use six wells per experimental condition, subdivided so that three wells provide triplicate measurements of the surface protein and the other three measure the total amount of expressed protein. After correcting for background, the data can provide an accurate measurement of the amount of surface protein relative to the total amount present in the cells.

To prepare for a surface ELISA, cells are grown in a 96-well plate and subjected to transfection for an appropriate length of time prior to beginning the assay. We routinely use Lipofectamine 2000 (Invitrogen) and transfect cells approximately 24 h prior to performing the assay. Each assay condition should be repeated over six wells, with an additional six wells receiving no treatment (or transfection with a blank plasmid, if appropriate). These wells will be used to measure the background signal.

An important consideration for the entire ELISA procedure is that variations in the number of cells per well can produce significant effects on the assay values. When plating the cells, care should be taken to utilize a uniform cell suspension so that the plated volume yields equivalent numbers of cells per well. In addition, cell lines such as human embryonic kidney cells (HEK293) that do not adhere strongly to plastic should be grown on PLL-coated wells in order to minimize cell loss during prefixation washes. Additionally, when washing the wells, the experimenter should be as consistent as possible about whether or where the suction pipette touches the bottom of the well during aspiration of media and wash liquids, since touching the cell layer will remove cells and affect the reading.

Once cells have been transfected and treated, all cells are carefully washed with cold PBS^{++}, after which the protocol for each set of wells (surface, total, and control) diverges for the labeling (Fig. 4). In general, 25 μl of diluted antibody is sufficient to cover the bottom surface in a 96-well plate.

Wells used to measure surface expression are first blocked at 4°C in blocking buffer for 30 min. Antibody against the extracellular epitope, diluted in blocking buffer, is then applied at 4°C for 1 h. An antibody dilution similar to that used for immunofluorescence experiments is generally suitable for this application. After incubation with the primary antibody, the cells are washed once in cold PBS^{++} and fixed for 20 min in 4% paraformaldeyhde. After fixation, the cells are washed three times with PBS and are then ready for the secondary antibody.

Wells used for the measurement of total PC1 protein are fixed in 4% paraformaldehyde after the initial wash with PBS^{++}. If the assay is being done in a single plate, these "total" wells will need to be allowed to fix while the "surface" wells are in the primary antibody at 4°C, so the fixation should last for 1 h. After three washes with PBS^{++}, the cells are permeabilized and blocked using a 30-min incubation in permeabilization buffer. The same primary antibody that is used for surface labeling is then diluted in the blocking buffer and applied for 1 h at room temperature. After this step, the cells are washed three times with permeabilization buffer and are then ready for the secondary antibody.

Control wells do not receive any primary antibody, but are fixed in 4% paraformaldehyde and then washed along with the other wells. The control wells measuring the surface background are left to sit in the blocking buffer until receiving the secondary antibody, but the control wells measuring the total background are permeabilized for 30 min after fixation and washing.

The secondary antibody is applied to all wells. The secondary antibody is HRP-conjugated antibody against the appropriate animal source of the primary antibody; in our case a goat anti-rabbit HRP is used to detect the polyclonal anti-Flag antibody. The dilution for the secondary antibody is 1:3000 for this rabbit HRP, but can be up to 1:10,000 for some mouse HRP conjugates. The secondary antibody is diluted in the blocking buffer and applied for 45 min at room temperature. After this incubation, the wells are washed three times with PBS and then 80 μl Ultra-TMB substrate, which has

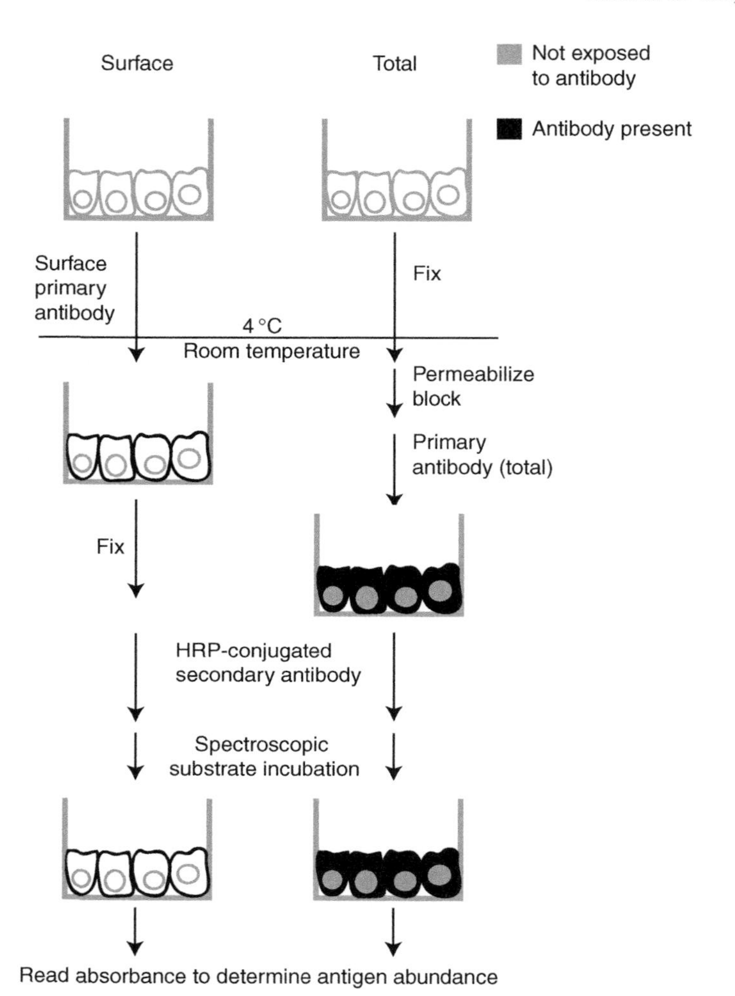

Fig. 4 Diagram of the ELISA protocol. A representation of cells grown in small-welled tissue culture plates illustrates the treatments for wells measuring the amount of PC1 protein on the surface and the total amount present in the cell. The requisite wash steps are not depicted here, and the protocol for control wells (without primary antibody) is also not shown.

been equilibrated to room temperature, is added for up to 15 min. The wells that have been permeabilized will react with this HRP substrate relatively quickly and care should be taken to stop the reaction before it proceeds far enough that the reaction product becomes saturating, after which the linearity of the assay is compromised. This time course is best determined empirically. The reaction is stopped after 10–15 min using 80 μl of 1 N sulfuric acid and the absorbance at 450 nm is measured using a spectrophotometric plate reader.

Due to endogenous peroxidase activity that is present to varying degrees in most cell types, there will be a small but measurable amount of background absorbance generated even in the untransfected control wells. Some published ELISA protocols use a mild hydrogen peroxide treatment before the secondary antibody incubation step to reduce this background, but we did not find this to be significantly effective in our system. To compensate for the background, we use the controls wells that do not receive the primary antibody. Utilizing a set of control wells for each treatment condition is especially useful if the experimental treatments change the total number of cells in any way, since that will alter background signal. If the experimental treatments have no effect on cell number, then it is feasible to do just one set of surface and total control wells for the entire plate. To analyze the data, we subtract the average of the appropriate control wells from each experimental reading, then average the readings across the experimental conditions, and take the ratio of surface:total absorbance signal.

C. Gal/VP Luciferase Assay

This PC1-Gal/VP-driven luciferase assay provides a tool to assay the nuclear translocation of the cleaved PC1 C-terminal tail and has the advantage that it permits simultaneous assessment of the total expressed protein in the cells by Western blot. Given the range of possible assays, the experiments can be set up in wells of almost any size, depending on the goals of a particular experiment. A 96-well plate can be used to effectively screen several conditions with biological replicates, while carrying out the experiment in 12-well or even 6-well plates allows one to both obtain a reading of luciferase while leaving enough lysate to be for Western blots or immunoprecipitations. If the assay results are to be read with a plate reader, then the cells should be plated in clear-bottomed, black-sided wells to minimize leakage of extraneous luciferase signal from adjoining wells.

The prerequisites for this assay include the cotransfection of cDNAs encoding UAS-promoted firefly luciferase, constituitively expressed *Renilla* luciferase, and PC1-Gal/VP. Once cells have reached an appropriate density in the plate of choice, they are transfected with the plasmids using Lipofectamine 2000 or a similar transfection agent. Since the plasmids are of different sizes and express at different levels, we routinely use a 1:20:60 ratio of *Renilla*:firefly:PC1-Gal4/VP16 in order to keep the luciferase signals within the linear detection range of our luminometer and maximize the signal from PC1-Gal4/VP16. The control condition for all assays is the transfection of only *Renilla* and firefly luciferase plasmids, replacing the PC1-Gal4/VP16 with an "empty" plasmid to keep the total micrograms of transfected DNA the same across all experimental conditions.

After transfection, the cells should be left for a minimum of 24 h, including the time required for any experimental treatment, before assaying. We routinely use the Dual-Luciferase kit from Promega, but alter the protocol for lysis depending on the goals of the assay. If the same lysate is to be used both for luciferase readings and for SDS-PAGE gel and Western blotting, then it is best to add protease inhibitors to the 1X passive lysis buffer provided with the luciferase kit. To detect the entire pool of protein, including any cleavage fragments that have translocated to the nucleus, we then sonicate the lysate and spin for 15 min at 18,000 g at 4°C before preparing an aliquot of the lysate for SDS-PAGE. The luciferase signal can be read either before or after sonication. Once the *Renilla* and luciferase readings are taken, the data are analyzed by dividing each sample's firefly luciferase signal by its corresponding *Renilla* luciferase value. This ratio per well is then normalized to the averaged ratio of all control wells, yielding a quantified measurement for each well that is effectively the fold-increase of the signal above the backgound. The ratioed values for each well can then be averaged for each experimental condition and statistically evaluated for significance.

V. Discussion

A reliable method to detect and quantify expression of proteins on the cell surface is of value in a wide range of applications. Measuring the size of the surface pool of a target protein can reveal cell biological processes that would not be seen with an immunofluorescence protocol that only recognizes the total pool of cell-associated protein. This is especially true in experiments that utilize cell-culture-based protein overexpression systems, since in these settings there may be dramatic and biologically interesting experimental effects on protein localization that would otherwise be missed. The visual signal produced by the population of protein trapped in the endoplasmic reticulum can mask the shift of a fraction of that protein to the cell surface, making a surface immunofluorescence protocol more illuminating than a technique that only sees the total pool. We have successfully used this surface immunofluorescence technique to demonstrate that PC1 is brought to the cell surface with different efficiencies as a function of the proteins with which it is coexpressed. A clean signal derived exclusively from the surface pool of protein is also useful in determining the protein's residence in specialized membrane subdomains. For example, pairing the surface antibody with an internal antibody against ciliary or junctional proteins may allow a more precise measurement of colocalization for proteins of interest that have both plasma membrane and intracellular distributions.

A key advantage of the surface immunofluorescence approach is that it is temporally specific. Since the surface labeling protocol is performed in the cold, and thus in the absence of ongoing membrane traffic, the technique effectively gives a snapshot of protein localization at a specific point in time, allowing the surface delivery to be analyzed over a time course. One hypothetical application of this capability involves examining several time points over the course of increasing cell confluency to assess the delivery of a protein of interest to junctions as they form. We have also used similarly designed time course experiments to assess the change in surface localization of PC1 protein following treatment with environmental stimuli and chemical compounds.

The surface labeling protocol described here is similar to that outlined by Bengtsson *et al.* (2008), with the added advantage that both secondary antibodies are employed in a single, simultaneous, postfixation incubation step. This cuts down on the time required for the overall protocol, and if greater speed is necessary, the times used here could be further optimized for a faster throughput. It should be noted that our protocol also allows for a blocking step before the surface antibody incubation to minimize the background signal that may improve the specificity of some primary antibodies.

While the immunofluorescence protocol is useful for detecting the changes in surface localization that occur at the level of individual cells, the ELISA protocol described here allows for a much more quantitative assessment of the amount of protein reaching the cell surface in a population of cells. While a whole-cell ELISA technique was developed quite some time ago to quantify surface proteins in bacteria (Elder *et al.* 1982), that uses a room temperature incubation to label the antigen and measures only the surface protein pool. Our protocol assays the ratio of the surface protein to the total amount expressed, thereby taking into account that the amount of surface protein may be affected by experimental treatments that alter the general stability or expression of the protein of interest. Additionally, by pairing each experimental condition with the appropriate controls, this protocol permits quantification of changes effected by a range of experimental conditions, including protein coexpression or drugs applied in the media. The miniaturization of the protocol allows for easy completion of a number of biological replicates, increasing confidence in the significance of experimental observations.

The biological importance of PC1's C-terminal cleavage has been recognized for several years, and the signaling effects of the soluble fragment have been suggested to be initiated by abnormal fluid flow past ciliated cells. Until the application of the Gal4/VP16 assay, though, there had been no way to quantify the nuclear translocation of the PC1 cleavage fragments. This Gal4/VP16 assay has now been utilized to demonstrate that the C-terminal cleavage of PC1 is enhanced by coexpression with PC2, a finding that had been suggested by Western blot and immunofluorescence, but had not been quantified using either technique. Furthermore, the relative ease and speed of assay completion allowed screening of conditions that altered intracellular calcium, revealing that the PC2 enhancement of PC1 cleavage is independent of intracellular calcium concentrations (Bertuccio *et al.*, 2009). These results show an interesting contrast between PC1 and fibrocystin, another ciliary protein that causes autosomal-recessive PKD when mutated. A luciferase assay based on a fibrocystin–Gal4/VP16 fusion protein revealed that fibrocystin undergoes a C-terminal cleavage, but that this cleavage requires activation of protein kinase C as well as increased cytoplasmic calcium concentrations (Hiesberger *et al.*, 2006). Similarly to PC1, the released C-terminal fragment of fibrocystin contains a nuclear localization sequence that mediates the nuclear translocation of the cleaved fragment. In combination, these results suggest a common mechanism of membrane-bound, ciliary proteins influencing intracellular signaling events through the cleavage and release of soluble, cytoplasmic fragments.

VI. Summary

The biology of ciliary proteins is fascinating and complex, and it is becoming clear that the localization and cleavage of these proteins play a key role in regulating their function. The techniques outlined here provide methods to visualize and quantify the localization and cleavage of membrane-bound proteins, allowing a more detailed understanding of how proteins such as PC1 are delivered to the cilia and plasma membrane, and what influences the cleavage that releases their cytoplasmic fragments. The PC1-Gal4/VP16 assay has been successfully used to show the calcium-independence of PC1 cleavage (Bertuccio *et al.*, 2009), laying the groundwork for subsequent exploration of the regulation of PC1 cleavage. In sum, these techniques contribute to the panoply of techniques available for the study of the dynamic cell biology of ciliary proteins.

References

Bakshi, P., Liao, Y. F., Gao, J., Ni, J., Stein, R., Yeh, L. A., and Wolfe, M. S. (2005). A high-throughput screen to identify inhibitors of amyloid beta-protein precursor processing. *J. Biomol. Screen.* **10**, 1–12.

Basavanna, U., Weber, K. M., Hu, Q., Ziegelstein, R. C., Germino, G. G., and Sutters, M. (2007). The isolated polycystin-1 COOH-terminal can activate or block polycystin-1 signaling. *Biochem. Biophys. Res. Commun.* **359**, 367–372.

Bengtsson, D. C., Sowa, K. M., and Arnot, D. E. (2008). Dual fluorescence labeling of surface-exposed and internal proteins in erythrocytes infected with the malaria parasite Plasmodium falciparum. *Nat. Protoc.* **3**, 1990–6.

Bertuccio, C. A., Chapin, H. C., Cai, Y., Mistry, K., Chauvet, V., Somlo, S., and Caplan, M. J. (2009). Polycystin-1 C-terminal cleavage is modulated by polycystin-2 expression. *J. Biol. Chem.* **284**, 21011–26.

Bukanov, N., Husson, H., Dackowski, W. R., Lawrence, B. D., Clow, P. A., Roberts, B. L., Klinger, K. W., and Ibraghimov-Beskrovnaya, O. (2002). Functional polycystin-1 expression is developmentally regulated during epithelial morphogenesis in vitro: Downregulation and loss of membrane localization during cystogenesis. *Hum. Mol. Genet.* **11**, 923–936.

Chauvet, V., Tian, X., Husson, H., Grimm, D. H., Wang, T., Hiesberger, T., Igarashi, P., Bennett, A. M., Ibraghimov-Beskrovnaya, O., Somlo, S., and Caplan, M. J. (2004). Mechanical stimuli induce cleavage and nuclear translocation of the polycystin-1 C terminus. *J. Clin. Invest.* **114**, 1433–1443.

Ebinu, J. O., and Yankner, B. A. (2002). A RIP tide in neuronal signal transduction. *Neuron* **34**, 499–502.

Elder, B. L., Boraker, D. K., and Fives-Taylor, P. M. (1982). Whole-bacterial cell enzyme-linked immunosorbent assay for Streptococcus sanguis fimbrial antigens. *J. Clin. Microbiol.* **16**, 141–4.

Gabow, P. A. (1993). Autosomal dominant polycystic kidney disease. *N. Engl. J. Med.* **329**, 332–342.

Guay-Woodford, L. M. (2004). RIP-ed and ready to dance: New mechanisms for polycystin-1 signaling. *J. Clin. Invest.* **114**, 1404–1406.

Hanaoka, K., Qian, F., Boletta, A., Bhunia, A. K., Piontek, K., Tsiokas, L., Sukhatme, V. P., Guggino, W. B., and Germino, G. G. (2000). Co-assembly of polycystin-1 and -2 produces unique cation-permeable currents. *Nature* **408**, 990–994.

Hiesberger, T., Gourley, E., Erickson, A., Koulen, P., Ward, C. J., Masyuk, T. V., Larusso, N. F., Harris, P. C., and Igarashi, P. (2006). Proteolytic cleavage and nuclear translocation of fibrocystin is regulated by intracellular Ca2+ and activation of protein kinase C. *J. Biol. Chem.* **281**, 34357–34364.

Hurley, W. L., Finkelstein, E., and Holst, B. D. (1985). Identification of surface proteins on bovine leukocytes by a biotin-avidin protein blotting technique. *J. Immunol. Methods* **85**, 195–202.

Lal, M., Song, X., Pluznick, J. L., Di Giovanni, V., Merrick, D. M., Rosenblum, N. D., Chauvet, V., Gottardi, C. J., Pei, Y., and Caplan, M. J. (2008). Polycystin-1 C-terminal tail associates with beta-catenin and inhibits canonical Wnt signaling. *Hum. Mol. Genet.* **17**, 3105–3117.

Lantinga-van Leeuwen, I. S., Dauwerse, J. G., Baelde, H. J., Leonhard, W. N., van de Wal, A., Ward, C. J., Verbeek, S., Deruiter, M. C., Breuning, M. H., de Heer, E., and Peters, D. J. (2004). Lowering of Pkd1 expression is sufficient to cause polycystic kidney disease. *Hum. Mol. Genet.* **13**, 3069–3077.

Low, S. H., Vasanth, S., Larson, C. H., Mukherjee, S., Sharma, N., Kinter, M. T., Kane, M. E., Obara, T., and Weimbs, T. (2006). Polycystin-1, STAT6, and P100 function in a pathway that transduces ciliary mechanosensation and is activated in polycystic kidney disease. *Dev. Cell* **10**, 57–69.

Lu, W., Peissel, B., Babakhanlou, H., Pavlova, A., Geng, L., Fan, X., Larson, C., Brent, G., and Zhou, J. (1997). Perinatal lethality with kidney and pancreas defects in mice with a targetted Pkd1 mutation. *Nat. Genet.* **17**, 179–181.

Maetzel, D., Denzel, S., Mack, B., Canis, M., Went, P., Benk, M., Kieu, C., Papior, P., Baeuerle, P. A., Munz, M., and Gires, O. (2009). Nuclear signalling by tumour-associated antigen EpCAM. *Nat. Cell Biol.* **11**, 162–171.

Nauli, S. M., Alenghat, F. J., Luo, Y., Williams, E., Vassilev, P., Li, X., Elia, A. E., Lu, W., Brown, E. M., Quinn, S. J., Ingber, D. E., and Zhou, J. (2003). Polycystins 1 and 2 mediate mechanosensation in the primary cilium of kidney cells. *Nat. Genet.* **33**, 129–137.

Pritchard, L., Sloane-Stanley, J. A., Sharpe, J. A., Aspinwall, R., Lu, W., Buckle, V., Strmecki, L., Walker, D., Ward, C. J., Alpers, C. E., Zhou, J., Wood, W. G., et al. (2000). A human PKD1 transgene generates functional polycystin-1 in mice and is associated with a cystic phenotype. *Hum. Mol. Genet.* **9**, 2617–2627.

Qian, F., Watnick, T. J., Onuchic, L. F., and Germino, G. G. (1996). The molecular basis of focal cyst formation in human autosomal dominant polycystic kidney disease type I. *Cell* **87**, 979–987.

Sadowski, I., Ma, J., Triezenberg, S., and Ptashne, M. (1988). GAL4-VP16 is an unusually potent transcriptional activator. *Nature* **335**, 563–564.

Streets, A. J., Wagner, B. E., Harris, P. C., Ward, C. J., and Ong, A. C. (2009). Homophilic and heterophilic polycystin 1 interactions regulate E-cadherin recruitment and junction assembly in MDCK cells. *J. Cell Sci.* **122**, 1410–1417.

Struhl, G., and Greenwald, I. (1999). Presenilin is required for activity and nuclear access of Notch in Drosophila. *Nature* **398**, 522–525.

CHAPTER 12

Assay for *In Vitro* Budding of Ciliary-Targeted Rhodopsin Transport Carriers

Dusanka Deretic[*,†] ***and*** **Jana Mazelova**[*]

[*]Department of Surgery, Division of Ophthalmology, University of New Mexico, Albuquerque, New Mexico 87131

[†]Department Cell Biology and Physiology, University of New Mexico, Albuquerque, New Mexico 87131

Abstract

Primary cilia and cilia-derived sensory organelles are cell's antennas that contain sensory receptors and signal transduction modules. Defects in the expression and targeting of ciliary proteins to this specialized cellular compartment lead to human disorders collectively known as ciliopathies. To examine the molecular basis for the

978-0-12-375024-2
DOI: 10.1016/S0091-679X(08)94012-7

ciliary targeting of the light receptor rhodopsin, we have developed a cell-free assay that reconstitutes its packaging into the specific post-Golgi rhodopsin transport carriers (RTCs). This assay accurately reproduces the *in vivo* process of carrier budding, while allowing examination of individual components of the macromolecular complexes, thus providing insight into a more general mechanism for the regulation of ciliary membrane targeting. Examples are shown for the use of this assay in rhodopsin trafficking. The cell-free assay is applicable to other ciliary-targeted sensory molecules.

I. Introduction

Almost all cells possess primary cilia that house an array of signal transduction modules. The molecular mechanisms underlying ciliary membrane synthesis and targeting have recently received increased attention, due to the ciliary involvement in a wide range of human diseases called ciliopathies, including retinal degeneration, polycystic kidney disease (PKD), Bardet–Biedl syndrome (BBS), and neural tube defects (Christensen *et al.*, 2007; Fliegauf *et al.*, 2007; Leroux, 2007; Nachury *et al.*, 2007; Singla and Reiter, 2006).

Retinal rod photoreceptor cells possess a specialized light-sensing organelle, the rod outer segment (ROS), which is filled with membranous disks that house the light-sensing machinery involved in visual transduction. ROS initially form from primary cilia, and a short "9 + 0" (nonmotile) connecting cilium remains as the sole path of communication between the ROS and the rod inner segment (RIS). The RIS houses biosynthetic membranes that continuously replenish ROS disk membranes, which are removed through daily shedding and phagocytosis by retinal pigment epithelial (RPE) cells. Photoreceptor connecting cilium corresponds to the transition zone of primary cilia, which is considered a gateway for the admission of specific proteins to this privileged intracellular compartment (Rosenbaum and Witman, 2002). Therefore, the ciliary targeting of light-sensing molecules underlies the process of ROS disk membrane renewal.

Photopigment rhodopsin represents 90% of the newly synthesized protein in rod photoreceptors and its biosynthetic pathway is best understood. Following synthesis and glycosylation in the Golgi, rhodopsin is incorporated into transport carriers (RTCs) that bud from the *trans*-Golgi network (TGN) and fuse with the specialized domain separating the ciliary membrane from the surrounding RIS plasma membrane (Deretic and Papermaster, 1991; Papermaster *et al.*, 1986). This specialized membrane domain for RTC docking and fusion, termed the periciliary ridge complex (PRC) in *Xenopus* (Peters *et al.*, 1983), houses an array of scaffold proteins that organize a ciliary and basal body protein network, which interacts with the Crumbs polarity complex (Gosens *et al.*, 2007; Maerker *et al.*, 2008). This is also the site of assembly of large molecular complexes involved in intraflagellar transport (IFT) (Baker *et al.*, 2003; Bhowmick *et al.*, 2009; Follit *et al.*, 2006; Pazour *et al.*, 2002; Rosenbaum and Witman, 2002; Rosenbaum *et al.*, 1999).

Successful isolation and purification of RTCs provided not only insight into their molecular composition, but was also the basis for the development of the retinal cell-free assay that reconstitutes RTC budding *in vitro* (Deretic, 2000; Deretic and Papermaster, 1993; Deretic *et al.*, 1996). Development of this assay led to the discovery that rhodopsin contains a sorting signal within its five C-terminal amino acids that regulates its incorporation into RTCs as they bud from the TGN (Deretic *et al.*, 1998). It has become increasingly clear that the VxPx motif in rhodopsin's C-terminal domain is a signal for sorting into RTCs and their targeting to the ROS (Deretic, 2006; Deretic *et al.*, 1998; Green *et al.*, 2000; Gross *et al.*, 2006; Shi *et al.*, 2004; Tam *et al.*, 2000). The rhodopsin C-terminal VxPx targeting motif is disrupted by the mutations that cause the most severe forms of the human disease autosomal dominant retinitis pigmentosa (adRP) (http://www.retina-international.com/sci-news/rhomut.htm). The VxPx targeting motif binds the small GTPase Arf4 and regulates incorporation of rhodopsin into RTCs at the TGN (Deretic *et al.*, 2005). Arf4, in turn, directs the assembly of a trafficking module comprised of the small GTPase Rab11, the Rab11/Arf effector FIP3, and an Arf-GAP/effector ASAP1 (Mazelova *et al.*, 2009a). Other ciliary membrane proteins also posses the VxPx motif (Geng *et al.*, 2006; Jenkins *et al.*, 2006); thus the Arf4-based targeting complex is most likely a part of the conserved machinery involved in the selection and packaging of the ciliary membrane cargo.

This chapter focuses on the experimental details of the frog retinal cell-free assay that reconstitutes the physiological process of RTC budding *in vitro*, in an ATP-, GTP-, and cytosol-dependent manner, as judged by the appropriate morphology, topology, and protein composition of budded carriers and by the retention of resident proteins in the TGN (Deretic, 2000). Using this assay we have identified the C-terminal VxPx ciliary targeting signal of rhodopsin and the essential components of the trafficking machinery that recognize ciliary cargo and regulate its incorporation into post-Golgi carriers and the carrier budding from the TGN (Deretic *et al.*, 1996, 1998, 2005; Mazelova *et al.*, 2009a).

II. Methods

A. Pulse-Labeling of Newly Synthesized Proteins in Isolated Frog Retinas

1. Rationale

All experiments are performed using frog retinas, because the large size of photoreceptor cells in the frog retina and the extensive turnover of its light-sensing membrane components, especially when compared to mammals (Besharse, 1986), provides an ideal system to define the roles of individual components of the membrane-trafficking machinery. Southern leopard frogs, *Rana berlandieri,* are used for the *in vitro* budding assay, because the kinetics of rhodopsin trafficking through the post-Golgi compartment of photoreceptor cells has been well established for this species (Deretic and Papermaster, 1991). Frogs are maintained at 22°C under dim white light in a 12 h light/dark cycle. Frogs are dark-adapted for 2 h before the experiment. Dark adaptation causes the retraction of pigment epithelium microvilli that are intercalated between the ROS, which facilitates retinal isolation. All experiments are conducted under dim red light, as illumination affects the buoyant density and subcellular distribution of membranes containing rhodopsin.

2. Procedure

Frogs are sacrificed under dim red light (Safelight Lamp, with Kodak 1A Safelight Filters), by pithing of the brain and spinal cord and decapitation using a small animal decapitator. Frog retinas are isolated using dissection tools that include two small forceps, fine scissors, and surgical blades. Seven retinas are incubated per 15 ml of oxygenated media (composition of which is described in the materials section), in 50 ml polycarbonate Erlenmeyer flasks with stoppers, connected to the 95%O_2/5%CO_2 tank. [^{35}S]-Express protein labeling mixture (1000 Ci/mmol) is added to a final concentration of 25 μCi per retina, and retinal cultures are incubated at 22°C for 1 h on the bench-top orbital shaker at 25 rpm. Following isotope incorporation, retinas are rinsed with 34% w/w ice-cold sucrose homogenizing medium containing protease inhibitors and transferred to a polycarbonate tube with fresh medium on ice (0.3 ml per retina).

B. Removal of the ROS from the Retina and Preparation of the Photoreceptor-Enriched PNS from Radiolabeled Cells

1. Rationale

Because of the layered structure of the retina, biosynthetic organelles in the RIS are accessible only upon removal of the ROS. To avoid extensive damage of retinal tissue, ROS are removed from the retina by low shear forces and separated from the remainder of the retina by flotation on high-density (34%) sucrose. Biosynthetic compartments from the RIS are then isolated from the remainder of the retina by homogenization of the retinal pellet in 0.25 M sucrose under conditions of slightly greater shear. The 0.25 M sucrose homogenizing medium is modified from that previously described (Deretic and Papermaster, 1993), so that NaCl is omitted to avoid the aggregation of membranes during *in vitro* incubation. This homogenization step preferentially releases RIS membranes from the retina, whereas the rest of the retina is relatively unbroken in large fragments. Retinal fragments and nuclei are then sedimented at low speed. This pellet contains ~85% of retinal tissue, but $<$10% of the newly synthesized rhodopsin (Papermaster *et al.*, 1975). The postnuclear supernatant (PNS), therefore, is highly enriched in photoreceptor organelles involved in ROS membrane biosynthesis. The retinal subcellular fractionation procedure and the preparation of the photoreceptor-enriched PNS are schematically outlined in Fig. 1.

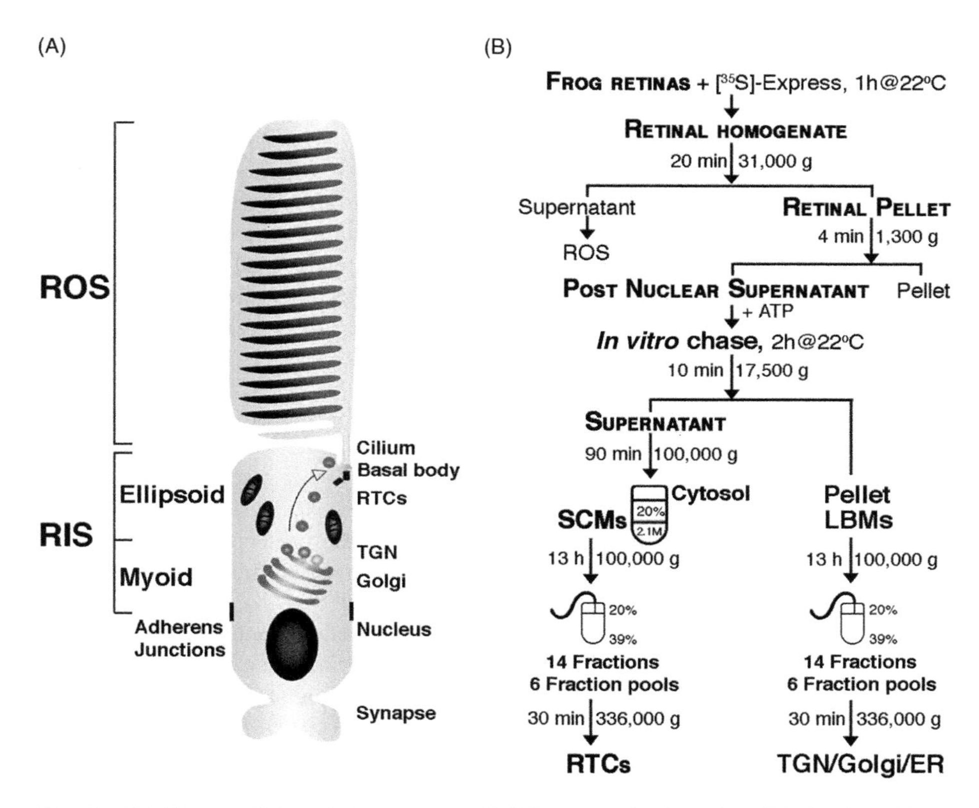

Fig. 1 (A) Diagram of the rod photoreceptor cell. Cilium protrudes from the cell body, the rod inner segment (RIS) and elaborates the rod outer segment (ROS) filled with membranous disks containing the photopigment rhodopsin and associated phototransduction machinery. Newly synthesized rhodopsin traverses the Golgi and the TGN, localized in the myoid region of the RIS, where it is incorporated into transport carriers (RTCs). RTCs travel from the TGN (arrow), though the mitochondria-laden ellipsoid region of the RIS, to the base of the cilium where they fuse with the RIS plasma membrane. (B) Schematic representation of the cell-free assay that reconstitutes RTC budding from the TGN. Following radiolabeling in cultured retinas, ROS are removed from the retina and floated on 34% sucrose. The retinal pellet is homogenized in 0.25 M sucrose and retinal fragments and nuclei are removed by low speed centrifugation. Postnuclear supernatant (PNS) is incubated in the presence of an ATP-regenerating or an ATP-depleting system for 2 h of cell-free chase. Upon termination of the assay, large biosynthetic membranes (LBMs) are sedimented and the supernatant containing the cytosol and the small carrier membranes (SCMs), which are predominantly RTCs, fractionated on a step sucrose gradient. SCMs are recovered from the step gradient and diluted with buffer. LBMs and SCMs are then fractionated on linear 20–39% sucrose density gradients. Fourteen fractions are pooled into six pools and membranes are sedimented and analyzed for the presence of radiolabeled rhodopsin.

2. Procedure

ROS are sheared by five passes through a 14G needle by gentle aspiration of retinas with a 10-ml syringe. The retinal homogenate in 34% sucrose homogenizing medium is overlaid with 1 ml of 1.10 g/ml sucrose solution and centrifuged at 31,000 g_{av}, for 20 min, at 4°C in a JA 25.50 rotor in a Beckman Avanti J-25 centrifuge. ROS are collected from the 1.10/34% sucrose interface. After removal of crude ROS, 1 ml aliquots of retinal pellets in 0.25 M sucrose, containing protease inhibitors (0.15 ml/retina), are homogenized with five passes of a loose-fitting Teflon-glass homogenizer (Thomas AA 831 with a clearance increased to 150 μm) driven by a Variable Speed Stirrer, model GT21 (G.K. Heller) with motor controller at setting 30, under dim red light in the cold room. Homogenized aliquots are pooled and the homogenate is centrifuged at 1300 g_{av}, for 4 min, at 4°C in a JA 25.50 rotor to pellet the nuclei and unbroken retinal fragments. The postnuclear supernatant obtained after this centrifugation (PNS) is the starting material for the cell-free assay.

C. *In Vitro* Incubation of Photoreceptor-Enriched PNS and Cell-Free Formation of Post-Golgi RTCs

1. Rationale

Radiolabeled PNS is incubated in the presence of ATP and an ATP-regenerating system for 2 h of chase *in vitro*, a chase period found to be optimal for radiolabeling of RTCs *in vivo*. At the onset of a typical *in vitro* chase experiment, after one hour of pulse labeling *in vivo*, rhodopsin is predominantly found in the Golgi/TGN (Deretic *et al.*, 1996). Therefore, rhodopsin trafficking during the cell-free chase reflects predominantly RTC budding from the Golgi/TGN.

2. Procedure

The standard assay for one cell-free reaction is as follows: to one ml of PNS in 0.25 M sucrose (obtained from seven radiolabeled retinas) 115 μl of 10 × concentrated buffer stock solution is added to give a final concentration of 25 mM Hepes-KOH pH 7.0, 25 mM KCl, and 2.5 mM MgAc. The 10 × concentrated buffer stock solution is designed to provide optimal conditions to achieve transport *in vitro*. The use of low-ionic strength medium is important because incubation of the PNS in the presence of salt causes membrane aggregation, which interferes with subsequent subcellular fractionation (Tooze and Huttner, 1992). The cell-fee transport assay is initiated by the addition of 50 μl of an ATP-regenerating system and by transfer to 22°C. To investigate the energy requirements of the transport *in vitro*, an ATP-depleting system is added to the PNS. Samples are incubated for 2 h at 22°C without agitation. The assay is terminated by the addition of 0.2 M EDTA to a final concentration of 3 mM. In addition to termination of the trafficking assay, EDTA also disrupts divalent cation cross-linking of unrelated membranes and permits their separation on sucrose gradients.

To examine the cytosol requirements for the cell-free trafficking, PNS is subfractionated into membrane-enriched and cytosolic fractions. One ml of PNS in 0.25 M sucrose (obtained from 14 retinas) is overlaid on a step sucrose gradient containing 1 ml of 0.5 M sucrose and 3 ml of 49% sucrose. Gradients are spun for 1 h at 75,000 g_{av} in an SW 50.1 rotor in an Optima L-90K ultracentrifuge (Beckman-Coulter, Fullerton, CA). Cytosol that remains in 0.25 M sucrose is further centrifuged at 336,000 g_{av} to remove residual membranes, and supernatants containing cytosolic proteins are collected. Membrane proteins that enter the 0.5 M sucrose layer are collected and divided in half. To one half of the membranes 0.5 ml of 5 mg/ml BSA in 0.25 M sucrose is added (–cytosol) and to the other half 0.5 ml of cytosolic proteins is added back (+cytosol). PNS is restored either to the original ratio of cytosol: membranes (full cytosol), or to the limited cytosol conditions (routinely 25% cytosol, diluted with 0.25 M sucrose). Reconstituted PNS (±cytosol) is assayed for RTC budding as described above. To study the role of specific cytosolic factors complete cytosol is further subfractionated, either biochemically or by immunodepletion. Cytosol immunodepletion is performed by incubation with Protein A beads (GE Healthcare Life Sciences, Piscataway, NJ) bound to specific antibodies, for 2 h at 4°C.

The standard cell-free assay is modified in the experiments where preincubation with antibodies, purified proteins, or synthetic peptides is used to perturb the function of the regulatory components. For example, the role of the C-terminal domain of rhodopsin in RTC budding was tested by preincubation with 100 μg of anti-rhodopsin C-terminal antibody mAb 11D5 (or its Fab fragments) for 30 min on ice, or by preincubation with 50 μM synthetic peptides that correspond to the C-terminal sequence of rhodopsin, for 30 min at 22°C. Control samples are preincubated with the buffer under identical conditions.

D. Subcellular Fractionation of PNS Following Cell-Free RTC Budding

1. Rationale

To monitor the cell-free formation of RTCs, assay mixtures are subjected to further subfractionation by sedimentation followed by equilibrium sucrose gradient centrifugation. After low speed centrifugation to sediment large biosynthetic membranes (LBMs), small carrier membranes (SCMs) are separated from the cytosol on step sucrose gradients (Morel *et al.*, 2000). The bulk of the retinal proteins are found in LBM fractions that contain the Golgi/TGN, the ER, the RIS plasma membrane, and fragmented mitochondria (Deretic, 2000; Deretic *et al.*, 2004; Mazelova *et al.*, 2009b). SCMs and LBMs are further subfractionated on linear 20–39% sucrose density gradients. The low-density SCM fractions contain RTCs. Other membranes, such as the vesiculated RIS plasma membrane, are present in SCM fractions only in minor amounts and their distribution does not follow the distribution of radiolabeled rhodopsin (Deretic *et al.*, 2004).

2. Procedure

Six different cell-free assays are routinely performed (a control and five experimental conditions). Upon completion of cell-free RTC budding, PNS is centrifuged to sediment LBMs at 17,500 g, for 10 min, at 4°C in a JA 25.50 rotor in a Beckman Avanti J-25 centrifuge. Supernatant (~3 ml) is loaded on a step sucrose gradient containing 2 ml of 2.1 M sucrose overlaid with 5 ml of 20% sucrose in 10 mM Tris acetate pH 7.4. Gradients are centrifuged for 90 min at 100,000 g_{av} in an SW 40 rotor (Beckman-Coulter) at 4°C. SCMs are recovered from the 20%/2.1 M sucrose interface in ~1.5 ml and diluted with an equal volume of 10 mM Tris acetate pH 7.4. Sedimented LBMs are ressuspended in 0.25 M sucrose. LBMs and SCMs are then overlayed on 10 ml linear 20–39% (w/w) sucrose gradients in 10 mM Tris-acetate pH 7.4 and 0.1 mM $MgCl_2$. Gradients are prepared as follows: a 0.5 ml cushion of 49% sucrose is prepared in the 13.5 ml ultraclear tube for the SW 40 rotor; 5 ml of 39% sucrose is pipeted into the near chamber and 5 ml of 20% sucrose into the far chamber of the gradient maker that is connected to an Auto Densi-Flow fractionator (Labconco, Kansas City, MO), which automatically deposits a linear gradient on the underlying cushion. LBMs and SCMs are centrifuged on linear density gradients at 100,000 g_{av} at 4°C, for 13 h in an SW 40 rotor in an Optima L-90K ultracentrifuge (Beckman-Coulter). Fractions (0.9 ml) are collected from the top of the gradient using the Auto Densi-Flow fractionator and a fraction collector (Gilson FC 205). A 50 μl aliquot of each fraction is used to measure the refractive index. Fourteen subcellular fractions are routinely pooled as follows: pool 1 = fractions 1–3, 2 = 4–6, 3 = 7–8, 4 = 9–10, 5 = 11–12 and 6 = 13–14. Pooled fractions are then diluted with 10 mM Tris acetate pH 7.4 and centrifuged at 336,000 g_{av} for 30 min in a 70.1 Ti rotor (Beckman-Coulter). Pellets are resuspended in 10 μl/retina of 10 mM Tris acetate pH 7.4 and aliquoted for further analysis.

a. Biochemical Analysis. To determine the efficiency of the cell-free post-Golgi membrane formation 20 μl aliquots from each fraction (equivalent to 2 retinas) are subjected to SDS-polyacrylamide gel electrophoresis (SDS-PAGE). Dried SDS gels are subjected to quantitative analysis of ^{35}S-labeled rhodopsin in retinal subcellular fractions in a PhosphorImager (Molecular Dynamics), whereas the images of the gels are generated by autoradiography at –85°C using Kodak BioMax MR film. To control for the possible fragmentation of the TGN, in some experiments the retention of the TGN markers in the TGN at the end of the reaction is determined by assaying the distribution of sialyltransferase activity across the sucrose density gradient. To ascertain the purity of RTCs, the distribution of a range of compartment-specific proteins is routinely determined by immunoblotting with specific antibodies using the ECL Western Blotting Detection System.

b. Electron Microscopy. Negatively stained images of retinal subcellular fractions are obtained by placing a 200 mesh carbon-coated EM grid on a 20 μl sample in sucrose for 15 min in a humidified chamber. Grids are blotted with filter paper, stained with 2% uranyl acetate for 2 min, air dried, and examined in a Philips CM-100 transmission electron microscope.

III. Materials

All Solutions Are Made Using Sterile DI Water

1. Stock Solutions
   1 M NaCl; 0.1 M $MgCl_2$; 1 M Tris-acetate, pH 7.4; 0.2 M EDTA (di Na^+ salt · $2H_2O$)

Stock sucrose solutions:

42% sucrose: 151.5 g sucrose/350 g solution in H_2O;
η (refractive index) = 1.4036; ρ (density) = 1.1870 g/ml at 20°C.
49% sucrose: 150 g sucrose/300 g solution in H_2O;
η = 1.4179; ρ = 1.2241 g/ml at 20°C

All sucrose concentrations are w/w so that solutions are prepared on a top-loading balance. Sucrose solutions are prepared on the day before the experiment as outlined in Table I.

Protease inhibitor stock solutions (1000×):

10 mg/ml antipain (10 μg/ml final)
1 mg/ml = 2 mM leupeptin (2 μM final)
10 mg/ml = 100,000 KIU/ml aprotinin (100 KIU/ml final)

Table I
Sucrose Solutions

	Homogenizing media		Step gradient solutions (g/ml)			Linear gradient solutions (%w/w)	
Solution name	34%	0.25M	1.10	1.11	1.13	20%	39%
% sucrose		(8.5%)	(24%)	(25%)	(30%)		
η (at 20°)	1.383	1.345	1.371	1.373	1.381	1.363	1.398
Components							
42% Sucrose	191 g	10.2 g	31.2 g	34.2 g	40.1 g	–	–
48.5% Sucrose	–	–	–	–	–	41.66 g	81 g
1 M NaCl	13 ml	–	–	–	–	–	–
0.1 M $MgCl_2$	0.4 ml	0.5 ml	50 μl	50 μl	50 μl	100 μl	100 μl
1 M Tris Acetate pH 7.0	1 ml	0.5 ml	50 μl	50 μl	50 μl	100 μl	100 μl
Add DD H_20 to weight	229.31 g	49.85 g	55.0 g	55.5 g	56.5 g	99.85 g	99.85 g
Day of the experiment add Protease inhibitors Antipain, Aprotinin, Leupeptin	230 μl each 690 μl total	50 μl each 150 μl total	–	–	–	50 μl each 150 μl total	50 μl each 150 μl total
Total weight	230 g	50 g	55.0 g	55.5 g	56.5 g	100 g	100 g

Protease inhibitor stock solutions are kept frozen at –20° in 400 μl aliquots and are added on the day of experiment. The volume added to the sucrose solutions is a sum of equal volumes of each protease inhibitor.

2. Medium for [^{35}S]-Express Incorporation

This medium formulation is based on the frog medium described by Wolf and Quimby (1964). Methionine and cysteine are added to the medium only during *in vivo* chase. Final concentrations of amino acids are as follows:

	mg/l		mg/l
L-Arginine	69.30	L-Methionine	8.25
L-Cysteine	13.20	L-Phenylalanine	17.60
L-Glutamine	160.60	L-Threonine	26.40
L-Histidine	23.10	L-Tryptophan	5.50
L-Isoleucine	28.60	L-Tyrosine	19.80
L-Leucine	28.60	L-Valine	25.85
L-Lysine	40.15		

Final concentrations of salts in the medium are according to Greenberger and Besharse (1983):

64 mM NaCl (3740 mg/l)
1 mM $MgCl_2$ (203 mg/l $MgCl_2 \cdot 6H_2O$)
35 mM $NaHCO_3$ (2940 mg/l)
2 mM KCl (150 mg/l)
1.8 mM $CaCl_2$ (18 μl of 100 mM sol.)
4.4 mM glucose (800 mg/l)

The pH of the medium is 7.5. This custom made medium is now available from (Millipore Bioscience Research Reagents, Temecula, CA).

3. ATP Regenerating and Depleting System

The ATP-regenerating system contains equal volumes of 100 mM ATP (Na salt, neutralized with NaOH), 800 mM creatine phosphate, and 4 mg/ml creatine phosphokinase in 50% w/v glycerol. The ATP-depleting system contains 10 mg/ml of hexokinase in 250 mM D-glucose.

4. The 10× Concentrated Buffer Stock Solution

250 mM Hepes-KOH pH 7.0, 250 mM KCl, and 15 mM MgAc.

IV. Results and Discussion

The cell-free system described here reconstitutes RTC budding in an ATP-, GTP-, cytosol-, Rab-, and Arf-dependent manner. Using this assay, we have recently determined that the association of Arf4 with TGN membranes is

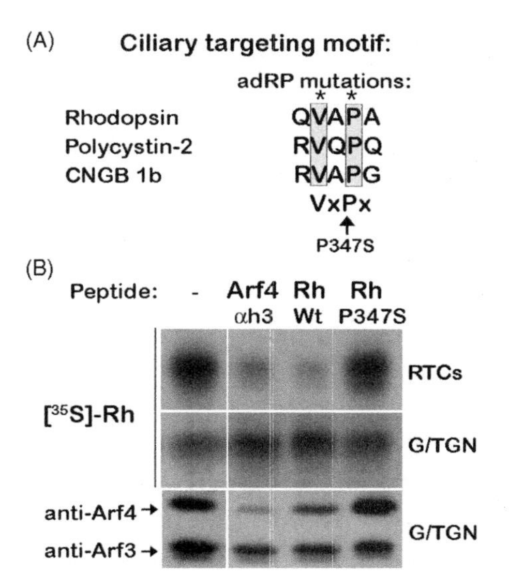

Fig. 2 Rhodopsin VxPx motif engages Arf4 at the TGN to regulate RTC budding. (A) ADRP mutations in the rhodopsin VxPx targeting motif are indicated with asterisks. (B) Frog retinas were pulse-labeled for 60 min and retinal PNS was incubated for 30 min with 50-μM peptides, as indicated in the panel, prior to a 2-h cell-free chase. Radiolabeled membrane proteins from two retinas were fractionated into Golgi/TGN and RTCs and analyzed by SDS-PAGE and autoradiography ([^{35}S]-Rh). Immunoblots of a duplicate gel were probed successively with anti-Arf3 and anti-Arf4, which specifically recognizes a ~20 kDa protein (left panel) and quantified (lower right panels). Modified from Mazelova *et al.* (2009a).

dependent on its access to the VxPx sorting motif present in rhodopsin (Mazelova *et al.*, 2009a). This is illustrated in Fig. 2. We found that the peptide corresponding to the α helix 3 unique to Arf4, and the rhodopsin C-terminal peptide inhibit RTC budding *in vitro*, while the peptide bearing the adRP mutation P347S in the VxPx motif has no effect. Unexpectedly, both inhibitory peptides nearly completely displaced Arf4 from the Golgi/TGN membranes, whereas the P347S peptide had no effect on Arf4 membrane association, indicating that P347S mutant rhodopsin is likely defective in interactions with Arf4 (Fig. 2).

A combination of biochemical and morphological assays was used to identify the Arf4-interacting proteins required for RTC budding. We first identified the Arf GAP that mediates GTP hydrolysis on Arf4 and regulates its function. Again, using the *in vitro* assay, we have determined that the Arf GAP ASAP1 is an essential regulator of Arf4 function and RTC budding (Mazelova *et al.*, 2009a), as illustrated in Fig. 3. The recombinant protein ASAP1 BAR-PZA (Mazelova *et al.*, 2009a) stimulates RTC budding. EM analysis of *in vitro* budded carriers demonstrated that RTCs formed in the presence of ASAP1

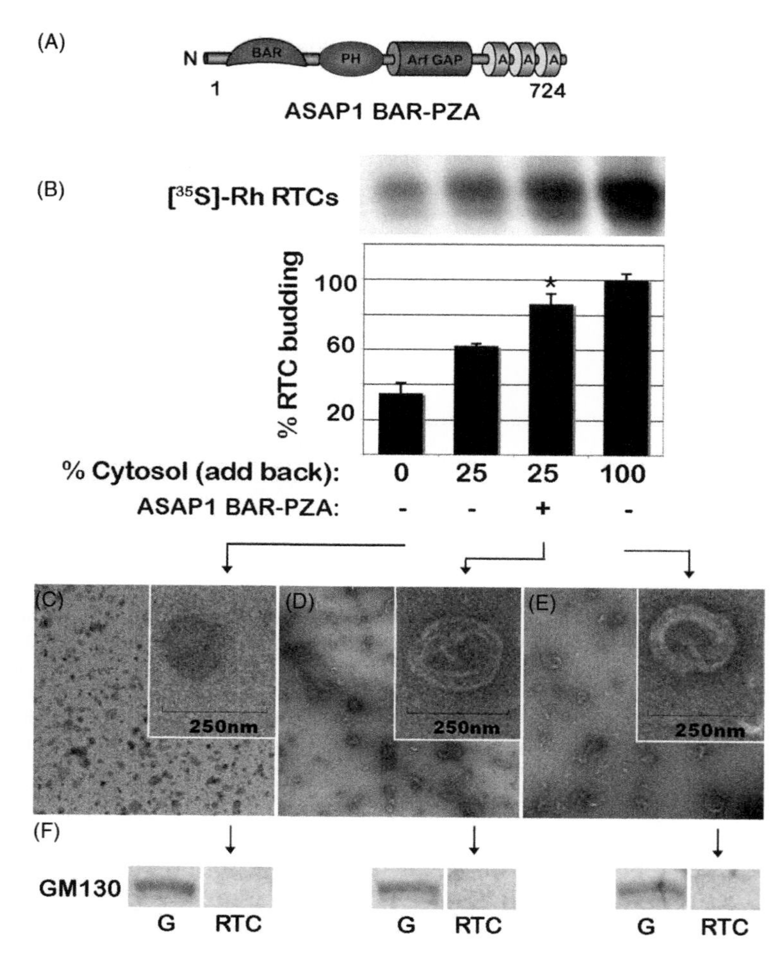

Fig. 3 Arf GAP ASAP1 and Rab11 regulate RTC budding from the TGN. (A) Schematic of the ASAP1 BAR-PZA construct. (B) Radiolabeled retinal PNS was separated into membrane and cytosolic fractions, which were reconstituted as indicated in the panel. 1 μM ASAP1-BAR-PZA was added to the assay in limited (25%) cytosol during the cell-free chase, and incorporation of rhodopsin into RTCs was determined ([^{35}S]-Rh). The quantification data are presented as the means ± SE of three separate experiments (*, $p = 0.01$). (C–E) Negatively stained *in vitro* budded RTCs (~250nm) examined by EM. (F) Following *in vitro* budding, Golgi and RTCs were immunoblotted with anti-GM130. (G) Cytosol was immunodepleted with anti-ASAP1 and Protein A beads, reducing ASAP1 to ~10% of control. (H) After *in vitro* budding in ASAP1-immunodepleted cytosol, membrane proteins from two retinas were separated by SDS-PAGE and autoradiographed ([^{35}S]-Rhodopsin, left panels). Duplicate gels were immunoblotted with ASAP1 (right panels). (I) Cytosol was immunodepleted with anti-Rab11, reducing it to nearly undetectable levels. (J) Following *in vitro* budding in Rab11-immunodepleted cytosol membrane proteins from two retinas were separated by SDS-PAGE and autoradiographed ([^{35}S]-Rhodopsin, left panels). Rab11 immunoblots of duplicate gels (right panels). Modified from Mazelova *et al.* (2009a).

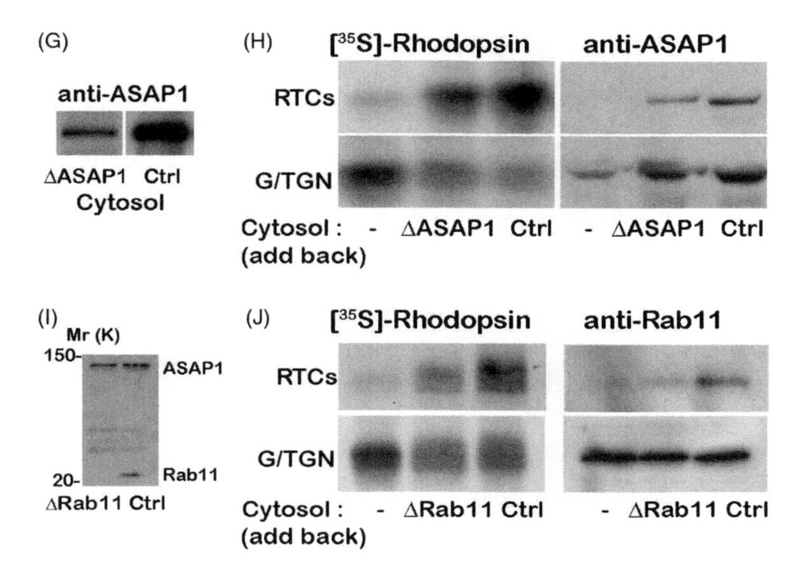

Fig. 3 (*Continued*)

BAR-PZA were morphologically identical to the ones formed in the full cytosol, and that RTC were undetectable when the cytosol was omitted from the assay (Fig. 3). Coimmunoprecipitation and pulldown experiments showed that Arf4, ASAP1, Rab11, and FIP3 form a protein complex (Mazelova *et al.*, 2009a). We then established that immunodepletion of the small GTPase Rab11 also inhibits RTC budding (Fig. 3). The fourth component of the trafficking complex, the Rab11/Arf effector FIP3, was also essential for RTC budding (Mazelova *et al.*, 2009a).

Finally, following *in vitro* RTC budding and subcellular fractionation on linear sucrose density gradients, we determined the distribution of Arf4, Rab11, FIP3, and ASAP1, and compared it to the distribution of radiolabeled rhodopsin incorporated into RTCs. As shown in Fig. 4, this analysis established that all components of the targeting complex are present on isolated RTCs, except for Arf4, which is present only in the Golgi/TGN-enriched fractions, but does not colocalize with Arf3. Localization of Arf4 is consistent with its inactivation by GTP hydrolysis and dissociation from the membrane, concomitant with RTC budding from the TGN. Moreover, by confocal microscopy we determined that TGN-derived nascent RTCs contain ASAP1, Rab11, and FIP3. Thus the cofractionation of the components of the complex with RTCs was corroborated by their colocalization in photoreceptor cells *in situ* (Mazelova *et al.*, 2009a). As the trafficking complex assembles at the TGN through Arf4, which directly binds the VxPx ciliary-targeting motif, this suggests that Arf4, Rab11, FIP3, and ASAP1 form a ciliary-targeting complex. These studies

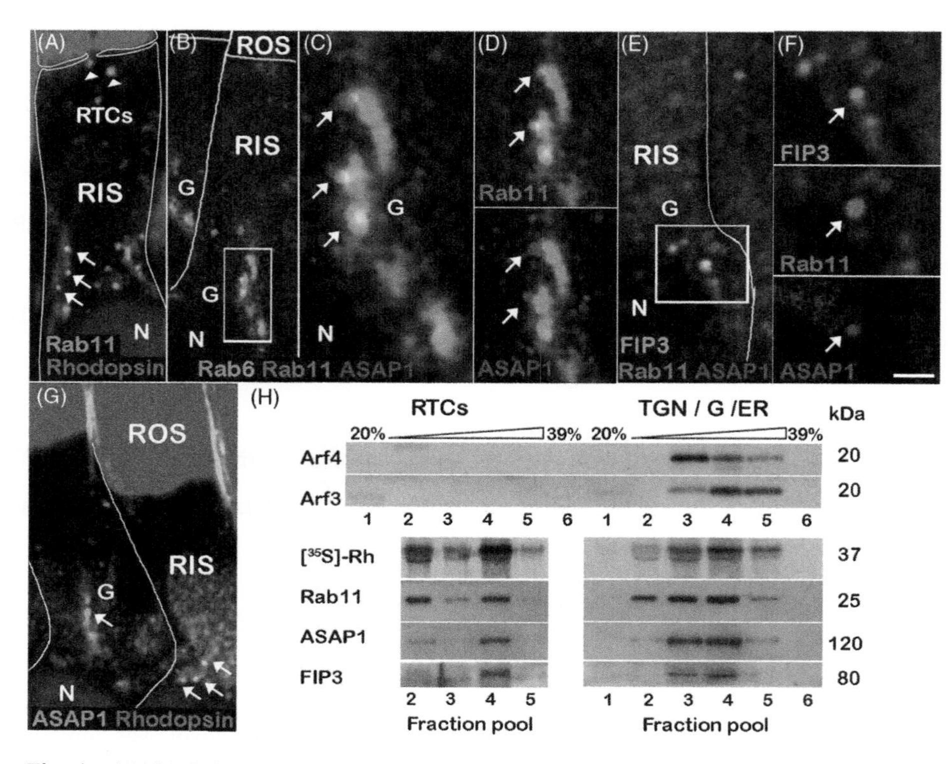

Fig. 4 ASAP1, Rab11, and FIP3 cofractionate and colocalize on nascent RTCs at the TGN. (A) A confocal optical section (0.7 μm) of frog retina. Rab11-positive puncta (yellow, arrows) aligned with rhodopsin-laden Golgi. Rab11 is also present on RTCs (arrowheads). (B) ASAP1 (blue) and Rab11 (red) colocalize in the bud-like profiles at the tips of the *trans*-Golgi (Rab6, green) (boxed area magnified in G). (C) Magnified *trans*-Golgi area from panel B, with ASAP1- and Rab11-positive buds (arrows). (D) Rab11 (red) and ASAP1 (blue) are shown separately. (E) FIP3 (green) overlaps with Rab11 (red) and ASAP1 (blue) in the punctate structures within the Golgi region (boxed area magnified in F). (F). FIP3, Rab11, and ASAP1 are shown in separate channels to better visualize their colocalization (arrows) (G) Rhodopsin C-terminal mAb 11D5 (red) labels the ROS and the Golgi (G), where ASAP1-positive puncta (yellow, arrows) line up with regular periodicity. (H) Following *in vitro* budding and subcellular fractionation, RTCs and Golgi/TGN were separated on linear sucrose density gradients. The distribution of rhodopsin was determined by autoradiography ([^{35}S]-Rh). Duplicate gels were immunoblotted as indicated in the panel. Arf4 and Arf3 immunoblots are from one experiment whereas [^{35}S]-Rh, Rab11, ASAP1, and FIP3 are from a different experiment. Bar, 3 μm in A, B, E, and G; 1 μm in D and F; 0.7 μm in C. Modified from Mazelova *et al.* (2009a). (See Plate no. 16 in the Color Plate Section.)

illustrate the suitability of the cell-free assay to reconstitute the physiological process of RTC budding, and to determine the role of specific regulators in the selection and packaging of sensory receptors and membrane cargo targeted to the primary cilia.

V. Summary

In summary, we have developed and validated a cell-free assay that reconstitutes the physiological process of cargo incorporation into specific ciliary-targeted post-TGN carriers involved in the delivery of rhodopsin to the cilia-derived sensory organelle of retinal rod photoreceptors. This assay is applicable for the studies of the molecular machinery involved in the targeting of sensory receptors to the primary cilia.

Acknowledgments

We thank Nancy Ransom, Lisa Astuto-Gribble, Hiroki Inoue, Beatrice Tam, Eric Schonteich, Rytis Prekeris, Orson Moritz, and Paul Randazzo for their invaluable contributions to the work presented here. Supported by the NIH grant EY-12421 to D.D.

References

Baker, S. A., Freeman, K., Luby-Phelps, K., Pazour, G. J., and Besharse, J. C. (2003). IFT20 links kinesin II with a mammalian intraflagellar transport complex that is conserved in motile flagella and sensory cilia. *J. Biol. Chem.* **278**, 34211–34218.

Besharse, J. C. (1986). Photosensitive membrane turnover: Differentiated membrane domains and cell–cell interaction. *In* "The Retina: A Model for Cell Biological Studies" (R. Adler and D. Farber, eds.), pp. 297–352. Academic Press, New York.

Bhowmick, R., Li, M., Sun, J., Baker, S. A., Insinna, C., and Besharse, J. C. (2009). Photoreceptor IFT complexes containing chaperones, guanylyl cyclase 1 and rhodopsin. *Traffic.* **10**, 648–663.

Christensen, S. T., Pedersen, L. B., Schneider, L., and Satir, P. (2007). Sensory cilia and integration of signal transduction in human health and disease. *Traffic.* **8**, 97–109.

Deretic, D. (2000). Rhodopsin trafficking in photoreceptors using retinal cell-free system. *Meth. Enzymol.* **315**, 77–88.

Deretic, D. (2006). A role for rhodopsin in a signal transduction cascade that regulates membrane trafficking and photoreceptor polarity. *Vision Res.* **46**, 4427–4433.

Deretic, D., and Papermaster, D. S. (1991). Polarized sorting of rhodopsin on post-Golgi membranes in frog retinal photoreceptor cells. *J. Cell Biol.* **113**, 1281–1293.

Deretic, D., and Papermaster, D. S. (1993). Isolation of post-Golgi membranes transporting newly synthesized rhodopsin. *In* "Methods for the Study of Photoreceptor Cells" (P. A. Hargrave, ed.), Vol. 15, pp. 108–120. Rockefeller University Press, New York.

Deretic, D., Puleo Scheppke, B., and Trippe, C. (1996). Cytoplasmic domain of rhodopsin is essential for post-Golgi vesicle formation in a retinal cell-free system. *J. Biol. Chem.* **271**, 2279–2286.

Deretic, D., Schmerl, S., Hargrave, P. A., Arendt, A., and McDowell, J. H. (1998). Regulation of sorting and post-Golgi trafficking of rhodopsin by its C-terminal sequence QVS(A)PA. *Proc. Natl. Acad. Sci. USA* **95**, 10620–10625.

Deretic, D., Traverso, V., Parkins, N., Jackson, F., Rodriguez De Turco, E. B., and Ransom, N. (2004). Phosphoinositides, ezrin/moesin and rac1 regulate fusion of rhodopsin transport carriers in retinal photoreceptors. *Mol. Biol. Cell* **15**, 359–370.

Deretic, D., Williams, A. H., Ransom, N., Morel, V., Hargrave, P. A., and Arendt, A. (2005). Rhodopsin C terminus, the site of mutations causing retinal disease, regulates trafficking by binding to ADP-ribosylation factor 4 (ARF4). *Proc. Natl. Acad. Sci. USA* **102**, 3301–3306.

Fliegauf, M., Benzing, T., and Omran, H. (2007). When cilia go bad: cilia defects and ciliopathies. *Nat. Rev. Mol. Cell Biol.* **8**, 880–893.

Follit, J. A., Tuft, R. A., Fogarty, K. E., and Pazour, G. J. (2006). The intraflagellar transport protein IFT20 is associated with the Golgi complex and is required for cilia assembly. *Mol. Biol. Cell* **17**, 3781–3792.

Geng, L., Okuhara, D., Yu, Z., Tian, X., Cai, Y., Shibazaki, S., and Somlo, S. (2006). Polycystin-2 traffics to cilia independently of polycystin-1 by using an N-terminal RVxP motif. *J. Cell Sci.* **119**, 1383–1395.

Gosens, I., van Wijk, E., Kersten, F. F., Krieger, E., van der Zwaag, B., Marker, T., Letteboer, S. J., Dusseljee, S., Peters, T., Spierenburg, H. A., Punte, I. M., Wolfrum, U. et al. (2007). MPP1 links the Usher protein network and the Crumbs protein complex in the retina. *Hum. Mol. Genet.* **16**, 1993–2003.

Green, E. S., Menz, M. D., LaVail, M. M., and Flannery, J. G. (2000). Characterization of rhodopsin mis-sorting and constitutive activation in a transgenic rat model of retinitis pigmentosa. *Invest. Ophthalmol. Vis. Sci.* **41**, 1546–1553.

Greenberger, L. M., and Besharse, J. C. (1983). Photoreceptor disc shedding in eye cups. Inhibition by deletion of extracellular divalent cations. *Invest. Ophthalmol. Vis. Sci.* **24**, 1456–1464.

Gross, A. K., Decker, G., Chan, F., Sandoval, I. M., Wilson, J. H., and Wensel, T. G. (2006). Defective development of photoreceptor membranes in a mouse model of recessive retinal degeneration. *Vision Res.* **46**, 4510–4518.

Jenkins, P. M., Hurd, T. W., Zhang, L., McEwen, D. P., Brown, R. L., Margolis, B., Verhey, K. J., and Martens, J. R. (2006). Ciliary targeting of olfactory CNG channels requires the CNGB1b subunit and the kinesin-2 motor protein, KIF17. *Curr. Biol.* **16**, 1211–1216.

Leroux, M. R. (2007). Taking vesicular transport to the cilium. *Cell.* **129**, 1041–1043.

Maerker, T., van Wijk, E., Overlack, N., Kersten, F. F., McGee, J., Goldmann, T., Sehn, E., Roepman, R., Walsh, E. J., Kremer, H., and Wolfrum, U. (2008). A novel Usher protein network at the periciliary reloading point between molecular transport machineries in vertebrate photoreceptor cells. *Hum. Mol. Genet.* **17**, 71–86.

Mazelova, J., Astuto-Gribble, L., Inoue, H., Tam, B. M., Schonteich, E., Prekeris, R., Moritz, O. L., Randazzo, P. A., Deretic, D. (2009a). Ciliary targeting motif VxPx directs assembly of a trafficking module through Arf4. *Embo J.* **28**, 183–92.

Mazelova, J., Ransom, N., Astuto-Gribble, L., Wilson, M. C., and Deretic, D. (2009b). Syntaxin 3 and SNAP-25 pairing, regulated by omega-3 docosahexaenoic acid, controls the delivery of rhodopsin for the biogenesis of cilia-derived sensory organelles, the rod outer segments. *J. Cell Sci.* **122**, 2003–2013.

Morel, V., Poschet, R., Traverso, V., and Deretic, D. (2000). Towards the proteome of the rhodopsin-bearing post-Golgi compartment of retinal photoreceptor cells. *Electrophoresis* **21**, 3460–3469.

Nachury, M. V., Loktev, A. V., Zhang, Q., Westlake, C. J., Peranen, J., Merdes, A., Slusarski, D. C., Scheller, R. H., Bazan, J. F., Sheffield, V. C., and Jackson, P. K. (2007). A Core Complex of BBS Proteins Cooperates with the GTPase Rab8 to Promote Ciliary Membrane Biogenesis. *Cell.* **129**, 1201–1213.

Papermaster, D. S., Converse, C. A., and Siu, J. (1975). Membrane biosynthesis in the frog retina: Opsin transport in the photoreceptor cell. *Biochemistry* **14**, 1343–1352.

Papermaster, D. S., Schneider, B. G., Defoe, D., and Besharse, J. C. (1986). Biosynthesis and vectorial transport of opsin on vesicles in retinal rod photoreceptors. *J. Histochem. Cytochem.* **34**, 5–16.

Pazour, G. J., Baker, S. A., Deane, J. A., Cole, D. G., Dickert, B. L., Rosenbaum, J. L., Witman, G. B., and Besharse, J. C. (2002). The intraflagellar transport protein, IFT88, is essential for vertebrate photoreceptor assembly and maintenance. *J. Cell Biol.* **157**, 103–113.

Peters, K. R., Palade, G. E., Schneider, B. G., and Papermaster, D. S. (1983). Fine structure of a periciliary ridge complex of frog retinal rod cells revealed by ultrahigh resolution scanning electron microscopy. *J. Cell Biol.* **96**, 265–276.

Rosenbaum, J. L., Cole, D. G., and Diener, D. R. (1999). Intraflagellar transport: The eyes have it. *J. Cell Biol.* **144**, 385–388.

Rosenbaum, J. L., and Witman, G. B. (2002). Intraflagellar transport. *Nat. Rev. Mol. Cell Biol.* **3**, 813–825.

Shi, G., Concepcion, F. A., and Chen, J. (2004). Targeting od visual pigments to rod outer segment in rhodopsin knockout mice. *In* "Photoreceptor Cell Biology and Inherited Retinal Degenerations" (D. S. Williams, ed.), pp. 93–109. World Scientific Publishing, Singapore.

Singla, V., and Reiter, J. F. (2006). The primary cilium as the cell's antenna: signaling at a sensory organelle. *Science.* **313**, 629–633.

Tam, B. M., Moritz, O. L., Hurd, L. B., and Papermaster, D. S. (2000). Identification of an outer segment targeting signal in the COOH terminus of rhodopsin using transgenic *Xenopus laevis. J. Cell Biol.* **151**, 1369–1380.

Tooze, S. A., and Huttner, W. B. (1992). Cell-free formation of immature secretory granules and constitutive secretory vesicles from trans-Golgi network. *Meth.Enzymol.* **219**, 81–93.

Wolf, K., and Quimby, M. C. (1964). Amphibian cell culture: Permanent cell line from the bullfrog (*Rana catesbeiana*). *Science* **144**, 1578–1580.

CHAPTER 13

Immunoelectron Microscopy of Vesicle Transport to the Primary Cilium of Photoreceptor Cells

Tina Sedmak, Elisabeth Sehn, *and* Uwe Wolfrum

Department of Cell and Matrix Biology, Institute of Zoology, Johannes Gutenberg University, Mainz D-55099, Germany

METHODS IN CELL BIOLOGY, VOL. 94

978-0-12-375024-2
DOI: 10.1016/S0091-679X(08)94013-9

Abstract

Cilia are organelles of high structural complexity. Since the biosynthetic machinery is absent from cilia all their molecular components must be synthesized in organelles of the cytoplasm and subsequently transported to the cilium. Ciliary cargos are thought to be translocated in the membrane of transport vesicles or association with these vesicles to the base of the cilium where the vesicles fuse with the periciliary target membrane for further delivery of their cargo into the ciliary compartment by the intraflagellar transport (IFT). Here we describe a modified preembedding labeling method as an alternative technique to conventional postembedding methods eligible for analyses of ciliary cargo vesicles and the distribution of ciliary molecules in subciliary compartments for immunoelectron microscopy. The preembedding labeling method preserves the antigenicity of ciliary antigens and its application reveals differential localization of individual IFT proteins in vertebrate photoreceptor cilia. Since membrane vesicles are conserved, the preembedding protocol additionally allows the identification of ciliary cargo vesicles by immunolabeling of individual IFT proteins and ciliary targeting molecules in ciliary photoreceptor cells. These results do not only confirm the central function of IFT molecules in ciliary transport, but further strengthen their role in transport processes in the cytoplasm. Furthermore, evidence for different alternative transport routes of cargo vesicles directed to different target membranes is gathered.

I. Introduction and Rationale

Cilia are polarized complex organelles that appear early in the evolution of the eukaryotic cell. Although, initially the main attention of the ciliary function was turned on cell and fluid propulsion, in recent years it has become clear that sensory reception is one of the major functions of cilia (Gerdes *et al.*, 2009; Pazour and Bloodgood, 2008). From this it is likely that all cilia perform sensory functions. In particular, primary cilia of mammals serve as multimodal antennae for signals from their extracellular environment (Pazour and Witman, 2003). Therefore, it is not a big surprise that major sensory receptor cells, like insect sensilla or mammalian olfactory cells, adapted cilia for sensory perception. The most drastic evolutionary modifications of cilia structure and function occurred in vertebrate photoreceptor cells (Pazour and Bloodgood, 2008; Roepman and Wolfrum, 2007) (Fig. 1).

In vertebrates, two types of ciliated photoreceptor cells, cones and rods, are arranged in the innermost layer of the neuronal retina of the eye. They are adapted to photopic vision allowing color perception and scotopic vision at low-light conditions, respectively. Cone and rod photoreceptor cells are highly polarized sensory neurons consisting of morphologically and functionally distinct cellular compartments (Fig. 1A). From their cell body an axon projects to the synaptic terminus, where ribbon synapses connect the photoreceptor cells with the second retinal neurons (tom Dieck and Brandstätter, 2006). At the other cell pole, a short dendrite named the inner segment terminates in a

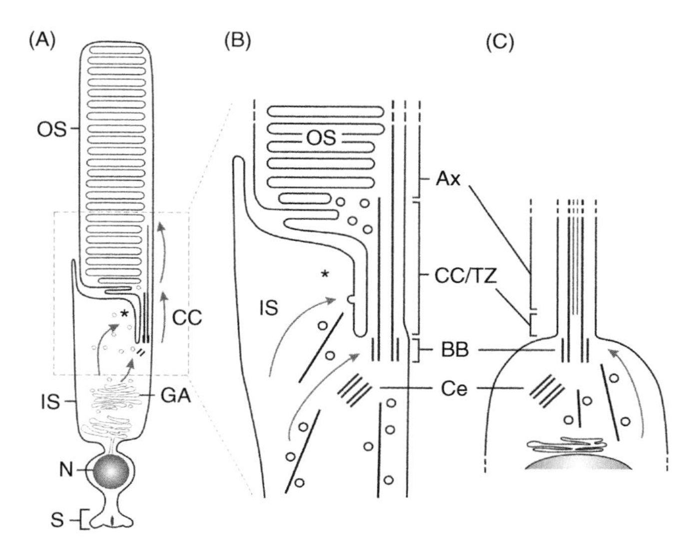

Fig. 1 Schemes of a ciliated photoreceptor cell compared with a prototypic cilium. (A) Scheme of a rod photoreceptor cell: a light-sensitive outer segment (OS) is bridged by the connecting cilium (CC) to an inner segment (IS) where all biosynthetic active cell organelles are located. (B) Schemes of the ciliary region of a rod cell and (C) of a prototypic cilium. *Arrows* in A–C indicate routs of ciliary cargo transport: post-Golgi vesicles bud from the Golgi apparatus (GA), translocate along microtubules through the inner segment (IS) to the ciliary base where they dock and fuse with the periciliary membrane. *Asterisk* indicates the region of the periciliary-targeting complex in the apical extension of the photoreceptor IS. Subsequently, membrane components translocate in the ciliary membrane to the ciliary compartment, the OS or the cilium/flagellum, respectively. Note that the photoreceptor OS resembles a highly modified primary cilium and the CC is homologue to the transition zone (TZ) of other cilia. Ax: axoneme; BB: basal body; Ce: cilia adjacent centriole.

light-sensitive outer segment which resembles a modified sensory cilium (Besharse and Horst, 1990; Pazour and Bloodgood, 2008; Roepman and Wolfrum, 2007). This outer segment is characterized by specialized flattened disk-like membranes where all components of the visual transduction cascade are arranged (Burns and Arshavsky, 2005). The visual signal transduction cascade in vertebrates is one of the best-studied examples of a G-protein transduction cascade. Photoexcitation leads to photoisomerization of the visual pigment rhodopsin which catalyzes GDP/GTP exchange at the visual heterotrimeric G-protein transducin, which in turn activates a phosphodiesterase, catalyzing cGMP hydrolysis in the cytoplasm which leads to the closure of cGMP-gated channels in the plasma membrane, finally resulting in photoreceptor cell hyperpolarization.

In addition to their sensory features, the photoreceptor cell outer segments are similar to other sensory cilia developmentally derived from primary cilia (Insinna and Besharse, 2008; Pazour and Bloodgood, 2008). As in other primary ciliary membranes, the phototransductive membranes of the outer segment are continually renewed throughout lifetime of the cell. Newly synthesized membranes are added at the base of the outer segment, whereas aged disks at the outer segment apex are phagocytosed by cells of the retinal pigment epithelium (Young, 1976). As it is the case in other cilia, there is no synthesis of biomolecules found in the outer segments and all outer segment components are continually synthesized in the endoplasmic reticulum (ER) organelle of the inner segment and transported through the slender connecting cilium to their destinations in the outer segment (Roepman and Wolfrum, 2007). The visual pigment rhodopsin is the most abundant protein of the outer segment (~85% of total outer segment protein) (Corless *et al.*, 1976), and an enormous number of rhodopsin molecules (10^7) per day are newly synthesized by photoreceptor cell (Besharse and Horst, 1990). These molecules have to traffic from the synthesizing organelles of the inner segment, through the connecting cilium, to their destination in the outer segment. Transport vesicles bearing rhodopsin in their membranes bud from the Golgi apparatus (Deretic, 2006) and are unidirectionally transported by cytoplasmic dynein along microtubules (Sung and Tai, 2000; Tai *et al.*, 1999) to a specific periciliary specialization (e.g., the periciliary ridge complex in amphibians or the membrane adhesion complex between the collar-like extension of the inner segment and the connecting cilium in rodents and mammals) at the base of the connecting cilium (Maerker *et al.*, 2008; Papermaster, 2002). Here at this specialization the vesicles dock and fuse with the target membranes of the apical inner segment and the membrane cargo is thought to move subsequently into the membrane domain of the connecting cilium (Roepman and Wolfrum, 2007). For the ciliary delivery two alternative processes have been suggested: (1) there is evidence for F-actin-based ciliary transport of rhodopsin by myosin VIIa (Liu *et al.* 1999; Wolfrum and Schmitt, 2000) and (2) based on mutational analyses the involvement of the intraflagellar transport (IFT) system including kinesin-II as transport motor along microtubules was favored (Insinna *et al.*, 2009; Marszalek *et al.*, 2000; Pazour *et al.*, 2002).

Although accumulating evidence indicates that vesicular transport of ciliary proteins from the *trans*-Golgi network is critical for proper development and maintenance of nonsensory cilia (Follit *et al.*, 2006), this process is much better understood in the sensory cilia of photoreceptor cells. This, as well as their clear morphological and functional compartmentalization, defines vertebrate photoreceptor cells as an excellent cellular system for study of the transport of cargo vesicles directed to cilia. In the present chapter we will focus on the visualization of vesicles containing ciliary cargo in photoreceptor cells by immunoelectron microscopy. We provide methodological instructions for a preembedding labeling procedure of vertebrate retina suitable for immunodetection of antigens in well-preserved transport vesicles.

II. Materials

A. Reagents and Solutions

Buffered solutions: Sorensen phosphate buffer (PB) 0.2 M (pH 7.4), 0.2 M $Na_2HPO_4 \cdot 2H_2O$ titrated with 0.2 M KH_2PO_4; phosphate-buffered saline (PBS), 137 mM NaCl, 3 mM KCl, 8 mM $Na_2HPO_4 \cdot 2H_2O$, 4 mM KH_2PO_4 (Ca^{2+} and Mg^{2+} free), pH 7.4; 0.05 M Tris-HCl buffer, pH 7.6; cacodylate buffer (CB), 0.2 M cacodylic acid·$3H_2O$ sodium salt, pH 7.4.

Fixatives: (1) Prefixation: 4% paraformaldehyde (PFA) in 0.1 M PB; PFA stock solution is freshly prepared just before use. PFA powder (extra pure Merck, Darmstadt, Germany) is dissolved in H_2O_{bidest} by constant stirring on a heating plate at 70°C (hood), droplets of 1 N NaOH are added until the solution clears. (2) Interim fixation: 2.5% glutaraldehyde (GA) (EM-grade 25% aqueous stock solution, Sigma-Aldrich, Deisenhofen, Germany) in 0.1 M CB. (3) Postfixation: 0.5% OsO_4 (stock).

Sucrose solutions 10, 20, and 30% in 0.1 M PB; 2% agar (Appichem, Darmstadt, Germany) solution in PBS.

Blocking solution: 10% normal goat serum (NGS) (Sigma-Aldrich), 1% bovine serum albumin (BSA) (Sigma-Aldrich) in PBS.

Antibody dilution solution: 3% NGS and 1% BSA in PBS. For 24-well plate at least 750 μl solution per well is necessary.

Antibodies: first antibodies: anti-IFT20, anti-IFT52, anti-IFT57 (Pazour *et al.*, 2002), and anti-MAGI-2 (Santa Cruz Inc., Santa Cruz, CA, USA); anti-mouse and anti-rabbit IgG biotin-conjugated second antibodies (Vector Laboratories Inc., Burlingame, CA, USA).

VECTASTAIN® ABC kit (Vector Laboratories Inc.): avidin (reagent A), biotinylated-horseradish-peroxidase (HRP) (reagent B).

1% 3,3′-diaminobenzidine tetrahydrochloride (DAB) (Sigma-Aldrich) in 0.05 M Tris-HCl used to prepare 0.05% DAB solution (hood).

Silver enhancement: solution A: 3% hexamethylenetetramine ($C_6H_{12}N_4$) in H_2O_{bidest}; solution B: 5% silver nitrate ($Ag\ NO_3$) in H_2O_{bidest}; solution C: 2.5% borax (disodium tetraborate) ($Na_2B_4O_7 \cdot 10H_2O$) in H_2O_{bidest}; solution D: 0.05% tetrachlorogold (III) acid trihydrate ($AuHCl_4 \cdot 3H_2O$) in H_2O_{bidest}; solution E: 2.5% sodium thiosulfate (Na_2SO_3) in H_2O_{bidest}.

Dehydration: graded series of 30–99.8% undenaturated ethanol; propylene oxide.

Embedding resin: Araldite® (ARALDITE CY212) and Araldite® hardener (ARALDITE HY 964) are mixed without air bubbles before accelerator (2,4,6,-Tris dimethylaminomethylphenol) is added (all compounds from Serva, Heidelberg, Germany).

B. Special Equipments and Tools

Dissection instruments, injection needle (ϕ 0.6 mm), copper block (size of object slide), super frost® object slides (Menzel, Braunschweig, Germany), superglue (UHU®), razor blades, a fine (size # 0) smooth art paint brush, toothpicks, 24-well plates, small homemade container with a nylon net (mesh size 80 μm^2) at the bottom, ACLAR® embedding film (Ted Pella Inc., Redding, CA, USA) plastic container with cover, copper grids (75 square mesh, diameter 3.05 mm) filmed with formvar support films, binocular, light microscope, vibratome, and ultramicrotome.

C. Animals

Present protocol is adapted to eyes of following vertebrates: *Xenopus laevis*, C57BL/6J mice, Wistar rats, domestic pig, and the Rhesus monkey, *Macaca mulatta*. Mice, rats, and

Xenopi are sustained on a 12-h light–dark cycle, with food and water *ad libitum* at local animal house. Pig eyes are obtained from the local slaughter house. *Macaca* eyes are provided from German Primate Center, Göttingen, Germany. All experiments conform with the statement by the Association for Research in Vision and Ophthalmology and by the local administration concerning the care and use of animals in research.

III. Methods and Procedures

A. Tissue Dissection and Prefixation (Day 1)

Although the present preembedding labeling protocol is described for retinas from eyes of diverse vertebrates (mouse, rat, pig, Rhesus monkey, and frog) it can be applied to other tissues. Eye balls are removed from sacrificed animals, dissected, and placed in fresh 4% PFA prefixative for 50 min at room temperature (RT). For better fixative penetration the cornea and the lens of the eyes are perforated with a needle (ϕ 0.6 mm). After 30 min prefixation sections along the *ora serrata* are performed so that the lens and the vitreous body can be removed. The remaining eye ball is placed into fresh prefixative for further 20 min. For removing the prefixative specimens are washed in 0.1 M PB for 4×15 min under constant rocking on ice. For cryoprotection the eye cups are infiltrated with 10% sucrose in 0.1 M PB for 2 h on ice. Subsequently, retinas are isolated from eye cups and placed in 20% sucrose (1–2 h; rocking on ice) followed by overnight incubation in 30% sucrose at 4°C.

B. Cracking, Embedding, and Vibratome Sectioning (Day 2)

To enable the penetration of antibodies into the tissue, plasma membranes are "cracked" by at least 3–4 freezing and thawing cycles. For this, sucrose-covered retinas are dissected in 3–4 pieces, which are spread on a super frost® slide and frozen on a copper block precooled by liquid nitrogen. Subsequently, frozen samples are melted on the heating plate (37°C). After cracking, specimens are rinsed with PBS (4×15 min, rocking on ice) and embedded in warm 2% agar in PBS (<50°C) in silicon-embedding molds. Soiled agar blocks are placed in ice-cold PBS and can be stored for a few days in PBS at 4°C.

Specimens are cut with a vibratome (Leica VT 1000 S). For this the specimen chamber is filled with ice-cold PBS. Agar blocks are fixed with superglue at the vibratome specimen holder, trimmed, and sectioned. 50-μm-thick sections are grabbed with a fine smooth art paint brush and transferred into a homemade container with a nylon net bottom located in a 24-well plate filled with ice-cold PBS.

C. Antibody Incubation (Day 2–7)

After washing in cold PBS for 15 min (rocking on ice), nonspecific binding of the first antibodies is prevented by floating sections in blocking solution for 2 h, on a shaker at RT. Subsequently, floating sections are incubated with first antibodies for 4 days on a shaker at 4°C. To prevent evaporation of the solution the 24-well plate is sealed with parafilm. After washing with PBS (4 × 15 min), sections are incubated with biotin-conjugated second antibodies for 2 h on a shaker at RT. Unbound second antibodies are washed away with PBS (4 × 15 min).

D. DAB Staining (Day 7)

The antibody staining is visualized using the VECTASTAIN® ABC kit according to manufacture instructions. VECTASTAIN® ABC kit provides the avidin-biotin-horseradish-peroxidase (HRP) complex (ABC) (Hsu *et al.*, 1981), which binds the biotin-conjugated second antibodies. PBS-washed sections are incubated with the premixed ABC of the VECTASTAIN® kit for 1.5 h in the dark.

After washes in PBS and then in 0.05 M Tris-HCl buffer, pH 7.6 (both 2 × 15 min, in the dark) the immune complex is visualized by a DAB reaction. For this, sections are preincubated with freshly prepared 0.05% DAB solution for 10 min at RT followed by incubation with 0.05% DAB solution containing 0.01% H_2O_2 until the sections change to brown (~5 min) indicating that DAB has been oxidized, converted into an insoluble brown polymer (Graham and Karnovsky, 1966). This reaction is stopped by the exchange of the DAB solution to 0.05 M Tris-HCl buffer, pH 7.6, and subsequent frequent washes in this Tris-HCl buffer for at least 30 min. After buffer change to 0.1 M CB (2 × 15 min) specimens are fixed in 2.5% GA in 0.1 M CB for 1 h at 4°C stabilizing the DAB staining as well as the cellular ultrastructure of the specimen. This interim fixation is stopped by washing with 0.1 M CB 15 min, before storing the samples in fresh 0.1 M CB overnight at 4°C.

Comment: For controls, first or second antibodies are omitted. Additional controls are preformed to analyze the endogenous peroxidase activity. For suppressing endogenous peroxidase activity sections are incubated with 3% H_2O_2 for 10 min before incubating with blocking solution.

E. Silver Enhancement of DAB Product (Day 8)

Silver enhancement is utilized to visualize the DAB product amplifying the staining for subsequent electron microscopic analyses (modified protocol from Leranth and Pickel, 1989). For this, sections are washed in H_2O_{bidest} (4 × 15 min, on a shaker at RT) to remove all available phosphates to prevent false labeling. For silver enhancement of the DAB product the three solutions A–C are mixed in a ratio of 5:0.25:0.5 and left on the specimens for 15 min at 60°C. After washing in H_2O_{bidest} (at least 3 × 3 min) sections are incubated with solution D for 4 min at RT and washed in H_2O_{bidest} again as above, before incubated with solution E for 4 min at RT. During H_2O_{bidest} washes (3 × 3 min) samples are transferred into smaller volume reaction tubes and H_2O_{bidest} is exchanged in two steps to 0.1 M CB.

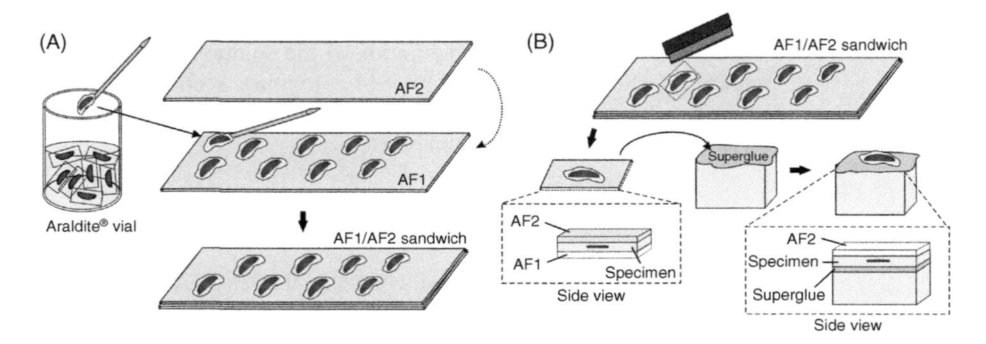

Fig. 2 Mounting and flat-embedding of vibratome sections. (A) Vibratome sections are transferred from the Araldite® containing glass vial using a toothpick to an ACLAR® film (AF1) and covered with a second ACLAR® film (AF2). (B) Selected sections are excised from the AF1/AF2 sandwich and the ACLAR® films are separated using a razor blade. The ACLAR® film containing the Araldite® flat-embedded sections are mounted with superglue onto an Araldite® block. Subsequently, ACLAR® film is removed and after trimming specimen blocks are read for ultrathin sectioning.

F. Postfixation, Flat-Embedding, and Polymerization (Day 8–12)

Postfixation of the samples is carried out in 0.5% OsO_4 in 0.1 M CB for 15 min at 4° C. After washing in ice-cold 0.1 M CB for 3 × 10 min, samples were dehydrated in a graded series of undenaturated ethanol (30–90%) (each dehydration step 10 min, 96 and 99.8% ethanol 2 × 10 min). Because Araldite® does not mingle with ethanol, propylene oxide is used as an intermedium for 2 × 5 min in small glass vials under the hood. Subsequently, samples are incubated in a 1:1 mixture of propylene oxide and Araldite® resin overnight. During this the propylene oxide is allowed to evaporate from the mixture and thereby, Araldite® resin slowly concentrates in the specimens. On day 9, each sample is transferred with a toothpick into a new glass vial containing pure Araldite® and placed again under the hood overnight before the sections are transferred with a toothpick on ACLAR® film and covered with a second ACLAR® film (Fig. 2A). Subsequently, sandwiches of Araldite® flat-embedded samples are polymerized for 48 h at 60°C (day 11, day 12).

G. Ultrathin Sections, Transmission Electron Microscopy Analyses, and Image Processing (Day 13)

Sections are selected by criteria "gross-morphology" changes of the retina sections by light microscopy. Selected sections are excised and the two ACLAR® films of the sandwich are separated by using a razor blade (Fig. 2B). The ACLAR® film containing the Araldite®-embedded section is fixed with superglue onto an Araldite® block oriented with the section "sunny side" down (Fig. 2B). After the glue is congealed the second ACLAR® film is detached and the specimen block is ultrathin sectioned (<60 nm) with an ultramicrotome Reichert Ultracut S. Ultrathin sections are collected on formvar-coated copper grids and are observed without or with heavy metal counter-staining (2% ethanolic uranyl acetate in 50% ethanol; 2% aqueous lead citrate) in a Technai 12 BioTwin transmission electron microscope (FEI, Eindhoven, the Netherlands). Images are obtained with a CCD camera (SIS MegaView3, Surface Imaging Systems, Herzogenrath, Germany) and processed with Adobe Photoshop CS (Adobe Systems, San Jose, CA, USA).

IV. Results and Discussion

Cilia are characteristic organelles of eukaryotes showing high structural complexity and diversity (600 to > 1200 polypeptides) (Pedersen and Rosenbaum, 2008; Roepman and Wolfrum, 2007). Since neither ribosomes nor membrane vesicles are present in cilia all their components must be synthesized in the cytoplasm of the cell body and subsequently delivered into the cilium. In the case of ciliary membrane proteins, electron microscopic analyses have indicated that proteins are synthesized on the rough ER and after passage through the Golgi apparatus they are incorporated in the membrane of post-Golgi vesicles and transported to the base of the cilium (Bouck, 1971; Papermaster, 2002; Sorokin, 1962). Here the vesicles dock and fuse with the periciliary membrane for the further delivery into the ciliary compartment. Furthermore, there is evidence that compounds of the axoneme of cilia and flagella, synthesized in the cytoplasm, associate with post-Golgi vesicles for their transport to the cilium (Rosenbaum and Witman, 2002). In either case, transport vesicles and their cargo get associated with the intra-flagellar transport (IFT) machinery at the base of the cilium prior to the further transport into the cilium. IFT protein complexes are thought to be assembled at the base of cilia and flagella. They are universal components of anterograde and retrograde microtubule-based transport processes in cilia required for the assembly and maintenance of cilia (Pedersen and Rosenbaum, 2008). In the present analysis of vesicular trafficking to the ciliary compartment we utilize the association of specific individual IFT proteins and of other molecules of ciliary-targeting machinery with transport vesicles for the visualization of ciliary cargo vesicles by immunoelectron microscopy. Fixation and embedding procedures always compromise between good morphology preservation and retention of antigenicity (Oliver and Jamur, 1999). Because fixations for postembedding procedures do not preserve membranes in general this constitutes an unacceptable condition for the analysis of cargo vesicles (Takamiya *et al.*, 1980), and because postembedding labeling of ciliary antigens on plastic sections failed or was so far of poor quality (Luby-Phelps *et al.*, 2008; T. Sedmak and U. Wolfrum, unpublished) we developed an alternative preembedding immunolabeling strategy to overcome these problems. For this purpose we modified previous preembedding protocols (Brandstätter *et al.*, 1996).

A. Expression and Subcellular Localization of Individual IFT in Photoreceptor Cells

Immunohistochemical analyses of cryosections through the retina of diverse vertebrate species by indirect immunofluorescence confirmed previous data obtained for the

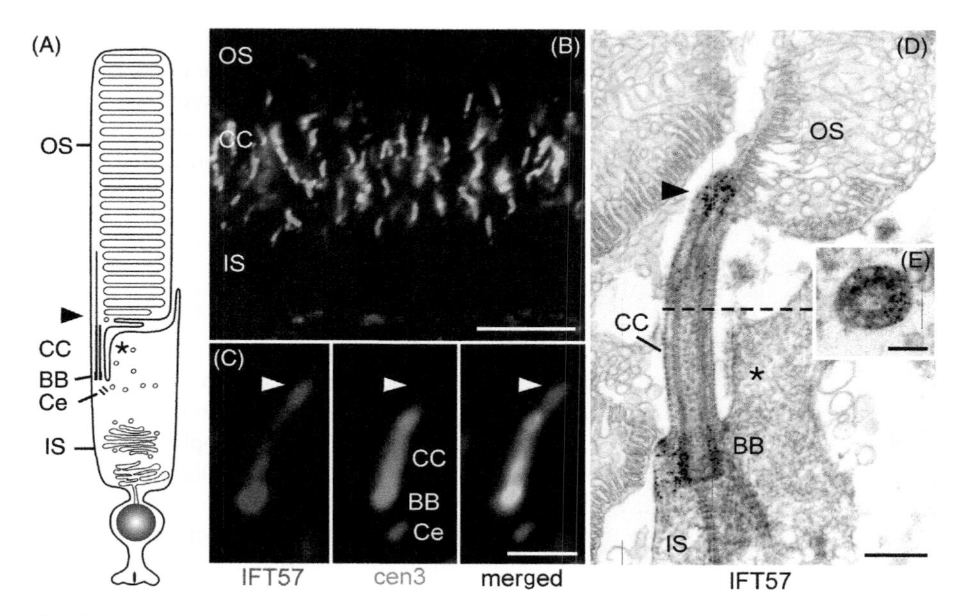

Fig. 3 Localization of IFT57 in the photoreceptor cilium. (A) Scheme of a rod photoreceptor cell. (B, C) Double immunofluorescence staining of IFT57 (red) and centrin3 (cen3) (green) in a cryosection through photoreceptor cell layer of a mouse retina. IFT57 partly colocalizes with centrin 3, a molecular marker for the ciliary apparatus (Trojan *et al.*, 2008). (C) High-magnification images of a single photoreceptor cilium from (B). IFT57 colocalizes with centrin 3 at the basal body (BB) and in the connecting cilium (CC), but not at the cilium adjacent centriole (Ce). In the absence of centrin 3, IFT57 is concentrated at the base of the outer segment (OS) (*arrow head*). (D) Preembedding immunoelectron microscopy labeling of IFT57 in a longitudinal section through part of a mouse rod cells. IFT57 is concentrated at the base of the OS (*arrow head*), sparsely decorated in the CC, and labeled in the periciliary region of the BB, but is absent from the periciliary apical extension of the IS (*asterisk*). (E) IFT57 labeling in a cross-section through the CC. Scale bars: B, 5 μm; C, 0.6 μm; D, 0.2 μm; E, 0.1 μm. (See Plate no. 17 in the Color Plate Section.)

ciliary localization of individual IFT proteins in the photoreceptor cell of the vertebrate retina (Luby-Phelps *et al.*, 2008; Pazour *et al.*, 2002) (Fig. 3B and C). However, our detailed analysis of the subciliary distribution of individual IFT polypeptides by high-resolution immunofluorescence microscopy combined with molecular markers of ciliary subcompartments determined a differential localization of individual IFT molecules in photoreceptor cells; a presence or an absence of the individual IFTs: at the cilia-adjacent centriole, at the basal body, in the periciliary membrane adhesion complex of the inner segment extension, in the connecting cilium, and in the axoneme (for details see Sedmak *et al.*, in preparation). Use of the preembedding methodology described above enabled us to demonstrate the subciliary localization of individual IFT proteins for the first time by the ultrastructural resolution of electron microscopy, exemplarily shown for IFT57 in Fig. 3D and E. In conclusion, our results further support the differential localization of individual IFT molecules in the subciliary compartment indicating slightly diverse but overlapping cellular functions.

B. Identification of Ciliary Cargo Vesicles in the Cytoplasm on Their Track to the Cilium

In addition to the subciliary analyses our preembedding method enabled us to stain the molecules associated with membrane vesicles present in the inner segment of photoreceptor cells. Using antibodies to label ciliary molecules and molecules involved in ciliary targeting (Maerker *et al.*, 2008; Bauß *et al.*, in preparation) we identified a set of inner segment vesicles as ciliary cargo vesicles independent of their cargo by immunoelectron microscopy (Fig. 4). Applying our preembedding protocol we detected individual IFT molecules, namely IFT20, IFT52, and IFT57, as well as molecules of the periciliary-targeting complex at the surface of vesicles present in the cytoplasm of the apical inner segment of photoreceptor cells (Figs. 4A/A′–C/C′). In

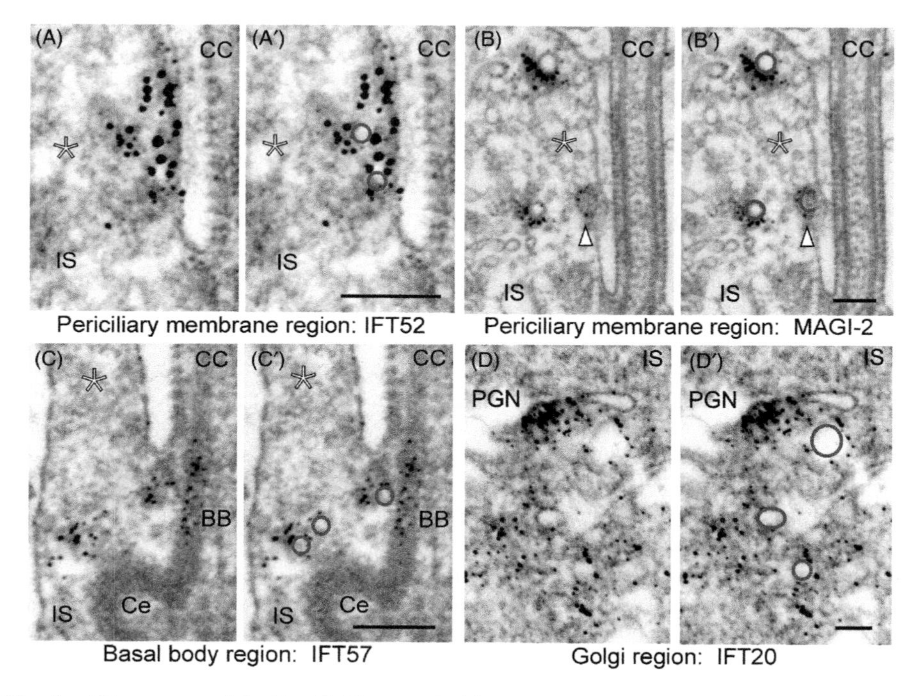

Fig. 4 Ciliary cargo vesicles identified by preembedding immunolabeling of individual IFT molecules and MAGI-2 in mouse photoreceptor cells. (A/A′, B/B′) Electron micrographs of longitudinal sections of the connecting cilium (CC) and the periciliary apical extension of the inner segment (IS) (*asterisk*) labeled with antibodies to IFT52 (A/A′) and to MAGI-2 (membrane associated guanylate kinase, WW and PDZ domain containing 2) protein (B/B′), a novel component of the periciliary targeting complex of the apical inner segment (Bauß *et al.*, in preparation; Maerker *et al.*, 2008). (A–B′) IFT52 and MAGI-2 are decorated at vesicles (circles in A′, B′) present at the docking zone of cargo vesicles. (C/C′) Immunoelectron microscopy labeling of IFT57 at cargo vesicles at the base of the photoreceptor cilium—in the region of the basal body (BB) and the adjacent centriole (Ce). *Arrowhead* indicates vesicle exocytosis in the periciliary targeting membrane. (D) Immunoelectron microscopy labeling of IFT20 in a cross-section through post-Golgi network (PGN) in the photoreceptor IS. IFT20 is associated with the membrane stacks of the PGN and present at budded post-Golgi vesicles. Scale bars: 0.2 μm. (See Plate no. 18 in the Color Plate Section.)

addition, these molecules were localized at vesicles during the processes of docking, fusion, and exocytosis (Fig. 4B/B′). IFT20 was further found to be associated with the membrane stacks of the post-Golgi network and with post-Golgi vesicles containing ciliary cargo (Fig. 4D/D′). Latter observation supported recent data (Follit *et al.*, 2006, 2008).

All in all, ciliary cargo vesicles were identified in all steps of the inner segment transport track traversed by the vesicles on their destination to the ciliary compartment, the photoreceptor outer segment (see detailed description above). These results confirm the data of previous studies on the visualization of cytoplasm or synaptic vesicles applying slightly different preembedding labeling techniques (Brandstätter *et al.*, 1996; Chuang et al., 2001; Melo *et al.*, 2009; Tanner *et al.*, 1996; van den Pol, 1986). Application of the protocol on retinas of different vertebrate species and preliminary data indicate that the present preembedding procedure described here is transferable to other cilia containing tissues and cells.

The present preembedding labeling method has certain advantages in comparison to conventional postembedding procedures: the rather mild prefixation is comparable to treatments used for immunohistochemical and cytochemical methods and preserves the antigenity of ciliary molecules much better than the processing in postembedding protocols. Although the preembedding protocol includes the rather crude cracking step, membranes are archived in an acceptable condition necessary for successful identification of cargo vesicles in the cytoplasm.

Acknowledgments

The present study was supported by the DFG (UW), the FAUN Stiftung (UW), and the Graduiertenförderung Rheinland-Pfalz (TS). We thank Dr. G. J. Pazour for kindly providing anti-IFT antibodies and Dr. K. Nagel-Wolfrum and Nora Overlack for critical comments on the manuscript.

References

Besharse, J.C., and Horst, C.J. (1990). The photoreceptor connecting cilium—a model for the transition zone. *In* "Ciliary and Flagellar Membranes" (R.A. Bloodgood, ed.), pp. 389–417. Plenum, New York.

Bauß, K., Maerker, T., Sehn, E., Kersten, F., Kremer, H., and Wolfrum, U. Cargo vesicle association and direct binding MAGI-2 to the of USH1G protein SANS. (in preperation).

Bouck, G.B. (1971). The structure, origin, isolation, and composition of the tubular mastigonemes of the Ochromas flagellum. *J. Cell Biol.* **50**, 362–384.

Brandstätter, J.H., Lohrke, S., Morgans, C.W., and Wassle, H. (1996). Distributions of two homologous synaptic vesicle proteins, synaptoporin and synaptophysin, in the mammalian retina. *J. Comp. Neurol.* **370**, 1–10.

Burns, M.E., and Arshavsky, V.Y. (2005). Beyond counting photons: Trials and trends in vertebrate visual transduction. *Neuron* **48**, 387–401.

Chuang, J.Z., Milner, T.A., and Sung, C.H. (2001). Subunit heterogeneity of cytoplasmic dynein: Differential expression of 14 kDa dynein light chains in rat hippocampus. *J. Neurosci.* **21**, 5501–5512.

Corless, J.M., Cobbs, W.H., III, Costello, M.J., and Robertson, J.D. (1976). On the asymmetry of frog retinal rod outer segment disk membranes. *Exp. Eye Res.* **23**, 295–324.

Deretic, D. (2006). A role for rhodopsin in a signal transduction cascade that regulates membrane trafficking and photoreceptor polarity. *Vision Res.* **46**, 4427–4433.

Follit, J.A., San Agustin, J.T., Xu, F., Jonassen, J.A., Samtani, R., Lo, C.W., and Pazour, G.J. (2008). The Golgin GMAP210/TRIP11 anchors IFT20 to the Golgi complex. *PLoS. Genet.* **4**, e1000315.

Follit, J.A., Tuft, R.A., Fogarty, K.E., and Pazour, G.J. (2006). The intraflagellar transport protein IFT20 is associated with the Golgi complex and is required for cilia assembly. *Mol. Biol. Cell* **17**, 3781–3792.

Gerdes, J.M., Davis, E.E., and Katsanis, N. (2009). The vertebrate primary cilium in development, homeostasis, and disease. *Cell* **137**, 32–45.

Graham, R.C., Jr. and Karnovsky, M.J. (1966). The early stages of absorption of injected horseradish peroxidase in the proximal tubules of mouse kidney: Ultrastructural cytochemistry by a new technique. *J. Histochem. Cytochem.* **14**, 291–302.

Hsu, S.M., Raine, L., and Fanger, H. (1981). Use of avidin-biotin-peroxidase complex (ABC) in immunoperoxidase techniques: A comparison between ABC and unlabeled antibody PAP procedures. *J. Histochem. Cytochem.* **29**, 577–580.

Insinna, C., and Besharse, J.C. (2008). Intraflagellar transport and the sensory outer segment of vertebrate photoreceptors. *Dev. Dyn.* **237**, 1982–1992.

Insinna, C., Humby, M., Sedmak, T., Wolfrum, U., and Besharse, J.C. (2009). Different roles for KIF17 and kinesin II in photoreceptor development and maintenance. *Dev. Dyn.* **238**, 2211–2222.

Leranth, C., and Pickel, V.M. (1989). Electron microscopic preembedding double immunostaining methods. *In* "Neuroanatomical Tract-Tracing Methods" (L. Heimer and L. Zaborszky, eds.), pp. 129–172. Plenum Press, New York.

Liu, X., Udovichenko, I.P., Brown, S.D., Steel, K.P., and Williams, D.S. (1999). Myosin VIIa participates in opsin transport through the photoreceptor cilium. *J. Neurosci.* **19**, 6267–6274.

Luby-Phelps, K., Fogerty, J., Baker, S.A., Pazour, G.J., and Besharse, J.C. (2008). Spatial distribution of intraflagellar transport proteins in vertebrate photoreceptors. *Vision Res.* **48**, 413–423.

Maerker, T., van Wijk, E., Overlack, N., Kersten, F.F., McGee, J., Goldmann, T., Sehn, E., Roepman, R., Walsh, E.J., Kremer, H., and Wolfrum, U. (2008). A novel Usher protein network at the periciliary reloading point between molecular transport machineries in vertebrate photoreceptor cells. *Hum. Mol. Genet.* **17**, 71–86.

Marszalek, J.R., Liu, X., Roberts, E.A., Chui, D., Marth, J.D., Williams, D.S., and Goldstein, L.S.B. (2000). Genetic evidence for selective transport of opsin and arrestin by kinesin-II in mammalian photoreceptors. *Cell* **102**, 175–187.

Melo, R.C., Spencer, L.A., Perez, S.A., Neves, J.S., Bafford, S.P., Morgan, E.S., Dvorak, A.M., and Weller, P.F. (2009). Vesicle-mediated secretion of human eosinophil granule-derived major basic protein. *Lab. Invest.* **89**, 769–781.

Oliver, C., and Jamur, M.C. (1999). Fixation and Embedding. *In* "Methods in Molecular Biology" (L.C. Lavois, ed.), pp. 319–326, Humana Press Inc., Totowa, NJ.

Papermaster, D.S. (2002). The birth and death of photoreceptors: The Friedenwald Lecture. *Invest. Ophthalmol. Vis. Sci.* **43**, 1300–1309.

Pazour, G.J., Baker, S.A., Deane, J.A., Cole, D.G., Dickert, B.L., Rosenbaum, J.L., Witman, G.B., and Besharse, J.C. (2002). The intraflagellar transport protein, IFT88, is essential for vertebrate photoreceptor assembly and maintenance. *J. Cell Biol.* **157**, 103–113.

Pazour, G.J., and Bloodgood, R.A. (2008). Targeting proteins to the ciliary membrane. *Curr. Top. Dev. Biol.* **85**, 115–149.

Pazour, G.J., and Witman, G.B. (2003). The vertebrate primary cilium is a sensory organelle. *Curr. Opin. Cell Biol.* **15**, 105–110.

Pedersen, L.B., and Rosenbaum, J.L. (2008). Intraflagellar transport (IFT) role in ciliary assembly, resorption and signalling. *Curr. Top. Dev. Biol.* **85**, 23–61.

Roepman, R., and Wolfrum, U. (2007). Protein networks and complexes in photoreceptor cilia. *In* "Subcellular Proteomics—From Cell Deconstruction to System Reconstruction" (M. Faupel and E. Bertrand, ed.), pp. 209–235. Springer, Dordrecht.

Rosenbaum, J.L., and Witman, G.B. (2002). Intraflagellar transport. *Nat. Rev. Mol. Cell Biol.* **3**, 813–825.

Sedmak, T., and Wolfrum, U. Differential ciliary and non-ciliary localization of intraflagellar transport molecules in the mammalian retina. (in preparation).

Sorokin, S. (1962). Centrioles and the formation of rudimentary cilia by fibroblasts and smooth muscle cells. *J. Cell Biol.* **15**, 363–377.
Sung, C.H., and Tai, A.W. (2000). Rhodopsin trafficking and its role in retinal dystrophies. *Int. Rev. Cytol.* **195**, 215–267.
Tai, A.W., Chuang, J.Z., Bode, C., Wolfrum, U., and Sung, C.H. (1999). Rhodopsin's carboxy-terminal cytoplasmic tail acts as a membrane receptor for cytoplasmic dynein by binding to the dynein light chain Tctex-1. *Cell* **97**, 877–887.
Takamiya, H., Batsford, S., and Vogt, A. (1980). An approach to postembedding staining of protein (immunoglobulin) antigen embedded in plastic: Prerequisites and limitations. *J. Histochem. Cytochem.* **28**, 1041–1049.
Tanner, V.A., Ploug, T., and Tao-Cheng, J.H. (1996). Subcellular localization of SV2 and other secretory vesicle components in PC12 cells by an efficient method of preembedding EM immunocytochemistry for cell cultures. *J. Histochem. Cytochem.* **44**, 1481–1488.
tom Dieck, S., and Brandstätter, J.H. (2006). Ribbon synapses of the retina. *Cell Tissue Res.* **326**, 339–346.
Trojan, P., Krauss, N., Choe, H.W., Giessl, A., Pulvermuller, A., and Wolfrum, U. (2008). Centrins in retinal photoreceptor cells: Regulators in the connecting cilium. *Prog. Retin. Eye Res.* **27**, 237–259.
van den Pol, A.N. (1986). Tyrosine hydroxylase immunoreactive neurons throughout the hypothalamus receive glutamate decarboxylase immunoreactive synapses: A double pre-embedding immunocytochemical study with particulate silver and HRP. *J. Neurosci.* **6**, 877–891.
Wolfrum, U., and Schmitt, A. (2000). Rhodopsin transport in the membrane of the connecting cilium of mammalian photoreceptor cells. *Cell Motil. Cytoskeleton* **46**, 95–107.
Young, R.W. (1976). Visual cells and the concept of renewal. *Invest. Ophthalmol. Vis. Sci.* **15**, 700–725.

CHAPTER 14

Polycystic Kidney Disease, Cilia, and Planar Polarity

Luis F. Menezes* *and* **Gregory G. Germino*,†**

*Department of Medicine, Johns Hopkins University School of Medicine, Baltimore, Maryland 21205

† National Institute of Diabetes and Digestive and Kidney Diseases, National Institutes of Health, Bethesda, Maryland 20892

Abstract

Cystic kidney diseases are characterized by dilated or cystic kidney tubular segments. Changes in planar cell polarity, flow sensing, and/or proliferation have been proposed to explain these disorders. Over the last few years, several groups have suggested that ciliary dysfunction is a central component of cyst formation. We review evidence for and against each of these models, stressing some of the inconsistencies

978-0-12-375024-2
DOI: 10.1016/S0091-679X(08)94014-0

that should be resolved if an accurate understanding of cyst formation is to be achieved. We also comment on data supporting a model in which ciliary function could play different roles at different developmental stages and on the relevance of dissecting potential differences between pathways required for tubule formation and/or maintenance.

I. Cystic Kidney Diseases

Cystic kidney disease is a heterogeneous group of genetic and acquired disorders characterized by dilated or cystic kidney tubular segments accompanied by a wide spectrum of abnormalities in both kidney and other organs. The cysts are lined by epithelial cells, are usually filled with fluid or amorphous material, and can originate in any segment, from Bowman's capsule to the collecting ducts.

For the purpose of this review, we are going to schematically separate genetic cystic kidney diseases in two broad groups: (1) the polycystic kidney disease (PKD) group, characterized by large, polycystic kidneys, including autosomal dominant polycystic kidney disease (ADPKD), autosomal recessive polycystic kidney disease (ARPKD); and (2) the group of hereditary cystic diseases with interstitial nephritis, characterized by small to normal size kidneys with tubular atrophy and significant interstitial fibrosis, including Bardet–Biedl syndrome (BBS), nephronophthisis (NPHP), and Alström syndrome (Fig. 1).

A. Polycystic Kidney Diseases

ADPKD is a common genetic disease, with a prevalence of ~1:1000, caused by mutations in *PKD1* or *PKD2* (Boucher and Sandford, 2004). It usually starts as a focal disease, but normal parenchyma is gradually replaced by cysts, resulting in end-stage kidney disease by the sixth decade of life in 50% of the affected patients. The focal nature of the disease is explained by a two-hit model, in which loss of the normal *PKD1* or *PKD2* alleles is a necessary pathogenic step leading to clonal expansion of cysts (Qian *et al*., 1996; Watnick *et al*., 1998). *PKD1* and *PKD2* encode polycystin-1 (PC1) and polycystin-2 (PC2), respectively.

ARPKD presents mostly in infants and young children with an estimated prevalence of 1:20,000 (Guay-Woodford and Desmond, 2003). It is caused by mutations in *PKHD1* and is characterized by cystic dilatations predominantly of collecting ducts, with few cysts originating in ascending loops of Henle and proximal tubules. Affected individuals universally develop hepatobiliary abnormalities as a consequence of ductal plate malformations of the liver. Bile duct proliferation and cystic expansion, choledochal cysts, ascending cholangitis, and congenital hepatic fibrosis are common complications, sometimes necessitating liver replacement. *PKHD1* encodes polyductin/fibrocystin (PD).

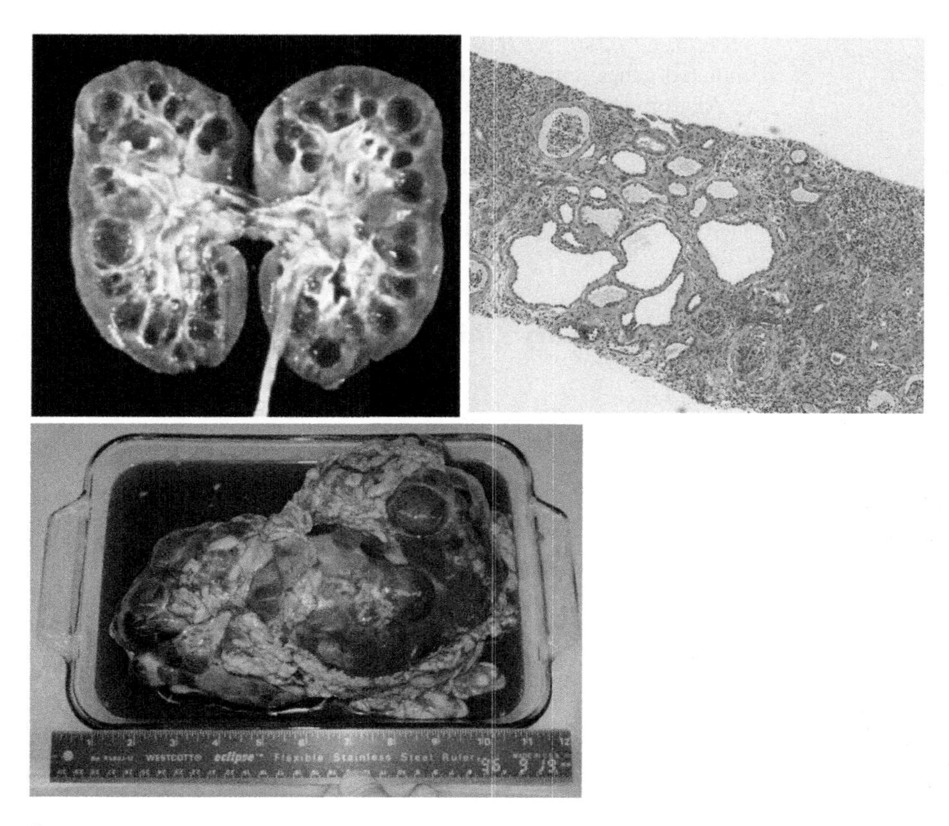

Fig. 1 On the top left, an example of an end-stage kidney from a patient with nephronophthisis (NPHP) and on the bottom left is an example of an autosomal dominant polycystic kidney disease (ADPKD). Note that the ADPKD kidney is over 3× the size of a normal kidney, and its gross structure is severely deformed by irregularly sized cysts. In contrast, the NPHP kidney is smaller than normal with macroscopic cysts clustered near the cortico-medullary junction. On microscopic examination (top right) fibrosis is usually a more prominent feature than cysts. (NPHP images used with permission of Drs. Hildebrandt, Bockenhauer, and Waldherr; from Hildebrandt and Zhou (2007)). (See Plate no. 19 in the Color Plate Section.)

B. Cystic Diseases with Interstitial Nephritis

BBS is a rare autosomal recessive syndrome characterized by atypical retinitis pigmentosa, postaxial polydactyly, obesity, mental retardation, hypogonadism in men, and a variety of renal abnormalities that include cysts, calyceal clubbing and blunting, tubulo-interstitial nephropathy, and dysplastic kidneys (Beales *et al.*, 1999). At least 12 genes have been implicated in BBS (Quinlan *et al.*, 2008).

Nephronophthisis is an autosomal recessive disease responsible for 10–20% of the cases of renal failure in children. It is characterized by cystic kidney tubules and interstitial fibrosis with tubulo-interstitial inflammatory infiltrate. So far, nine genes have been implicated in NPHP (*NPHP1, NPHP2/inversin, NPHP3, NPHP4, NPHP5,*

NPHP6/CEP290, NPHP7/GLIS2, NPHP8/RPGRIP1L, and *NPHP9/NEK8*), but the mutated genes remain unidentified in ~70% of the cases (Hildebrandt *et al.*, 2009).

Alström syndrome is a rare autosomal recessive disease, with fewer than 450 reported cases, caused by mutations in *Alms1* (Marshall *et al.*, 2007). Its main features are usually observed in children and include atypical retinitis pigmentosa, sensorineural hearing loss, insulin resistance, and obesity. In adult patients, kidney disease, characterized by hyalinization of tubules and interstitial fibrosis, may be observed.

II. The Primary Cilium and Cystic Kidney Diseases

A. Morphological Evidence

The first observation suggesting a link between PKD and cilia came from work in *Caenorhabditis elegans* showing colocalization of GFP-tagged LOV-1 and PKD-2—the homologues of PC1 and PC2, respectively—in the cytoplasm and cilium of male-specific sensory neurons (Barr and Sternberg, 1999). Subsequently, cloning of the intraflagellar transport (IFT) gene, *IFT88*, in *Chlamydomonas* showed that it is homologous to mouse and human *Tg737*, a gene previously linked to renal cystic disease (Pazour *et al.*, 2000). In *Chalamydomonas*, IFT88 is required for the assembly of flagella; similarly, *Ift88*$^{Orpk/Orpk}$ (*Tg737*) mutant mice have shorter primary cilia in collecting ducts at postnatal days 4 and 7 (Pazour *et al.*, 2000). Previous results had shown that *Tg737* mutant mice have an ARPKD-like phenotype, initially with proximal tubule cysts with a gradual shift toward a predominantly collecting duct cyst phenotype and ductal plate malformation in the liver (Moyer *et al.*, 1994).

The stage was now set to place this previously obscure organelle at the center of PKD research. Prompted by these results, a series of studies localized various cysto-proteins to cilia (Fig. 2). PC2 was detected in cilia of both wild type and *Tg737* mutant mice (Pazour *et al.*, 2002). PC1, PC2, polaris (the protein encoded by *Tg737*), and cystin (the protein encoded by the gene mutated in *cpk* mice, another ARPKD-like mouse model (Hou *et al.*, 2002)) were colocalized in cilia of mouse collecting duct cell lines (Yoder *et al.*, 2002), and polyductin/fibrocystin was also found in the primary cilium (Gallagher *et al.*, 2008; Kaimori *et al.*, 2007; Menezes *et al.*, 2004; Wang *et al.*, 2004; Ward *et al.*, 2003; Zhang *et al.*, 2004). In *Drosophila*, PC2 has been localized to the distal tip of the sperm tail (Watnick *et al.*, 2003).

Contemporary with these studies, BBS proteins were also shown to colocalize in cilia-associated structures. Myc-tagged BBS8 was localized to the basal body in cell lines and to ciliated neurons in *C. elegans*, where it has the same distribution pattern as homologues of BBS1, BBS2, BBS7, and IFT88 (Ansley *et al.*, 2003). Subsequently, comparative genomics approaches to identify flagellar and basal body proteins picked out BBS5 and showed that it is expressed exclusively at the base of the cilia in *C. elegans* (Li *et al.*, 2004). In addition, BBS1, BBS2, BBS4, BBS5, BBS7, BBS8, and BBS9 were shown to form a stable complex, the BBSome, which localizes to the centrosome or basal body (Nachury *et al.*, 2007).

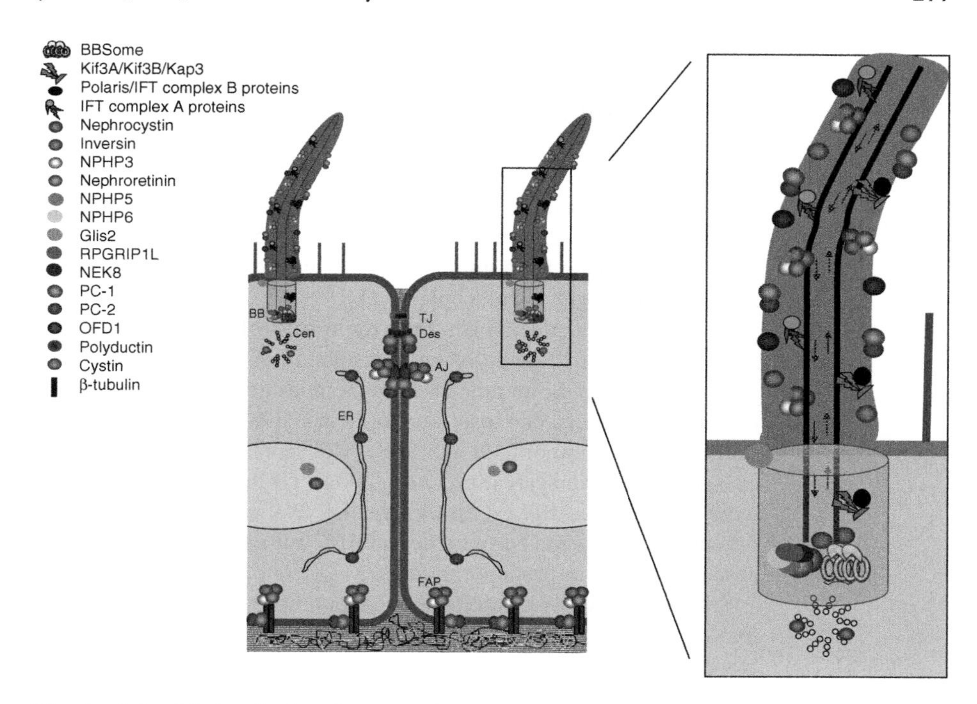

Fig. 2 Subcellular localization of cystoproteins. Over 20 cystoproteins have been localized to the primary cilium/basal body complex (legend on top left) but many also map to other intracellular domains. AJ, adherens junction; BB, basal body; Cen, centriole; ER, endoplasmic reticulum; FAP, focal adhesion plaque; TJ, tight junction. Adapted from Watnick and Germino (2003). (See Plate no. 20 in the Color Plate Section.)

NPHP proteins also seem to be present in cilia. Following the immunodetection studies showing that NPHP2/inversin and NPHP1 colocalized in a punctuated pattern in cilial axoneme (Otto *et al.*, 2003), NPHP4 was seen in ciliated neurons in *C. elegans* (Wolf *et al.*, 2005); NPHP5 in cilia of renal tubular epithelial cells (Otto *et al.*, 2005); NPHP6 in centrosomes (Sayer *et al.*, 2006); GLIS2 in nuclei and cilial axoneme (Attanasio *et al.*, 2007); RPGRIP1L in centrosomes, colocalized with NPHP4 (Delous *et al.*, 2007); and NEK8 in the proximal region cilia (Mahjoub *et al.*, 2005). Alms1 was also detected in cilia (Hearn *et al.*, 2005).

B. Functional Evidence

The first functional evidence linking the primary cilium to cystic disease came from the aforementioned *Tg737* mouse studies. The *ORPKD* mouse line, which has a hypomorphic allele of the Tg737 locus produced by a random transgene insertion into the locus, has morphologically abnormal cilia, liver abnormalities, and renal cystic disease very similar to that seen in human ARPKD (Moyer *et al.*, 1994). Complete inactivation of the locus ($Tg737^{\Delta2-3\beta\text{-Gal}}$) results in an even more severe phenotype.

Mutant embryos develop left/right axis abnormalities and ventral node cells lacking a central cilium (Murcia *et al.*, 2000). Similar findings have been described for murine models of *NPHP2*. The *inv* mouse mutant is the result of a random transgene insertion into the *inv* (*Nphp2*) locus. Homozygotes develop large cystic kidneys and situs inversus (Mochizuki *et al.*, 1998; Morgan *et al.*, 1998). Interestingly, the cilia of mutant mice appear to have normal morphology, suggesting that grossly normal appearing cilia can nonetheless have altered ciliary function (Phillips *et al.*, 2004). This has relevance for a number of other cystic models where ciliary structure is also normal on the light microscopic level.

Additional evidence comes from the results of an insertional mutagenesis screen in zebrafish. Zebrafish embryos are transparent animals with simple pronephric kidneys, composed of two glomeruli, fused in the midline, each connected to a pronephric duct. Pronephric kidneys can also develop cysts and have been used as a model to understand PKD (Drummond *et al.*, 1998). The screen identified 15 loci that when mutated caused pronephric cysts (Sun *et al.*, 2004). Of the ten genes that were identified, eight were further investigated, showing that five were involved in cilia formation and function, three were homologues of IFT components, and one was associated with absence of cilia in pronephric ducts.

Studies of a *Kif3A* conditional knockout mouse model (Lin *et al.*, 2003) also indicate that cystic disease can result from disrupting ciliary function. KIF3A forms a heterotrimeric KIF3 kinesin motor complex with KIF3B and KAP3 and is thought to mediate plus-end-directed microtubule-based transport of protein complexes and membrane-bound vesicles, and cilia assembly (Hirokawa, 2000). Following distal tubule-specific *Ksp-cre* conditional inactivation of *Kif3A*, mutant mice displayed progressive kidney cyst formation, having a few cystic segments at postnatal day (P5) (roughly 10 days post-inactivation) but massive kidney enlargement and diffuse distal tubule involvement at P35. Cyst formation was accompanied by loss of primary cilia and increased levels of nuclear β-catenin, a marker of increased canonical Wnt activity, in cystic epithelium (Lin *et al.*, 2003). Consistent with this finding, *Kif3a*$^{-/-}$ embryos were found to have increased canonical Wnt activity using the BAT-gal reporter line (Corbit *et al.*, 2008). Since dysregulated Wnt signaling had been previously implicated as a possible cause of renal cystic disease, these studies suggest a possible mechanism linking ciliary dysfunction to renal cyst formation. This topic is discussed further below.

Despite the interesting correlation with cilia assembly, it is important to note that many of the proteins associated with cystic disease have multiple functions within the cell and localize to a number of different subcellular compartments. The multiplicity of functions for individual proteins within cells, which may well be interdependent, makes it difficult to ascribe specific consequences to a protein's ciliary localization (Corbit *et al.*, 2008). For example, KIF3 mediates microtubule transport within the cell to more than just the primary cilium. PC2 appears to be an integral protein of the endoplasmic reticulum (ER) where it may function as a calcium-release channel, and nephrocystin-1 has been reported to form a complex with Pyk2, p130(Cas), and tensin in cell-matrix adhesions at the basolateral membrane (Benzing *et al.*, 2001; Koulen *et al.*, 2002). Interestingly, conditional inactivation of KAP3 results in trapping of

N-cadherin in the Golgi, reduction in the cell-boundary pool of N-cadherin and consequent increase in cytoplasmic β-catenin (Teng *et al.*, 2005). These data prompt one to ask whether the effects of disrupting KIF3A are primarily the result of ciliary dysfunction or loss of other functions.

The IFT mutants provide the strongest evidence linking the ciliary function of the various cystogenes to the cystic phenotype. Currently, IFT proteins are thought to function solely in ciliary formation and function. If future studies suggest additional roles for this class of proteins, additional strategies may be required to discriminate between their ciliary and nonciliary properties and their relationship to cystic disease.

Recent study of one of the genes identified in the zebrafish screen may be particularly informative. *Seahorse* mutants look very similar to IFT mutants, with L/R axis abnormalities, altered body curvature, and pronephric cysts (Kishimoto *et al.*, 2008). Interestingly, mutant pronephric cystic epithelial cells have normal ciliary structure, length, and motility. The protein appears to be present solely in the cytoplasm where it forms a complex with Disheveled. Disheveled is a key component of the Wnt signaling system, playing a critical role in both the canonical and noncanonical/planar cell polarity (PCP) pathways. These studies suggest that seahorse may function downstream of the primary cilium, serving as a link between it and the Wnt pathway. A number of cystoproteins may similarly serve either to transduce ciliary signals or to function in effector pathways.

III. Explanatory Models for Cystic Kidney Diseases

Kidneys seem to have few phenotypic responses to pathological insults. Among these, cystic change and interstitial fibrosis are particularly common. Therefore, it should come as no surprise that disruption of developmental or homeostatic process results in cystic kidney disease. We shall briefly consider the role of cilia in (1) establishment and (2) maintenance of kidney architecture and how disrupting these functions may result in cystic kidney diseases.

A. Establishing Kidney Architecture

In nephrogenesis, Wnt signaling is of paramount importance. Wnt9b is expressed in the stalk of the ureteric bud (UB) as it invades and branches in the metanephric mesenchyme (MM); acting as a paracrine signal, it seems to be essential for the expression of early markers of the tubulogenic pathway, such as Pax8, Fgf8, and Wnt4 (Carroll *et al.*, 2005). Wnt4, on its turn, can be detected in condensing mesenchyme, pretubular aggregates, comma- and S-shaped bodies, and in epithelial tubular derivatives (Stark *et al.*, 1994). Furthermore, *Wnt4*$^{-/-}$ mice have small kidneys with morphologically undifferentiated MM (Stark *et al.*, 1994). Both these molecules seem to employ primarily the canonical, β-catenin-dependent, pathway. Actually, studies using loss- and gain-of-function β-catenin conditional alleles showed that β-catenin activation is necessary and sufficient to initiate the tubulogenic program and can induce MM in *Wnt4*$^{-/-}$ and *Wnt9b*$^{-/-}$ mice (Park *et al.*, 2007).

It is clear, however, that a proper balance of canonical Wnt activity is required for normal morphogenesis. In transgenic mice, overexpression of mutant, constitutively active β-catenin results in focal cyst formation in all tubular segments, defective kidney maturation, persistence of immature tubules, and globally increased proliferation and apoptosis rates that do not correlate with cystic change (Saadi-Kheddouci *et al.*, 2001). Similarly, *Ksp-cre* conditional inactivation of *APC*, which also increases cellular β-catenin activity, results in severely cystic kidneys by postnatal day 2. Mutant kidneys have cysts derived from all tubular segments, hyper-proliferative epithelium, and occasional adenomas (Qian *et al.*, 2005).

As noted earlier, the observation of increased nuclear β-catenin levels in cystic epithelium of Kif3a mutants suggested a possible link between ciliary function and Wnt signaling. Further evidence came from studies of the *inv* mutant mouse. Loss of *Nphp2* function results in a cystic mouse kidney phenotype similar to the one in β-catenin overexpression models (Simons *et al.*, 2005). Conversely, Simon *et al.* found that overexpression of the *Nphp2* gene product, inversin, in cultured cells increased β-catenin degradation, possibly by targeting Disheveled (Dvl1) for proteosomal degradation. Importantly, the authors found that knock down of *Nphp2* in *Xenopus laevis* resulted in convergent extension phenotypes characteristic of PCP defects, suggesting that inversin might work as both a negative modulator of the canonical and a positive modulator of the noncanonical Wnt pathways (Fig. 3). Later studies by Corbit *et al.* indicate that this may be a general property of ciliary signaling. They studied tissues and cultured cells from three different mutants that have disrupted ciliogenesis (*Kif3a, Ift88 and Ofd1*) and found them to be hyper-responsive to Wnt stimulation (Corbit *et al.*, 2008). Their data suggest that in contrast to its role as a promoter of hedgehog signaling, the primary cilium instead appears to restrict the activity of the canonical Wnt pathway, possibly by enhancing the activity of PCP pathways. Seahorse appears to be an important cytoplasmic regulator of Wnt activity that like inversin may help determine the balance between canonical and noncanonical signaling.

B. Planar Cell Polarity and Cystic Kidney Disease

The PCP pathway was first defined for its role in wing hair and bristle orientation in *Drosophila*. Two groups of evolutionarily conserved core PCP factors have been identified. The first includes Frizzled (Fz), Flamingo (Fmi), Disheveled (Dsh/Dvl), Prickle (Pk), Strabismus/Van Gogh (Stbm/Vang), and Diego (Dgo; Diversin and Inversin); and the second Fat (Ft), Dachsous (Ds), Four-jointed (Fj), and Atrophin (Atro). Other factors that seem to play a role in PCP are *fuzzy, inturned, nemo*, some Rho GTPases, and the PAR proteins (Simons and Mlodzik, 2008). Several of the vertebrate homologues of PCP genes were linked to convergent extension during gastrulation and neurulation; and to stereocilia(inner ear)/hair(epidermis) orientation (Simons and Mlodzik, 2008).

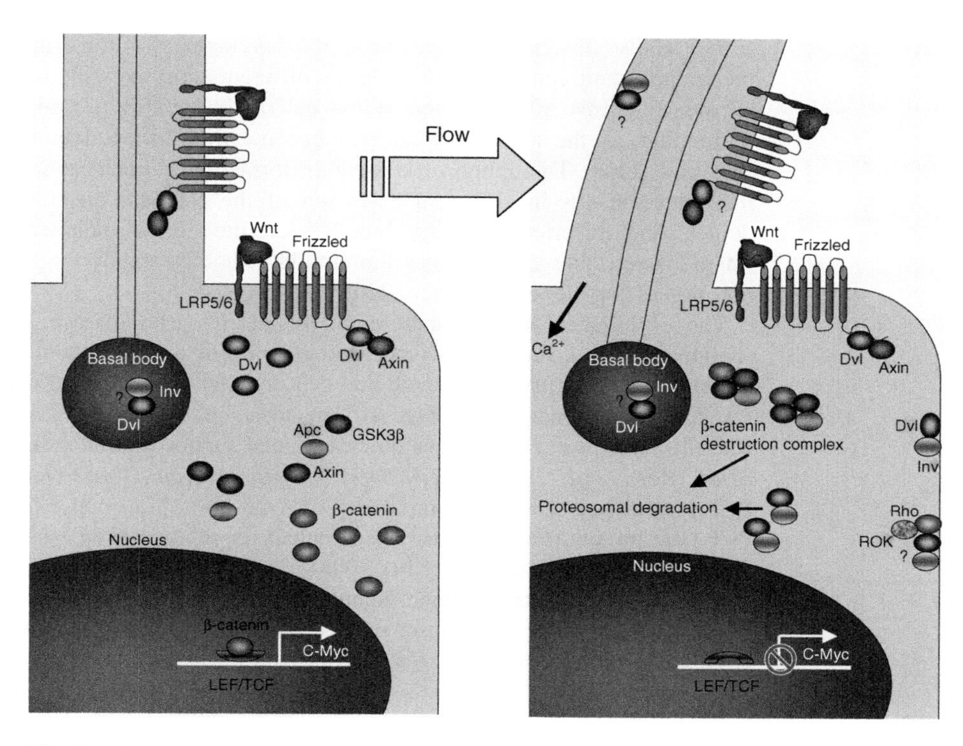

Fig. 3 Flow-based model of Wnt regulation in the kidney. Simon *et al.* (2005) suggest that Wnt signals primarily via β-catenin-dependent pathways in the absence of flow. Ligand binding by the Frizzled (Fz)/LRP complex results in inactivation of the β-catenin destruction complex, increased cytoplasmic and nuclear β-catenin levels, and upregulation of effector gene expression (left). Stimulation of the primary cilium by flow is postulated to increase expression of inversin (inv) and reduce levels of cytoplasmic Dvl by increasing its proteosomal degradation. Flow sensing by the primary cilium is thought to function as a switch, flipping Wnt signaling from the canonical to noncanonical pathway. It is not known whether the inv/Dvl or Fz/LRP complexes are present in the cilium/basal body complex (indicated by question marks). From Germino (2005). (See Plate no. 21 in the Color Plate Section.)

One model proposes that PCP is established through oriented cell division. In *Drosophila*, cell fate is in part determined by unequal segregation of signaling molecules after cell division. This is accomplished by asymmetric cellular distribution followed by oriented cell division, a process controlled by noncanonical Wnt signaling in *Drosophila* neuroblast (Bellaiche *et al.*, 2001). It was also proposed that oriented cell division could control organ shape in *Drosophila*, as a clear correlation between the axis of cell division and the shape of wing was noticed (Baena-Lopez *et al.*, 2005). In zebrafish, injection of mRNA encoding a mutant form of Dsh results in randomization of the axis of cell division and disrupts convergent extension, providing evidence that these two processes are linked (Gong *et al.*, 2004).

PCP is also observed in mice during the development of the organ of Corti, where the auditory hair cells are located. In the differentiating hair cell, a single kinocilium transiently forms at the center of the cell and migrates toward its final cellular destination, at the apex of what will become the V-shaped bundle of stereocilia (Axelrod, 2008). Disruption of kinocilium formation in cochlear-specific conditional *Ift88* mice results in abnormal extension of the cochlear duct, mislocalized basal bodies, and misoriented ciliary bundles, despite normal polarization of Fz3 and Vangl2, suggesting that the kinocilium might be a relay station for positional information processing (Jones and Chen, 2008).

The observation that inversin might result in cystic disease by affecting PCP signaling led to a model in which dysregulation of the noncanonical Wnt signaling that normally oriented cell growth and cell division might instead result in renal cyst formation (Germino, 2005) (Fig. 4). The idea that altered PCP signaling might be causally linked to cystic disease quickly gained support as mutations in *BBS1, BBS4, BBS6* (Ross *et al.*, 2005), and *NPHP3* (Bergmann *et al.*, 2008) were associated with PCP phenotypes. The recent discovery of renal cysts in a mouse line with a mutant PCP core protein (Fat4) suggests that primary disruption of PCP signaling can cause the cystic phenotype (Saburi *et al.*, 2008).

The model predicts that there should be evidence for altered cellular orientation immediately preceding the onset of cyst formation. To test this hypothesis, Fischer *et al.* analyzed the axis of mitosis in mice with renal-specific *HNF1β* inactivation and

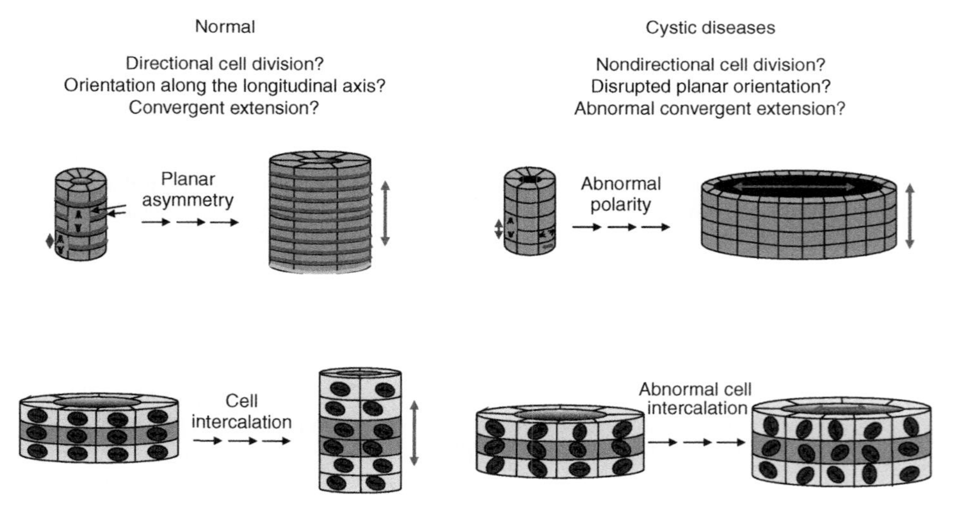

Fig. 4 Noncanonical Wnt signaling and tubular morphogenesis. Planar cell polarity might be involved in the establishment of normal tubular architecture in several different ways, including directional cell division, preservation of cellular spatial orientation information, and/or convergent extension/directional cell migration. Disruption of any of these processes could potentially result in cyst formation. Adapted from Germino (2005). (See Plate no. 22 in the Color Plate Section.)

in the *Pck* ARPKD rat model (Fischer *et al.*, 2006). *HNF1β* mutant mice develop cystic kidneys by P8, a process thought to result from decreased expression of *Pkhd1, Pkd2*, and possibly other cystogenes (Gresh *et al.*, 2004). Analysis of the axis of cell division in P3–P4 *HNF1β* mutants revealed randomization, in contrast to control animals where the axis was parallel with that of the tubular lumen. Young *Pck* rats (P8–P12), which are usually cystic by 3 weeks of age, had roughly 23% of mitoses in tubular segments not aligned with the extension axis, versus 5% in control animals. The authors propose that randomized cell division precede and could cause cystogenesis in ARPKD models and that cilia could provide the orientation cues necessary for proper tubular elongation.

Further support for this model is provided by the *Fat4* knockout mouse line (Saburi *et al.*, 2008). These animals have cyst formation, starting at embryonic day 16.5 (E16.5), accompanied by change in the distribution of the angle between the axis of cell division and tubular elongation. In addition, the cystic disease is aggravated by disruption of the PCP genes *Vangl2* and *Fjx1*. Though Fat4 localizes to cilia, mutant animals seem to have morphologically normal cilia. Evidence linking ciliary dysfunction directly to altered PCP signaling and renal cyst formation is provided by studies of *Ift20 Hoxb7Cre* conditional mice. The mutants almost completely lack cilia in their collecting duct cells at E18 and have obvious cystic disease at P10 (Jonassen *et al.*, 2008). Cyst formation in this model was preceded by increased canonical Wnt signaling and misoriented axis of cell division.

A recent study of a mouse *Wnt9b* mutant more directly links noncanonical Wnt signaling to this process (Karner *et al.*, 2009). Because germline *Wnt9b* null homozygotes fail to form kidneys, the investigators conditionally inactivated the gene in the kidney by generating floxed *Wnt9b* mice expressing Ksp-Cre, which restricts gene inactivation to collecting duct stalks. *KspCre; Wnt9b$^{-/flox}$* mice have few visible cysts at P1 but many by P10. The authors examined the mitotic axes of tubular epithelial cells in controls and mutants at P5 and also found that they were mostly oriented parallel to the long axis of the tubule in controls and random in the precystic mutant specimens.

Several lines of evidence suggest that the link between mitotic axis orientation and cyst formation might not be as direct as the prior studies suggest, however. In the case of the *Pck* rat, for example, it is unclear how the *HNF1β* mutants form a normal kidney despite the abnormal orientation of its mitotic axes that then proceeds to become cystic (Fischer *et al.*, 2006). Furthermore, if the consequences of PCP defects are exacerbated by increased rates of proliferation, as suggested by the authors, it would be expected that most of the cysts would be present by the time nephrogenesis is complete. Even more perplexing, Bonnett *et al.* recently reported abnormal mitotic spindle orientation at P2 in mouse *Pkd1* heterozygotes, which develop rare cysts (Bonnet *et al.*, 2009). This result would suggest that an altered axis of cell division can be completely unlinked to renal cyst formation.

The *Wnt9b* studies also uncovered some surprising findings. One of the Wnt9b lines has a hypomorphic allele that results in more severe cystic disease beginning at E15.5 (Karner *et al.*, 2009). Interestingly, the authors found that the mitotic axes of the mutant and normal specimens at E15.5 were similar and much more random than in the normal P5 samples, with over 60% having an axis orthogonal (60–90°) to the luminal axis. This prompted Karner *et al.* to examine the mitotic axis in normal specimens at a range of other developmental time points and found that it was random through birth, in transition from P0 to P5, and finally oriented approximately parallel to the tubules' long axis at P5. They concluded that tubule elongation in the latter stages of renal development was likely mediated by oriented cell division but must be determined by some other mechanism at earlier time points. In looking for other explanations, they found that both the diameter of tubules and the number of cells lining them decreased by ~50% from E13.5 to P1 in normals but significantly less than this in mutants. Moreover, they noted that the longitudinal axis of individual cells was generally perpendicular to the axis of tubular extension at E15.5 but more random in mutants. Collectively, these data suggest that convergent extension processes drive the early stages of tubular elongation, which is later maintained by oriented cell division. These processes may require the Rho/Jnk effector branch of the PCP pathway. Activated Rho and phospho-Jnk2 were present in normal P0 kidneys but not in mutants.

In sum, these data provide strong support for the hypothesis that defective PCP signaling can play a pathogenic role in cyst formation during renal development. It is likely that defective convergent extension, possibly coupled to disordered cellular division, may be sufficient to drive cyst formation in some cases. Neither process is likely to explain the later onset of cystic disease in the *Pck* rat, however, indicating that other processes likely are important in at least some cystic diseases. The temporal disconnect between the disorientation of the mitotic axis and the cystic phenotype in the *Pck* rat also suggests that alterations in the axis of cell division may not necessarily be a driving force behind cyst growth but rather a marker of dysregulated cellular orientation. In the case of *Pkd1*, it is clear that some other second step must be required for cysts to form if the mitotic axis data are accurate. How the processes that cause these forms of dysregulated cellular orientation result in a delayed cystic phenotype remains to be determined.

C. Proliferation and Cystic Kidney Disease

There is little question that cellular proliferation must contribute to the massively enlarged organs observed in many individuals with ADPKD. This has led many to conclude that hyper-proliferation must be a critical factor in the pathogenesis of the disease and a suitable target for therapeutic intervention. Multiple studies of human and mouse cystic specimens have provided supportive evidence, reporting high rates of cellular proliferation and occasional foci of hyperplasia and polyps in cystic epithelia (Bukanov *et al.*, 2006; Chang *et al.*, 2006; Lanoix *et al.*, 1996; Nadasdy *et al.*, 1995). Further support comes from mice models that develop cystic disease either as a consequence of transgenic expression of oncogenes (*c-myc*, SV40 large T-antigen, *β-catenin*) or as an inactivation of various tumor suppressors (*APC, Tsc1, Tsc2*) (Kelley *et al.*, 1991; Onda *et al.*, 1999; Qian *et al.*, 2005; Wilson *et al.*, 2005; Trudel *et al.*,

1991). Based on such evidence, several groups have proposed antiproliferative agents as a viable therapeutic approach, and various studies have addressed the efficacy of antiproliferative drugs in different cystic models, consistently showing a correlation between reduction of proliferation with slower cyst formation (Bukanov *et al.*, 2006; Shillingford *et al.*, 2006; Sweeney *et al.*, 2003).

Despite such apparent success, studies of orthologous models of ADPKD suggest that the relationship between cell proliferation and cystic disease may be more complicated than once thought. In one study of *Pkd2* mutant mice, investigators reported a 10-fold increase in proliferation rates in cystic mice compared to controls (Chang *et al.*, 2006). Curiously, the authors also found a 10-fold increase in the proliferation rate of noncystic epithelial cells in *Pkd2*$^{+/-}$ mice, which develop rare renal cysts. These data suggest that either an increased proliferation rate itself is not sufficient to cause cystic disease or the markers used to track proliferation were not sufficiently specific.

The relationship between cyst formation and cellular proliferation in *Pkd1* is not much clearer. Lantinga-van Leuwen *et al.* reported a proliferation index that was 5 to 20-fold higher in cystic kidney tissues of adult mice with conditional inactivation of *Pkd1* and 3× higher in noncystic renal epithelium of *Pkd1*$^{+/-}$ mice (which develop rare cysts) than in the corresponding control groups (Lantinga-van Leeuwen *et al.*, 2007). The data appear to be consistent with what was reported for *Pkd2* but the apparently increased rate of proliferation in haploinsufficient, phenotypically normal tissue raises the same issues regarding the relationship between proliferation rates and cyst induction. Shibazaki *et al.* used a different *Pkd1* model and a different method of measuring cell proliferation and achieved somewhat similar results. They reported BrdU incorporation rates twice normal in DBA-positive cyst-lining cells at P7 (15% vs 8%) and P12 (8% vs 5%) in a mouse line with conditional inactivation of *Pkd1* restricted to the kidney (Shibazaki *et al.*, 2008). More striking was the difference at P24 where the rate was 4% in the mutants but undetectable in the controls.

Our group had used a different *Pkd1* model, two complementary methods of assaying cellular proliferation and a comprehensive, unbiased method of quantifying the results and reached a different conclusion. We examined the relationship between the timing of *Pkd1* inactivation and cyst formation using a tamoxifen-inducible Cre recombinase (Piontek *et al.*, 2007). We found that inactivation of *Pkd1* in young mice before P13 resulted in severely cystic kidneys within 3 weeks, whereas inactivation at day 14 and later resulted in cysts only after 5 months (Fig. 5). This abrupt change in response to *Pkd1* inactivation coincided with a previously undefined brake point in renal growth. We found that the very high proliferation rates that prevailed during fetal and postnatal renal development dropped precipitously around the same time that the kidney became relatively insensitive to *Pkd1* loss (Fig. 6). Importantly, however, we did not see a significant difference in the proliferation rate of cystic versus control epithelia at any time point, though the proliferation rate of the adult cystic trended slightly higher (<2×) than the controls (Fig. 7). The lack of significance may be due to the relatively small sample size that we analyzed.

We interpreted the results to suggest that the defective growth regulation is not likely to be the primary defect and that the relationship between cellular proliferation and cyst

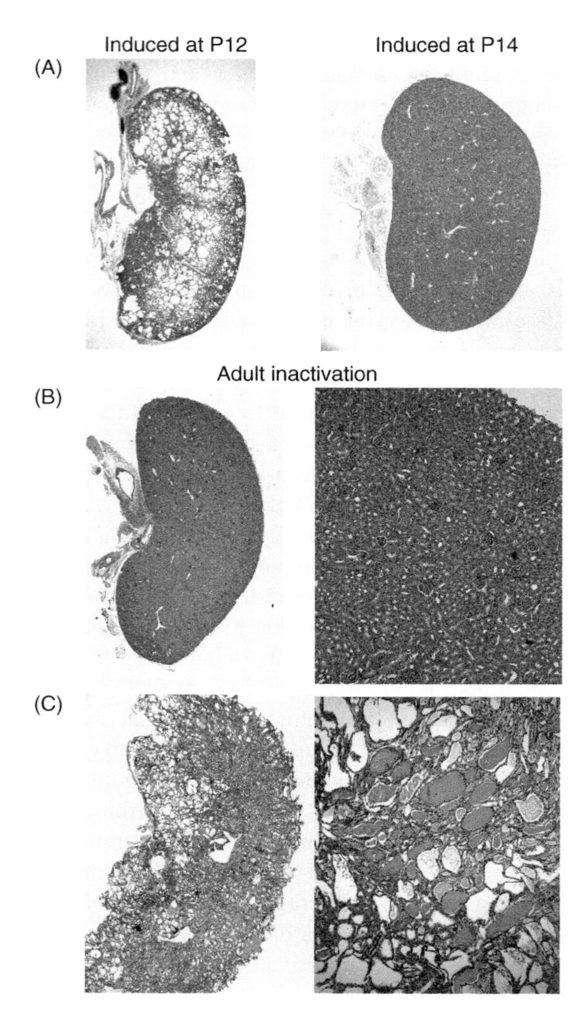

Fig. 5 Phenotypic response to acquired *Pkd1* loss is exquisitely sensitive to developmental life stage. (A) Kidneys of *Pkd1*$^{cond/cond}$; tamoxifen-Cre+ mice induced at P12 were cystic within 3 weeks (left), whereas those of mice induced at P14 remained normal 3 months later (right). (B/C) Adult inactivation results in late-onset renal cystic disease. Shown is H&E staining of kidneys harvested 3 months (B) or 6 months (C) after *Pkd1* inactivation induced in 6-week-old *Pkd1*$^{cond/cond}$; tamoxifen-Cre+ mice. From Piontek *et al.* (2007). (See Plate no. 23 in the Color Plate Section.)

formation must be indirect, recognizing the experimental limitations of our study. We could not exclude the possibility that *Pkd1* mutants have slightly higher proliferation rates and that the tools used in our study were not sensitive enough to detect them, nor that proliferation could occur in bursts that were missed. We note, however, that recent studies of other cystic models have also found no differences in the proliferation rate

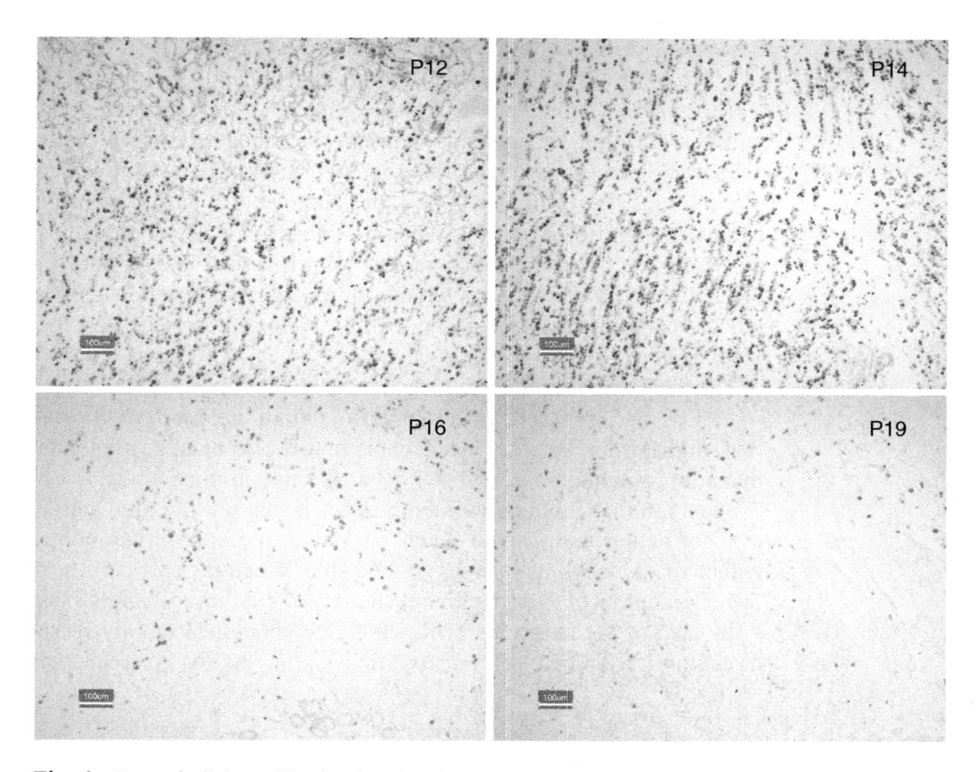

Fig. 6 Rates of cellular proliferation drop abruptly around the time that kidneys become less sensitive to acquired *Pkd1* loss. Ki67-stained sections of uninduced $Pkd1^{cond/cond}$; tamoxifen-Cre–; ROSA26R+ kidneys harvested on P12–P19. Similar results were obtained for kidneys of uninduced $Pkd1^{cond/cond}$; tamoxifen-Cre+ ROSA26R+ mice (data not shown). Scale bar, 100 μm. From Piontek *et al.* (2007). (See Plate no. 24 in the Color Plate Section.)

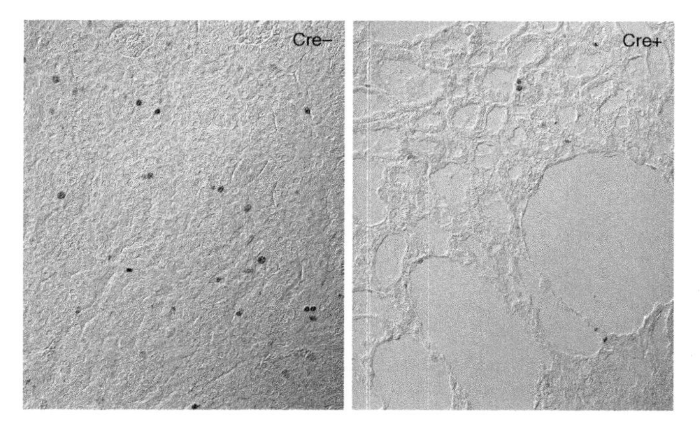

Fig. 7 Proliferation rates in mouse *Pkd1*-associated cystic kidneys are not appreciably higher than in control specimens. Ki67-stained sections of $Pkd1^{cond/cond}$; tamoxifen-Cre+ (left) and $Pkd1^{cond/cond}$; tamoxifen-Cre– (right) kidneys 3 weeks after induction at P12. From Piontek *et al.* (2007). (See Plate no. 25 in the Color Plate Section.)

between control and mutant samples during cystogenesis (Karner *et al.*, 2009; Saburi *et al.*, 2008; Sullivan-Brown *et al.*, 2008).

The apparent lack of a primary role for hyper-proliferation in cyst formation together with the intriguing relationship between the change in susceptibility to *Pkd1* loss and the abrupt decrease in cell proliferation rates prompted us to consider other mechanistic explanations. We examined gene expression patterns during this temporal window and identified significant changes that we think mark a developmental switch that signals the end of the terminal renal maturation process (Fig. 8). These findings led us to conclude that *Pkd1* regulates tubular morphology in both developing and adult kidney, but the pathologic consequences of inactivation are defined by the organ's developmental status.

The fact that the temporal window marking the kidneys' exquisite sensitivity to *Pkd1* loss also is marked by a high proliferation rate prompts consideration of how the two might be linked. Any mechanistic model also must accommodate the observation that antiproliferative agents have been effective in mice models of cystic disease. One explanation that could address both issues is that proliferation acts as an accelerant for cyst formation and growth when present in the appropriate context. A situation in which the kidney must undergo dynamic structural reorganization, such as during development or in its reparative response to renal injury, might be such a condition.

In each of the examples cited, epithelial cells must rapidly reestablish their planar orientation after cell division (Ciruna *et al.*, 2006). A recent study of neuroblast

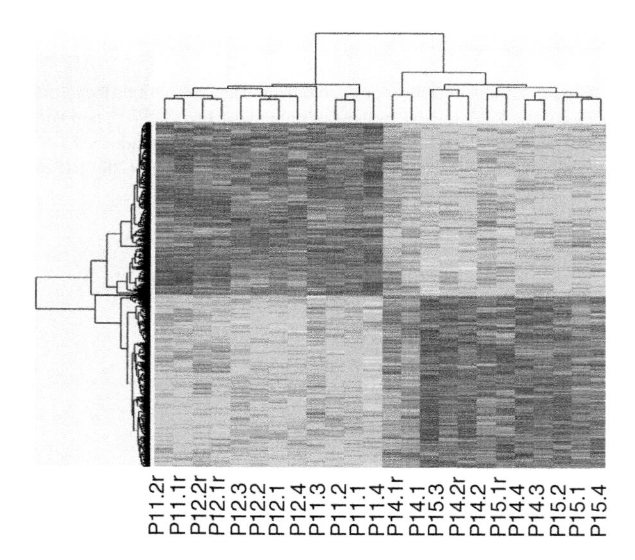

Fig. 8 Abrupt brake point in rate of renal proliferation parallels dramatic changes in gene expression. Heatmap plot of genes ($n = 827$) that varied between the P11 + P12 and P14 + P15 groups, showing a clear switch between the two groups (yellow, greater expression than the mean expression at P11; red, less expression than the mean at P11). Details including lists of differentially expressed genes can be found in Piontek *et al.* (2007). From Piontek *et al.* (2007).

development in zebrafish implicates PCP components in this process. In zebrafish, proper neural tube formation requires that, as cells divide in the neural plate, one daughter cell remain on the ipsilateral side, whereas the other intercalates across the midline and integrates into the contralateral side. Analyzing the morphology of neural anlage during development, the authors showed that in *Vangl2* mutant embryos the neural tube failed to form. This phenotype was explained by a failure of the daughter cells that lost contact with the cell membrane during cell division to reintegrate. Using GFP-tagged Prickle as a marker of polarization, the authors further demonstrated that following cell division both wild-type (WT) and mutant cells lose the asymmetric Prickle localization, which is recovered in WT, but not in mutant cells. Remarkably, the neural tube defect was rescued by inhibition of proliferation. The authors propose that PCP signaling is required after cell division in order to reestablish polarity.

We speculate that the *Pkd1* gene product, polycystin-1, may play a similar role during development or in responding to renal injury. This model may also help to explain why antiproliferative agents have been effective in rodent models of cystic disease. What it doesn't readily explain is why all cellular proliferation in mouse kidneys with induced loss of *Pkd1* at > P14 does not initiate cyst formation. Proliferation rates at P16 in both control and mutant samples are severalfold higher than in adult mice yet *Pkd1* inactivation at P14 does not result in cysts for many months. We offer two different possible explanations. Firstly, the cellular context might be important. Proliferation in the setting of developmental or reparative processes (which are thought to recapitulate some aspects of renal development) might be very different than that which sporadically occurs under static conditions. The second possibility is suggested by the aforementioned zebrafish studies. Experiments using WT-mutant chimaeras suggested that both cell-autonomous and noncell-autonomous components are important in establishing planar polarity. These studies suggest that positional information may be established and stabilized by a field of cells sensing gradients. In the setting of isolated cell division, the neighboring community may help the cell reestablish its orientation, even in the absence of cell guidance cues like *Pkd1*.

D. Flow and Cystic Kidney Disease

As mentioned in a previous section, cilia bending is thought to trigger transient calcium influx, release of nucleotides and activation of epithelial P2 receptors. After a cell-type specific lag phase, these steps ultimately result in a significant release in intracellular calcium and a subsequent refractory period. (Praetorius and Leipziger, 2009; Praetorius and Spring, 2001, 2003). How cilia sense mechanical stimuli is still an ongoing research topic. However, some evidence would suggest that PC1 and PC2 could play a role (Nauli *et al.*, 2003, 2006). In epithelial kidney cells, fluid shear stress-induced ciliary bending was shown to result in immediate increase intracellular calcium levels that were not present in $Pkd1^{del34/del34}$ cells.

The flow hypothesis for cystogenesis asserts that flow sensing by primary cilia results in increased activity of inversin, enhanced degradation of disheveled, and a switch from canonical to noncanonical Wnt signaling. The net result of this is a decrease in cell proliferation and changes in cell polarity. Disruption of ciliary structure/function would then result in cyst formation. Since several "cystoproteins" localize to cilia, this could provide a unifying mechanism for cyst formation. Closer analysis, however, reveals certain inconsistencies that have to be resolved: (1) as discussed earlier, the kidney cystic diseases encompass a vast spectrum of different phenotypes, reasonably conserved within each disease, from primarily fibrotic to mostly cystic and it is difficult to imagine that the same defect could account for all of them; (2) *in vitro* models show that in the absence of flow *Pkd1* expression can induce tubulogenesis in cyst-forming MDCK cells (Boletta *et al.*, 2000); (3) metanephric kidneys in organ culture do not spontaneously become cystic; in fact, even *Pkd* mutant organs require additional stimulation to become cystic in this system (Magenheimer *et al.*, 2006); (4) as discussed in the next session, there is a considerable delay between loss of ciliary integrity and cyst formation in the adult kidney, which is difficult to explain if the primary role of cilia in cystogenesis is to decode dynamic changes of flow (Davenport *et al.*, 2007; Piontek *et al.*, 2007); (5) recent studies have reported that TRPV4 and PC2 form a mechano- and thermosensitive molecular sensor in the cilium that responds to flow with transient calcium currents with kinetics similar to those reported by Praetorius *et al.*, yet TRPV4-deficient zebrafish and mice lack renal cysts (Kottgen *et al.*, 2008). Collectively, these observations challenge the concept that defective ciliary flow sensing constitutes a fundamental mechanism of cystogenesis

E. Maintaining Kidney Architecture and Function

One interesting open question in kidney biology is whether the pathways required for kidney development are also necessary to maintain kidney architecture. A weaker version could be formulated as: does the time of gene inactivation influence the biology of cystic diseases? A corollary is that different molecular mechanisms could be responsible for cyst formation in early (i.e., during development/maturation) versus late (i.e., postmaturation) stages in life, and respond to drug therapy differently.

Our study examining the consequences of *Pkd1* inactivation at different time points has begun addressing this question (Piontek *et al.*, 2007). We had found that time course of cyst formation after *Pkd1* inactivation is extremely time sensitive. Though these studies did not address the potential differences in the mechanisms of cyst formation in early- versus late-onset disease, it is interesting to note that, in both cases, the interval between identification of dilated tubules and globally cystic kidneys seemed similar, suggesting that in the late-onset model there might be a lag-phase followed by rapid cyst development. Another remarkable finding is that the cystic disease explodes globally and essentially simultaneously throughout the organ in both cases.

Similar studies using conditional alleles of *Tg737* and *Kif3a* addressed the related question of whether cilia are required in postnatal life (Davenport *et al.*, 2007). Analogous to the *Pkd1* results, this study also demonstrated a 16+ week lag phase between gene inactivation and cyst formation in animals induced after 8 weeks of life, contrasting to rapid onset ($<$2 weeks) in animals induced *in utero*.

Together, these results demonstrated that inactivation of *Pkd1* and other ciliary factors in adult animals can cause cystic disease and suggest that cilia are required not only for proper kidney development, but also for maintenance of normal function/ morphology. Several models could explain this observation:

1. Cilia could maintain the cells in a low proliferative and/or more differentiated state, and loss of cilia would push the cell toward a high proliferative, less differentiated state. How cilia could control proliferation is still not completely clear. However, it was shown that inversin binds to Apc2, a subunit of the anaphase-promoting complex, and has a dynamic expression pattern during the cell cycle, which could support a model in which information sensed by the primary cilia could be used to modulate the cell cycle (Morgan *et al.*, 2002). Other models of cell cycle control by cilia involve PDGF, hedgehog, and Wnt signaling (Christensen *et al.*, 2008). Arguing against the applicability of this model to PKD is the very long lag phase between the time of loss of ciliary function and the onset of cystic disease. In fact, the mouse studies challenge any role for cilia in regulating cell proliferation in the context of a stable, well-differentiated system.
2. Cilia could sense chemical gradients or mechanical stress/flow and confer polarity to the tissue; loss of cilia would result in disruption of normal polarity of the tissue and cyst formation. This model can nicely explain the rapid onset of cyst formation during development but must assume a more tonic role for ciliary signaling in the mature organ.
3. Cilia could release molecules that relay information to other cells, controlling tissue polarity and/or proliferative status (Kaimori *et al.*, 2007); the same caveat applies here as it did for item #2.
4. Finally, cilia may be correlated to some other biological property which is the underlying cause of cyst formation, and would be therefore a marker of some other process.

The exquisite sensitivity to developmental stage could additionally be explained by different models: (1) cilia (or some other underlying property that correlates with cilia presence) have different functions during different developmental stages, and the kinetics of cyst formation would depend on these functions; (2) cilia have the same functions at different stages of life but these functions are required differently as the cells develop; (3) cells in different developmental stages could respond differently to signals that are relayed through cilia; (4) field effects of signaling pathways could induce mutant cells to behave normally and vice versa; in the early induction models, the clonal expansion of a growing organ would result in fields of cells of certain size behaving abnormally and forming cysts. Once the number of cells is below a certain threshold, there would be less recruitment of normal cells to behave abnormally and mutant cells would be kept under some tonic control. A corollary would be that higher inactivation rates would tend to push the late-onset models toward the early-onset.

An interesting prediction of these models is that insults that result in disruption of cilia, increased proliferation, loss of tissue polarity, and/or reversal of differentiation status by genetic manipulation and/or ischemic/chemical would cause cyst formation. As several studies showed, ischemic and toxic kidney injury seem to result in all such changes and do indeed accelerate cyst formation (Happe *et al.*, 2009; Patel *et al.*, 2008; Takakura *et al.*, 2009). Curiously, *TamCre* induced inactivation of *Kif3a* as early as P10 did not result in cyst formation within the 2 months of observation, despite loss of cilia (Patel *et al.*, 2008). It is possible that longer follow-up would have identified kidney cysts, but it is still relevant to notice the significant lag between loss of cilia and morphological changes. Cyst formation could be induced in these mutant animals by ischemic injury in what seemed to be a proliferation-dependent process. However, these studies cannot unambiguously prove which mechanisms are responsible for cyst formation nor address whether cystic phenotype is the outcome of an abnormal response and/or if mutant cells are more susceptible to injury. The latter is a relevant concern since it is well recognized that even genetically normal kidneys will develop cystic changes in response to severe renal injury.

IV. Final Remarks

In the last few years, cilia have emerged from obscurity into the spotlight of kidney cystic disease research. There is an overwhelming abundance of data linking cilia to renal cystic disease but the causal nature of such relationships is still being defined. As with most new ideas, the initial acceptance into mainstream science was based on studies trying to prove its importance. Whether all forms of cystic disease arise as a direct or indirect consequence of defective ciliary function remains to be determined. As the glare of novelty starts to dim, it is now time to design experiments that will rigorously test the ciliary model of cystogenesis by seeking to disprove it. This will not only help us better understand ciliary function, but also yield a more comprehensive understanding of cyst formation. A particularly interesting challenge will be in determining how loss of ciliary function, or more precisely, factors known to be important for its function, result in abnormal tubule morphology months later.

Acknowledgments

The authors would like to thank Dr. Klaus Piontek and other members of the Germino, Watnick, and Qian laboratories for helpful discussions. This work was supported in part by the PKD Foundation, the NKF of Maryland, R37DK48006, and the intramural program of the NIDDK.

References

Ansley, S.J., Badano, J.L., Blacque, O.E., Hill, J., Hoskins, B.E., Leitch, C.C., Kim, J.C., Ross, A.J., Eichers, E.R., Teslovich, T.M., Mah, A.K., Johnsen, R.C., *et al.*, (2003). Basal body dysfunction is a likely cause of pleiotropic Bardet–Biedl syndrome. *Nature* **425**, 628–633.

Attanasio, M., Uhlenhaut, N.H., Sousa, V.H., O'Toole, J.F., Otto, E., Anlag, K., Klugmann, C., Treier, A.C., Helou, J., Sayer, J.A., Seelow, D., Nurnberg, G., *et al.*, (2007). Loss of GLIS2 causes nephronophthisis in humans and mice by increased apoptosis and fibrosis. *Nat. Genet.* **39**, 1018–1024.

Axelrod, J.D. (2008). Basal bodies, kinocilia and planar cell polarity. *Nat. Genet.* **40**, 10–11.

Baena-Lopez, L.A., Baonza, A., and Garcia-Bellido, A. (2005). The orientation of cell divisions determines the shape of *Drosophila* organs. *Curr. Biol.* **15**, 1640–1644.

Barr, M.M., and Sternberg, P.W. (1999). A polycystic kidney-disease gene homologue required for male mating behaviour in *C. Elegans. Nature* **401**, 386–389.

Beales, P.L., Elcioglu, N., Woolf, A.S., Parker, D., and Flinter, F.A. (1999). New criteria for improved diagnosis of Bardet–Biedl syndrome: Results of a population survey. *J. Med. Genet.* **36**, 437–446.

Bellaiche, Y., Gho, M., Kaltschmidt, J.A., Brand, A.H., and Schweisguth, F. (2001). Frizzled regulates localization of cell-fate determinants and mitotic spindle rotation during asymmetric cell division. *Nat. Cell Biol.* **3**, 50–57.

Benzing, T., Gerke, P., Hopker, K., Hildebrandt, F., Kim, E., and Walz, G. (2001). Nephrocystin interacts with Pyk2, p130(Cas), and tensin and triggers phosphorylation of Pyk2. *Proc. Natl. Acad. Sci. USA* **98**, 9784–9789.

Bergmann, C., Fliegauf, M., Bruchle, N.O., Frank, V., Olbrich, H., Kirschner, J., Schermer, B., Schmedding, I., Kispert, A., Kranzlin, B., Nurnberg, G., Becker, C., *et al.*, (2008). Loss of nephrocystin-3 function can cause embryonic lethality, Meckel-Gruber-like syndrome, situs inversus, and renal-hepatic-pancreatic dysplasia. *Am. J. Hum. Genet.* **82**, 959–970.

Boletta, A., Qian, F., Onuchic, L.F., Bhunia, A.K., Phakdeekitcharoen, B., Hanaoka, K., Guggino, W., Monaco, L., and Germino, G.G. (2000). Polycystin-1, the gene product of PKD1, induces resistance to apoptosis and spontaneous tubulogenesis in MDCK cells. *Mol. Cell* **6**, 1267–1273.

Bonnet, C.S., Aldred, M., von Ruhland, C., Harris, R., Sandford, R., and Cheadle, J.P. (2009). Defects in cell polarity underlie TSC and ADPKD-associated cystogenesis. *Hum. Mol. Genet.* **18**, 2166–2176.

Boucher, C., and Sandford, R. (2004). Autosomal dominant polycystic kidney disease (ADPKD, MIM 173900, PKD1 and PKD2 genes, protein products known as polycystin-1 and polycystin-2). *Eur. J. Hum. Genet.* **12**, 347–354.

Bukanov, N.O., Smith, L.A., Klinger, K.W., Ledbetter, S.R., and Ibraghimov-Beskrovnaya, O. (2006). Long-lasting arrest of murine polycystic kidney disease with CDK inhibitor roscovitine. *Nature* **444**, 949–952.

Carroll, T.J., Park, J.S., Hayashi, S., Majumdar, A., and McMahon, A.P. (2005). Wnt9b plays a central role in the regulation of mesenchymal to epithelial transitions underlying organogenesis of the mammalian urogenital system. *Dev. Cell* **9**, 283–292.

Chang, M.Y., Parker, E., Ibrahim, S., Shortland, J.R., Nahas, M.E., Haylor, J.L., and Ong, A.C. (2006). Haploinsufficiency of Pkd2 is associated with increased tubular cell proliferation and interstitial fibrosis in two murine Pkd2 models. *Nephrol. Dial. Transplant.* **21**, 2078–2084.

Christensen, S.T., Pedersen, S.F., Satir, P., Veland, I.R., and Schneider, L. (2008). The primary cilium coordinates signaling pathways in cell cycle control and migration during development and tissue repair. *Curr. Top. Dev. Biol.* **85**, 261–301.

Ciruna, B., Jenny, A., Lee, D., Mlodzik, M., and Schier, A.F. (2006). Planar cell polarity signalling couples cell division and morphogenesis during neurulation. *Nature* **439**, 220–224.

Corbit, K.C., Shyer, A.E., Dowdle, W.E., Gaulden, J., Singla, V., Chen, M.H., Chuang, P.T., and Reiter, J.F. (2008). Kif3a constrains beta-catenin-dependent wnt signalling through dual ciliary and non-ciliary mechanisms. *Nat. Cell Biol.* **10**, 70–76.

Davenport, J.R., Watts, A.J., Roper, V.C., Croyle, M.J., van Groen, T., Wyss, J.M., Nagy, T.R., Kesterson, R.A., and Yoder, B.K. (2007). Disruption of intraflagellar transport in adult mice leads to obesity and slow-onset cystic kidney disease. *Curr. Biol.* **17**, 1586–1594.

Delous, M., Baala, L., Salomon, R., Laclef, C., Vierkotten, J., Tory, K., Golzio, C., Lacoste, T., Besse, L., Ozilou, C., Moutkine, I., Hellman, N.E., *et al.*, (2007). The ciliary gene RPGRIP1L is mutated in cerebello-oculo-renal syndrome (Joubert syndrome type B) and Meckel syndrome. *Nat. Genet.* **39**, 875–881.

Drummond, I.A., Majumdar, A., Hentschel, H., Elger, M., Solnica-Krezel, L., Schier, A.F., Neuhauss, S.C., Stemple, D.L., Zwartkruis, F., Rangini, Z., Driever, W., and Fishman, M.C. (1998). Early development of the zebrafish pronephros and analysis of mutations affecting pronephric function. *Development* **125**, 4655–4667.

Fischer, E., Legue, E., Doyen, A., Nato, F., Nicolas, J.F., Torres, V., Yaniv, M., and Pontoglio, M. (2006). Defective planar cell polarity in polycystic kidney disease. *Nat. Genet.* **38**, 21–23.

Gallagher, A.R., Esquivel, E.L., Briere, T.S., Tian, X., Mitobe, M., Menezes, L.F., Markowitz, G.S., Jain, D., Onuchic, L.F., and Somlo, S. (2008). Biliary and pancreatic dysgenesis in mice harboring a mutation in Pkhd1. *Am. J. Pathol.* 172, 417–429.

Germino, G.G. (2005). Linking cilia to Wnts. *Nat. Genet.* **37**, 455–457.

Gong, Y., Mo, C., and Fraser, S.E. (2004). Planar cell polarity signalling controls cell division orientation during zebrafish gastrulation. *Nature* **430**, 689–693.

Gresh, L., Fischer, E., Reimann, A., Tanguy, M., Garbay, S., Shao, X., Hiesberger, T., Fiette, L., Igarashi, P., Yaniv, M., and Pontoglio, M. (2004). A transcriptional network in polycystic kidney disease. *EMBO J.* **23**, 1657–1668.

Guay-Woodford, L.M., and Desmond, R.A. (2003). Autosomal recessive polycystic kidney disease: The clinical experience in North America. *Pediatrics* **111**, 1072–1080.

Happe, H., Leonhard, W.N., van der Wal, A., van de Water, B., Lantinga-van Leeuwen, I.S., Breuning, M.H., de Heer, E., and Peters, D.J. (2009). Toxic tubular injury in kidneys from Pkd1-deletion mice accelerates cystogenesis accompanied by dysregulated planar cell polarity and canonical Wnt signaling pathways. *Hum. Mol. Genet.* **18**, 2532–2542.

Hearn, T., Spalluto, C., Phillips, V.J., Renforth, G.L., Copin, N., Hanley, N.A., and Wilson, D.I. (2005). Subcellular localization of ALMS1 supports involvement of centrosome and basal body dysfunction in the pathogenesis of obesity, insulin resistance, and type 2 diabetes. *Diabetes* **54**, 1581–1587.

Hildebrandt, F., Attanasio, M., and Otto, E. (2009). Nephronophthisis: Disease mechanisms of a ciliopathy. *J. Am. Soc. Nephrol.* **20**, 23–35.

Hildebrandt, F., and Zhou, W. (2007). Nephronophthisis-associated ciliopathies. *J. Am. Soc. Nephrol.* **18**, 1855–1871.

Hirokawa, N. (2000). Stirring up development with the heterotrimeric kinesin KIF3. *Traffic* **1**, 29–34.

Hou, X., Mrug, M., Yoder, B.K., Lefkowitz, E.J., Kremmidiotis, G., D'Eustachio, P., Beier, D.R., and Guay-Woodford, L.M. (2002). Cystin, a novel cilia-associated protein, is disrupted in the Cpk mouse model of polycystic kidney disease. *J. Clin. Invest.* **109**, 533–540.

Jonassen, J.A., San Agustin, J., Follit, J.A., and Pazour, G.J. (2008). Deletion of IFT20 in the mouse kidney causes misorientation of the mitotic spindle and cystic kidney disease. *J. Cell Biol.* **183**, 377–384.

Jones, C., and Chen, P. (2008). Primary cilia in planar cell polarity regulation of the inner ear. *Curr. Top. Dev. Biol.* **85**, 197–224.

Kaimori, J.Y., Nagasawa, Y., Menezes, L.F., Garcia-Gonzalez, M.A., Deng, J., Imai, E., Onuchic, L.F., Guay-Woodford, L.M., and Germino, G.G. (2007). Polyductin undergoes notch-like processing and regulated release from primary cilia. *Hum. Mol. Genet.* **16**, 942–956.

Karner, C.M., Chirumamilla, R., Aoki, S., Igarashi, P., Wallingford, J.B., and Carroll, T.J. (2009). Wnt9b signaling regulates planar cell polarity and kidney tubule morphogenesis. *Nat. Genet.* **41**, 793–799.

Kelley, K.A., Agarwal, N., Reeders, S., and Herrup, K. (1991). Renal cyst formation and multifocal neoplasia in transgenic mice carrying the simian virus 40 early region. *J. Am. Soc. Nephrol.* **2**, 84–97.

Kishimoto, N., Cao, Y., Park, A., and Sun, Z. (2008). Cystic kidney gene seahorse regulates cilia-mediated processes and Wnt pathways. *Dev. Cell* **14**, 954–961.

Kottgen, M., Buchholz, B., Garcia-Gonzalez, M.A., Kotsis, F., Fu, X., Doerken, M., Boehlke, C., Steffl, D., Tauber, R., Wegierski, T., Nitschke, R., Suzuki, M., *et al.*, (2008). TRPP2 and TRPV4 form a polymodal sensory channel complex. *J. Cell Biol.* **182**, 437–447.

Koulen, P., Cai, Y., Geng, L., Maeda, Y., Nishimura, S., Witzgall, R., Ehrlich, B.E., and Somlo, S. (2002). Polycystin-2 is an intracellular calcium release channel. *Nat. Cell Biol.* **4**, 191–197.

Lanoix, J., D'Agati, V., Szabolcs, M., and Trudel, M. (1996). Dysregulation of cellular proliferation and apoptosis mediates human autosomal dominant polycystic kidney disease (ADPKD). *Oncogene* **13**, 1153– 1160.

Lantinga-van Leeuwen, I.S., Leonhard, W.N., van der Wal, A., Breuning, M.H., de Heer, E., and Peters, D.J. (2007). Kidney-specific inactivation of the Pkd1 gene induces rapid cyst formation in developing kidneys and a slow onset of disease in adult mice. *Hum. Mol. Genet.* **16**, 3188–3196.

Li, J.B., Gerdes, J.M., Haycraft, C.J., Fan, Y., Teslovich, T.M., May-Simera, H., Li, H., Blacque, O.E., Li, L., Leitch, C.C., Lewis, R.A., Green, J.S., *et al.*, (2004). Comparative genomics identifies a flagellar and basal body proteome that includes the BBS5 human disease gene. *Cell* **117**, 541–552.

Lin, F., Hiesberger, T., Cordes, K., Sinclair, A.M., Goldstein, L.S., Somlo, S., and Igarashi, P. (2003). Kidney-specific inactivation of the KIF3A subunit of kinesin-II inhibits renal ciliogenesis and produces polycystic kidney disease. *Proc. Natl. Acad. Sci. USA* **100**, 5286–5291.

Magenheimer, B.S., St. John, P.L., Isom, K.S., Abrahamson, D.R., De Lisle, R.C., Wallace, D.P., Maser, R.L., Grantham, J.J., and Calvet, J.P. (2006). Early embryonic renal tubules of wild-type and polycystic kidney disease kidneys respond to cAMP stimulation with cystic fibrosis transmembrane conductance regulator/Na(+), K(+), 2Cl(–) Co-transporter-dependent cystic dilation. *J. Am. Soc. Nephrol.* **17**, 3424–3437.

Mahjoub, M.R., Trapp, M.L., and Quarmby, L.M. (2005). NIMA-related kinases defective in murine models of polycystic kidney diseases localize to primary cilia and centrosomes. *J. Am. Soc. Nephrol.* **16**, 3485–3489.

Marshall, J.D., Beck, S., Maffei, P., and Naggert, J.K. (2007). Alstrom syndrome. *Eur. J. Hum. Genet.* **15**, 1193–1202.

Menezes, L.F., Cai, Y., Nagasawa, Y., Silva, A.M., Watkins, M.L., Da Silva, A.M., Somlo, S., Guay-Woodford, L.M., Germino, G.G., and Onuchic, L.F. (2004). Polyductin, the PKHD1 gene product, comprises isoforms expressed in plasma membrane, primary cilium, and cytoplasm. *Kidney Int.* **66**, 1345–1355.

Mochizuki, T., Saijoh, Y., Tsuchiya, K., Shirayoshi, Y., Takai, S., Taya, C., Yonekawa, H., Yamada, K., Nihei, H., Nakatsuji, N., Overbeek, P.A., Hamada, H. *et al.*, (1998). Cloning of inv, a gene that controls left/right asymmetry and kidney development. *Nature* **395**, 177–181.

Morgan, D., Eley, L., Sayer, J., Strachan, T., Yates, L.M., Craighead, A.S., and Goodship, J.A. (2002). Expression analyses and interaction with the anaphase promoting complex protein Apc2 suggest a role for inversin in primary cilia and involvement in the cell cycle. *Hum. Mol. Genet.* **11**, 3345–3350.

Morgan, D., Turnpenny, L., Goodship, J., Dai, W., Majumder, K., Matthews, L., Gardner, A., Schuster, G., Vien, L., Harrison, W., Elder, F.F., Penman-Splitt, M., *et al.*, (1998). Inversin, a novel gene in the vertebrate left–right axis pathway, is partially deleted in the inv mouse. *Nat. Genet.* **20**, 149–156.

Moyer, J.H., Lee-Tischler, M.J., Kwon, H.Y., Schrick, J.J., Avner, E.D., Sweeney, W.E., Godfrey, V.L., Cacheiro, N.L., Wilkinson, J.E., and Woychik, R.P. (1994). Candidate gene associated with a mutation causing recessive polycystic kidney disease in mice. *Science* **264**, 1329–1333.

Murcia, N.S., Richards, W.G., Yoder, B.K., Mucenski, M.L., Dunlap, J.R., and Woychik, R.P. (2000). The Oak Ridge polycystic kidney (Orpk) disease gene is required for left–right axis determination. *Development* **127**, 2347–2355.

Nachury, M.V., Loktev, A.V., Zhang, Q., Westlake, C.J., Peranen, J., Merdes, A., Slusarski, D.C., Scheller, R.H., Bazan, J.F., Sheffield, V.C., and Jackson, P.K. (2007). A core complex of BBS proteins cooperates with the GTPase Rab8 to promote ciliary membrane biogenesis. *Cell* **129**, 1201–1213.

Nadasdy, T., Laszik, Z., Lajoie, G., Blick, K.E., Wheeler, D.E., and Silva, F.G. (1995). Proliferative activity of cyst epithelium in human renal cystic diseases. *J. Am. Soc. Nephrol.* **5**, 1462–1468.

Nauli, S.M., Alenghat, F.J., Luo, Y., Williams, E., Vassilev, P., Li, X., Elia, A.E., Lu, W., Brown, E.M., Quinn, S.J., Ingber, D.E., and Zhou, J. (2003). Polycystins 1 and 2 mediate mechanosensation in the primary cilium of kidney cells. *Nat. Genet.* **33**, 129–137.

Nauli, S.M., Rossetti, S., Kolb, R.J., Alenghat, F.J., Consugar, M.B., Harris, P.C., Ingber, D.E., Loghman-Adham, M., and Zhou, J. (2006). Loss of polycystin-1 in human cyst-lining epithelia leads to ciliary dysfunction. *J. Am. Soc. Nephrol.* **17**, 1015–1025.

Onda, H., Lueck, A., Marks, P.W., Warren, H.B., and Kwiatkowski, D.J. (1999). Tsc2(±) mice develop tumors in multiple sites that express gelsolin and are influenced by genetic background. *J. Clin. Invest.* **104**, 687–695.

Otto, E.A., Loeys, B., Khanna, H., Hellemans, J., Sudbrak, R., Fan, S., Muerb, U., O'Toole, J.F., Helou, J., Attanasio, M., Utsch, B., Sayer, J.A., *et al.*, (2005). Nephrocystin-5, a ciliary IQ domain protein, is mutated in Senior–Loken syndrome and interacts with RPGR and calmodulin. *Nat. Genet.* **37**, 282–288.

Otto, E.A., Schermer, B., Obara, T., O'Toole, J.F., Hiller, K.S., Mueller, A.M., Ruf, R.G., Hoefele, J., Beekmann, F., Landau, D., Foreman, J.W., Goodship, J.A., *et al.*, (2003). Mutations in INVS encoding inversin cause nephronophthisis type 2, linking renal cystic disease to the function of primary cilia and left–right axis determination. *Nat. Genet.* **34**, 413–420.

Park, J.S., Valerius, M.T., and McMahon, A.P. (2007). Wnt/beta-catenin signaling regulates nephron induction during mouse kidney development. *Development* **134**, 2533–2539.

Patel, V., Li, L., Cobo-Stark, P., Shao, X., Somlo, S., Lin, F., and Igarashi, P. (2008). Acute kidney injury and aberrant planar cell polarity induce cyst formation in mice lacking renal cilia. *Hum. Mol. Genet.* **17**, 1578–1590.

Pazour, G.J., Dickert, B.L., Vucica, Y., Seeley, E.S., Rosenbaum, J.L., Witman, G.B., and Cole, D.G. (2000). *Chlamydomonas* IFT88 and its mouse homologue, polycystic kidney disease gene tg737, are required for assembly of cilia and flagella. *J. Cell Biol.* **151**, 709–718.

Pazour, G.J., San Agustin, J.T., Follit, J.A., Rosenbaum, J.L., and Witman, G.B. (2002). Polycystin-2 localizes to kidney cilia and the ciliary level is elevated in orpk mice with polycystic kidney disease. *Curr. Biol.* **12**, R378–R380.

Phillips, C.L., Miller, K.J., Filson, A.J., Nurnberger, J., Clendenon, J.L., Cook, G.W., Dunn, K.W., Overbeek, P.A., Gattone, V.H., II, and Bacallao, R.L. (2004). Renal cysts of inv/inv mice resemble early infantile nephronophthisis. *J. Am. Soc. Nephrol.* **15**, 1744–1755.

Piontek, K., Menezes, L.F., Garcia-Gonzalez, M.A., Huso, D.L., and Germino, G.G. (2007). A critical developmental switch defines the kinetics of kidney cyst formation after loss of Pkd1. *Nat. Med.* **13**, 1490–1495.

Praetorius, H.A., and Leipziger, J. (2009). Released nucleotides amplify the cilium-dependent, flow-induced [Ca(2)] response in MDCK cells. *Acta. Physiol.* (Oxf). 2009 Nov; **197(3)**, 241–251.

Praetorius, H.A., and Spring, K.R. (2001). Bending the MDCK cell primary cilium increases intracellular calcium. *J. Membr. Biol.* **184**, 71–79.

Praetorius, H.A., and Spring, K.R. (2003). Removal of the MDCK cell primary cilium abolishes flow sensing. *J. Membr. Biol.* **191**, 69–76.

Qian, C.N., Knol, J., Igarashi, P., Lin, F., Zylstra, U., Teh, B.T., and Williams, B.O. (2005). Cystic renal neoplasia following conditional inactivation of Apc in mouse renal tubular epithelium. *J. Biol. Chem.* **280**, 3938–3945.

Qian, F., Watnick, T.J., Onuchic, L.F., and Germino, G.G. (1996). The molecular basis of focal cyst formation in human autosomal dominant polycystic kidney disease type I. *Cell* **87**, 979–987.

Quinlan, R.J., Tobin, J.L., and Beales, P.L. (2008). Modeling ciliopathies: Primary cilia in development and disease. *Curr. Top. Dev. Biol.* **84**, 249–310.

Ross, A.J., May-Simera, H., Eichers, E.R., Kai, M., Hill, J., Jagger, D.J., Leitch, C.C., Chapple, J.P., Munro, P.M., Fisher, S., Tan, P.L., Phillips, H.M., *et al.*, (2005). Disruption of Bardet–Biedl syndrome ciliary proteins perturbs planar cell polarity in vertebrates. *Nat. Genet.* **37**, 1135–1140.

Saadi-Kheddouci, S., Berrebi, D., Romagnolo, B., Cluzeaud, F., Peuchmaur, M., Kahn, A., Vandewalle, A., and Perret, C. (2001). Early development of polycystic kidney disease in transgenic mice expressing an activated mutant of the beta-catenin gene. *Oncogene* **20**, 5972–5981.

Saburi, S., Hester, I., Fischer, E., Pontoglio, M., Eremina, V., Gessler, M., Quaggin, S.E., Harrison, R., Mount, R., and McNeill, H. (2008). Loss of Fat4 disrupts PCP signaling and oriented cell division and leads to cystic kidney disease. *Nat. Genet.* **40**, 1010–1015.

Sayer, J.A., Otto, E.A., O'Toole, J.F., Nurnberg, G., Kennedy, M.A., Becker, C., Hennies, H.C., Helou, J., Attanasio, M., Fausett, B.V., Utsch, B., Khanna, H., *et al.*, (2006). The centrosomal protein nephrocystin-6 is mutated in Joubert syndrome and activates transcription factor ATF4. *Nat. Genet.* **38**, 674–681.

Shibazaki, S., Yu, Z., Nishio, S., Tian, X., Thomson, R.B., Mitobe, M., Louvi, A., Velazquez, H., Ishibe, S., Cantley, L.G., Igarashi, P., and Somlo, S. (2008). Cyst formation and activation of the extracellular regulated kinase pathway after kidney specific inactivation of Pkd1. *Hum. Mol. Genet.* **17**, 1505–1516.

Shillingford, J.M., Murcia, N.S., Larson, C.H., Low, S.H., Hedgepeth, R., Brown, N., Flask, C.A., Novick, A.C., Goldfarb, D.A., Kramer-Zucker, A., Walz, G., Piontek, K.B., *et al.*, (2006). The mTOR pathway is regulated by polycystin-1, and its inhibition reverses renal cystogenesis in polycystic kidney disease. *Proc. Natl. Acad. Sci. USA* **103**, 5466–5471.

Simons, M., and Mlodzik, M. (2008). Planar cell polarity signaling: From fly development to human disease. *Annu. Rev. Genet.* **42**, 517–540.

Simons, M., Gloy, J., Ganner, A., Bullerkotte, A., Bashkurov, M., Kronig, C., Schermer, B., Benzing, T., Cabello, O.A., Jenny, A., Mlodzik, M., Polok, B., *et al.*, (2005). Inversin, the gene product mutated in nephronophthisis type II, functions as a molecular switch between wnt signaling pathways. *Nat. Genet.* **37**, 537–543.

Stark, K., Vainio, S., Vassileva, G., and McMahon, A.P. (1994). Epithelial transformation of metanephric mesenchyme in the developing kidney regulated by Wnt-4. *Nature* **372**, 679–683.

Sullivan-Brown, J., Schottenfeld, J., Okabe, N., Hostetter, C.L., Serluca, F.C., Thiberge, S.Y., and Burdine, R.D. (2008). Zebrafish mutations affecting cilia motility share similar cystic phenotypes and suggest a mechanism of cyst formation that differs from pkd2 morphants. *Dev. Biol.* **314**, 261–275.

Sun, Z., Amsterdam, A., Pazour, G.J., Cole, D.G., Miller, M.S., and Hopkins, N. (2004). A genetic screen in zebrafish identifies cilia genes as a principal cause of cystic kidney. *Development* **131**, 4085–4093.

Sweeney, W.E., Jr., Hamahira, K., Sweeney, J., Garcia-Gatrell, M., Frost, P., and Avner, E.D. (2003). Combination treatment of PKD utilizing dual inhibition of EGF-receptor activity and ligand bioavailability. *Kidney Int.* **64**, 1310–1319.

Takakura, A., Contrino, L., Zhou, X., Bonventre, J.V., Sun, Y., Humphreys, B.D., and Zhou, J. (2009). Renal injury is a third hit promoting rapid development of adult polycystic kidney disease. *Hum. Mol. Genet.* **18**, 2523–2531.

Teng, J., Rai, T., Tanaka, Y., Takei, Y., Nakata, T., Hirasawa, M., Kulkarni, A.B., and Hirokawa, N. (2005). The KIF3 motor transports N-cadherin and organizes the developing neuroepithelium. *Nat. Cell Biol.* **7**, 474–482.

Trudel, M., D'Agati, V., and Costantini, F. (1991). C-Myc as an inducer of polycystic kidney disease in transgenic mice. *Kidney Int.* **39**, 665–671.

Wang, S., Luo, Y., Wilson, P.D., Witman, G.B., and Zhou, J. (2004). The autosomal recessive polycystic kidney disease protein is localized to primary cilia, with concentration in the basal body area. *J. Am. Soc. Nephrol.* **15**, 592–602.

Ward, C.J., Yuan, D., Masyuk, T.V., Wang, X., Punyashthiti, R., Whelan, S., Bacallao, R., Torra, R., LaRusso, N.F., Torres, V.E., and Harris, P.C. (2003). Cellular and subcellular localization of the ARPKD protein; fibrocystin is expressed on primary cilia. *Hum. Mol. Genet.* **12**, 2703–2710.

Watnick, T.J., and Germino, G. (2003). From cilia to cyst. *Nat. Genet.* **34**, 355–366.

Watnick, T.J., Jin, Y., Matunis, E., Kernan, M.J., and Montell, C. (2003). A flagellar polycystin-2 homolog required for male fertility in drosophila. *Curr. Biol.* **13**, 2179–2184.

Watnick, T.J., Torres, V.E., Gandolph, M.A., Qian, F., Onuchic, L.F., Klinger, K.W., Landes, G., and Germino, G.G. (1998). Somatic mutation in individual liver cysts supports a two-hit model of cystogenesis in autosomal dominant polycystic kidney disease. *Mol. Cell* **2**, 247–251.

Wilson, C., Idziaszczyk, S., Parry, L., Guy, C., Griffiths, D.F., Lazda, E., Bayne, R.A., Smith, A.J., Sampson, J.R., and Cheadle, J.P. (2005). A mouse model of tuberous sclerosis 1 showing background specific early post-natal mortality and metastatic renal cell carcinoma. *Hum. Mol. Genet.* **14**, 1839–1850.

Wolf, M.T., Lee, J., Panther, F., Otto, E.A., Guan, K.L., and Hildebrandt, F. (2005). Expression and phenotype analysis of the nephrocystin-1 and nephrocystin-4 homologs in *Caenorhabditis elegans. J. Am. Soc. Nephrol.* **16**, 676–687.

Yoder, B.K., Hou, X., and Guay-Woodford, L.M. (2002). The polycystic kidney disease proteins, polycystin-1, polycystin-2, polaris, and cystin, are co-localized in renal cilia. *J. Am. Soc. Nephrol.* **13**, 2508–2516.

Zhang, M.Z., Mai, W., Li, C., Cho, S.Y., Hao, C., Moeckel, G., Zhao, R., Kim, I., Wang, J., Xiong, H., Wang, H., Sato, Y., *et al.*, (2004). PKHD1 protein encoded by the gene for autosomal recessive polycystic kidney disease associates with basal bodies and primary cilia in renal epithelial cells. *Proc. Natl. Acad. Sci. USA* **101**, 2311–2316.

CHAPTER 15

Constructing and Deconstructing Roles for the Primary Cilium in Tissue Architecture and Cancer

E. Scott Seeley[*,†] ***and*** **Maxence V. Nachury**[†]

[*]Department of Pathology, Stanford University, Stanford, California 94305

[†]Department of Molecular and Cellular Physiology, Stanford University, Stanford, California 94305

Abstract

Primary cilia are exquisitely designed sensory machines that have evolved at least three distinct sensory modalities to monitor the extracellular environment. The presence and activation of growth factor, morphogen, and hormone receptors within the confines of the ciliary membrane, the intrinsic physical relationship between the ciliary axoneme and the centriole, and the preferential assembly of primary cilia on the apical surfaces of tissue epithelia highlight the importance of this

978-0-12-375024-2
DOI: 10.1016/S0091-679X(08)94015-2

organelle in the establishment and maintenance of tissue architecture and homeostasis. Accordingly, recent studies begin to suggest roles for these organelles in oncogenesis and tumor suppression. Here, we review the sensory properties of primary cilia, assess the "history" of the primary cilium in cancer, and draw upon recent findings in a discussion of how the primary cilium may influence tissue architecture and neoplasia.

I. Introduction

Primary cilia are assembled by many cell types within the human body (http://www.bowserlab.org/primarycilia/cilialist.html) and in some cases during select stages of development (Marion *et al.*, 2009; see also Chapter 1 by Bloodgood, this volume). Their unique topology provides a highly specialized surface area which may be fully dedicated to the detection, transformation, and relay of external cues to the cell body. In addition, there are likely to exist important relationships between the cilium and the cell cycle, as would be suggested by the dynamic assembly and resorption patterns seen during the cell cycle (Fonte *et al.*, 1971; Tucker *et al.*, 1979) and by the intrinsic relationship between cilium and centriole (reviewed in Seeley and Nachury, 2009). Accordingly, defective cilium assembly or function can lead to pleiotropic disorders of tissue architecture and proliferation, including polycystic kidney disease (Pazour *et al.*, 2000; Qin *et al.*, 2001; Yoder *et al.*, 2002), Bardet–Biedl syndrome (BBS) (Ansley *et al.*, 2003; Nachury *et al.*, 2007), and Alstrom syndrome (Hearn *et al.*, 2005; Li *et al.*, 2007). While there have been few thorough studies of the behavior of this organelle in the setting of cancer, interest in the relationship between primary cilia and oncogenesis is growing. Here, we review the remarkable sensory properties of primary cilia, the classes of receptor complexes that are found within the ciliary membrane, and how primary cilium function and dysfunction influences tissue homeostasis, architecture, and potentially, the development of cancer.

II. Sensory Modalities

Primary cilia have evolved distinct structural properties and at least three distinct sampling modalities to detect and transmit a wide variety of different stimuli to the interior of the cell. The primary cilium has the capacity to sense flow and mechanical stress via classes of mechanosensory calcium channels present in the ciliary membrane (Praetorius and Spring, 2001; Yoder *et al.*, 2002), low-abundance ligands via resident, noncycling receptors (Huang *et al.*, 2007), and changes in ligand concentrations by utilizing intraflagellar transport (IFT), a bidirectional motility seen along the length of cilia and flagella, or other transport machinery to provide continuous sampling (Huang *et al.*, 2007; Rohatgi *et al.*, 2007). Additional mechanisms for "tuning" might occur through regulated delivery of ciliary signals at the level of retrograde trafficking or cilium resorption. Together, these properties render the primary cilium exceptionally well equipped to enforce tissue homeostasis and to stimulate adaptive changes in tissue architecture.

III. Flow and Mechanosensation

The ability to detect and monitor flow rates and shear forces permits the function and adaptation of kidney tubules and the ducts of liver and pancreas. The conduction of fluids from one location to another by these cellular tubes is life-sustaining and must be preserved in the face of occlusion. How then, might these systems adapt to preserve flow should blockage arise? Ideally, a luminal detection system would be in place to detect diminished flow rates and increase tubule caliber. The primary cilium contains flow-responsive calcium channels that, when activated, flux calcium ions to the cytoplasm (Praetorius and Spring, 2001). The many functions of intracellular calcium ions include stimulation of cytoskeletal remodeling and cell division, two activities which can lead to increased tubule caliber. In the setting of constant unidirectional flow, such as that of urine in the nephron, the flow-responsive calcium channels on the upstream face of the cilium, those that are constantly subjected to lateral tension and shear forces, are likely silent secondary to stimulation-dependent desensitization (Fig. 1). In contrast, those of the downstream face are likely silent due to lack of lateral tension and exposure to flow (Fig. 1). Thus, during continual unidirectional flow, the primary cilia of the nephron are inactive with regard to calcium flux. In the setting of occlusion, fluid backup, reversal of flow, and increased fluid pressure within the system would reverse cilium orientation and increase lateral tension along the newly upstream face of the ciliary membrane, triggering calcium influx (Fig. 1). Depending on context, such an event may lead to cellular lengthening and cell division. In this way, the cilium may increase tubule caliber and restore flow in the face of occlusion. Similarly, stimulation of the primary cilia of the vasculature leads to calcium influx and to nitric oxide production, two second messengers that increase vessel caliber (Abou Alaiwi *et al.*, 2009). In accord with these concepts, the loss of the regulatory environment provided by the cilium results in unabated calcium influx in renal tubule cells and to tissue pathology otherwise seen following irreversible tubule occlusion (Cano *et al.*, 2004; Pazour *et al.*, 2000; Zhang *et al.*, 2005).

IV. Low-Abundance Ligands

In many circumstances, trace amounts of ligand are utilized to coordinate homeostatic signaling. Receptors, including growth factor and hormone receptors, are often highly concentrated within the ciliary membrane (Ma *et al.*, 2005;

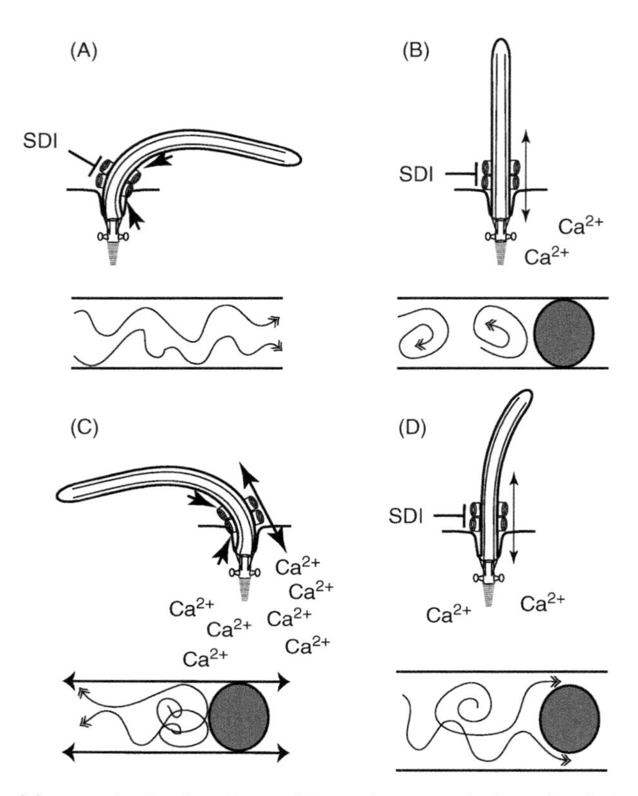

Fig. 1 Model of flow sensing by the primary cilium and response in face of occlusion. Upper images depict cilium orientation and lower images depict direction of flow in presence and absence of occlusion. (A) In setting of continuous flow, downstream calcium channels are inactive due to minimal lateral membrane tension (inward arrows) and upstream calcium channels are inactive due to stimulus-dependent inhibition (SDI). (B, C) Following tubule occlusion, flow ceases, primary cilia begin to deflect and increased lateral membrane tension (outward arrows) activates calcium signaling, lengthening of the tubule epithelium, increasing tubule caliber, and leading to restoration of flow (D).

Schneider *et al.*, 2005), permitting the accumulation of ligand–receptor complexes on the ciliary membrane. Further, there is evidence to suggest that the cilium can immobilize certain receptor populations within its membrane while subjecting others to continuous trafficking in and out of the organelle (Huang *et al.*, 2007). The ability to concentrate and immobilize receptors within the ciliary membrane maximizes sensitivity and at the same time distances high concentrations of catalytic receptor complexes from important cytoplasmic effectors, minimizing the potential consequences of aberrant receptor "firing." To activate their cytoplasmic or nuclear effectors, activated ciliary receptors must be delivered to the cell body by IFT, other ciliary trafficking systems, or cilium resorption, offering additional steps that may increase fidelity and minimize aberrant signal transduction (Fig. 2).

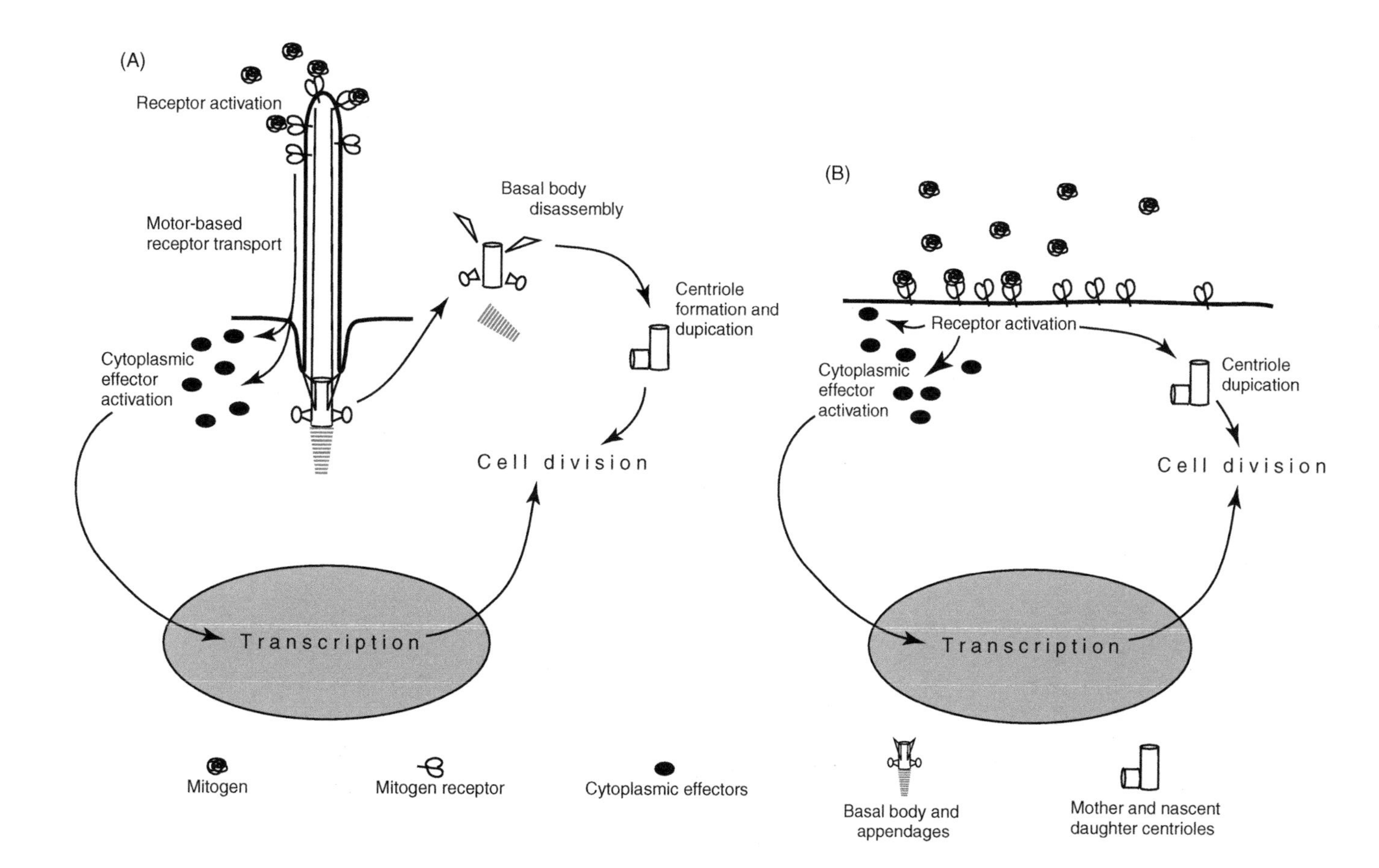

Fig. 2 Model of cell cycle events in relation to presence or absence of the primary cilium. (A) In the presence of a primary cilium mitogens bind to ciliary receptors and activated receptor complexes are trafficked the length of the cilium into the cell body where they activate cytoplasmic effectors and subsequently, the transcription of genes required for cell cycle progression. Simultaneously, basal body appendages are disassembled to permit centriole duplication and mitosis. (B) In the absence of a primary cilium, activated mitogenic receptor complexes require minimal trafficking to activate their cytoplasmic effectors and basal body disassembly is not required for centriole duplication to proceed.

V. Concentration Shifts

A third sensory modality made available by the primary cilium relates to IFT and the presence of receptors which cycle in and out of the ciliary membrane, permitting continuous sampling of the extracellular environment (Huang *et al.*, 2007; Rohatgi *et al.*, 2007). Here, the rate-of-receipt of occupied or activated receptor complexes relayed from the cilium and detected in the cell body can serve to adjust or activate signaling pathways important for adaptive or developmental programs (Fig. 2). Intriguingly, IFT is quite dynamic and the rate of entry and exit of IFT particles in and out of the cilium can vary markedly (Pan and Snell, 2005), providing for continuous tuning of ciliary sampling rates according to ligand and physiological context.

VI. Ciliary Receptors

Clearly, the structural and biochemical features of the primary cilium convey a marked capacity for sensory activities. What types of receptors take advantage of the flexibility and fidelity conveyed by this unique signaling platform? Growth factor, morphogen, odorant, and hormone receptors, ion channels, and multipass G-protein-coupled receptors (GPCRs) have all been localized to the ciliary membrane (Marshall and Nonaka, 2006). At present, due to the lack of a primary cilium preparation of high purity and yield, the full complement of ciliary receptor modules has yet to be enumerated. However, targeted analyses have yielded an already remarkable diversity, and new methods are being developed for primary cilia isolation and purification (see Chapter 4 by Mitchell *et al.* and Chapter 5 by Huang *et al.*, this volume).

VII. Growth Factor Receptors

Growth factor receptors are coupled to mitogenic signaling cascades and ion channels which regulate cell growth and proliferation. Both epidermal growth factor receptor (EGFR) and platelet-derived growth factor receptor-alpha alpha (PDGFR-$\alpha\alpha$) are known to localize to the ciliary membrane (Schneider *et al.*, 2005; Ma *et al.*, 2005). Following serum starvation, PDGFR-$\alpha\alpha$ is present nearly exclusively (Schneider *et al.*, 2005) and EGFR markedly enriched (Ma *et al.*, 2005) in the ciliary membrane.PDGFR-$\alpha\alpha$ ligation leads to the rapid appearance of phosphotyrosine and phospho-Mek1/2 along along the length of the cilium (Schneider *et al.*, 2005). In the absence of the primary cilium, PDGFR-$\alpha\alpha$ fails to accumulate in any membrane compartment (Schneider *et al.*, 2005). Together, these findings suggest that the cilium functions as the exclusive site of PDGFR-$\alpha\alpha$ activation. In addition, the appearance of phospho-Mek1/2 in the cilium suggests that intermediate components of the PDGFR signaling cascade, such as Ras, Raf, and PKC, are also present within the primary cilium.

While EGFR has also been shown to localize to primary cilia (Ma *et al.*, 2005), the significance of its ciliary localization is unclear at present. At a minimum, there is evidence to suggest that stimulation of cilium-localized EGFR leads to calcium flux via the polycystin calcium channel (Ma *et al.*, 2005). Whether downstream kinase activation occurs within the cilium or whether ciliary activation of the EGFR is sufficient to induce cytoplasmic signal transduction remain to be seen, however. In this regard, it would be interesting to learn whether ciliary EGFR signaling induces cellular responses distinguishable from those of plasma membrane-associated signaling.

While other canonical growth factor receptors have not yet been found within primary cilia, fibroblast growth factor has been shown to modulate cilium length (Neugebauer *et al.*, 2009). It will no doubt be of interest to learn the complete repertoire of ciliary growth factor receptors and to develop a better understanding of how these proteins are trafficked to and within the cilium.

VIII. Morphogens

Morphogens such as those of the Wnt and Hedgehog (Hh) families mediate important developmental switches and can influence cell fate and nascent tissue architecture over long distances (see also Chapter 10 by Goetz *et al.*, this volume). While not appreciated for many years, the discovery and biochemical characterization of IFT led rapidly to the realization that primary cilia have central roles in mammalian morphogen signaling (Huangfu, 2003).

In a phenotypic screen designed to identify genes important for hedgehog signaling in mice, Anderson and coworkers found that mutation of IFT172 gave rise to neural tube and other defects seen in hedgehog mutants (Huangfu, 2005). Subsequent work from many additional groups has shown that Smoothened (Smo), Patched (Ptch), and Gli family transcription factors all localize to the primary cilium and seem to traffic dynamically in and out of the organelle in a Hh-dependent manner (Corbit *et al.*, 2005; Haycraft *et al.*, 2005; Rohatgi *et al.*, 2007). Defects in cilium assembly prevent Gli processing and activation of Gli activator and repressor functions in response to Hh signals (Haycraft *et al.*, 2005; Liu *et al.*, 2005). A current model of the ciliary Hh signaling posits that Hh–Ptch interactions relieve Ptch-mediated inhibition of Smo, leading to ciliary Smo accumulation and to the activation of Gli processing and function (Rohatgi *et al.*, 2007). While the molecular details of the mechanisms of ciliary Hh pathway activation remain to be fully elucidated, they clearly evidence the remarkable dynamics of regulated membrane traffic occurring between the primary cilium and other cellular compartments. Important issues to be further addressed are whether alternative pathways of Gli processing might function in the absence of a primary cilium *in vivo*, which transport systems traffic Hh pathway components to and within the primary cilium, and why primary cilia are required to modulate Hh signaling in vertebrates but not in lower organisms such as insects.

In contrast to the hedgehog pathway, the relationship between primary cilia and Wnt signaling is less clearly resolved. On one hand, Wnt receptors, APC (a negative regulator of Wnt signaling), and β-catenin (a Wnt effector) have all been localized along the length and at the base of primary cilia (Marion *et al.*, 2009; Ross *et al.*, 2005). Interestingly, the loss of the primary cilium via mutations affecting the basal body or IFT has been suggested to augment canonical Wnt signaling, induce increased nuclear accumulation of β-catenin, and increase cellular responsiveness to soluble Wnts (Cano *et al.*, 2004; Corbit *et al.*, 2008). Further, the expression of Wnt ligands is increased markedly, possibly via an autocrine mechanism, in cilium assembly mutants (Jonassen *et al.*, 2008). On the other hand, more recent studies in murine embryos and zebrafish do not support a role for the cilium in the transmission of canonical Wnt signals (Huang and Schier, 2009; Ocbina *et al.*, 2009). Overall, it is likely that a certain degree of context dependence accounts for these diverse findings and much work remains to be done.

IX. Hormone and G-Protein-Coupled Receptors

While the relationship between the cilium and growth factor and morphogen receptors is becoming clear, less is known regarding the influence of ciliary localization on signaling by the olfactory and hormone receptors that are found there. While serotonin and somatostatin receptors localize to the ciliary membrane, it is unknown whether cilium dysfunction perturbs signaling by their effector complexes. While leptin signaling is perturbed in BBS mutants, it is unknown whether the leptin receptor localizes to cilia. Overall, there should be a great deal to be learned through the identification of ciliary receptor complexes, including the uncharacterized GPCRs that are known to reside in the ciliary membrane (Marshall and Nonaka, 2006).

X. Tissue Homeostasis and Architecture

The important relationship between primary cilia and development can be deduced from the complement of signaling receptors found within the ciliary membrane, the stimulus-dependent trafficking of morphogen receptors and their downstream effectors in and out of the primary cilium, the deregulation of growth and developmental pathways in cilium assembly mutants, and the remarkable pleiotropism of heritable developmental defects that are seen in the setting of cilium dysfunction in humans.

As described earlier, the sensory capacities of the primary cilium provide means to shape tissue architecture in response to environmental challenge and developmental cues. In humans, heritable mutations in genes required for cilium assembly or in those encoding ciliary receptors leads to disordered epithelial proliferation and to the kidney, pancreas, and liver cysts seen in polycystic kidney disease (see also Chapter 14 by Menezes and Germino, this volume). Mutations affecting cilium function also lead to BBS and Alstrom syndrome, two pleiotropic disorders encompassing cyst formation, dilated cardiomyopathy, short stature, type II diabetes, morbid obesity, hypertension, mental retardation, and skeletal abnormalities, and to a host of other rare heritable disorders of development (Sharma *et al.*, 2008). What general concepts may explain how cilium dysfunction might lead to such developmental defects?

Growth-promoting ligand receptors which are normally sequestered within the primary cilium accumulate in other cellular compartments or are destabilized in cilium assembly mutants (Lin *et al.*, 2003; Ma *et al.*, 2005; Schneider *et al.*, 2005). EGFR is known to accumulate within primary cilia and the loss of Kif3a, a component of the anterograde IFT motor complex, prevents cilium assembly and leads to EGFR accumulation and persistent mitogen-activated protein kinase (MAPK) activity (Cano *et al.*, 2006; Lin *et al.*, 2003). It is likely that, in addition to participating in calcium signaling, growth factor receptors such as EGFR and PDGFR-$\alpha\alpha$ are sequestered to the cilium during quiescence to prevent aberrant signaling, to couple receptor activation to one or more of the sensory modalities of the cilium, and to impose a requirement for cilium resorption or retrograde ciliary trafficking on the activation of cytoplasmic and nuclear targets. Similarly, in the absence of a primary cilium, polycystin 2 accumulates in the apical plasma membrane of kidney tubule cells, leading to unabated and deregulated calcium signaling (Siroky, 2006). Together, these findings suggest that the loss of the cilium can lead to redistribution of its membrane effectors and subsequently, to aberrant signal transduction. It is tempting to speculate that signaling via this mechanism may account for the increased levels of c-Myc that are also seen in cilium assembly mutants (Jonassen *et al.*, 2008; Lin *et al.*, 2003).

The cilium may also function to coordinate the downregulation of certain signaling molecules. Proteasomes localize about basal bodies and centrosomes and a number of ciliary proteins, including β-catenin, are known to be downregulated via proteasomal degradation (Aberle *et al.*, 1997; Fabunmi *et al.*, 2000; Gerdes *et al.*, 2007). It is possible that the cilium provides a conduit for trafficking to the proteasome and, in this way, the loss of the cilium could lead to persistent signaling due to the accumulation of proteasome targets such as β-catenin. Such a mechanism could contribute to the increase in Wnt signaling seen in cilium assembly mutants.

Overall, while it is clear that defects of cilium assembly or ciliary receptors lead to dramatic alterations in tissue architecture, the specific mechanisms involved have yet to be revealed and we are left to speculate. It is intriguing however, that the loss of the primary cilium seems to recapitulate architectural phenotypes that might be expected were certain of the known receptors of its membrane to signal continuously.

XI. Cancer

There are compelling reasons to believe that primary cilia can stimulate or enhance cancer development. Because primary cilia are required for Hh to activate downstream Gli transcription factors, the loss of this organelle should repress the development of hedgehog-dependent cancers bearing mutations that lie upstream of the Glis but not affect the development of tumors bearing activating mutations in the Gli transcription factors themselves. Indeed, a pair of recent studies (Han *et al.*, 2009; Wong *et al.*,

2009) has shown that the development of Smo-dependent neoplasms requires the presence of primary cilia while those that depend on mutant Gli2 are enhanced by mutations that preclude primary cilium assembly. In the case of medulloblastoma, the presence of cilia amongst a group of human tumors correlated with the presence of Wnt or Hh pathway mutations, in accord with roles for the cilium in modulation of these pathways (Han *et al.*, 2009).

While the primary cilium can modulate the development of hedgehog-dependent neoplasms, the relationship between this organelle and the development of more common malignancies is less clear. The loss of genes encoding proteins required directly for cilium assembly leads to epithelial hyperplasia, metaplasia, and cystic disease; defects in apico-basal cell and planar cell polarity; persistent Wnt, TGF-β, and MAPK signaling; EGFR and c-Myc accumulation; and deregulated Hh signaling (Cano *et al.*, 2004, 2006; Corbit *et al.*, 2008; Jonassen *et al.*, 2008; Lin *et al.*, 2003; Pazour *et al.*, 2000). Further, the loss of the cilium may remove requirements for cilium resorption, basal body disassembly, and receptor transport steps from the cell division cycle (Fig. 2). As such, it would be expected that loss or dysfunction of the cilium may stimulate the development of more common epithelial malignancies.

What then, is the fate and significance of the primary cilium during cancer development? No clear picture has yet emerged. The majority of instances in which this topic is treated in the literature occur as reports of pathological oddities rather than careful analyses involving significant numbers of patient specimens. While it has been suggested that the loss of the primary cilium is a well-known consequence of cellular transformation (Plotnikova *et al.*, 2008; Pugacheva *et al.*, 2007), an extensive review of this topic (Wheatley, 1995) states, "it is already well known that a wide variety of tumours possess cells with primary cilia…several cases have been reported of malignant cell types apparently producing cilia after malignant transformation when their 'normal counterparts' were supposedly unciliated, as in the case of hepatocytes of the LI-IO line," and, "…transformed variants of the otherwise well-behaved cell line, 3T3 still exhibit, however, the same high incidence of ciliation at confluency as the progenitor, untransformed 3T3 line."

An additional problem relates to the significance of a loss of primary cilia, were it to occur as a regular feature of neoplastic transformation. Primary cilia exhibit dynamic patterns of assembly and disassembly as they progress through the cell cycle (Tucker *et al.*, 1979). There exist examples of cell types in which ciliation seems to occur only during cell cycle arrest (Dingemans, 1969) and others in which cilia are present during interphase (Fonte *et al.*, 1971; Tucker *et al.*, 1979). As cilium resorption can thus occur as a physiologic consequence of cell cycle progression, any loss of this organelle in cancer may merely represent the enhanced rates of cell division that are seen in this setting. This issue has been almost completely ignored by recent studies, including those attempting to link mutations of the Von Hipple Lindau (VHL) tumor suppressor to defects in ciliogenesis (Esteban, 2006; Lutz, 2006) and more recent studies examining the presence of cilia in clinical medulloblastoma and basal cell carcinoma specimens (Han *et al.*, 2009; Wong *et al.*, 2009). Indeed, careful studies demonstrate that VHL deletion alone has no effect on the ciliation of renal epithelial cells and that phosphatase and tensin homologue (PTEN) or glycogen synthase kinase 3 beta (GSK-β) must also be deleted to block cilium assembly (Frew *et al.*, 2008; Thoma *et al.*, 2007). While the potential involvement of proliferative deciliation was not addressed for GSK3-β/VHL codeletion, PTEN/VHL codeletion lead to a five- to sixfold increase in proliferation (Frew *et al.*, 2008). Thus, it is very possible that the loss of primary cilia in these settings is a physiological consequence of enhanced proliferation.

To address these issues, we recently examined the fate of the primary cilium in the setting of pancreatic ductal adenocarcinoma (PDA) (Seeley *et al.*, 2009). Here, in 17 patients and in murine developmental models, we found that primary cilia were completely absent from the neoplastic epithelium (Fig. 3) (Seeley *et al.*, 2009) but

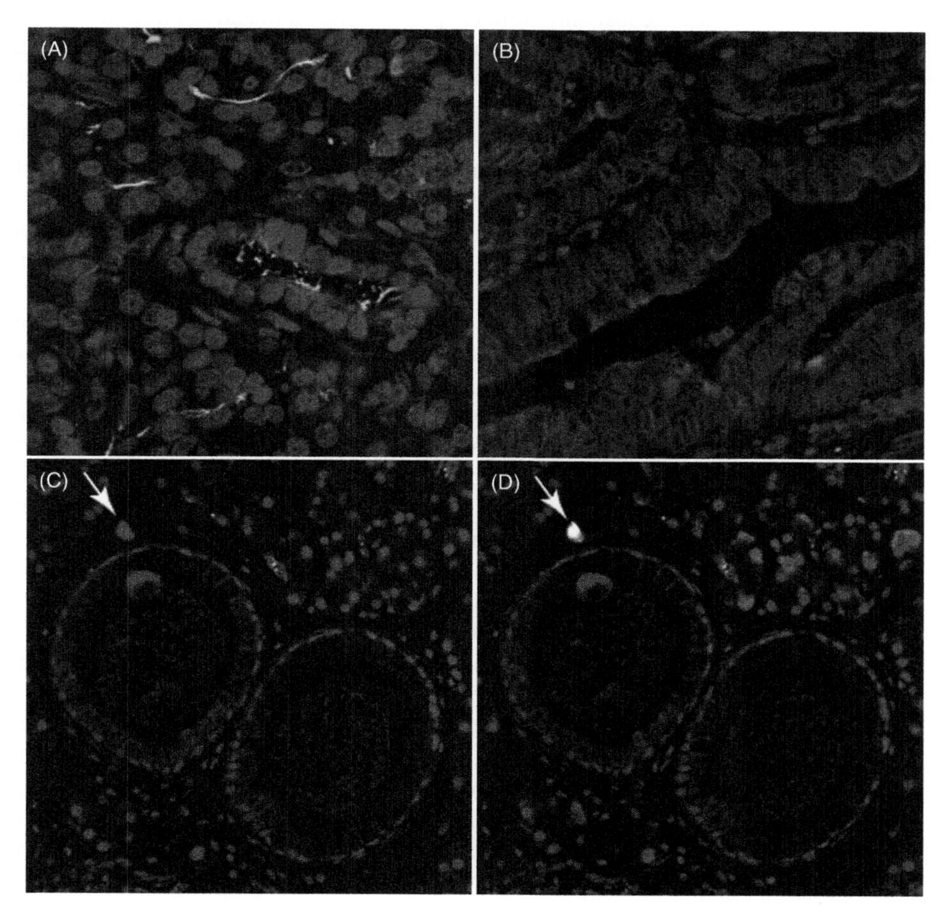

Fig. 3 Primary cilium assembly is suppressed in pancreatic cancer. (A) Normal pancreatic duct. (B) Well-differentiated pancreatic ductal adenocarcinoma in adjacent region of same section as shown in "B." In "B" and "C", acetylated tubulin (primary cilia) staining is shown in white. (C) Pair of murine PanIN-1 complexes, acetylated tubulin (cilia) shown in white. (D) Same lesion as in C but with Ki67 staining shown in white and indicated by arrow.

not the stroma of PDA lesions (E. S. Seeley, unpublished observation). The majority of cases of PDA are thought to initiate via the formation of preinvasive neoplasms termed Pancreatic Intraepithelial Neoplasia (PanIN) and we were likewise unable to identify a single ciliated neoplastic epithelium in any of these types of lesions. In agreement with several prior studies, 97 and 60% of the neoplastic epithelial cells in the earliest appearing PanINs and in PDA, respectively, were quiescent as assessed by Ki67, proliferating cell nuclear antigen (PCNA), and BrdU labelling. In contrast, the epithelia of chronic pancreatitis and acinar to ductal metaplasia were not associated with primary cilium loss but were instead associated with highly elongated cilia.

Interestingly, when oncogenic Kras was expressed in all pancreatic lineages, primary cilia were lost only from neoplastic epithelium and not from untransformed islets, ducts, centroacinar cells, or acinar–ductal metaplasms. Finally, while confluence and serum deprivation-induced quiescence did not restore cilium assembly in murine PDA cells, the addition of PI3 kinase inhibitors induced robust ciliation in both contexts. We draw several conclusions from this study. One, a pathologic and reversible suppression of cilium assembly occurs as a constant feature of pancreatic cancer. Two, primary cilia depletion can occur as a consequence of oncogenic signaling in the setting of cell cycle arrest and independent of dominant mutations in genes required for organelle assembly. Three, cilium depletion is likely to represent an intrinsic feature of pancreatic epithelium transformation. While it is not known whether the loss of genes required for cilium assembly might enhance PDA development or how the assembly of the organelle is directly suppressed in PDA, our findings suggest the primary cilium functions as a tumor suppressor organelle.

While it has been shown that certain Hh-dependent cancers may require primary cilia, more likely the primary cilium functions as a general tumor suppressor in common epithelial malignancies such as those of pancreas, liver, and prostate. The primary cilium contains growth-promoting receptor complexes which, if not segregated from their cytoplasmic effectors by placement in the cilium, would likely end up in cellular compartments such as the recycling endocome or plasma membrane. Here, these normally cilium-resident proteins would be more prone to aberrant activation and likely to activate downstream cytoplasmic effectors. In the setting of a 'stimulatory' cytoplasm, as in the context of Ras or other pro-oncogenic mutations, this scenario may have dire consequences for the cell. In this way, mistargeting of cilium-resident proteins would provide easy access to substrates necessary for transformation by Ras oncogenes. Accordingly, cilium depletion and neoplasm formation are inseparable when oncogenic Kras is expressed throughout the pancreas.

Finally, akin to the studies of Wong and Han, the pattern of primary cilium loss in PDA provides predictions regarding the influence of Hh and Wnt signaling in this malignancy. Primary cilia are present on stromal cells (E. S. Seeley, unpublished observation), absent from the cancerous epithelium, and are required for Hh stimulation of the Hh pathway (Haycraft *et al.*, 2005). Thus, Hh pathway activation should occur only in the stromal compartment of PDA. Indeed, several recent studies have shown that, while Hh is expressed in the neoplastic epithelium of PDA, pathway activation seems to occur solely in the stroma (Theunissen and de Sauvage, 2009). In contrast, the fact that cilium dysfunction leads to enhanced Wnt activation and secretion (Cano *et al.*, 2004; Corbit *et al.*, 2008; Jonassen *et al.*, 2008) predicts that Wnt signaling would be enhanced in both the stroma and neoplastic epithelium of PDA. This too has recently been demonstrated (Pasca di Magliano *et al.*, 2007; Wang *et al.*, 2009). While the significance of these geographical patterns of activity and their potential interdependence remains unclear, they fit well with the pattern of cilium loss in PDA.

XII. Summary

In all, it is not difficult to imagine that primary cilium dysfunction may play a role in the development of malignancies which arise from ductular epithelia such as those of the liver, kidney, prostate, and pancreas. Either directly or indirectly, the assembly and function of this organelle is likely, like other organelles, to require the function of several hundred gene products. Thus, a great number of transcriptional perturbations or mutations or combinations thereof could lead to frequent cilium dysfunction in the setting of cancer. From this perspective, it would be perhaps surprising if primary cilium function was not frequently altered in some way during oncogenesis. In this regard, it will be particularly interesting to learn if cilium integrity is required only for cancers which depend on hedgehog signaling or instead, whether the genetic diversity of cancers is reflected by distinct patterns of ciliation across the spectrum of human malignancy.

References

Aberle, H., Bauer, A., Stappert, J., Kispert, A., and Kemler, R. (1997). Beta-catenin is a target for the ubiquitin-proteasome pathway. *EMBO J.* **16**, 3797–3804.

Abou Alaiwi, W.A., Takahashi, M., Mell, B.R., Jones, T.J., Ratnam, S., Kolb, R.J., and Nauli, S.M. (2009). Ciliary polycystin-2 is a mechanosensitive calcium channel involved in nitric oxide signaling cascades. *Circ. Res.* **104**, 860–869.

Ansley, S.J., Badano, J.L., Blacque, O.E., Hill, J., Hoskins, B.E., Leitch, C.C., Kim, J.C., Ross, A.J., Eichers, E.R., Teslovich, T.M., Mah, A.K., Johnsen, R.C., Cavender, J.C., Lewis, R.A., Leroux, M.R., Beales, P.L., and Katsanis, N. (2003). Basal body dysfunction is a likely cause of pleiotropic Bardet-Biedl syndrome. *Nature* **425**, 628–633.

Cano, D.A., Murcia, N.S., Pazour, G.J., and Hebrok, M. (2004). Orpk mouse model of polycystic kidney disease reveals essential role of primary cilia in pancreatic tissue organization. *Development* **131**, 3457–3467.

Cano, D.A., Sekine, S., and Hebrok, M. (2006). Primary cilia deletion in pancreatic epithelial cells results in cyst formation and pancreatitis. *Gastroenterology* **131**, 1856–1869.

Corbit, K., Shyer, A., Dowdle, W., Gaulden, J., Singla, V., and Reiter, J. (2008). Kif3a constrains β-catenin-dependent Wnt signalling through dual ciliary and non-ciliary mechanisms. *Nat. Cell Biol.* **10**, 70–76.

Corbit, K.C., Aanstad, P., Singla, V., Norman, A.R., Stainier, D.Y., and Reiter, J.F. (2005). Vertebrate Smoothened functions at the primary cilium. *Nature* **437**, 1018–1021.

Dingemans, K.P. (1969). The relation between cilia and mitoses in the mouse adenohypophysis. *J. Cell Biol.* **43**, 361–367.

Esteban, M. (2006). Formation of primary cilia in the renal epithelium is regulated by the von Hippel-Lindau tumor suppressor protein. *J. Am. Soc. Nephrol.* **17**, 1801–1806.

Fabunmi, R.P., Wigley, W.C., Thomas, P.J., and DeMartino, G.N. (2000). Activity and regulation of the centrosome-associated proteasome. *J. Biol. Chem.* **275**, 409–413.

Fonte, V.G., Searls, R.L., and Hilfer, S.R. (1971). The relationship of cilia with cell division and differentiation. *J. Cell Biol.* **49**, 226–229.

Frew, I.J., Thoma, C.R., Georgiev, S., Minola, A., Hitz, M., Montani, M., Moch, H., and Krek, W. (2008). pVHL and PTEN tumour suppressor proteins cooperatively suppress kidney cyst formation. *EMBO J.* **27**, 1747–1757.

Gerdes, J., Liu, Y., Zaghloul, N., Leitch, C., Lawson, S., Kato, M., Beachy, P., Beales, P., Demartino, G., Fisher, S., Badano, J., and Katsanis, N. (2007). Disruption of the basal body compromises proteasomal function and perturbs intracellular Wnt response. *Nat. Genet.* **39**, 1350–1360.

Han, Y.G., Kim, H.J., Dlugosz, A.A., Ellison, D.W., Gilbertson, R.J., and Alvarez-Buylla, A. (2009). Dual and opposing roles of primary cilia in medulloblastoma development. *Nat. Med.* **15**, 1062–1065.

Haycraft, C., Banizs, B., Aydin-Son, Y., Zhang, Q., Michaud, E., and Yoder, B. (2005). Gli2 and Gli3 localize to cilia and require the intraflagellar transport protein polaris for processing and function. *PLoS Genet.* **1**, e53.

Hearn, T., Spalluto, C., Phillips, V.J., Renforth, G.L., Copin, N., Hanley, N.A., and Wilson, D.I. (2005). Subcellular localization of ALMS1 supports involvement of centrosome and basal body dysfunction in the pathogenesis of obesity, insulin resistance, and type 2 diabetes. *Diabetes* **54**, 1581–1587.

Huang, K., Diener, D., Mitchell, A., Pazour, G., Witman, G., and Rosenbaum, J. (2007). Function and dynamics of PKD2 in Chlamydomonas reinhardtii flagella. *J. Cell Biol.* **179**, 501–514.

Huang, P., and Schier, A.F. (2009). Dampened Hedgehog signaling but normal Wnt signaling in zebrafish without cilia. *Development* **136**, 3089–3098.

Huangfu, D., Liu, A., Rakeman, A.S., Murcia, N.S., Niswander, L., and Anderson, K.V. (2003). Hedgehog signalling in the mouse requires intraflagellar transport proteins. *Nature* **426**, 83–87.

Jonassen, J., San Agustin, J., Follit, J., and Pazour, G. (2008). Deletion of IFT20 in the mouse kidney causes misorientation of the mitotic spindle and cystic kidney disease. *J. Cell Biol.* **183**, 377–384.

Li, G., Vega, R., Nelms, K., Gekakis, N., Goodnow, C., Mcnamara, P., Wu, H., Hong, N., and Glynne, R. (2007). A role for Alström syndrome protein, Alms1, in kidney ciliogenesis and cellular quiescence. *PLoS Genet.* **3**, e8.

Lin, F., Hiesberger, T., Cordes, K., Sinclair, A.M., Goldstein, L.S., Somlo, S., and Igarashi, P. (2003). Kidney-specific inactivation of the KIF3A subunit of kinesin-II inhibits renal ciliogenesis and produces polycystic kidney disease. *Proc. Natl. Acad. Sci. USA* **100**, 5286–5291.

Liu, A., Wang, B., and Niswander, L.A. (2005). Mouse intraflagellar transport proteins regulate both the activator and repressor functions of Gli transcription factors. *Development* **132**, 3103–3111.

Lutz, M. (2006). Primary cilium formation requires von Hippel-Lindau gene function in renal-derived cells. *Cancer Res* **66**, 6903–6907.

Ma, R., Li, W.P., Rundle, D., Kong, J., Akbarali, H.I., and Tsiokas, L. (2005). PKD2 functions as an epidermal growth factor-activated plasma membrane channel. *Mol. Cell. Biolchem.* **25**, 8285–8298.

Marion, V., Stoetzel, C., Schlicht, D., Messaddeq, N., Koch, M., Flori, E., Danse, J.M., Mandel, J.L., and Dollfus, H. (2009). Transient ciliogenesis involving Bardet-Biedl syndrome proteins is a fundamental characteristic of adipogenic differentiation. *Proc. Natl. Acad. Sci. USA* **106**, 1820–1825.

Marshall, W.F., and Nonaka, S. (2006). Cilia: Tuning in to the cell's antenna. *Curr. Biol.* **16**, R604–R614.

Nachury, M.V., Loktev, A.V., Zhang, Q., Westlake, C.J., Peränen, J., Merdes, A., Slusarski, D.C., Scheller, R.H., Bazan, J.F., Sheffield, V., and Jackson, P.K. (2007). A core complex of BBS proteins cooperates with the GTPase Rab8 to promote ciliary membrane biogenesis. *Cell* **129**, 1201–1213.

Neugebauer, J.M., Amack, J.D., Peterson, A.G., Bisgrove, B.W., and Yost, H.J. (2009). FGF signalling during embryo development regulates cilia length in diverse epithelia. *Nature* **458**, 651–654.

Ocbina, P.J., Tuson, M., and Anderson, K.V. (2009). Primary cilia are not required for normal canonical Wnt signaling in the mouse embryo. *PLoS One* **4**, e6839.

Pan, J., and Snell, W.J. (2005). Chlamydomonas shortens its flagella by activating axonemal disassembly, stimulating IFT particle trafficking, and blocking anterograde cargo loading. *Dev. Cell.* **9**, 431–438.

Pasca di Magliano, M., Biankin, A.V., Heiser, P.W., Cano, D.A., Gutierrez, P.J., Deramaudt, T., Segara, D., Dawson, A.C., Kench, J.G., Henshall, S.M., Sutherland, R.L., Dluosz, A., Rustgi, A.K., and Hebrok, M. (2007). Common activation of canonical Wnt signaling in pancreatic adenocarcinoma. *PLoS One* **2**, e1155.

Pazour, G.J., Dickert, B.L., Vucica, Y., Seeley, E.S., Rosenbaum, J.L., Witman, G.B., and Cole, D.G. (2000). Chlamydomonas IFT88 and its mouse homologue, polycystic kidney disease gene tg737, are required for assembly of cilia and flagella. *J. Cell Biol.* **151**, 709–718.

Plotnikova, O., Golemis, E., and Pugacheva, E. (2008). Cell cycle-dependent ciliogenesis and cancer. *Cancer Res* **68**, 2058–2061.

Praetorius, H.A., and Spring, K.R. (2001). Bending the MDCK cell primary cilium increases intracellular calcium. *J. Membr. Biol.* **184**, 71–79.

Pugacheva, E.N., Jablonski, S.A., Hartman, T.R., Henske, E.P., and Golemis, E.A. (2007). HEF1-dependent Aurora A activation induces disassembly of the primary cilium. *Cell* **129**, 1351–1363.

Qin, H., Rosenbaum, J.L., and Barr, M.M. (2001). An autosomal recessive polycystic kidney disease gene homolog is involved in intraflagellar transport in C. elegans ciliated sensory neurons. *Curr. Biol.* **11**, 457–461.

Rohatgi, R., Milenkovic, L., and Scott, M. (2007). Patched1 regulates Hedgehog signaling at the primary cilium. *Science* **317**, 372–376.

Ross, A., May-Simera, H., Eichers, E., Kai, M., Hill, J., Jagger, D., Leitch, C., Chapple, J., Munro, P., Fisher, S., Phillips, H., Leroux, M., Henderson, D., Murdoch, J., Copp, A., Eliot, M., Lupski, J., Kemp, D., Dolifus, H., Tada, M., Katsanis, N., Forge, A., and Beales, P. (2005). Disruption of Bardet-Biedl syndrome ciliary proteins perturbs planar cell polarity in vertebrates. *Nat. Genet.* **37**, 1135–1140.

Schneider, L., Clement, C., Teilmann, S.C., Pazour, G., Hoffmann, E.K., Satir, P., and Christensen, S. (2005). PDGFRalphaalpha signaling is regulated through the primary cilium in fibroblasts. *Curr. Biol.* **15**, 1861–1866.

Seeley, E.S., Carrière, C., Goetze, T., Longnecker, D.S., and Korc, M. (2009). Pancreatic cancer and precursor pancreatic intraepithelial neoplasia lesions are devoid of primary cilia. *Cancer Res* **69**, 422–430.

Seeley, E.S., and Nachury, M.V. (2009). The perennial organelle: Assembly and resorption of the primary cilium. *J. Cell Sci.*, in press.

Sharma, N., Berbari, N.F., and Yoder, B.K. (2008). Ciliary dysfunction in developmental abnormalities and diseases. *Curr. Top. Dev. Biol.* **85**, 371–427.

Siroky, B. (2006). Loss of primary cilia results in deregulated and unabated apical calcium entry in ARPKD collecting duct cells. *AJP: Renal Physiol.* **290**, F1320–F1328.

Theunissen, J.W., and de Sauvage, F.J. (2009). Paracrine Hedgehog signaling in cancer. *Cancer Res* **69**, 6007–6010.

Thoma, C., Frew, I., Hoerner, C., Montani, M., Moch, H., and Krek, W. (2007). pVHL and GSK3β are components of a primary cilium-maintenance signalling network. *Nat. Cell Biol.* **9**, 588–595.

Tucker, R.W., Pardee, A.B., and Fujiwara, K. (1979). Centriole ciliation is related to quiescence and DNA synthesis in 3T3 cells. *Cell* **17**, 527–535.

Wang, L., Heidt, D.G., Lee, C.J., Yang, H., Logsdon, C.D., Zhang, L., Fearon, E.R., Ljungman, M., and Simeone, D.M. (2009). Oncogenic function of ATDC in pancreatic cancer through Wnt pathway activation and beta-catenin stabilization. *Cancer Cell* **15**, 207–219.

Wheatley, D.N. (1995). Primary cilia in normal and pathological tissues. *Pathobiology* **63**, 222–238.

Wong, S.Y., Seol, A.D., So, P.L., Ermilov, A.N., Bichakjian, C.K., Epstein, E.H., Dlugosz, A.A., and Reiter, J.F. (2009). Primary cilia can both mediate and suppress Hedgehog pathway-dependent tumorigenesis. *Nat. Med.* **15**, 1055–1061.

Yoder, B.K., Hou, X., and Guay-Woodford, L.M. (2002). The polycystic kidney disease proteins, polycystin-1, polycystin-2, polaris, and cystin, are co-localized in renal cilia. *J. Am. Soc. Nephrol.* **13**, 2508–2516.

Zhang, Q., Davenport, J.R., Croyle, M.J., Haycraft, C.J., and Yoder, B.K. (2005). Disruption of IFT results in both exocrine and endocrine abnormalities in the pancreas of Tg737(orpk) mutant mice. *Lab. Invest.* **85**, 45–64.

SECTION IV

Posttranslational Modifications

CHAPTER 16

Polyglutamylation and the *fleer* Gene

Narendra H. Pathak ***and*** **Iain A. Drummond**

Department of Medicine and Genetics, Harvard Medical School and Nephrology Division, Massachusetts General Hospital, Charlestown, Massachusetts 02129

I. Introduction

The zebrafish *fleer* gene is a homolog of *Caenorhabditis elegans dyf-1* and encodes a tetratricopeptide repeat (TPR) repeat protein that is a component of the cilia proteome. *fleer* is essential for cilia function and its deficiency has been shown to reduce cilia tubulin glutamylation in zebrafish (Pathak *et al.*, 2007), disrupt OSM-3 kinesin-dependent assembly of the singlet microtubule of *C. elegans* amphid cilia (Ou *et al.*, 2005) and induce B-tubule structural defects (Pathak *et al.*, 2007). *fleer* is

METHODS IN CELL BIOLOGY, VOL. 94

978-0-12-375024-2
DOI: 10.1016/S0091-679X(08)94016-4

highly expressed in ciliated cells, localizes to cilia, and regulates tubulin glutamylation selectively within the cilia axoneme (Pathak *et al.*, 2007).

Polyglutamylation is a posttranslational tubulin modification that modulates charge-based protein interactions of microtubules by enhancing their electronegative character (Janke *et al.*, 2008). Polyglutamylated tubulins are abundant in stable microtubule-rich organelles; cilia, centrioles, and neuronal axons. Tubulin glutamylases and glycylases have been recently identified as related members of the Tubulin Tyrosine Ligase Like (TTLL) protein family. Individual TTLLs show distinct substrate preferences for tubulin subtypes and vary in their ability to initiate or elongate the amino acid side chains. Different TTLL proteins contribute to the microtubule heterogeneity proposed to be important for organelle-specific functions (van Dijk *et al.*, 2007). The Fleer protein lacks homology to TTLL proteins and must regulate tubulin glutamylation indirectly, possibly by acting as an adaptor protein that facilitates access or transport of TTLL protein(s) to the cilia compartment. In this review, we describe tubulin polyglutamylation with respect to cilia structure and function. We also provide a synopsis of zebrafish cilia defects in the *fleer* mutant as a reference for using zebrafish to study the in *vivo* significance and enzymatic mechanisms of tubulin glutamylation.

II. Significance of Polyglutamylation

Polyglutamylation is a posttranslational modification predominantly associated with tubulins in stable microtubule structures such as cilia, centrioles, neuronal axons, and mitotic spindles. Cilia are useful organelles for *in vivo* functional analysis of tubulin glutamylation since axonemal tubulin is highly modified. In addition, cilia are easily accessible, have a defined, elaborate microtubule architecture and a wealth of knowledge has been built through biochemical and genetic analysis of their formation and motility.

Cilia occur in a wide range of eukaryotic species from simple unicellular ciliates to the most complex vertebrates, and glutamylated tubulins have been detected in all ciliated species examined thus far. Glutamylated tubulin in cilia appears as a gradient with its maxima towards the basal bodies and decrease gradually toward the distal cilia tips (Pathak *et al.*, 2007). At the level of cilia ultrastructure, glutamylated tubulin antibodies locate near the B-tubules of the outer doublet microtubules and appear to be excluded from the transition zone and central pair (Lechtreck and Geimer, 2000). A functional role for tubulin glutamylation is indicated by the fact that hypoglutamylation induced by genetic defects of tubulins in *Tetrahymena* (Redeker *et al.*, 2005) or regulatory proteins like *fleer* (Pathak *et al.*, 2007) significantly decreases cilia motility and beat amplitude. Hyperglutamylation-induced arrest of cilia motility observed in *Tetrahymena* overexpressing the tubulin glutamylase Ttll6Ap indicates that a critical balance in the level of tubulin glutamylation is essential (Janke *et al.*, 2005). Inhibition of cilia motility by injection of antiglutamyl antibodies demonstrates that protein interactions of glutamylated tubulins control cilia beat amplitude (Gagnon *et al.*, 1996; Vent *et al.*, 2005). Glutamylation modulates tubulin–protein interactions by addition of acidic charges. Dynamic changes in tubulin glutamylation may be necessary for microtubule interactions with Microtubule Associated Proteins (MAPs), Tau, and kinesins to regulate cellular processes such as neurite extension, synaptic receptor redistribution, and processive movement of kinesin motors (Janke *et al.*, 2008).

Glutamylation is not exclusively associated with tubulins and has been recently implicated as a general modulator of protein function. A large repertoire of cellular proteins has recently been purified on the basis of their affinity to the glutamylation-specific antibody GT335 (van Dijk *et al.*, 2008). While the functional significance of nontubulin protein glutamylation is not well understood, putative glutamylation sites have been identified in nucleosome-associated protein NAP1 (Regnard *et al.*, 2000). Association of glutamylated cytoplasmic microtubules with proteins such as septins indicates these might be involved in vectorial vesicle-mediated delivery of proteins (Spiliotis *et al.*, 2008).

III. Biochemistry of Tubulin Glutamylation

Tubulin glutamylation is catalyzed by distinct enzymes on alpha- and beta-tubulin subtypes at a site near their C-terminal tails. These sites contain a stretch of glutamates that can be coupled to either glutamate or glycine residues by distinct glutamylases or glycylases. This overlap in glutamylation and glycylation sites, raises the possibility of competition between their distinct enzymes for access to tubulin substrates. Tubulin glutamylation proceeds by a two step process: (1) Initiation involves a gamma-carboxy linkage between two glutamate residues, one intrinsic to tubulin and the other in the side chain. (2) Elongation involves an alpha-carboxy linkage between the side chain glutamate to additional glutamates (Janke *et al.*, 2008).

Tubulin glutamylases are related members of the TTLL protein family and most of these genes have homologs in other species. Six of the 13 murine TTLLs (*ttll* 1, 4, 5, 6, 7, and 9) encode tubulin glutamylases, while three others (*ttll* 3, 8, and 10) encode tubulin glycine ligases (Rogowski *et al.*, 2009; van Dijk *et al.*, 2007; Wloga *et al.*, 2009). Individual tubulin glutamylases differ with respect to their substrate preferences for either alpha- or beta-tubulin subtypes and participate distinctly in the catalysis of initiation or elongation process. Only one beta-tubulin glutamylase, TTLL7 has been shown to be capable of initiation as well as extension of glutamates (Mukai *et al.*, 2009). Homologs of *ttll* genes in different species have been shown to possess divergent properties. For example, murine TTLL6 prefers alpha–tubulin, whereas its *Tetrahymena* homolog Ttll6Ap shows preference toward beta-tubulins. It is possible that multiple splice variants within a species could also have divergent substrate preferences (Janke *et al.*, 2005). Distinct activities of TTLL proteins could thus generate an extremely diverse range of tubulin protein variants, well beyond the complex repertoire of tubulin gene variants. Diversity in tubulin codes has been put forward as a hypothetical explanation for how organelle-specific organization and functions of microtubules could be specified (Verhey and Gaertig, 2007).

IV. Functional Significance of Tubulin Glutamylase

Functional analysis of individual *ttll* genes has revealed that tubulin glutamylases affect diverse microtubule-based cellular processes. Loss of the pGS1 subunit TTLL1 in ROSA 22 mice disrupts normal motility of KIF1A (Ikegami *et al.*, 2007). Loss of ttll6 affects olfactory cilia formation in zebrafish (Pathak *et al.*, 2007), whereas over-expression of Ttll6Ap in *Tetrahymena* induces hyperglutamylation and arrest of their cilia motility (Janke *et al.*, 2005). Knockdown of ttll7 in PC12 cells affects extension of their neurite processes (Ikegami *et al.*, 2006). Functions of tubulin glycylases have also begun to emerge from similar analyses. For instance, knockdown of *ttll3* affects cilia assembly in *Tetrahymena* and zebrafish (Rogowski *et al.*, 2009; Wloga *et al.*, 2009).

As mentioned before, cilia are rich in glutamylated tubulins. Importantly, green fluorescent protein (GFP) reporter fusions of all TTLLs have been localized to cilia of Madine-Darby Canine Kidney (MDCK) cells (van Dijk *et al.*, 2007) and *Tetrahymena* (Janke *et al.*, 2005; Wloga *et al.*, 2009). Moreover, distinct microtubule glutamylation patterns within substructures of cilia show preferential association of specific TTLL proteins. For example, in *Tetrahymena*, TTLL1 localizes to basal bodies in contrast to the lengthwise distribution of TTLL6Ap (Janke *et al.*, 2005).

Cilia defects are a common consequence of alterations in tubulin glutamylation deficiencies and are demonstrated to occur from knockdown of *ttll6* in zebrafish and *Tetrahymena* (Janke *et al.*, 2005; Pathak *et al.*, 2007). The existence of distinct cilia subtypes within model organisms such as *Tetrahymena* (Wloga *et al.*, 2008), *C. elegans* (Mukhopadhyay *et al.*, 2007), and zebrafish (Pathak *et al.*, 2007) will no doubt help dissect contributions of TTLLs to the generation of microtubule heterogeneity.

Importantly, analysis of cilia formation and function has received a strong stimulus due to the emergence of cilia defects as the central etiology in various inherited human disorders with pathological manifestations involving kidneys and sensory organs (Marshall, 2008). Independent bioinformatic compilations from these studies have indicated that over 600 proteins may be required for ciliogenesis (Avidor-Reiss *et al.*, 2004; Ostrowski *et al.*, 2002). Since localized protein synthesis does not occur within the cilium, all proteins must be imported into the cilium by the microtubule and motor protein-dependent process of intraflagellar transport (Pedersen and Rosenbaum, 2008). Given the association between tubulin glutamylation and kinesin function (Ikegami *et al.*, 2007), failure to properly glutamylate axonemal tubulins could adversely affect the delivery of many cilia proteins.

V. The Zebrafish *fleer* Mutant as a Paradigm for Analysis of Cilia Tubulin Glutamylation

The zebrafish system offers several advantages for genetic analysis of ciliogenesis on account of its short generation time, rapid external development, optical clarity, sequenced genome, the ability to induce antisense-mediated gene knockdowns, and ease of expressing transgenes (Drummond, 2005). Zebrafish larvae display morphologically distinct cilia subtypes in both sensory and epithelial organs. These cilia are easily visualized with antibodies against acetylated tubulin and distinctions with respect to their arrangement, length, motility, and number can be easily quantified. *fleer (flr)* is a zebrafish cystic mutation isolated during the Boston ENU mutagenesis screen (Drummond *et al.*, 1998). Loss of *fleer* function results in pleiotropic phenotypes (Pathak *et al.*, 2007) including curved body axis, hydrocephalus, pronephric cysts, *situs inversus*, and abnormal otolith number. This spectrum of defects is similar to other zebrafish cilia mutants and to human ciliopathies (Pathak *et al.*, 2007). Positional cloning revealed that the *fleer* gene is orthologous to the *C. elegans* gene *dyf-1*and encodes a TPR repeat protein identified within the cilia proteome (Fig. 1) (Ou *et al.*, 2005; Pathak *et al.*, 2007). *fleer* is a single-copy gene in zebrafish, localized to chromosome 3. The nonsense point mutation in *fleer* is predicted to truncate the mutant polypeptide to only the first 265 of its 681 amino acids Fig. 1 (Pathak *et al.*, 2007). The C-terminus of Fleer/Dyf-1 appears to be functionally important as the transposon inserted sequence in the *C. elegans dyf-1* mutant is at the extreme C-terminus and the *dyf-1* mutant gene is predicted to encode a nearly complete polypeptide (Ou *et al.*, 2005).

B-subfiber defects observed in *fleer* mutant cilia microtubule doublets offered an initial clue that *fleer* plays a role in tubulin posttranslational modifications (Fig. 2). Similar ultrastructural defects were observed in *Tetrahymena* tubulin mutants that are

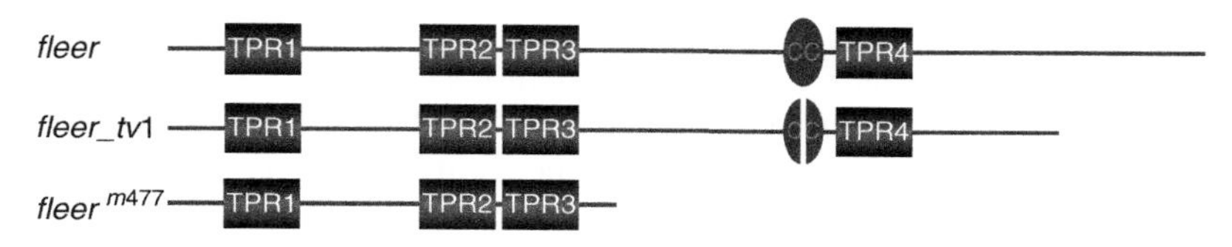

Fig. 1 Fleer and Fleer transcript variant 1 are predicted to contain four TPR motifs and a coiled coil domain (CC). The predicted fleer m477 mutant protein is truncated by a stop codon after TPR 3.

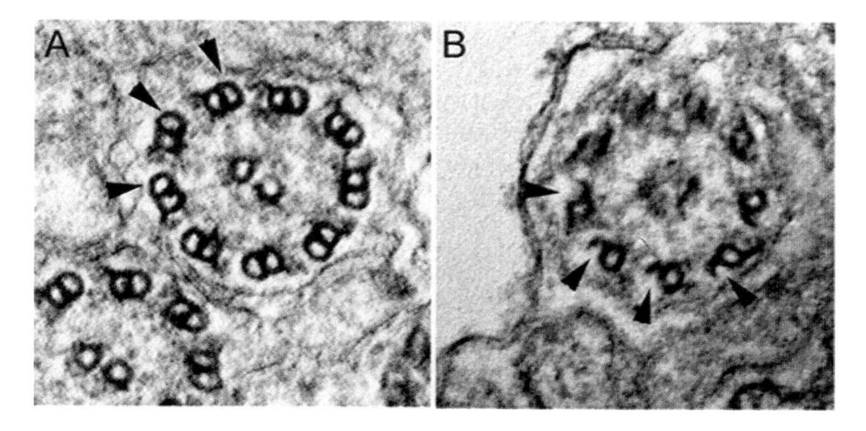

Fig. 2 *fleer* loss of function causes structural defects in the ciliary axoneme. (A) Ultrastructure of normal single pronephric cilia showing 9+2 microtubule architecture at 56 hpf and intact B-tubules of the microtubule doublets (arrowheads). (B) *fleer* mutant cilia in cross section show a specific loss of structure in the outer aspect of the B-tubules (arrowheads) with gaps in microtubules and accumulation of some electron-dense material between the doublets and the cilia outer membrane.

deficient in tubulin glutamylation and glycylation (Redeker *et al.*, 2005). Also, the B-subfiber is the localization site of the antiglutamylated antibody GT335 (Lechtreck and Geimer, 2000). Electron microscopic analysis of pronephric cilia in normal zebrafish indicates characteristic $9 + 2$ arrangement of microtubules (Fig. 2A). In contrast, in the *fleer* mutant, consistent gaps occur within the outer aspect of the B-tubule (Fig. 2B) (Pathak *et al.*, 2007). Confirmation that *fleer* specifically regulates tubulin glutamylation in cilia axonemes was provided by confocal immunofluorescence studies. Double staining embryos with antiacetylated tubulin (to reveal cilia structure) and the antiglutamyl tubulin antibody GT335 demonstrated that glutamyl tubulin is absent in *fleer* mutant axonemes (Fig. 3A–F). Similarly, *C. elegans* dyf-1 mutant outer labial cilia lack glutamyl-modifed tubulin (Fig. 3G–L). Thus fleer acts as a critical regulator of tubulin glutamylation and its loss of function can be exploited to analyze the consequences of loss of this tubulin posttranslational modification.

VI. Situs Inversus in *fleer* is Caused by Motility and Length Defects of Kupffer's Vesicle Cilia

Kupffer's vesicle (KV) is a disc-shaped, ciliated structure, present transiently in the zebrafish embryos between the 7 and 15 somite segmentation stages. Single, motile cilia in the lumen of wild-type Kupffer's vesicle are radially arranged (Fig. 4A). The length of these cilia is approximately 4.5 μm. Counterclockwise fluid flow generated by motion of the normal KV cilia signals asymmetric organ laterality (Kramer-Zucker *et al.*, 2005). At the molecular level, left-sided expression of the nodal-related gene *southpaw (spw), lefty*, and *pitx2A* in lateral plate mesoderm is a reliable indicator of normal KV cilia function (Kramer-Zucker *et al.*, 2005). Crosses of fleer heterozygotes produce clutches of embryos with Mendelian ratios of mutant embryos (25%). Of these mutants, 10–12% display absent or reversed *spw* expression, indicating randomization of left–right asymmetry. Measurements of KV cilia in *fleer* knockdown embryos demonstrated a reduction in cilia length to ~3 μm or about 65% of the normal value (Fig. 4B) (Pathak *et al.*, 2007).

VII. Pronephric Cysts in *fleer* Result from Defects of Cilia Motility and Length

Bilateral pronephroi in zebrafish larvae are organs of osmoregulation. Single or multicilia project from distinct pronephric cell types that are interspersed in a salt-and-pepper manner within the pronephric nephron. Motility of pronephric cilia is important for elimination of high volumes of dilute urine outside the body (Kramer-Zucker *et al.*, 2005). The length of these cilia is a function of their location along the anterio-posterior axis of the pronephros. In the anterior or proximal segment, cilia are on average 12 μm long while in the most posterior

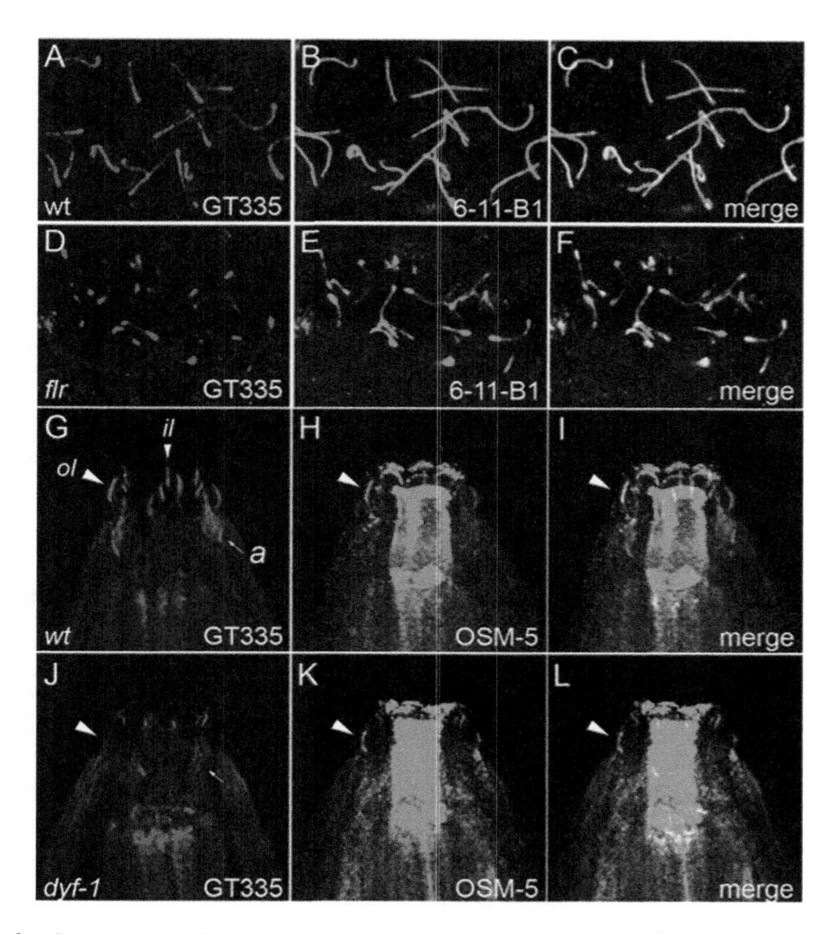

Fig. 3 *fleer* is required for axonemal tubulin polyglutamylation. (A–F) Tubulin polyglutamylation and acetylation was visualized in zebrafish pronephric cilia by sequential labeling with antibodies mAb GT335 and mAb 6-11B-1. (A) In wild-type pronephric cilia at 56 hpf, polyglutamylated tubulin is distributed in a decreasing gradient from the basal body to the cilia tip. (B) Acetylated tubulin immunostaining of wild-type cilia reveals the total length of the cilia. (C) Merge of A and B. (D) In the *fleer* mutant pronephric cilia, polyglutamylated tubulin is restricted to the basal body region and severely reduced in axonemes. (E) Acetylated tubulin staining could still be observed in axonemes devoid of the polyglutamylated tubulin. (F) Merge of D and E shows that polyglutamylated tubulin is found only in the proximal segment of *fleer* cilia. (G–L): Loss of tubulin polyglutamylation in *dyf-1 C. elegans* mutant cilia. Tubulin polyglutamylation in *C. elegans* hermaphrodite sensory cilia was detected with the monoclonal antibody GT335 and overall cilia structure was detected with anti-OSM-5 antibody. Anti-OSM-5 staining also gave nonspecific labeling of the pharynx, unrelated to cilia. (G) GT335 immunoreactivity identifies polyglutamylated tubulin in multiple wild-type hermaphrodite cilia: outer labial (large arrowheads; *ol*), inner labial (small arrowheads; *il*), and amphid cilia (small arrow; *a*). (H) OSM-5 positive outer labial cilia (arrowheads), (I) merge of A and B; arrowheads denote outer labial cilia positive for both GT335 and OSM-5. (J) Polyglutamylated tubulin in *dyf-1* mutant hermaphrodites is reduced. Large arrowheads: position of outer labial cilia. (K) OSM-5 positive outer labial cilia in *dyf-1* mutants. (L) Merge of D and E; *dyf-1* outer labial cilia are negative for GT335. (See Plate no. 26 in the Color Plate Section.)

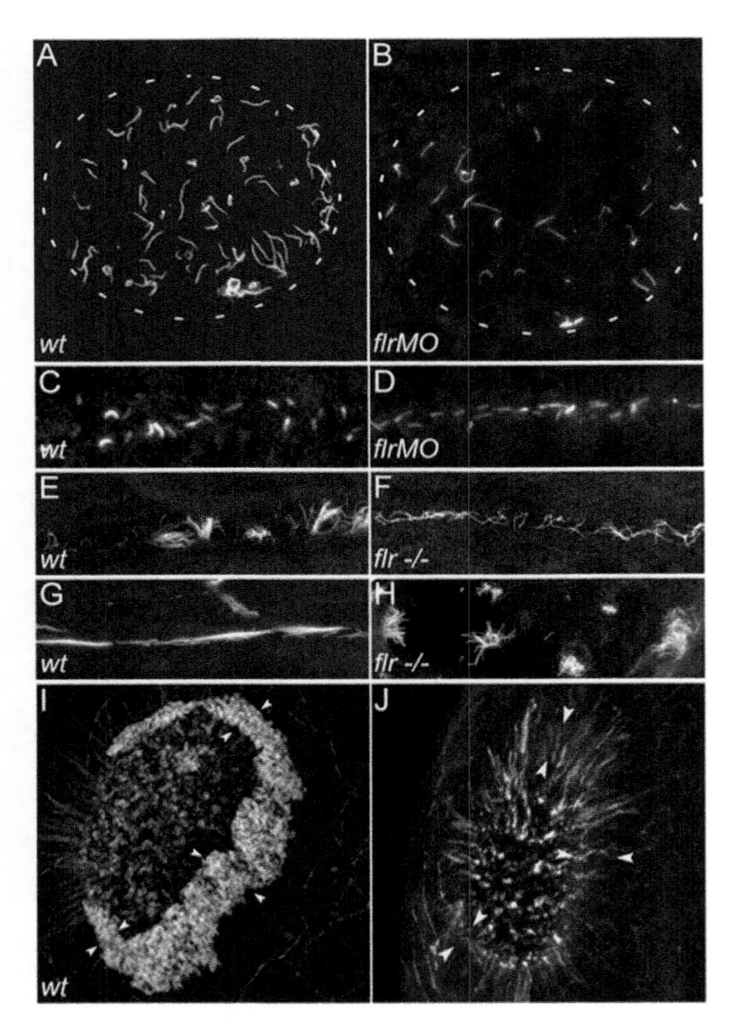

Fig. 4 *fleer* loss of function differentially affects cilia in different organs. Cilia were visualized in wild-type and *fleer* mutant or morphant larvae stained with an antibody to acetylated tubulin. (A) Kupffer's vesicle contains radially arranged cilia of uniform length in wild-type embryos. White dashed outline represents the borders of Kupffer's vesicle in A and B. (B) *fleer* morphant embryos exhibit severe shortening or absence of cilia preferentially at the periphery of Kupffer's vesicle. (C) In the presumptive pronephric duct at the 10 somite stage, short cilia were apparent in both (C) wild-type and (D) *fleer* morphant embryos. (E) Later at 48 hpf, single cilia and multiciliated cells are observed in wild-type embryos while in the *fleer* mutant pronephros, (F) cilia were shorter and multiciliated cells were not observed. (G) In 88 hpf wild-type larvae, multiciliated cells project dense, brightly staining bundles of cilia into the narrow lumen of the pronephros. (H) *fleer* mutant pronephroi at 88 hpf contain dispersed bundles of cilia that are nonuniform in length. (I) The dense ring of cilia on the epithelium surrounding the olfactory placode (arrowheads) of 60hpf wild-type larvae was absent in 60 h *fleer* mutant larvae (J). Residual staining for acetylated tubulin in *fleer* mutant olfactory placodes was associated with intracellular projections of olfactory neurons.

or distal segment they measure 6 μm. Cilia in the midsegment reach an intermediate length of 9 μm (Pathak *et al.*, 2007). Variation of cilia length as a function of their position necessitates that comparisons be made from corresponding pronephric segments. Measurements of cilia length are straightforward in the distended lumens of cystic mutants or morphants and it is convenient to experimentally induce such distension in wild-type pronephros by mechanical obstruction (Kramer-Zucker *et al.*, 2005). Although cells bearing distinct single and multicilia are interspersed, views of the pronephric midsegment are dominated by bold bundles of multicilia (Fig. 4E) (Liu *et al.*, 2007). In stark contrast, only single cilia appear in the distended midsegment of *fleer* mutants, with the multicilia strikingly absent or regressed (Fig. 4F and H) (Pathak *et al.*, 2007). Comparisons of single cilia length in wild-type and *fleer* larvae indicated that loss of fleer causes a consistent shortening of these cilia to 60–70% of the normal values in all segments.

VIII. Olfactory Placode Cilia

Olfactory multicilia are motile and appear in a dense C-shape pattern by 48 hpf on epithelial cells surrounding a central disk with numerous short sensory cilia (Fig. 4I). The prominent, superficial nature of olfactory cilia offers an easy landmark for observing cilia defects. The motile multicilia are thought to be nonsensory and function to direct water over single, chemosensory cilia on olfactory neurons emerging from in small pits near the center of the placode at about 4 dpf (Hansen and Zeiske, 1993; Hansen and Zielinski, 2005). Consistent with the loss of pronephric multicilia, olfactory multicilia are completely absent in *fleer* mutants (Fig. 4J).

IX. Abnormal Otolith Numbers are Caused by Defects of Motile Otic Placode Cilia

Otic placodes harbor two distinct cilia subtypes at slightly different developmental stages. Numerous short, motile cilia occur on most of the surface between 18 and 27 hpf and help to create fluid vortices that crystallize two prominent otololiths from particulate biomineralized secretions. Formation of longer tether cilia around 25 hpf anchors the otoliths (Riley, 2003; Wiederhold *et al.*, 2003).

Genetic defects affecting cilia motility typically cause abnormal otolith number as the biomineralized particles move in Brownian manner that coalesce by gravity either into one large or multiple small otoliths. Defective tether cilia render one or more otoliths to appear loose upon tilting the larvae. Absence of otoliths, however, is not a characteristic cilia defect.

X. Retinal Degeneration in *fleer* Results from Defects of the Connecting Cilium in Photoreceptors

Transport of visual pigments to the outer segments of photoreceptors occurs via a modified primary cilium or connecting cilium (Insinna *et al.*, 2008). The missing outer segments, observed in the histological stained plastic sections of *fleer* mutants occur due to defects in the transporting function of the connecting cilium (Pathak *et al.*, 2007).

XI. *fleer*-Like Phenotypes in Other Species

The Fleer polypeptide shows high degree of evolutionary conservation across invertebrate and vertebrate species. Surprisingly, in mammalian genomes, *fleer* homologs appear to be expressed pseudogenes that lack introns. However, these sequences have no premature stop codons and the encoded protein has been detected in proteomic analysis (2D protein gels) of human lung cilia (Ostrowski *et al.*, 2002). Mutation in the *C. elegans* fleer ortholog *dyf1* affects the ciliated endings of chemosensory neurons, making the amphid neurons defective in the uptake of a fluorescent lipophillic dye (Ou *et al.*, 2005). The amphid cilia in the *C. elegans dyf-1* mutant cilia are short and lack distal singlet microtubule containing sections. In addition, we showed that tubulin glutamylation is absent from the cilia of outer labial neurons (Pathak *et al.*, 2007) despite relatively normal structure.

Dyf-1 mutations engineered in *Tetrahymena* cause a severe shortening of cilia and manifest defects in the B-tubule and central pair formation similar to the zebrafish *fleer* mutant. These defects are similar to the previously reported *fleer* olfactory placode cilia phenotype where cilia fail to form (Pathak *et al.*, 2007). Analysis of cilia remnant stubs in *Tetrahymena* DYF-1 mutants using an antipolyglutamine antibody (ID5) suggested that these cilia were hyperglutamyled, the opposite of what has been reported in *fleer* mutants (Dave *et al.*, 2009). The main difference here may be that hyperglutamylation may occur in *Tetrahymena* as a response to failed cilia assembly, while cilia analyzed for glutamylation in *fleer* mutants (single pronephric cilia) did not show severe ciliogenesis defects, allowing for analysis of glutamylation in near-normal length cilia. Also, it will be important to further characterize tubulin glutamylation in fish and other organisms using antibodies that detect both glutamylation branch point (GT335) and polyglutamine chains (ID5) to extend and confirm these results.

Finally, B-subfiber defects similar to those seen in *fleer* mutants have also been reported in *C. elegans* mutants of the transition zone proteins *nphp1* and *nphp4* (Jauregui *et al.*, 2008), and the mouse mutant hennin which harbors a mutation in Arl13b (Caspary *et al.*, 2007). Unlike *fleer*, the B-tubule defects of these mutants are located at the inner seam between the A and B tubule. Assessment of glutamylation defects induced by these mutations could address whether B-tubule defects could also induce glutamylation defects.

XII. Methods

A. Cilia Motility Defects in *fleer* Mutants

The optically clear zebrafish larvae present an excellent opportunity to visualization of cilia function by light microscopy. Recording cilia motility in the pronephric and olfactory cilia is performed in live 2.5 dpf zebrafish larvae using a 40× or 63× water-dipping objective and differential interference contrast (DIC or Nomarski) optics. Larvae suspended in an anesthetic mix (dilute tricaine egg water containing 20 mM 2,3-butanedionemonoxime or BDM) make observation convenient due to their immobility and elimination of background blood flow induced by reversible blockage of the heart beat by BDM. The immobilized larvae are suspended laterally in a clear depression slide or a glass bottom plate on a supporting base of 3% methylcellulose solution. Cilia motility is recorded with a high-speed video camera capable of recording images between 200 and 250 frames/s. Movies acquired in QuickTime format can be imported for analysis by open source software such as Image J. Detailed protocols for approaches to high-speed video analysis of cilia in zebrafish are presented in Chapter 11 by Drummond in Methods in Cell Biology, volume 93 "Studying Cilia in Zebrafish".

In the *fleer* mutant, motile cilia are absent in the olfactory placode. Single pronephric cilia move with markedly reduced beat amplitude but a normal frequency (Pathak *et al.*, 2007). Similar high-speed video analysis of the otic placode between 21 and 25 hpf allows observation of the oscillating otoliths in normal zebrafish. In contrast, otic placodes of *fleer* mutants appear filled with numerous small particles in Brownian motion.

B. Whole-mount Confocal Analysis of Cilia Tubulin Posttranslational Modifications

For comprehensive analysis of cilia tubulin posttranslational modifications, it is necessary to combine the use of an antibody that reports overall cilia length with a second antibody that distinguishes mono- or polyglutamyl tubulin modifications. This approach is required to distinguish between overall shortening of cilia (seen for instance in IFT mutants) and specific defects in tubulin modifications. The commonly used antiacetylated tubulin monoclonal antibody 6-11-B1 (Sigma-Aldrich, USA; T6793) works well to visualize overall cilia structure in zebrafish. Other useful monoclonals that work well in zebrafish include antipolyglutamyl tubulin (GT335 or B3) and antigamma tubulin (GTU-88). Tubulin glutamylation (GT335 immunoreactivity) in wild-type cilia appears strongest at the basal body and is progressively reduced towards the distal tips of cilia. In striking contrast, GT335 staining appears largely excluded from the axonemes of *fleer* single cilia while antiacetylated tubulin of the same cilia confirmed this results was not due to overall cilia shortening. This approach requires the combined use of two mouse monoclonal antibodies. The following protocol details how to detect two different mouse monoclonal antibodies in the same tissue.

1. Materials

Phosphate-buffered saline (PBS)
DiMethyl Sulfoxide (DMSO)
Tween20
Normal goat serum (Sigma G9023)
Bovine serum albumin (BSA) (Sigma A8022) or gelatin from cold water fish skin (Sigma G7765)
Normal mouse serum (Sigma M5905)
Goat antimouse Fab fragments (JacksonImmunoResearch, USA 115-007-003)
Methanol
Benzyl alcohol
Glycerol
N-propyl gallate
Formaldehyde
Hydrogen peroxide
SDS

2. Solutions

PBST:	PBS + 0.5% Tween 20
Blocking solution:	PBS with 1% DMSO 0.5 % Tween 20 1% BSA or 0.3% gelatin from cold water fish skin 10% normal goat serum
Incubation solution:	PBS with 1% DMSO 0.5 % Tween 20 2% normal goat serum
Dent's fixative:	80% methanol 20% DMSO
Rehydration solutions:	75:25 MeOH/PBST 50:50 MeOH/PBST 25:75 MeOH/PBST
Mounting medium:	53% benzyl alcohol (by weight) 45% glycerol (by weight) and 2% *N*-propyl gallate

3. Methods

One of the main advantages of using zebrafish immunofluorescence is the transparency of their embryos. To achieve maximum embryo transparency, development of pigmentation can be blocked by raising embryos in 0.003% phenylthiourea egg water.

4. Fixation

1. Fix embryos in Dent's fixative for 3 h to overnight at room temperature. After fixation embryos can be stored in 100% methanol at –20°C.
2. Rehydrate Dent's fixed embryos with graded changes of methanol PBT:
   75:25 MeOH/PBST 15 min
   50:50 MeOH/PBST 15 min
   25:75 MeOH/PBST 15 min
   PBT

5. Antibody Staining

3. Block nonspecific binding by incubating fixed embryos for 2 h to overnight in blocking solution. Incubations are done in Eppendorf tubes at 4°C on a nutator rocking platform.
4. Incubate with primary antibody (6-11-B1 at 1:1000) in incubation solution overnight at 4°C. If necessary to economize on primary antibodies, the incubation can be done in 50–100 μl without agitation.
5. After incubation with primary antibody, wash at least 4× 30 min with incubation solution. A volume of 2% normal goat serum is included in all steps to reduce background.
6. Incubate with fluorophore-coupled secondary antibody in incubation solution overnight. Alexa antimouse secondary antibodies (Invitrogen, USA) work well at 1:1000 dilution. Following incubation, wash at least 4× 30 min with Incubation solution on a rocking platform.
7. Incubate embryos in 10% normal mouse serum (Sigma M5905) in incubation solution for 2 h at 4°C.
8. Wash with incubation solution 4× 30 min at 4°C
9. Incubate embryos with goat antimouse Fab fragments at 1:20 dilution in incubation buffer; (Jackson-ImmunoResearch 115-007-003) overnight at 4°C.
10. Wash with incubation solution 4× 30 min at 4°C.
11. Block with blocking solution 2 h at 4°C
12. Incubate with antiglutamylated tubulin (GT335 or B3; 1:400), detect with a second, different fluorescent antibody, and wash as described above for the primary incubation with antiacetylated tubulin (steps 1–6). Wash twice with PBS and mount in mounting medium for confocal microscopy.

C. Mounting the Sample for Confocal Microscopy

To minimize optical distortion caused by mismatch in refractive index of the sample, coverslips, and immersion oil, use a mounting medium that has the same refractive index (1.513) as the immersion oil. We make a mounting medium developed by Gustafsson *et al.* (1999), that is, a mixture of glycerol and benzyl alcohol and contains an antifade compound (*N*-propyl gallate) that is essential for preventing signal bleaching, especially when using 488 nM fluorophores (FITC, Alexa 488) on large Z image stacks.

Embryos can be placed in mounting medium directly after washing. The difference in density of the mounting medium and PBS is significant and causes turbulence but does not damage the sample. It is best to transfer embryos to mounting medium in a depression slide or directly on a microscope slide since the embryos become essentially invisible and can be hard to find in an Eppendorf tube. Change to fresh mounting medium and transfer embryos to a standard microscope slide. Using small balls of modeling clay, support the edges of a coverslip to provide space for the embryo and coverslip the sample. Alternatively, make a coverslip bridge with additional coverslips as spacers. The orientation of the embryo can often be shifted by moving the coverslip. This sample configuration is suited for viewing with an upright microscope using a 63× oil immersion objective.

Acknowledgments

We would like to thank all members of the Drummond lab and our collaborators for their contributions to this work. This work was supported by NIH Grants DK053093 to I. A. D. and DK078741 to N. P.

References

Avidor-Reiss, T., Maer, A.M., Koundakjian, E., Polyanovsky, A., Keil, T., Subramaniam, S., and Zuker, C.S. (2004). Decoding cilia function: Defining specialized genes required for compartmentalized cilia biogenesis. *Cell* **117**, 527–539.

Caspary, T., Larkins, C.E., and Anderson, K.V. (2007). The graded response to Sonic Hedgehog depends on cilia architecture. *Dev. Cell* **12**, 767–778.

Dave, D., Wloga, D., Sharma, N., and Gaertig, J. (2009). DYF-1 is required for assembly of the axoneme in Tetrahymena. *Eukaryot. Cell* **8**, 1397–1406.

Drummond, I.A. (2005). Kidney development and disease in the zebrafish. *J. Am. Soc. Nephrol.* **16**, 299–304.

Drummond, I.A., Majumdar, A., Hentschel, H., Elger, M., Solnica-Krezel, L., Schier, A.F., Neuhauss, S.C., Stemple, D.L., Zwartkruis, F., Rangini, Z., Driever, W., and Fishman, M.C. (1998). Early development of the zebrafish pronephros and analysis of mutations affecting pronephric function. *Development* **125**, 4655–4667.

Gagnon, C., White, D., Cosson, J., Huitorel, P., Edde, B., Desbruyeres, E., Paturle-Lafanechere, L., Multigner, L., Job, D., and Cibert, C. (1996). The polyglutamylated lateral chain of alpha-tubulin plays a key role in flagellar motility. *J. Cell Sci.* **109**(Pt 6), 1545–1553.

Gustafsson, M.G., Agard, D.A., and Sedat, J.W. (1999). I5M: 3D widefield light microscopy with better than 100 nm axial resolution. *J. Microsc.* **195**, 10–16.

Hansen, A., and Zeiske, E. (1993). Development of the olfactory organ in the zebrafish, Brachydanio rerio. *J. Comp. Neurol.* **333**, 289–300.

Hansen, A., and Zielinski, B.S. (2005). Diversity in the olfactory epithelium of bony fishes: Development, lamellar arrangement, sensory neuron cell types and transduction components. *J. Neurocytol.* **34**, 183–208.

Ikegami, K., Heier, R.L., Taruishi, M., Takagi, H., Mukai, M., Shimma, S., Taira, S., Hatanaka, K., Morone, N., Yao, I., Campbell, P.K., Yuasa, S., et al. (2007). Loss of alpha-tubulin polyglutamylation in ROSA22 mice is associated with abnormal targeting of KIF1A and modulated synaptic function. *Proc. Natl. Acad. Sci. USA* **104**, 3213–3218.

Ikegami, K., Mukai, M., Tsuchida, J., Heier, R.L., Macgregor, G.R., and Setou, M. (2006). TTLL7 is a mammalian beta-tubulin polyglutamylase required for growth of MAP2-positive neurites. *J. Biol. Chem.* **281**, 30707–30716.

Insinna, C., Pathak, N., Perkins, B., Drummond, I., and Besharse, J.C. (2008). The homodimeric kinesin, Kif17, is essential for vertebrate photoreceptor sensory outer segment development. *Dev. Biol.* **316**, 160–170.

Janke, C., Rogowski, K., and van Dijk, J. (2008). Polyglutamylation: A fine-regulator of protein function? "Protein Modifications: Beyond the usual suspects" review series. *EMBO Rep.* **9**, 636–641.

Janke, C., Rogowski, K., Wloga, D., Regnard, C., Kajava, A.V., Strub, J.M., Temurak, N., van Dijk, J., Boucher, D., van Dorsselaer, A., Suryavanshi, S., Gaertig, J., et al. (2005). Tubulin polyglutamylase enzymes are members of the TTL domain protein family. *Science* **308**, 1758–1762.

Jauregui, A.R., Nguyen, K.C., Hall, D.H., and Barr, M.M. (2008). The Caenorhabditis elegans nephrocystins act as global modifiers of cilium structure. *J. Cell. Biol.* **180**, 973–988.

Kramer-Zucker, A.G., Olale, F., Haycraft, C.J., Yoder, B.K., Schier, A.F., and Drummond, I.A. (2005). Cilia-driven fluid flow in the zebrafish pronephros, brain and Kupffer's vesicle is required for normal organogenesis. *Development* **132**, 1907–1921.

Lechtreck, K.F., and Geimer, S. (2000). Distribution of polyglutamylated tubulin in the flagellar apparatus of green flagellates. *Cell Motil. Cytoskeleton* **47**, 219–235.

Liu, Y., Pathak, N., Kramer-Zucker, A., and Drummond, I.A. (2007). Notch signaling controls the differentiation of transporting epithelia and multiciliated cells in the zebrafish pronephros. *Development* **134**, 1111–1122.

Marshall, W.F. (2008). The cell biological basis of ciliary disease. *J. Cell Biol.* **180**, 17–21.

Mukai, M., Ikegami, K., Sugiura, Y., Takeshita, K., Nakagawa, A., and Setou, M. (2009). Recombinant mammalian tubulin polyglutamylase TTLL7 performs both initiation and elongation of polyglutamylation on beta-tubulin through a random sequential pathway. *Biochemistry* **48**, 1084–1093.

Mukhopadhyay, S., Lu, Y., Qin, H., Lanjuin, A., Shaham, S., and Sengupta, P. (2007). Distinct IFT mechanisms contribute to the generation of ciliary structural diversity in C. elegans. *EMBO J.* **26**, 2966–2980.

Ostrowski, L.E., Blackburn, K., Radde, K.M., Moyer, M.B., Schlatzer, D.M., Moseley, A., and Boucher, R.C. (2002). A proteomic analysis of human cilia: Identification of novel components. *Mol. Cell Proteomics* **1**, 451–465.

Ou, G., Blacque, O.E., Snow, J.J., Leroux, M.R., and Scholey, J.M. (2005). Functional coordination of intraflagellar transport motors. *Nature* **436**, 583–587.

Pathak, N., Obara, T., Mangos, S., Liu, Y., and Drummond, I.A. (2007). The zebrafish fleer gene encodes an essential regulator of cilia tubulin polyglutamylation. *Mol. Biol. Cell* **18**, 4353–4364.

Pedersen, L.B., and Rosenbaum, J.L. (2008). Intraflagellar transport (IFT) role in ciliary assembly, resorption and signalling. *Curr. Top. Dev. Biol.* **85**, 23–61.

Redeker, V., Levilliers, N., Vinolo, E., Rossier, J., Jaillard, D., Burnette, D., Gaertig, J., and Bre, M.H. (2005). Mutations of tubulin glycylation sites reveal cross-talk between the C termini of alpha- and beta-tubulin and affect the ciliary matrix in Tetrahymena. *J. Biol. Chem.* **280**, 596–606.

Regnard, C., Desbruyeres, E., Huet, J.C., Beauvallet, C., Pernollet, J.C., and Edde, B. (2000). Polyglutamylation of nucleosome assembly proteins. *J. Biol. Chem.* **275**, 15969–15976.

Riley, B.B. (2003). Genes controlling the development of the zebrafish inner ear and hair cells. *Curr. Top. Dev. Biol.* **57**, 357–388.

Rogowski, K., Juge, F., van Dijk, J., Wloga, D., Strub, J.M., Levilliers, N., Thomas, D., Bre, M.H., Van Dorsselaer, A., Gaertig, J., and Janke, C. (2009). Evolutionary divergence of enzymatic mechanisms for posttranslational polyglycylation. *Cell* **137**, 1076–1087.

Spiliotis, E.T., Hunt, S.J., Hu, Q., Kinoshita, M., and Nelson, W.J. (2008). Epithelial polarity requires septin coupling of vesicle transport to polyglutamylated microtubules. *J. Cell Biol.* **180**, 295–303.

van Dijk, J., Miro, J., Strub, J.M., Lacroix, B., van Dorsselaer, A., Edde, B., and Janke, C. (2008). Polyglutamylation is a post-translational modification with a broad range of substrates. *J. Biol. Chem.* **283**, 3915–3922.

van Dijk, J., Rogowski, K., Miro, J., Lacroix, B., Edde, B., and Janke, C. (2007). A targeted multienzyme mechanism for selective microtubule polyglutamylation. *Mol. Cell* **26**, 437–448.

Vent, J., Wyatt, T.A., Smith, D.D., Banerjee, A., Luduena, R.F., Sisson, J.H., and Hallworth, R. (2005). Direct involvement of the isotype-specific C-terminus of beta tubulin in ciliary beating. *J. Cell Sci.* **118**, 4333–4341.

Verhey, K.J., and Gaertig, J. (2007). The tubulin code. *Cell Cycle* **6**, 2152–2160.

Wiederhold, M.L., Harrison, J.L., and Gao, W. (2003). A critical period for gravitational effects on otolith formation. *J. Vestib. Res.* **13**, 205–214.

Wloga, D., Rogowski, K., Sharma, N., Van Dijk, J., Janke, C., Edde, B., Bre, M.H., Levilliers, N., Redeker, V., Duan, J., Gorovsky, M.A., Jerka-Dziadosz, M., et al. (2008). Glutamylation on alpha-tubulin is not essential but affects the assembly and functions of a subset of microtubules in Tetrahymena thermophila. *Eukaryot. Cell* **7**, 1362–1372.

Wloga, D., Webster, D.M., Rogowski, K., Bre, M.H., Levilliers, N., Jerka-Dziadosz, M., Janke, C., Dougan, S.T., and Gaertig, J. (2009). TTLL3 Is a tubulin glycine ligase that regulates the assembly of cilia. *Dev. Cell* **16**, 867–876.

CHAPTER 17

Regulation of Cilia assembly, Disassembly, and Length by Protein Phosphorylation

Muqing Cao, Guihua Li, *and* Junmin Pan
School of Life Sciences, Tsinghua University, Beijing 100084, China

Abstract

The exact mechanism by which cells are able to assemble, regulate, and disassemble cilia or flagella is not yet completely understood. Recent studies in several model systems, including *Chlamydomonas, Tetrahymena, Leishmania, Caenorhabditis elegans*, and mammals, provide increasing biochemical and genetic evidence that

METHODS IN CELL BIOLOGY, VOL. 94

978-0-12-375024-2
DOI: 10.1016/S0091-679X(08)94017-6

phosphorylation of multiple protein kinases plays a key role in cilia assembly, disassembly, and length regulation. Members of several protein kinase families—including aurora kinases, never in mitosis A (NIMA)-related protein kinases, mitogen-activated protein (MAP) kinases, and a novel cyclin-dependent protein kinase—are involved in the ciliary regulation process. Among the newly identified protein kinase substrates are *Chlamydomonas* kinesin-13 (CrKinesin13), a microtubule depolymerizer, and histone deacetylase 6 (HDAC6), a microtubule deacetylase. *Chlamydomonas* aurora/Ipl1p-like protein kinase (CALK) and CrKinesin13 are two proteins that undergo phosphorylation changes correlated with flagellar assembly or disassembly. CALK becomes phosphorylated when flagella are lost, whereas CrKinesin13 is phosphorylated when new flagella are assembled. Conversely, suppressing CrKinesin13 expression results in cells with shorter flagella.

I. Introduction

The microtubule-based cellular organelles, cilia and eukaryotic flagella, serve as motors and cellular antennae (Fliegauf *et al.*, 2007; Gerdes *et al.*, 2009; Pan, 2008). They are ancient organelles, well conserved throughout the biological kingdom, and present in organisms ranging from protists to mammals. For historical reasons, cilia and eukaryotic flagella actually refer to the same cellular organelle; thus, these two terms are used interchangeably.

A cilium is anchored to the cell body through a structure called the basal body, which is derived from the centriole. The length of a cilium is cell-type-specific and is maintained during G1 or G0 phase of the cell cycle so that the cilium can properly perform its function. Cilia may be lost in response to environmental stress or development signals by two distinct mechanisms: gradually shortening from the cilia tip (resorption) or detachment from the cilia base (deflagellation). When cilia are lost, cells are able to regenerate new cilia of the same length. Cilia are resorbed before mitosis, thereby releasing the centrioles to participate in spindle formation. After cytokinesis, cilia are formed on the centrioles. In vertebrates, a primary cilium is typically formed on the mother centriole in uniciliated cells. In the eukaryotic organism *Chlamydomonas*, however, two flagella are assembled by nucleating microtubules from both the mother and daughter centrioles. Cilia formation in multiciliated cells involves the generation of multiple centrioles either spontaneously (acentriolar pathway) or through duplication of existing centrioles (centriolar pathway) (Hagiwara *et al.*, 2000; Sorokin, 1962, 1968). The assembly, length regulation, and loss of cilia are complex cellular processes involving gene expression, protein transport, remodeling of protein complexes, and signal transduction.

Although cilia structure, function, and consequences of their defects are largely known, less is understood about the regulatory network involved in cilia assembly, disassembly, and length maintenance. Protein phosphorylation has emerged as a key mechanism for regulating these biological processes. We will discuss recent evidence on the function of protein phosphorylation in the control of each step of the cilia "life cycle": assembly, disassembly, and length control.

II. Protein Phosphorylation in Cilia Assembly

Cilia formation requires multiple cellular processes: signaling to initiate cilia assembly, expression of ciliary genes, remodeling of cytoplasmic microtubules, protein transport to the cilia assembly site, and posttranslational modification of axonemal microtubules. Exciting discoveries have recently been made regarding phosphorylation control of cilia assembly, which will open new research opportunities for studying ciliary regulation.

A. CrKinesin13 Phosphorylation

Chlamydomonas offers unique advantages for studying cilia assembly; for example, its flagella can be amputated experimentally and they will reassemble synchronously within 2 h. Analysis of fully grown flagella in *Chlamydomonas* has revealed enriched protein kinase and phosphatase activities (Bloodgood, 1992; Boesger *et al.*, 2009). In addition, protein kinase inhibitor added before flagellar regeneration impairs flagellar assembly (Harper *et al.*, 1993). Recent findings indicate that protein kinase activities predominantly occurring in the cell body may play a pivotal role in regulating cilia assembly and disassembly (Berman *et al.*, 2003; Tam *et al.*, 2007).

To identify novel phosphorylated proteins involved in flagellar regulation, Piao *et al.* (2009) took a biochemical approach to identify and purify phosphoproteins from *Chlamydomonas*. By using mitotic protein monoclonal 2 (MPM-2), a monoclonal antibody that recognizes pSer/pThr-Pro (Harper *et al.*, 1993; Westendorf *et al.*, 1994), a 70-kDa phosphoprotein was found to appear only during flagellar assembly. This phosphoprotein was subsequently identified as a member of the microtubule-depolymerizing kinesin13 gene family and therefore named "CrKinesin13." CrKinesin13 is phosphorylated when flagella assembly begins under all conditions tested: after deflagellation by pH shock, after transferring aflagellated cells from agar plate to liquid medium, and after reversing flagellar shortening to elongation. In contrast to the weak phosphorylation of glycogen synthase kinase (GSK)-3β during flagellar assembly (Wilson and Lefebvre, 2004), CrKinesin13 is vigorously phosphorylated. Confirming this enzyme's critical role, researchers have found that suppressing CrKinesin13 expression by RNA interference (RNAi) results in cells with shorter flagella, indicating defects in flagellar assembly (Piao *et al.*, 2009). The participation of a microtubule-depolymerizing kinesin in flagellar assembly raises the intriguing question of whether the cytoplasmic microtubules serve as a tubulin pool for assembling axonemal microtubules (Piao *et al.*, 2009). Adding a protein synthesis inhibitor before flagellar assembly of the CrKinesin13 RNAi strains does indeed block flagellar assembly, which is consistent with the hypothesis that CrKinesin13 is critical to assembly (unpublished data). Additional research is needed to identify the protein kinase responsible for CrKinesin13 phosphorylation and further our understanding of how cytoplasmic microtubules contribute to cilia assembly.

In addition to CrKinesin13, Salisbury and colleagues have reported that two phosphoproteins of 90 and 34 kDa, both recognized by the MPM-2 antibody, appeared during flagellar assembly in *Chlamydomonas* (Harper *et al.*, 1993). These two proteins are phosphorylated immediately after flagellar regeneration and dephosphorylated about 30 min thereafter. However, the nature of these two phosphoproteins is not clear, nor is their relationship to CrKinesin13.

B. Role of GSK-3β

GSK-3β regulates microtubule dynamics via phosphorylation of microtubule-associated proteins (Buttrick and Wakefield, 2008; Forde and Dale, 2007); therefore, it is unsurprising to find that GSK-3β is involved in ciliogenesis and cilia length control. In *Chlamydomonas*, flagellar assembly requires GSK-3β activity. Adding lithium chloride (LiCl), an inhibitor of GSK-3β, to cells undergoing flagellar assembly results in flagella less than half the normal length (Williams *et al.*, 2004; Wilson and Lefebvre, 2004). Blocking protein synthesis by cycloheximide also allows assembly of half-length flagella. However, LiCl in combination with cycloheximide blocks flagellar assembly almost completely (Wilson and Lefebvre, 2004), indicating that the shorter flagella phenotype caused by LiCl results from more than the inhibitory effects of protein synthesis. We propose that there is an inactive pool of flagellar precursors that must become active during flagellar assembly. GSK-3β may be involved in activating this inactive pool, and LiCl blocks this step. Because GSK-3β is known to be involved in regulating microtubule dynamics, GSK-3β may function by inducing the disassembly of cytoplasmic microtubules to release free tubulin dimers for the assembly of axonemal microtubules.

As discussed above, we have hypothesized that CrKinesin13 is involved in converting cytoplasmic microtubules to tubulin subunits during cilia assembly; therefore, it will be interesting to test whether GSK-3β phosphorylates CrKinesin13. If GSK-3β promotes the disassembly of microtubules, then LiCl treatment of cells with full-grown flagella may stimulate further flagellar growth. In fact, several studies have reported that finding (Nakamura *et al.*, 1987; Periz *et al.*, 2007; Wilson and Lefebvre, 2004), in addition to our own unpublished data. It appears, therefore, that GSK-3β may use its activity in promoting microtubule depolymerization to regulate flagellar assembly as well as flagellar length control.

Similarly, the regulation of ciliogenesis by GSK-3β has been demonstrated in mammalian cells. GSK-3β and the von Hippel–Lindau protein (pVHL), a tumor suppressor, have been found to regulate ciliogenesis cooperatively in kidney epithelial cells. The assembly of cilia requires both proteins. The inhibitory phosphorylation state of GSK-3β correlates with the reduced frequency of primary cilia formation in renal cysts (Thoma *et al.*, 2007).

C. Protein Phosphorylation of Intraflagellar Transport Particle Proteins

Cilia assembly requires transporting newly synthesized ciliary protein precursors to the cilia assembly site. This transport involves two-way traffic of protein complexes, called intraflagellar transport (IFT), between the cell body and the tip of the cilium

(Cole, 2003; Rosenbaum and Witman, 2002; Scholey, 2003). The anterograde transport of protein complexes (IFT particles) from the cell body to the cilia tip is driven by kinesin-2, whereas the retrograde transport from the tip of cilia to the cell body is mediated by cytoplasmic dynein. This IFT process has been found to be essential for cilia assembly and maintenance in almost all ciliated cells. IFT particle proteins were first isolated from *Chlamydomonas* (Cole *et al.*, 1998; Piperno and Mead, 1997). Each particle is composed of multiple copies of two protein complexes designated A and B, which contain at least 6 and 13 proteins, respectively (Cole and Snell, 2009). Among these proteins, IFT 74/72 and IFT 57/55 each exhibit two isoforms by sodium dodecyl sulfate polyacrylamide gel electrophoresis (SDS-PAGE) analysis; these were subsequently shown to be encoded by the same gene (Cole *et al.*, 1998; Qin *et al.*, 2004). Although the nature of these different isoforms is not known, it is likely that the larger molecular weight form is phosphorylated. The recently identified protein IFT 25 has been shown to be phosphorylated (Lechtreck *et al.*, 2009; Wang *et al.*, 2009). IFT particles undergo remodeling as well as loading or unloading cargo at the cilia tip and base; therefore, protein phosphorylation may play a key role during these processes. The identification of the protein kinases involved will provide further insight into how IFT and cilia assembly are regulated.

D. Regulation of Acetylation of Axonemal Microtubules

Cilia are acetylated during assembly and deacetylated prior to disassembly (L'Hernault and Rosenbaum, 1983, 1985). In *Chlamydomonas*, it has been shown that the acetylation occurs on K40 of α-tubulin; however, ectopic overexpression of α-tubulin with a K-to-R mutation does not affect flagellar assembly, disassembly, or apparent flagellar motility (Kozminski *et al.*, 1993). Although its exact role is not understood, the acetylation process is controlled by phosphorylation. In mammalian cells, phosphatase inhibitor-2 is enriched in primary cilia and knockdown of this inhibitor by RNAi reduces cilia acetylation levels. Furthermore, treatment with calyculin A, a phosphatase inhibitor, partially recovers the reduced acetylation level of cilia in knockdown cells. Thus, protein phosphorylation positively regulates acetylation of cilia (Wang and Brautigan, 2008).

E. Regulation of Ciliary Gene Expression

The transcription of ciliary genes is under active control. In *Chlamydomonas*, when flagella are regenerated after *in vitro* deflagellation, a set of specific genes including those encoding tubulin and flagellar axonemal components are highly induced (Lefebvre and Rosenbaum, 1986; Li *et al.*, 2004; Stolc *et al.*, 2005). Cilia are separated from the cell body by the proposed "flagellar pore" at the ciliary base (Rosenbaum and Witman, 2002). The protein components of the ciliary membrane, matrix, axoneme, and accessory structures are either unique to or highly enriched in the cilia. All of the genes involved in cilia structure, function, or regulation are termed ciliary genes. It is likely that these ciliary genes are under the same transcriptional control. Indeed, in

C. elegans, a bioinformatic approach reveals that about 750 ciliary genes contain the X-box motif in the promoter region. The transcription of these genes is postulated to be under the control of members of the forkhead transcription factors (Blacque *et al.*, 2005; Efimenko *et al.*, 2005). Several studies have demonstrated that forkhead transcription factor hepatocyte nuclear factor-3/forkhead homologue 4 (HFH-4) (also known as "Foxj1" or "RFX-3") from either mouse or zebrafish is involved in ciliogenesis (Chen *et al.*, 1998; Stubbs *et al.*, 2008; Yu *et al.*, 2008). The signaling cascade leading to the activation of transcription factors is not well established. In *Chlamydomonas*, when expression of flagellar-specific genes is induced during flagellar regeneration, a few phosphoproteins appear as mentioned above. Evidence is still needed to show a direct link between protein phosphorylation and transcription of ciliary genes. Recently, it has been shown that disrupting fibroblast growth factor (FGF) signaling in zebrafish results in downregulation of the transcription of Foxj1 transcription factors and genes required for cilia assembly (Neugebauer *et al.*, 2009). Thus, protein kinase activity involved in FGF signaling may regulate transcription of ciliary genes (Eswarakumar *et al.*, 2005).

F. Protein Phosphorylation Control over Centrioles for Ciliogenesis

The fact that no cilia are assembled during mitosis, even though centrioles are available, indicates the presence of a ciliogenesis suppressor. The initial step of cilia assembly after cytokinesis involves converting a mitotic centriole into a cilia assembly-competent centriole. What proteins are involved and how are they controlled?

Centriolar protein CP110 has been found to interact with centrosomal protein of 97 kDa (Cep97) to form a complex that suppresses cilia formation (Spektor *et al.*, 2007). Downregulation of CP110 expression results in aberrant cilia formation in cells undergoing mitosis, whereas overexpression of CP110 suppresses cilia formation in quiescent cells. Interestingly, in quiescent cells, this complex is localized to the daughter centriole, from which no cilium is formed. This protein complex likely functions by blocking microtubule polymerization on the centriole. Because protein phosphorylation exerts tight control over cell cycle progression (Hunter, 1987), studying the conversion of different functional centrioles may reveal the mechanisms by which the cell cycle regulates ciliary assembly.

III. Protein Phosphorylation in Cilia Disassembly

Before mitosis, primary cilia are lost in order to release the basal body to engage in mitosis (Pan and Snell, 2007; Quarmby and Parker, 2005). The precise phase of the cell cycle during which cilia are lost has not been definitively determined; different researchers have proposed different answers, from the S phase, to the G2/M phase, to prophase (Pugacheva *et al.*, 2007; Rieder *et al.*, 1979; Tucker *et al.*, 1979). Cells may lose cilia by deflagellation or cilia resorption. The current evidence favors the view that cilia resorption occurs prior to cell division (Cavalier-Smith, 1974; Marshall *et al.*, 2005; Rieder *et al.*, 1979; Tucker *et al.*, 1979; our unpublished data). In *Chlamydomonas* studies, flagella can occasionally be seen in cells undergoing mitosis, thereby raising the question of whether flagellar loss is a necessary step for cell division (Piasecki *et al.*, 2008). Because deflagellation and flagellar resorption share common signaling components (Pan *et al.*, 2004; Parker and Quarmby, 2003), it is likely that when flagellar resorption fails to proceed properly, the deflagellation mechanism is improperly activated. Thus, the flagella, though separated from the basal body by activation of the deflagellation mechanism, may still be attached to the cell body through the flagellar membrane.

A. *Chlamydomonas* Aurora-Like Protein Kinase in Cilia Disassembly

How are cilia disassembled? What are the molecular players? *Chlamydomonas* aurora-like protein kinase (CALK), an aurora-like protein kinase, was the first enzyme shown to be phosphorylated during flagellar resorption (Pan *et al.*, 2004). CALK is phosphorylated immediately after flagellar shortening has been induced. Simple treatment of cells with sodium pyrophosphate, which triggers flagellar shortening (Lefebvre *et al.*, 1978), induces CALK phosphorylation. During zygote maturation, when flagella undergo gradual shortening, CALK is phosphorylated. Suppression of CALK expression by RNAi impairs flagellar shortening, demonstrating that CALK is required for flagellar disassembly.

Our unpublished data show that CALK is phosphorylated at multiple sites. However, the protein kinase(s) that phosphorylate(s) CALK and its substrates is still elusive. The flagellar length is determined by the balance of assembly and disassembly activities; therefore, CALK may be involved in controlling flagellar length. CALK is the fundamental member in the flagellar disassembly pathway, and the discovery of its phosphorylation is one step to revealing the complete signaling cascade that leads to flagellar disassembly.

B. The Aurora-A–HDAC Pathway in Primary Cilia Disassembly

Shortly after the discovery of the role of CALK during flagellar disassembly in *Chlamydomonas*, aurora-A was demonstrated to play a pivotal role in primary cilia disassembly in mammalian cells (Pugacheva *et al.*, 2007). Aurora-A undergoes protein phosphorylation during cilia resorption prior to mitosis. Once aurora-A has been activated, resorption of primary cilia begins. Microtubule deacetylase HDAC6 is found to be a target of aurora-A, and the activation of HDAC6 by aurora-A phosphorylation is essential for cilia resorption. The cilia resorption pathway is extremely well conserved from unicellular green algae to mammalian cells.

C. Flagellar Targeting of CrKinesin13 and Regulation of IFT

The disassembly of the flagellar axoneme starts from the tip (Marshall and Rosenbaum, 2001). Surprisingly, CrKinesin13, which participates in flagellar assembly as discussed above, is transported to the flagella soon after flagellar resorption has been induced (Piao *et al.*, 2009). Suppressing CrKinesin13 expression by RNAi inhibits flagellar shortening. Thus, CrKinesin13 functions in disassembling axonemal microtubules during flagellar shortening. In *Leishmania* and *Giardia*, homologues of CrKinesin13 have been reported to be involved in flagellar length control (Blaineau *et al.*, 2007; Dawson *et al.*, 2007). The working mechanisms and regulation of these homologues may differ from those of CrKinesin13, since the *Leishmania* homologue is constitutively present in the flagella. Knowing that CALK is phosphorylated during flagellar disassembly, researchers must still determine whether CALK is involved in flagellar targeting of CrKinesin13.

The disassembled flagellar components are carried back via IFT (Qin *et al.*, 2004), which is deliberately regulated to coordinate the disassembly process. The transport of CrKinesin13 requires IFT because its transport is inhibited at 32°C in the temperature-sensitive mutant of IFT motor protein FLA10. During flagellar disassembly, IFT trafficking increases severalfold, while IFT anterograde cargo loading is suppressed and retrograde cargo loading is activated (Pan and Snell, 2005). Thus, the disassembled flagellar components can be transported back to the cell body rapidly and efficiently. It remains to be shown whether the protein phosphorylation cascade regulates IFT during flagellar resorption.

Apart from aurora kinases, NIMA-related protein kinases (Nrks) may also actively participate in flagellar resorption, since overexpression of Nrks results in cells with shorter cilia (Wloga *et al.*, 2006). Biochemical data showing changes in the properties of Nrks during flagellar resorption may provide key evidence to support this view (more in the following section).

IV. Protein Phosphorylation in Cilia Length Control

Members of divergent protein kinase families play a role in controlling cilia length (Wilson *et al.*, 2008). The ciliary function of these protein kinases has largely been revealed by mutant analysis, RNAi suppression, and/or gene overexpression.

A. LF2 and LF4 Protein Kinases and Long-Flagellar Mutants of *Chlamydomonas*

In *Chlamydomonas*, cells assemble flagella to a constant length of approximately 12 μm. The same flagellar length is maintained until the beginning of the next cell cycle or cellular stress. Four long-flagellar (lf) mutants have been identified (Asleson and Lefebvre, 1998; Barsel *et al.*, 1988). Defects in any of four long-flagella genes result in loss of flagellar length control, creating cells with flagella whose lengths vary and can measure more than 20 μm. Interestingly, two of these four genes encode protein kinases, demonstrating that protein phosphorylation plays a pivotal role in controlling flagellar length. The first protein kinase identified is a homologue of a mitogen-activated protein (MAP) kinase, and it is encoded by long-flagella gene *LF4* (Berman *et al.*, 2003). Though LF4p is barely present in the flagella, the *LF4* transcript is upregulated during flagellar regeneration, implicating its possible role in flagellar assembly. Surprisingly, *lf4* null mutants can assemble flagella during the cell cycle and after deflagellation by pH shock. Since the majority of *lf4* null mutant cells possess flagella longer than 12 μm, LF4p apparently maintains flagellar length by shortening (Berman *et al.*, 2003).

The second protein kinase identified is a novel member of the cyclin-dependent protein kinase family, which lacks the cyclin-binding motif (Tam *et al.*, 2007). The *lf2* null mutant produces multiple phenotypes indicative of defects in flagellar length regulation. They have stumpy or unequal length flagella, accumulate IFT particles at the flagellar tips, and are unable to regenerate flagella after deflagellation by pH shock. In contrast, an allelic mutant *lf2-3*, in which the protein kinase activity of LF2p is supposedly disrupted because of an aberrant insertion of a short amino acid sequence within the protein kinase domain, produces more than 50% cells with abnormally long flagella. Furthermore, *lf2* null mutant cells transformed with a kinase dead construct display a long-flagella phenotype and are able to regenerate flagella, though slowly. Thus, protein kinase activity may be required to enforce normal flagellar length. The fact that *lf2-3* mutant cells, but not *lf2* null mutant cells, are able to regenerate flagella after pH shock indicates that the intact protein structure —independent of protein kinase activity—may be required for flagellar assembly. Biochemical analysis including yeast two-hybrid assay and sucrose gradient separation followed by Western blotting demonstrates that LF2p forms a protein complex named the length regulation complex (LRC). This complex includes two long-flagella proteins, LF1p and LF3p (Nguyen *et al.*, 2005; Tam *et al.*, 2003, 2007). Because complete knockout of either component of the LRC complex results in flagellar regeneration defects, the major role of the LRC complex may involve flagellar assembly rather than length control.

B. NIMA-Related Protein Kinases

Evidence obtained from mammals, worms, and *Chlamydomonas* implicates Nrks in ciliary control. In kat and jck murine models of polycystic kidney disease, the mutated genes have been identified as *Nek*1 and *Nek*8, respectively (Guay-Woodford, 2003; Liu *et al.*, 2002; Upadhya *et al.*, 2000). Both are members of the Nrk family. The involvement of Nrks in cystic kidney disease indicates their possible connection with ciliary regulation. Indeed, *Nek1* is localized in the basal body of primary cilia of inner medullary collecting duct (IMCD) cells and at the centrosome during cell division, but not found in the cilia themselves (Bradley and Quarmby, 2005). *Nek8* is localized in the proximal region of primary cilia from cultured cells as well as those from kidney tissue (Mahjoub *et al.*, 2005; Sohara *et al.*, 2008). It also forms a protein complex with polycystin-2 (PC-2), a calcium channel whose mutation results in polycystic kidney disease (Nauli *et al.*, 2003). *Nek8* mutation results in mislocalization of *Nek8* along the full length of cilia, which induces phosphorylation of PC-2 and increased expression of PC-1 and PC-2. In addition, *Nek8* mutant mice seem to have abnormally long cilia. *Nek8* in jck mice has a missense mutation, resulting in the substitution of a conserved glycine by valine at position 448, outside the protein kinase domain (Sohara *et al.*,

2008). Thus, the conserved glycine may be involved in the cellular localization control of *Nek8*. In addition, the proper localization of a protein kinase in cilia may regulate ciliary gene expression and the functionality of the intact cilium.

Evidence from *Chlamydomonas* and *Tetrahymena* also supports the role of Nrks in ciliary regulation. In *Tetrahymena*, several members of the Nrk family, including *Nrk1, Nrk2, Nrk17*, and *Nrk30*, are present in cilia and/or basal bodies. Overexpression of *Nrk2, Nrk17*, or *Nrk30* under the control of inducible promoters results in cilia shortening or disassembly, whereas overexpression of a kinase-dead *Nrk2* induces cilia lengthening (Wloga *et al.*, 2006). In *Chlamydomonas, FA2* and *CNK2*, two members of the Nrk family, are involved in flagellar function. *FA2* and its protein kinase activity are essential for deflagellation (Mahjoub *et al.*, 2004). Overexpression of *Cnk2* results in cells with shorter flagella, whereas RNAi knockdown of *Cnk2* expression results in longer flagella (Bradley and Quarmby, 2005).

C. Mitogen-Activated Protein Kinases

In *Leishmania,* MAP kinases have been shown to regulate flagellar length. There are two cell stages in the life cycle of *Leishmania*: promastigote and amastigote. The promastigote has long flagella and the ability to swim, whereas the amastigote has short flagella buried in the flagellar pockets and cannot swim. The two stages may differentiate into each other due to variations in environmental factors such as temperature and pH (Zilberstein and Shapira, 1994). Promastigotes with the deletion of a mitogen-activated protein (MAP) kinase kinase from *Leishmania mexicana*, (*LmxMKK*) encoding a MAP kinase kinase or *LmxMPK3* encoding a MAP kinase exhibit flagella with an average length of 2 μm, about one-fifth the flagellar length of the wild type. It has been noticed that 11% of the flagella in the *LmxMKK* mutant lack central pair microtubules, indicating that LmxMKK-mediated signaling may be involved in flagellar assembly (Erdmann *et al.*, 2006; Wiese *et al.*, 2003). However, deletion of the MAP kinase gene *LmxMPK9* induces longer flagella in promastigotes, in sharp contrast to the phenotype of an *LmxMPK3*-deleted mutant (Bengs *et al.*, 2005). This comparison indicates a complex pathway regulating flagellar assembly and length control by MAP kinase signaling in *Leishmania*. The conservation of flagellar length control by MAP kinases in other model organisms remains to be shown.

V. Conclusion

The protein kinases and/or their substrates that have been implicated in ciliary regulation are depicted in Fig. 1. The phosphoproteins involved in assembly and/or disassembly of cilia (e.g., LF2p) may control ciliary length as well. However, phosphoproteins that regulate flagellar length may not be required for flagellar assembly (e.g., LF4p). Genetic evidence supports the idea that most of these phosphoproteins play a role in ciliary regulation; however, relevant biochemical evidence or cellular property changes must also be demonstrated to pinpoint the roles of these

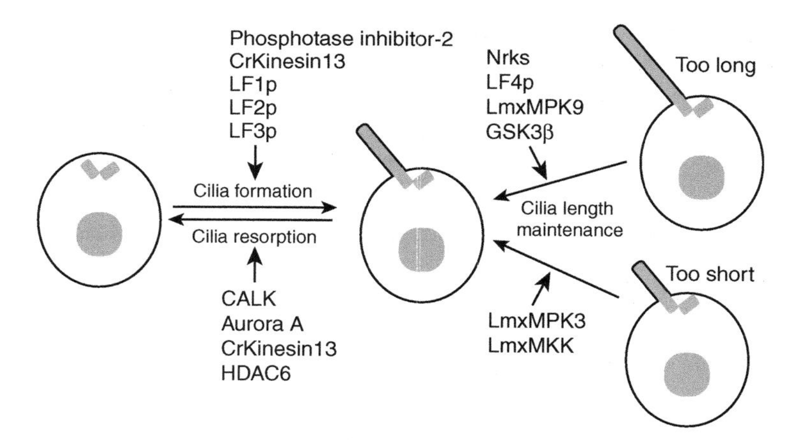

Fig. 1 A drawing of a simple cell with one round nucleus and one primary cilium is used as the prototype of a ciliated cell. Phosphoproteins involved in phosphorylation control of different stages of the "life cycle" of cilia are depicted. Although only positive regulation by these proteins is shown, negative regulation may also occur. (Please refer to the text for the description of individual proteins.)

phosphoproteins in a particular step of ciliary control. The participation of multiple protein kinases in controlling cilia length raises new questions for future research, including whether and how these kinases cross talk and what their substrates are. The answers to these questions will facilitate our understanding of the molecular mechanisms underlying ciliary regulation.

Acknowledgments

We are indebted to Dr. William Snell from UT Southwestern Medical Center, Dallas, for insightful suggestion and comments on the manuscript. This work was supported in part by the National Natural Science Foundation of China (Nos. 30671090, 30830057, and 30771084) and the National Basic Research Program of China (No. 2007CB914401) (to JP).

References

Asleson, C.M., and Lefebvre, P.A. (1998). Genetic analysis of flagellar length control in *Chlamydomonas reinhardtii*: A new long-flagella locus and extragenic suppressor mutations. *Genetics* **148**, 693–702.

Barsel, S.E., Wexler, D.E., and Lefebvre, P.A. (1988). Genetic analysis of long-flagella mutants of *Chlamydomonas reinhardtii. Genetics* **118**, 637–648.

Bengs, F., Scholz, A., Kuhn, D., and Wiese, M. (2005). LmxMPK9, a mitogen-activated protein kinase homologue affects flagellar length in *Leishmania mexicana. Mol. Microbiol.* **55**, 1606–1615.

Berman, S.A., Wilson, N.F., Haas, N.A., and Lefebvre, P.A. (2003). A novel MAP kinase regulates flagellar length in *Chlamydomonas. Curr. Biol.* **13**, 1145–1149.

Blacque, O.E., Perens, E.A., Boroevich, K.A., Inglis, P.N., Li, C., Warner, A., Khattra, J., Holt, R.A., Ou, G., Mah, A.K., McKay, S.J., Huang, P., et al. (2005). Functional genomics of the cilium, a sensory organelle. *Curr. Biol.* **15**, 935–941.

Blaineau, C., Tessier, M., Dubessay, P., Tasse, L., Crobu, L., Pages, M., and Bastien, P. (2007). A novel microtubule-depolymerizing kinesin involved in length control of a eukaryotic flagellum. *Curr. Biol.* **17**, 778–782.

Bloodgood, R.A. (1992). Calcium-regulated phosphorylation of proteins in the membrane-matrix compartment of the *Chlamydomonas* flagellum. *Exp. Cell Res.* **198**, 228–236.

Boesger, J., Wagner, V., Weisheit, W., and Mittag, M. (2009). Analysis of flagellar phosphoproteins from *Chlamydomonas reinhardtii. Eukaryot. cell* **8**, 922–932.

Bradley, B.A., and Quarmby, L.M. (2005). A NIMA-related kinase, Cnk2p, regulates both flagellar length and cell size in *Chlamydomonas. J. Cell Sci.* **118**, 3317–3326.

Buttrick, G.J., and Wakefield, J.G. (2008). PI3-K and GSK-3: Akt-ing together with microtubules. *Cell Cycle* **7**, 2621–2625.

Cavalier-Smith, T. (1974). Basal body and flagellar development during the vegetative cell cycle and the sexual cycle of *Chlamydomonas reinhardtii. J. Cell Sci.* **16**, 529–556.

Chen, J., Knowles, H.J., Hebert, J.L., and Hackett, B.P. (1998). Mutation of the mouse hepatocyte nuclear factor/forkhead homologue 4 gene results in an absence of cilia and random left-right asymmetry. *J. Clin. Invest.* **102**, 1077–1082.

Cole, D.G. (2003). The intraflagellar transport machinery of *Chlamydomonas reinhardtii. Traffic* **4**, 435–442.

Cole, D.G., Diener, D.R., Himelblau, A.L., Beech, P.L., Fuster, J.C., and Rosenbaum, J.L. (1998). *Chlamydomonas* kinesin-II-dependent intraflagellar transport (IFT): IFT particles contain proteins required for ciliary assembly in *Caenorhabditis elegans* sensory neurons. *J. Cell Biol.* **141**, 993–1008.

Cole, D.G., and Snell, W.J. (2009). SnapShot: Intraflagellar transport. *Cell.* **137**, 784–784.e1.

Dawson, S.C., Sagolla, M.S., Mancuso, J.J., Woessner, D.J., House, S.A., Fritz-Laylin, L., and Cande, W.Z. (2007). Kinesin-13 regulates flagellar, interphase, and mitotic microtubule dynamics in *Giardia intestinalis. Eukaryot. cell* **6**, 2354–2364.

Efimenko, E., Bubb, K., Mak, H.Y., Holzman, T., Leroux, M.R., Ruvkun, G., Thomas, J.H., and Swoboda, P. (2005). Analysis of xbx genes in *C. elegans. Development* **132**, 1923–1934.

Erdmann, M., Scholz, A., Melzer, I.M., Schmetz, C., and Wiese, M. (2006). Interacting protein kinases involved in the regulation of flagellar length. *Mol. Biol. Cell.* **17**, 2035–2045.

Eswarakumar, V.P., Lax, I., and Schlessinger, J. (2005). Cellular signaling by fibroblast growth factor receptors. *Cytokine Growth Factor Rev.* **16**, 139–149.

Fliegauf, M., Benzing, T., and Omran, H. (2007). When cilia go bad: Cilia defects and ciliopathies. *Nat. Rev. Mol. Cell Biol.* **8**, 880–893.

Forde, J.E., and Dale, T.C. (2007). Glycogen synthase kinase 3: A key regulator of cellular fate. *Cell. Mol. Life Sci.* **64**, 1930–1944.

Gerdes, J.M., Davis, E.E., and Katsanis, N. (2009). The vertebrate primary cilium in development, homeostasis, and disease. *Cell* **137**, 32–45.

Guay-Woodford, L.M. (2003). Murine models of polycystic kidney disease: Molecular and therapeutic insights. *Am. J. Physiol. Renal. Physiol.* **285**, F1034–F1049.

Hagiwara, H., Ohwada, N., Aoki, T., and Takata, K. (2000). Ciliogenesis and ciliary abnormalities. *Med. Electron. Microsc.* **33**, 109–114.

Harper, J.D., Sanders, M.A., and Salisbury, J.L. (1993). Phosphorylation of nuclear and flagellar basal apparatus proteins during flagellar regeneration in *Chlamydomonas reinhardtii. J. Cell Biol.* **122**, 877–886.

Hunter, T. (1987). A thousand and one protein kinases. *Cell.* **50**, 823–829.

Kozminski, K.G., Diener, D.R., and Rosenbaum, J.L. (1993). High level expression of nonacetylatable alpha-tubulin in *Chlamydomonas reinhardtii. Cell Motil. Cytoskeleton* **25**, 158–170.

L'Hernault, S.W., and Rosenbaum, J.L. (1983). *Chlamydomonas* alpha-tubulin is posttranslationally modified in the flagella during flagellar assembly. *J. Cell Biol.* **97**, 258–263.

L'Hernault, S.W., and Rosenbaum, J.L. (1985). *Chlamydomonas* alpha-tubulin is posttranslationally modified by acetylation on the epsilon-amino group of a lysine. *Biochemistry (Mosc).* **24**, 473–478.

Lechtreck, K.F., Luro, S., Awata, J., and Witman, G.B. (2009). HA-tagging of putative flagellar proteins in *Chlamydomonas reinhardtii* identifies a novel protein of intraflagellar transport complex B. *Cell Motil. Cytoskeleton* **66**, 469–482

Lefebvre, P.A., Nordstrom, S.A., Moulder, J.E., and Rosenbaum, J.L. (1978). Flagellar elongation and shortening in *Chlamydomonas*. IV. Effects of flagellar detachment, regeneration, and resorption on the induction of flagellar protein synthesis. *J. Cell Biol.* **78**, 8–27.

Lefebvre, P.A., and Rosenbaum, J.L. (1986). Regulation of the synthesis and assembly of ciliary and flagellar proteins during regeneration. *Annu. Rev. Cell Biol.* **2**, 517–546.

Li, J.B., Gerdes, J.M., Haycraft, C.J., Fan, Y., Teslovich, T.M., May-Simera, H., Li, H., Blacque, O.E., Li, L., Leitch, C.C., Lewis, R.A., Green, J.S., et al. (2004). Comparative genomics identifies a flagellar and basal body proteome that includes the BBS5 human disease gene. *Cell.* **117**, 541–552.

Liu, S., Lu, W., Obara, T., Kuida, S., Lehoczky, J., Dewar, K., Drummond, I.A., and Beier, D.R. (2002). A defect in a novel Nek-family kinase causes cystic kidney disease in the mouse and in zebrafish. *Development* **129**, 5839–5846.

Mahjoub, M.R., Qasim Rasi, M., and Quarmby, L.M. (2004). A NIMA-related kinase, Fa2p, localizes to a novel site in the proximal cilia of *Chlamydomonas* and mouse kidney cells. *Mol. Biol. Cell.* **15**, 5172–5186.

Mahjoub, M.R., Trapp, M.L., and Quarmby, L.M. (2005). NIMA-related kinases defective in murine models of polycystic kidney diseases localize to primary cilia and centrosomes. *J. Am. Soc. Nephrol.* **16**, 3485–3489.

Marshall, W.F., Qin, H., Rodrigo Brenni, M., and Rosenbaum, J.L. (2005). Flagellar length control system: Testing a simple model based on intraflagellar transport and turnover. *Mol. Biol. Cell.* **16**, 270–278.

Marshall, W.F., and Rosenbaum, J.L. (2001). Intraflagellar transport balances continuous turnover of outer doublet microtubules: Implications for flagellar length control. *J. Cell Biol.* **155**, 405–414.

Nakamura, S., Tabino, H., and Kojima, M.K. (1987). Effect of lithium on flagellar length in *Chlamydomonas reinhardtii. Cell Struct. Funct.* **12**, 369–374.

Nauli, S.M., Alenghat, F.J., Luo, Y., Williams, E., Vassilev, P., Li, X., Elia, A.E., Lu, W., Brown, E.M., Quinn, S.J., Ingber, D.E., and Zhou, J. (2003). Polycystins 1 and 2 mediate mechanosensation in the primary cilium of kidney cells. *Nat. Genet.* **33**, 129–137.

Neugebauer, J.M., Amack, J.D., Peterson, A.G., Bisgrove, B.W., and Yost, H.J. (2009). FGF signalling during embryo development regulates cilia length in diverse epithelia. *Nature* **458**, 651–654.

Nguyen, R.L., Tam, L.W., and Lefebvre, P.A. (2005). The LF1 gene of *Chlamydomonas reinhardtii* encodes a novel protein required for flagellar length control. *Genetics* **169**, 1415–1424.

Pan, J. (2008). Cilia and ciliopathies: From *Chlamydomonas* and beyond. *Sci. China. C. Life Sci.* **51**, 479–486.

Pan, J., and Snell, W.J. (2007). The primary cilium: Keeper of the key to cell division. *Cell* **129**, 1255–1257.

Pan, J., and Snell, W.J. (2005). *Chlamydomonas* shortens its flagella by activating axonemal disassembly, stimulating IFT particle trafficking, and blocking anterograde cargo loading. *Dev. Cell* **9**, 431–438.

Pan, J., Wang, Q., and Snell, W.J. (2004). An aurora kinase is essential for flagellar disassembly in *Chlamydomonas. Dev. Cell.* **6**, 445–451.

Parker, J.D., and Quarmby, L.M. (2003). *Chlamydomonas* fla mutants reveal a link between deflagellation and intraflagellar transport. *BMC Cell Biol.* **4**, 11.

Periz, G., Dharia, D., Miller, S.H., and Keller, L.R. (2007). Flagellar elongation and gene expression in *Chlamydomonas reinhardtii. Eukaryot. cell* **6**, 1411–1420.

Piao, T., Luo, M., Wang, L., Guo, Y., Li, D., Li, P., Snell, W.J., and Pan, J. (2009). A microtubule depolymerizing kinesin functions during both flagellar disassembly and flagellar assembly in *Chlamydomonas. Proc. Natl. Acad. Sci. USA* **106**, 4713–4718.

Piasecki, B.P., LaVoie, M., Tam, L.W., Lefebvre, P.A., and Silflow, C.D. (2008). The Uni2 phosphoprotein is a cell cycle regulated component of the basal body maturation pathway in *Chlamydomonas reinhardtii. Mol. Biol. Cell.* **19**, 262–273.

Piperno, G., and Mead, K. (1997). Transport of a novel complex in the cytoplasmic matrix of *Chlamydomonas* flagella. *Proc. Natl. Acad. Sci. USA* **94**, 4457–4462.

Pugacheva, E.N., Jablonski, S.A., Hartman, T.R., Henske, E.P., and Golemis, E.A. (2007). HEF1-dependent Aurora A activation induces disassembly of the primary cilium. *Cell* **129**, 1351–1363.

Qin, H., Diener, D.R., Geimer, S., Cole, D.G., and Rosenbaum, J.L. (2004). Intraflagellar transport (IFT) cargo: IFT transports flagellar precursors to the tip and turnover products to the cell body. *J. Cell Biol.* **164**, 255–266.

Quarmby, L.M., and Parker, J.D. (2005). Cilia and the cell cycle? *J. Cell Biol.* **169**, 707–710.

Rieder, C.L., Jensen, C.G., and Jensen, L.C. (1979). The resorption of primary cilia during mitosis in a vertebrate (PtK1) cell line. *J. Ultrastruct. Res.* **68**, 173–185.

Rosenbaum, J.L., and Witman, G.B. (2002). Intraflagellar transport. *Nat. Rev. Mol. Cell Biol.* **3**, 813–825.

Scholey, J.M. (2003). Intraflagellar transport. *Annu. Rev. Cell Dev. Biol.* **19**, 423–443.

Sohara, E., Luo, Y., Zhang, J., Manning, D.K., Beier, D.R., and Zhou, J. (2008). Nek8 regulates the expression and localization of polycystin-1 and polycystin-2. *J. Am. Soc. Nephrol.* **19**, 469–476.

Sorokin, S.P. (1962). Centrioles and the formation of rudimentary cilia by fibroblasts and smooth muscle cells. *J. Cell Biol.* **15**, 363–377.

Sorokin, S.P. (1968). Reconstructions of centriole formation and ciliogenesis in mammalian lungs. *J. Cell Sci.* **3**, 207–230.

Spektor, A., Tsang, W.Y., Khoo, D., and Dynlacht, B.D. (2007). Cep97 and CP110 suppress a cilia assembly program. *Cell* **130**, 678–690.

Stolc, V., Samanta, M.P., Tongprasit, W., and Marshall, W.F. (2005). Genome-wide transcriptional analysis of flagellar regeneration in *Chlamydomonas reinhardtii* identifies orthologs of ciliary disease genes. *Proc. Natl. Acad. Sci. USA* **102**, 3703–3707.

Stubbs, J.L., Oishi, I., Izpisua Belmonte, J.C., and Kintner, C. (2008). The forkhead protein Foxj1 specifies node-like cilia in Xenopus and zebrafish embryos. *Nat. Genet.* **40**, 1454–1460.

Tam, L.W., Dentler, W.L., and Lefebvre, P.A. (2003). Defective flagellar assembly and length regulation in LF3 null mutants in *Chlamydomonas. J. Cell Biol.* **163**, 597–607.

Tam, L.W., Wilson, N.F., and Lefebvre, P.A. (2007). A CDK-related kinase regulates the length and assembly of flagella in *Chlamydomonas. J. Cell Biol.* **176**, 819–829.

Thoma, C.R., Frew, I.J., Hoerner, C.R., Montani, M., Moch, H., and Krek, W. (2007). pVHL and GSK3beta are components of a primary cilium-maintenance signalling network. *Nat. Cell Biol.* **9**, 588–595.

Tucker, R.W., Scher, C.D., and Stiles, C.D. (1979). Centriole deciliation associated with the early response of 3T3 cells to growth factors but not to SV40. *Cell.* **18**, 1065–1072.

Upadhya, P., Birkenmeier, E.H., Birkenmeier, C.S., and Barker, J.E. (2000). Mutations in a NIMA-related kinase gene, Nek1, cause pleiotropic effects including a progressive polycystic kidney disease in mice. *Proc. Natl. Acad. Sci. USA* **97**, 217–221.

Wang, W., and Brautigan, D.L. (2008). Phosphatase inhibitor 2 promotes acetylation of tubulin in the primary cilium of human retinal epithelial cells. *BMC Cell Biol.* **9**, 62.

Wang, Z., Fan, Z.C., Williamson, S.M., and Qin, H. (2009). Intraflagellar transport (IFT) protein IFT25 is a phosphoprotein component of IFT complex B and physically interacts with IFT27 in *Chlamydomonas. PLoS One* **4**, e5384.

Westendorf, J.M., Rao, P.N., and Gerace, L. (1994). Cloning of cDNAs for M-phase phosphoproteins recognized by the MPM2 monoclonal antibody and determination of the phosphorylated epitope. *Proc. Natl. Acad. Sci. USA* **91**, 714–718.

Wiese, M., Kuhn, D., and Grunfelder, C.G. (2003). Protein kinase involved in flagellar-length control. *Eukaryot. Cell* **2**, 769–777.

Williams, R., Ryves, W.J., Dalton, E.C., Eickholt, B., Shaltiel, G., Agam, G., and Harwood, A.J. (2004). A molecular cell biology of lithium. *Biochem. Soc. Trans.* **32**, 799–802.

Wilson, N.F., Iyer, J.K., Buchheim, J.A., and Meek, W. (2008). Regulation of flagellar length in *Chlamydomonas. Semin. Cell Dev. Biol.* **19**, 494–501.

Wilson, N.F., and Lefebvre, P.A. (2004). Regulation of flagellar assembly by glycogen synthase kinase 3 in *Chlamydomonas reinhardtii. Eukaryot. Cell* **3**, 1307–1319.

Wloga, D., Camba, A., Rogowski, K., Manning, G., Jerka-Dziadosz, M., and Gaertig, J. (2006). Members of the NIMA-related kinase family promote disassembly of cilia by multiple mechanisms. *Mol. Biol. Cell* **17**, 2799–2810.

Yu, X., Ng, C.P., Habacher, H., and Roy, S. (2008). Foxj1 transcription factors are master regulators of the motile ciliogenic program. *Nat. Genet.* **40**, 1445–1453.

Zilberstein, D., and Shapira, M. (1994). The role of pH and temperature in the development of *Leishmania* parasites. *Annu. Rev. Microbiol.* **48**, 449–470.

CHAPTER 18

Posttranslational Protein Modifications in Cilia and Flagella

Roger D. Sloboda

Biological Sciences, Dartmouth College, Hanover, New Hampshire 03755 and The Marine Biological Laboratory, Woods Hole, Massachusetts 02543

Abstract

Tubulin and other flagellar and ciliary proteins are the substrates for a host of posttranslational modifications (PTMs), many of which have been highly conserved over evolutionary time. In addition to the binding of MAPs (microtubule-associated proteins) that provide a specific functionality, or the use of different tubulin isotypes to convey a specific function, most cells rely on an array of PTMs. These include phosphorylation, acetylation, glycylation, glutamylation, and methylation. The first

978-0-12-375024-2
DOI: 10.1016/S0091-679X(08)94018-8

and the last of this list are not unique to the tubulin in cilia and flagella, while the others are. This chapter will review briefly these varying modifications and will conclude with detailed methods for their detection and localization at the limit of resolution provided by electron micrscopy.

I. Introduction

The basic structure of the subunit protein of microtubules, the tubulin dimer composed of α- and β-tubulin, is essentially unchanged over evolutionary time. For example, yeast tubulin and mammalian brain tubulin will assemble into hybrid copolymers (Clayton *et al.*, 1979; Kilmartin, 1981), and a chimeric β-tubulin comprised of chick and yeast sequences when transfected into mouse cells produces a β-tubulin that incorporates efficiently into all of the microtubules of the host cell (Bond *et al.*, 1986). Such data suggest the microtubule subunit lattice is generic and does not itself alone specify all of the varied functions of microtubules in cells, which can have different tasks and characteristics in different types of cells or even the same cell type. For example, microtubules can be stable or dynamic in interphase and in the mitotic apparatus; assemble into single microtubules in the cytoplasm or complex arrays in cilia, flagella, and axostyles; comprise the circular marginal band; and organize into triplets in basal bodies and centrioles, all belying the seemingly universal evolutionary conservation of the basic structure of the tubulin dimer.

In vitro, purified tubulin assembles into microtubules having 12–15 subunits in cross section. If the lateral contacts between protofilaments are between α- and β-tubulin (i.e., αβ), the resulting subunit lattice is termed the A lattice. If the lateral contacts are αα and ββ, the B-lattice results. This terminology arose from the initial observation that the A and B subfibers of flagellar outer doublets had different lattices (Amos and Klug, 1974). However, more recent data indicate that the outer doublet microtubules display only the B lattice, as do microtubules assembled *in vitro* from purified tubulin. This conclusion was drawn from diffraction data performed on microtuble samples decorated by the binding of the head of squid kinesin to flagellar outer doublets or to *in vitro* assembled microtubules (Song and Mandelkow, 1993). Kinesin binds only to β-tubulin, and this binding provides a strong reflection that clearly reveals via X-ray diffraction the lattice of the decorated microtubule. Song and Mandelkow argue that reflections indicating the A lattice in the A tubule of the outer doublets originally reported by Amos and Klug may have resulted from the presence of microtubule associated proteins (MAPs) (Sloboda *et al.*, 1975) in that preparation, a conclusion suggested by the following observations. The presence of MAPs such as end-binding protein 1 (EB1) (Vitre *et al.*, 2008) or doublecortin (Moores *et al.*, 2004) has been reported to constrain the assembly of tubulin *in vitro* into microtubules having predominantly 13 protofilaments and the A-lattice conformation.

Thus, with respect to overall structure, microtubules alone are relatively generic, and specific functionality is determined in one of three ways. First, the binding of specific MAPs to microtubules conveys specific function, and this has been shown quite clearly by the Dis1/TOG family of MAPs (Ohkura *et al.*, 2001), members of which have been found in yeast to mammals, including the model plant *Arabidopsis*. Via their interaction with various microtubule containing structures, and their ability to bind other classes of regulatory and motor molecules, the Dis1/TOG MAPs help to convey specific functionalities on different microtubule-based organelles. They are thus important in spindle, centrosome, and kinetochore function, spindle assembly, cortical microtubule assembly and hence cell morpohogenesis, and so on. Specificity of MAP binding may arise from tubulin isotype variations in the populations of microtubules present in a given cell or tissue (Cowan *et al.*, 1988).

Second, there are a number of different isoforms of tubulin encoded by the genomes of various organisms, and certainly, specific microtubule functionality may derive from microheterogeneity in the sequences of specific tubulins. For example, the dynamic instability of microtubules is altered by the tubulin isotype composition of the polymer (Panda *et al.*, 1994), and axonemal microtubules in *Chlamydomonas* differ in isotype content (Johnson, 1998). Indeed, flagellar assembly in *Drosophila* is exquisitely sensitive to the tubulin isotypes available (Raff *et al.*, 2000). For example, β2-tubulin is required for normal assembly of the $9+2$ axoneme. In male flies expressing only β1-tubulin, the axonemes are $9+0$; hence β2-tubulin in some manner specifies central pair assembly, a functionality that β1-tubulin lacks. Similarly, the cold stability characteristic of the microtubules of Antarctic fishes results not from the binding of specific MAPs (Detrich *et al.*, 1990) but is rather due to intrinsic differences in the primary sequence of Antarctic fish β-tubulin (Detrich, 1997). For example, eight amino acid differences distinguish cold-adapted β-tubulin from other vertebrate β-tubulins, and the majority of these are in a structural domain that lies at the dimer–dimer interface in the microtubule. Finally, note also that α-tubulin undergoes an interesting, perhaps unique, modification. The genome encodes a Tyr residue at the C-terminus. This Tyr is removed soon after assembly into the polymer (Gundersen *et al.*, 1987) by tubulin tyrosine carboxypeptidase (the enzyme prefers tubulin in the polymer form) to reveal the penultimate Glu residue at the C-terminus. The Tyr residue can be added back by tubulin tyrosine ligase (which prefers dimers). Although glu-tubulin is a marker for stable microtubules, this modification alone does not convey stability (Webster *et al.*, 1990).

A third basis for the generation of functional variety in microtubules is derived from posttranslational protein modifications. Such modifications presumably in turn either promote the association of specific protein complexes (molecular motors, plus end tracking proteins, severing proteins, MAPs) with the microtubule, promoting or enhancing a specific functionality (or preventing another) or alter the stability or assembly characteristics of the resulting polymer. With respect to posttranslational modifications (PTMs), tubulin sets the paradigm not only for the variety of modifications it undergoes, but also for the conservation of many of these modifications over evolutionary time. This chapter begins with a brief overview of PTMs of tubulin and other proteins that are known to occur in cilia and flagella, and then concludes with some methodologies to identify and localize specific modifications. For more extensive details on tubulin PTMs, there are some excellent recent reviews as well (Gaertig and Wloga, 2008; Hammond *et al.*, 2008; Verhey and Gaertig, 2007).

II. Posttranslational Protein Modification in Cilia and Flagella

In relative order of their discovery in flagella or cilia, the posttranslational protein modifications (PTMs) to be reviewed here include phosphorylation, acetylation, glycylation, glutamylation, and methylation. The myriad number of PTMs that occur in flagella speaks directly to the molecular complexity of the axonemal microtubules with respect to the variety of different proteins with which they interact.

A. Phosphorylation

Chlamydomonas flagella contain a number of different phosphorylated proteins, including α-tubulin (Piperno and Luck, 1976), outer arm dynein (King and Witman, 1994), proteins of the radial spokes (Piperno *et al.*, 1981), and components of the membrane/matrix fraction (Bloodgood, 1992). Metabolic labeling with ^{32}P (Bloodgood and Salomonsky, 1995) indicates clearly that these modifications are functionally important, as protein phosphorylation levels change as flagellar activities change (Bloodgood, 1992; Bloodgood and Salomonsky, 1991, 1994). There are ~600 proteins in the flagellar proteome (Pazour *et al.*, 2005), and a recent analysis of the *Chlamydomonas* flagellar phosphoproteome via immobilized metal affinity chromatography identified 141 phosphopeptides representing 32 flagellar proteins (Boesger *et al.*, 2009). This number represents 5% of the total estimated number of flagellar proteins and is hence likely to be an underestimate as previous data derived from two-dimensional (2-D) gel analyses of *Chlamydomonas* flagella identified 80 phosphoproteins in the axoneme fraction alone (Piperno *et al.*, 1981). Many of these proteins are of unknown function. However, note that changes in the phosphorylation state of inner arm dynein clearly correlate with changes in outer doublet sliding (Habermacher and Sale, 1995, 1996, 1997). More recently, Smith and coworkers have reported that calmodulin-mediated calcium-regulated events play a role in the signal transduction that occurs between the central pair microtubules and the dynein arms, which is transduced by the radial spokes (Dymek and Smith, 2007). These interactions are important in determining which dynein arms are active, and therefore the waveform of the resulting flagellar beat (Wargo and Smith, 2003; Wargo *et al.*, 2005). This signal transduction mechanism very likely involves protein phosphorylation events, but this has not been directly proved as yet. Also, a recent report (Piao *et al.*, 2009) has shown that kinesin-13, a kinesin that induces microtubule depolymerization, is phosphorylated as cells grow flagella and dephosphorylated upon completion of flagellar generation. In addition, kinesin-13 is transported into flagella during resorption.

Various protein kinases have been implicated in flagellar and ciliary length control, including glycogen synthase kinase 3 (Thoma *et al.*, 2007; Wilson and Lefebvre, 2004), a NIMA-related kinase (Bradley and Quarmby, 2005), a novel MAP kinase (Berman *et al.*, 2003; Mahjoub *et al.*, 2004), and a cyclin-related kinase (Tam *et al.*, 2007). Furthermore, the activity and/or localization to the flagellum of some of these kinases are themselves regulated by phosphorylation. More details on the role of protein phosphorylation and protein kinases in flagellar dynamics can be found in Chapter 17 by Cao *et al.*, this volume.

A protein kinase first described in *Chlamydomonas* flagella has recently been shown to operate in primary cilia as well. CALK (a *Chlamydomonas* aurora-like kinase) translocates into the flagella in a kinesin-2-dependent manner (Pan and Snell, 2003) when cells receive a signal telling them to shorten their flagella; CALK also becomes phosphorylated (Pan and Snell, 2005; Pan *et al.*, 2004). The signal transduction mechanism controlling regulated flagella shortening by CALK is different than that involved in the transient flagellar shortening that occurs during the process of flagellar length control (Pan and Snell, 2005). In yet another example of the amazing conservation of enzymatic activities and protein function over evolutionary time, injection of Aurora-A, the vertebrate homolog of CALK, into tissue culture cells containing primary cilia induced cilia disassembly within a few minutes (Pugacheva *et al.*, 2007). Pugacheva and coworkers also showed that RNAi depletion of Aurora-A or inhibition of the activity of Aurora-A prevented the cilia resorption that is normally induced by serum addition. How might a protein kinase induce ciliary disassembly? We will return to this question after discussing a second flagellar protein PTM, acetylation.

B. Acetylation

Acetylation occurs on two residues in α-tubulin: on a highly reactive lysine (HRL, Lys 394; Szasz *et al.*, 1986) in brain microtubules and on the ϵ-amino group of Lys 40 in α-tubulin (Greer *et al.*, 1985; L'Hernault and Rosenbaum, 1985a,b). Acetylation of Lys 40 is a marker of stable microtubules, such as those that comprise the outer doublet microtubules of cilia and flagella, although this modification alone does not appear to increase the stability of microtubules, at least *in vitro* (Maruta *et al.*, 1986). In addition, acetylation of Lys 40 is nonessential in *Tetrahymena* cilia (Gaertig *et al.*, 1995) and *Caenorhabditis elegans* sensory neurons (Fukushige *et al.*, 1999), and probably in *Chlamydomonas* flagella as well (Kozminski *et al.*, 1993), as cells in which Lys 40 has been mutated have no noticeable phenotype. Based on the structure of the dimer, Lys 40 is predicted to face the lumen of the microtubule. Hence it is unclear how the enzymes that acetylate and deacetylate Lys 40 would have access to that residue in a polymerized microtubule. Yet a specific deacetylase removes this group from Lys 40 during flagellar resorption (L'Hernault and Rosenbaum, 1985b), a process that requires disassembly of the microtubules of the outer doublets. It is not clear, however, if deacetylation is required for a net dissociation of tubulin dimers from the microtubule or if it occurs after tubulin subunits are removed from the microtubule lattice. At the very least, acetylation of Lys 40 likely serves as a marker of microtubules that have been stabilized by some other mechanism, for example, via a plus end capping protein (Palazzo *et al.*, 2003).

The importance of deacetylation and phosphorylation reactions has recently been demonstrated in the disassembly of *Chlamydomonas* flagella and primary cilia in tissue culture cells. HDAC6, a histone deacetylase, can remove the methyl group from Lys 40 *in vitro* (Hubbert *et al.*, 2002; Matsuyama *et al.*, 2002), as can a second deacetylase, Sir2 (North *et al.*, 2003). The aurora-like kinase CALK, in *Chlamydomonas*, is itself activated by phosphorylation and this modification occurs upon induction of flagellar resorption (Pan *et al.*, 2004), although the kinase responsible for the phosphorylation of CALK has not yet been identified. In tissue culture cells, aurora-A phosphorylation occurs coincident with resorption of the primary cilium prior to mitosis. Activated aurora-A then phosphorylates HDAC6, the microtubule deacetylase, a process required for ciliary disassembly (Pugacheva *et al.*, 2007). More detailed information on the role of phosphorylation and deacetylation can be found in Chapter 17 by Cao *et al.* and Chapter 7 by Plotnikova *et al.*, this volume.

C. Polyglycylation

Polyglycylation involves the addition of one or more short chains of Gly residues via an isopeptide linkage to the carboxyl group of Glu residues in the C-terminal tail of α- and β-tubulin (Redeker *et al.*, 1994); up to 34 Gly residues can be added enzymatically in this manner. The extent of polyglycylation of tubulin in tracheal cilia varies with position along the axoneme, increasing in amount as one moves from the tip to the base (Dossou *et al.*, 2007). Although *Tetrahymena* cilia lacking all α-tubulin glycylation sites are normal, glycylation of β-tubulin is an essential modification in ciliary axonemes. Cilia lacking three of the five β-tubulin glycylation sites are nonmotile due to lack of the central pair microtubules (Xia *et al.*, 2000). The enzymes responsible for adding glycyl residues to α- and β-tubulin (one enzyme to initiate the isopeptide branch point, a second to elongate the glycyl side chain) have very recently been identified (Rogowski *et al.*, 2009; Wloga *et al.*, 2009). In these studies, knockdown of the glycyl ligase via RNAi or expression of a dominant negative form of the enzyme demonstrated that polyglycylation is necessary for the assembly of ciliary and flagellar axonemes. Surprisingly, these data also indicated that reduction of polyglycylation was correlated with an increase in polyglutamylation (see Section II.D). Because both glycylation and glutamylation modifications occur on residues in the C-terminal tail of α-tubulin, it is possible that one modification produces steric hindrance that decreases the accessibility of the tail residues to the other set of enzymes. Hence, the two modifying systems oppose one another, and it has recently been suggested that the role of polyglycylation in promoting axonemal microtubule assembly is due to its ability to inhibit polyglutamylation (Wloga *et al.*, 2009).

D. Polyglutamylation

This PMT is associated with stable microtubules in structures such as the axoneme, the basal body/centriole, and the axons of neurons (by contrast, polyglycylation occurs predominantly in flagellated and ciliated cells). The extent of polyglutamylation among ciliates, however, can vary considerably. For example, tubulin in the *Paramecium* axoneme is glutamylated to a small extent, while tubulin in the axonemes of *Tetrahymena* is extensively glutamylated (Redeker *et al.*, 2005). Glutamylation occurs on the B subfiber of the outer doublets, and both α- and β-tubulin are substrates for this PTM. Polyglutamylation of α-tubulin (Edde *et al.*, 1990; Fouquet *et al.*, 1994) is catalyzed by a specific polyglutamylase (Tt116), and the extent to which tubulin becomes modified is modulated by the *fleer* gene (Pathak *et al.*, 2007). More detailed information on polyglutamylation can be found in Chapter 16 by Pathak and Drummond, this volume.

As with polyglycylation (Dossou *et al.*, 2007), the extent of polyglutamylation is highest at the base and decreases as one moves toward the ciliary tip (Pathak *et al.*, 2007). The reason for this gradient in distribution of these PTMs is currently unknown. One possibility is that the gradient relates to the "age" of the modified microtubules, as the base of the cilium is older than the tip from an assembly point of view, and the tip continually turns over subunits (Marshall and Rosenbaum, 2001). Alternatively, the enzymes responsible for these modifications might themselves be distributed unevenly in the flagellum (Bre *et al.*, 1996), and this distribution is reflected by the level of PTM along the length of the organelle. Finally, perhaps these enzymes associate with IFT particles and enter the flagellum only during the events of flagellar assembly but do not associate with IFT particles once assembly is complete.

E. Methylation

We have recently identified protein methylations that occur on Arg residues in six specific flagellar proteins (Schneider *et al.*, 2008; Sloboda and Howard, 2009). Protein arginine methyltransferases (PRMTs) are the class of enzymes responsible for these methylations, which occur in several forms. The first is the addition of two methyl groups to one of the guanidino nitrogens of Arg, producing asymmetric dimethyl arginine (aDMA). The second places a single methyl group on each of the two guanidino nitrogens, producing symmetric dimethyl arginine (sDMA). The third, less common modification, methylates only one of the guanidino nitrogens, producing monomethyl arginine (MMA). The fourth class, thus far only reported in yeast, methylates the nitrogen of the Arg guanidino group that is bonded to the δ carbon. The methylations we have discovered in flagella are sDMA in full-length flagella and aDMA in resorbing flagella. A type I PRMT produces aDMA and a type II PRMT produces sDMA (Bedford, 2007).

Protein methylation is critical not only for gene transcription, but also for regulating steps in signal transduction and protein targeting (McBride and Silver, 2001). Several small guanosine triphosphatases (GTPases) and the catalytic subunit of protein phosphatase 2A are methylated during cell signaling events (Mumby, 2001); the latter event then enhances the association of the phosphatase with various interacting proteins. In flagella, previous work has indicated that histone H1, itself a substrate for methylation, is a component of the flagellar axoneme where it plays a role in microtubule stability (Multigner *et al.*, 1992). We have not detected aDMA residues in a protein with the size of H1 in resorbing flagella; however, we have identified in full-length flagella a protein modified with sDMA that is in the size range of H1 (Sloboda and Howard, 2009).

Methylation modifications can affect protein–protein interactions or enzymatic activities. Depending on the methylated protein and its interacting partner(s), methylation could inhibit protein–protein interaction by removing one or more required H-bond donors. Alternatively, methylation could promote protein–protein interaction (Mumby, 2001), as methylated Arg has been demonstrated to have an increased affinity for the indole ring of tryptophan residues (Hughes and Waters, 2006). Protein methylation may be a necessary step in the disassembly of axonemal components, for example, the dynein arms, radial spokes, outer doublet MTs, and so on, or methylation may be required to promote the association of disassembled axonemal proteins with the retrograde IFT machinery. Elucidating the precise role of protein methylation in flagellar dynamics will require identification and characterization of not only the methylated proteins but also the PRMTs responsible for carrying out the methylation reactions.

III. Methods for the Detection of PTM in Cilia and Flagella using the Electron Microscope

A. Solutions Required

All solutions are made with ultrapure (18 MΩ) water.

10 mM HEPES, pH 7.5
1 M ethanolamine, pH 8
Elution buffer: EB (100 mM glycine, 150 mM NaCl, pH 2.4)
Neutralizing buffer: NB (1 M Tris, brought to pH 8.0 with HCl)
PBS: phosphate-buffered saline (for a 10× stock solution, combine 80 g NaCl, 2 g KCl, 11.5 g Na_2HPO_4, and 2 g KH_2PO_4; dissolve and add water to 1 l. The pH of the stock solution will be ~6.8–6.9. When diluted 10-fold to make a working solution, the pH will be ~7.5.)
HMDEK: 10 mM HEPES, pH 7.5, 5 mM $MgSO_4$, 1 mM DTT, 0.5 mM EDTA, and 25 mM KCl.)
3.7% Formaldehyde: 10-fold dilution of formaldehyde stock (purchased as a 37% solution) in HMDEK
0.1% PEI-coated cover slips: Add a solution of 0.1% polyethylene imine (PEI) in water to the surface of a coverslip. Let sit for 5 min at room temperature. Aspirate off the PEI and replace with water. Repeat a total of three times. Air dry. Keep coverslips in a dust free container (i.e., in a Petri dish lined with filter paper) during the wash procedures and until use.
4% Paraformaldehyde in HMDEK: Vortex 0.4 g paraformaldehyde in HMDEK until dissolved. Alternatively, add 0.4 g paraformaldehyde to 6.0 ml of 60°C water containing 0.1 ml of 0.5 N NaOH. When dissolved, add the remaining components of HMDEK from concentrated stocks, check the pH, and bring to a final volume of 10.0 ml.
TEM Blocking Buffer: PBS containing 2% bovine serum albumin, 0.1% gelatin, and 0.05% Tween-20.

The procedures outlined below have been used to detect methylated proteins in *Chlamydomonas* flagella. The steps should be readily adapted to other PTMs, the primary caution being that the use of well-characterized, highly specific, affinity-purified antibodies is the main key to success.

B. Antibody Affinity Purification

We have used earlier two different methods to generate affinity-purified polyclonal antibodies. The first uses antigen immobilized on nitrocellulose as the affinity matrix, following the procedure of Talian *et al.* (1983) exactly as published. Most often, however, for localization of proteins with PTMs, synthetic peptide antibodies coupled to Keyhole Limpet Hemocyanin (KLH) are used as the immunogen. Thus, we use as the affinity matrix a synthetic peptide representing the PTM that was initially used as the antigen, and we bind this peptide to Affi-gel 15 (Bio-Rad, Richmond, CA) to generate the affinity matrix. To do this:

1. Determine the pI of the peptide. There are a number of web-based tools for calculating this based on the known sequence of the peptide, for example, http://ca.expasy.org/tools/pi_tool.html.
2. Avoid using buffers such as Tris-HCl or Tris/glycine, that is, those containing primary amines, as these will couple to the Affigel. Rather, suspend the peptide to be coupled in bicarbonate, acetate, or a sulfonic acid-based buffer such as HEPES, at low ionic strength (10 mM) and suitable pH.
3. Use Affi-gel 10 if the coupling is to be carried out in a buffer whose pH is at or below the pI of the peptide. Use Affi-gel 15 if working at a pH near or above the pI of the peptide.
4. Prepare the Affi-gel and carry out the coupling reaction at 4°C, according to the manufacturer's instructions. Under these conditions, most coupling reactions near completion within 30–60 min and are complete within 4 h.
5. Coupling efficiency of course varies with the characteristics of the peptide or protein being coupled to the Affi-gel. When working with peptides, we routinely allow the reaction to proceed for the full 4 h.
6. After the binding reaction is complete and excess antigen has been removed, block any remaining binding sites on the beads by adding 1 M ethanolamine, pH 8, to a final concentration of 100 mM. Agitate gently for 1 h at 4°C.
7. Transfer beads to a suitable column, and wash with PBS. Clarify 1–2 ml of antiserum by centrifugation in a microfuge for 10 min.
8. Apply the antibody supernatant to the column and allow the column to flow at a slow rate. Collect the outflow and reapply it to the column. Collect and reapply the outflow to the column a second time.
9. Wash the column with 10 ml PBS and then elute with elution buffer (100 mM glycine, 150 mM NaCl, pH 2.4). Collect 1 ml fractions equal to at least 10 column volumes (i.e., collect 10, 1 ml fractions for an affinity column whose volume is 1 ml) and immediately neutralize each fraction by the addition of 200 μl 1 M Tris-HCl at pH 8.0.

10. Depending on the ratio of loading volume to column volume, affinity-purified antibodies will be found in the first third, or so, of the elution profile (i.e., fractions 1 through 3 or 4, for a 1 ml column). However, for some of our polyclonal antibodies, we have noted a variation in binding affinities such that antibodies to the same antigenic peptide elute at different positions in the eluant, due apparently to different binding affinities for the peptide. Thus, it is good practice the first time a particular serum is used for affinity purification to collect 20–30 column volumes and assay each fraction for the presence of peptide specific antibodies. To do this, spot 1 μl of each fraction on nitrocellulose, block, and process for detection of primary antibodies as one would an immunoblot. Pool the fractions based on the elution profile, and test them for specificity via immunoblotting and immunofluorescence.

We have used both whole-mount scanning EM and thin section EM to localize antigens (methylated Arg peptide antigens and others) with great success. The procedures provided below most likely will work with any well-characterized, affinity-purified antibodies purified as outlined above.

C. Immunogold Scanning EM

The procedures outlined here are based on previously published work (Sloboda and Howard, 2007, 2009); readers may wish to consult these manuscripts for more detailed information and relevant figures. All solution changes are made by sedimentation of the axonemes in a swinging bucket rotor (e.g., Sorvall HB-4) at 12,000 × *g* for 15 min.

1. Isolate flagellar axonemes via previously published procedures (Pazour *et al.*, 2005; Pedersen *et al.*, 2006; Schneider *et al.*, 2008; Sloboda and Howard, 2007) or isolate the axonemes from primary cilia as outlined in this volume (see Chapter 4 by Mitchell *et al.* and Chapter 5 by Huang *et al.*, this volume). Apportion into 50–100 μl aliquots depending on the number of primary antibodies being assayed.
2. If desired, the axonemes can be fixed in 3.7% formaldehyde in HMDEK, followed by a wash into fresh HMDEK by sedimentation and resuspension. Repeat the wash a second time.
3. Add an appropriate dilution of primary antibody to each tube and incubate at room temperature for 2–4 h with gently rocking.
4. Wash out the primary antibody by sedimentation and resuspension in HMDEK. Repeat this step once to remove any primary antibodies that may be trapped in the pellet.
5. Add to the resuspended axonemes an appropriate dilution of secondary antibodies labeled with colloidal gold. Note that these antibodies can be purchased with gold tags of various diameters, depending on the manufacturer, making double (or even triple) labeling experiments possible (Sloboda and Howard, 2007).
6. Incubate with gentle rocking at room temperature for 2 h.

7. Collect and wash the axonemes as in step 4; add a third wash to ensure complete removal of unbound secondary antibodies.
8. Our scanning EM (FEI XL-30, FEI Company, Salem, MA) accepts 12-mm diameter glass coverslips. Coat these with 0.1% PEI on the day of use. Place the labeled axonemes from step 7 on a PEI-coated coverslip and allow to adhere for 10 min at room temperature.
9. Withdraw the sample with an autopipettor or aspirator and fix by immersion of the coverslip in a small (35×10 mm) disposable Petri dish containing 1% glutaraldehyde in HMDEK. Incubate at 4 C overnight.
10. Prepare samples for the scanning EM: critical point dry the samples (we use a Samdri 795) and coat them with 2–3 nm of osmium. We use a Samdri 795 and an SPI plasma coater, respectively, for these purposes. It is important to use a plasma coater and not a sputter coater for the osmium coat as the plasma results in a metal coat that is almost perfectly amorphous and thus well below the resolution limit for scanning electron microscopy (SEM) specimens such as those described here. Moreover, the stabilizing coat of metal does not preclude detection of the gold label on the antibodies. When the SEM is operated to detect backscattered electrons, a clear view of the gold (as white dots) on the antibodies is obtained. This field can then be compared to a conventional three-dimensional (3-D) view of the same field obtained from operation of the SEM in the conventional mode, that is, to detect secondary electrons. Figure 1 provides an example.

D. Thin Section Immunogold EM

The procedures outlined here are based on previously published work on the localization of methyl Arg residues in flagella (Sloboda and Howard, 2007, 2009), in which can be found more detailed information and relevant figures.

1. Isolate flagellar axonemes via previously published procedures (Pazour *et al.*, 2005; Pedersen *et al.*, 2006; Schneider *et al.*, 2008; Sloboda and Howard, 2007) or isolate the axonemes from primary cilia as outlined in this volume (see Chapter 4 by Mitchell *et al.* and Chapter 5 by Huang *et al.*, this volume).
2. Gently overlay the final flagellar or ciliary pellet with 4% paraformaldehyde in HMDEK. Incubate at room temperature for 1 h and overnight at 4°C.
3. Dehydrate and embed in LR White resin, section, and place sections on Ni grids.
4. To process the sections on grids for immunolocalization, arrange a piece of Parafilm on the bench, covered with a cover taken from a large Petri dish. Place drops of PBS on the Parafilm and float one grid section side down on each drop for 15 min to hydrate the samples.
5. Move the grids to new drops of TEM blocking buffer, cover, and incubate at room temperature for 30 min.
6. Prepare new drops on Parafilm of affinity purified primary antibody at the appropriate dilution in TEM blocking buffer and float the grids on these drops. Cover and incubate at room temperature for 2 h.

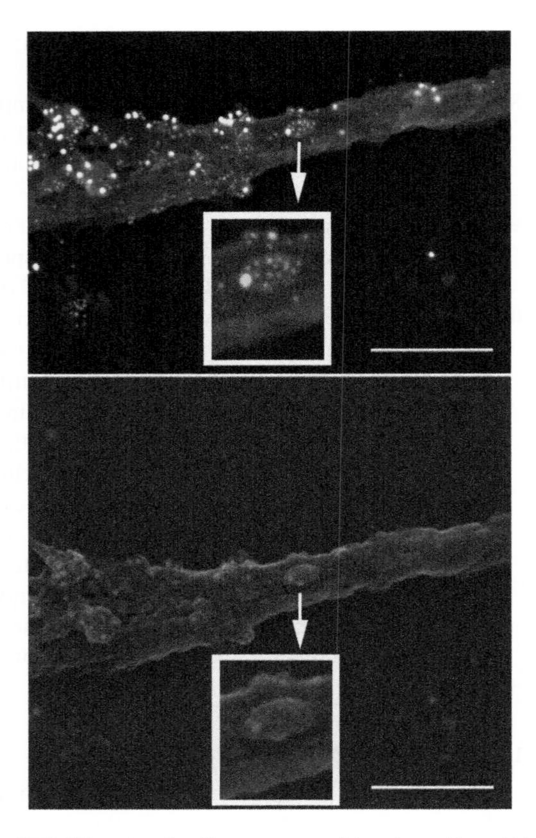

Fig. 1 Localization of IFT 139, an intraflagellar transport particle polypeptide, and kinesin-2 via immunogold scanning EM. IFT 139 is revealed by small (12 nm) gold particles, and kinesin-2 by large (25 nm) gold particles. The image generated by backscattered electrons (top) clearly reveals the gold particles. The corresponding conventional SEM image generated by secondary electrons is shown in the bottom panel. The scale bar is 500 nm, and the insets (showing a single IFT particle labeled with one 25-nm gold particle and multiple 12 nm particles) are shown at twice that magnification. This figure is a portion of Fig. 4 of Sloboda and Howard (2007). *Cell Motil. Cytoskeleton* **64**, 446–460 and is reproduced with permission.

7. Wash the grids by floating on successive drops (three total) of PBS, 5 min each.
8. Place the grids on drops of appropriately diluted gold-labeled secondary antibodies in blocking buffer for 1 h at room temperature. For this step we use 25-nm diameter gold-labeled antibodies from Electron Micrsocopy Sciences, diluted 1:20 in TEM blocking buffer.
9. Wash the grids as in step 7 followed by two, 5 min washes in water, and then allow them to air dry.
10. Stain the grids with 2% aqueous uranyl acetate for 7 min followed by Reynold's lead citrate for a maximum of 10 s. Do not overstain the samples, as this will tend to occlude the detection of the gold label. The times listed here are sufficient to

stain the axonemes in the section to reveal their characteristic ultrastructure while leaving the gold particles clearly visible. Figure 2 shows a low-power view of a field of labeled flagella demonstrating good localization with essentially no background.

Note that each gold particle used in the procedures above is surrounded by a halo of antibody molecules bound to the gold by their Fc tails, as shown in the inset of Fig. 6B of Sloboda and Howard (2007). Note also that each antibody molecule has a length of 14.5 nm and hence a single 25-nm gold particle has an effective diameter of roughly 50 nm (14.5 + 25 + 14.5 nm). In a thin section sample, a given gold particle is bound in turn to the Fc portion of the primary antibody, itself with a length of 14.5 nm. Thus, the center of mass of a given gold particle in an EM localization experiment of the type outlined in Sections III.C and III.D could be up to as much as ~40 nm away from the actual antigen (14.5 + 14.5 + 12.5 nm), assuming both antibodies are fully extended to their maximum lengths.

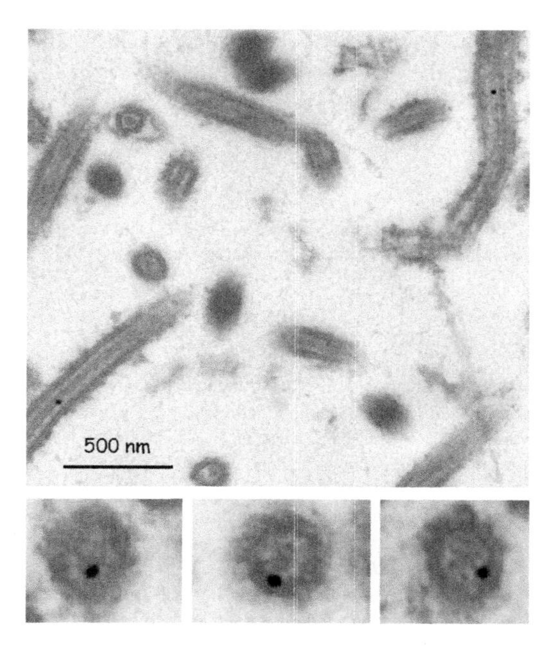

Fig. 2 Top: localization of a protein posttranslationally modified with symmetric dimethyl arginine (sDMA) via immunogold transmission electron microscopy. This low-power field demonstrates the lack of background labeling obtained via the procedures outlined here. The gold particles (25 nm, black dots) are restricted to the sectioned flagella and localize to the central pair of the axoneme. No gold particles are visible on the section where there is no sample. Bottom: collage showing cross sections of axonemes demonstrating the localization of asymmetric dimethyl arginine (aDMA) residues. This posttranslational modification (PTM) appears to be associated with the outer doublet array and not the central pair microtubules. This figure is derived from Figs. 6 and 7 of Sloboda and Howard (2009). *Cell Motil. Cytoskeleton* **66**, 650–660 and is reproduced with permission.

E. Concluding Remarks

The array of PTMs for which tubulin and other flagella proteins are substrates is remarkable not only for the diversity of these modifications and their conservation over evolutionary time, but also for the relative lack of information available as to their specific functions in flagellar motility and dynamics. Many of these PTMs have been observed to occur during a specific cellular event, for example, deacetylation of Lys 40 in α-tubulin or methylation of other proteins during flagellar disassembly. Yet Lys 40 itself, for example, is not an essential residue: cells lacking Lys at position 40 in α-tubulin have no detectable phenotype. It would appear, then, that a single approach, for example, mutation of a specific residue in tubulin or another posttranslationally modified flagellar protein, might not always provide insight into function. Future understanding of the role of PTMs in flagellar and ciliary dynamics will more often than not require combined approaches that innovatively use techniques of molecular genetics, protein biochemistry, and cell biology to determine the function of a specific PTM. Some insight is beginning to be obtained through analysis and characterization of the enzymes involved in producing and removing a specific PTM, followed by knockout, knockdown, or dominant-negative approaches to test function. Such experiments set the paradigm for future analyses of the role of PTM function in the biology of cilia and flagella.

Acknowledgments

The author thanks the NSF (MCB0418877) and the NIH (DK071720) for their support of the work from the author's laboratory that is summarized here.

References

Amos, L., and Klug, A. (1974). Arrangement of subunits in flagellar microtubules. *J. Cell Sci.* **14**, 523–549.

Bedford, M.T. (2007). Arginine methylation at a glance. *J. Cell Sci.* **120**, 4243–4246.

Berman, S.A., Wilson, N.F., Haas, N.A., and Lefebvre, P.A. (2003). A novel MAP kinase regulates flagellar length in *Chlamydomonas*. *Curr. Biol.* **13**, 1145–1149.

Bloodgood, R.A. (1992). Calcium-regulated phosphorylation of proteins in the membrane-matrix compartment of the *Chlamydomonas* flagellum. *Exp. Cell Res.* **198**, 228–236.

Bloodgood, R.A., and Salomonsky, N.L. (1991). Regulation of flagellar glycoprotein movements by protein phosphorylation. *Eur. J. Cell Biol.* **54**, 85–89.

Bloodgood, R.A., and Salomonsky, N.L. (1994). The transmembrane signaling pathway involved in directed movements of *Chlamydomonas* flagellar membrane glycoproteins involves the dephosphorylation of a 60-kD phosphoprotein that binds to the major flagellar membrane glycoprotein. *J. Cell Biol.* **127**, 803–811.

Bloodgood, R.A., and Salomonsky, N.L. (1995). Phosphorylation of Chlamydomonas flagellar proteins. *Methods Cell Biol.* **47**, 121–127.

Boesger, J., Wagner, V., Weisheit, W., and Mittag, M. (2009). Analysis of flagellar phosphoproteins from Chlamydomonas reinhardtii. *Eukaryot Cell* **8**, 922–932.

Bond, J.F., Fridovich-Keil, J.L., Pillus, L., Mulligan, R.C., and Solomon, F. (1986). A chicken-yeast chimeric beta-tubulin protein is incorporated into mouse microtubules in vivo. *Cell* **44**, 461–468.

Bradley, B.A., and Quarmby, L.M. (2005). A NIMA-related kinase, Cnk2p, regulates both flagellar length and cell size in Chlamydomonas. *J. Cell Sci.* **118**, 3317–3326.

Bre, M.H., Redeker, V., Quibell, M., Darmanaden-Delorme, J., Bressac, C., Cosson, J., Huitorel, P., Schmitter, J.M., Rossler, J., Johnson, T., Adoutte, A., and Levilliers, N. (1996). Axonemal tubulin polyglycylation probed with two monoclonal antibodies: Widespread evolutionary distribution, appearance during spermatozoan maturation and possible function in motility. *J. Cell Sci.* **109**(Pt. 4), 727–738.

Clayton, L., Pogson, C.I., and Gull, K. (1979). Microtubule proteins in the yeast, *Saccharomyces cerevisiae. FEBS Lett.* **106**, 67–70.

Detrich, H.W., III (1997). Microtubule assembly in cold-adapted organisms: Functional properties and structural adaptations of tubulins from antarctic fishes. *Comp. Biochem. Physiol. A.* **118**, 501–513.

Detrich, H.W., III, Neighbors, B.W., Sloboda, R.D., and Williams, R.C., Jr. (1990). Microtubule-associated proteins from Antarctic fishes. *Cell Motil. Cytoskeleton* **17**, 174–186.

Dossou, S.J., Bre, M.H., and Hallworth, R. (2007). Mammalian cilia function is independent of the polymeric state of tubulin glycylation. *Cell Motil. Cytoskeleton* **64**, 847–855.

Dymek, E.E., and Smith, E.F. (2007). A conserved CaM- and radial spoke associated complex mediates regulation of flagellar dynein activity. *J. Cell Biol.* **179**, 515–526.

Edde, B., Rossier, J., Le Caer, J.P., Desbruyeres, E., Gros, F., and Denoulet, P. (1990). Posttranslational glutamylation of alpha-tubulin. *Science* **247**, 83–85.

Fouquet, J.P., Edde, B., Kann, M.L., Wolff, A., Desbruyeres, E., and Denoulet, P. (1994). Differential distribution of glutamylated tubulin during spermatogenesis in mammalian testis. *Cell Motil. Cytoskeleton* **27**, 49–58.

Fukushige, T., Siddiqui, Z.K., Chou, M., Culotti, J.G., Gogonea, C.B., Siddiqui, S.S., and Hamelin, M. (1999). MEC-12, an alpha-tubulin required for touch sensitivity in *C. elegans. J. Cell Sci.* **112**(Pt. 3), 395–403.

Gaertig, J., Cruz, M.A., Bowen, J., Gu, L., Pennock, D.G., and Gorovsky, M.A. (1995). Acetylation of lysine 40 in alpha-tubulin is not essential in *Tetrahymena thermophila. J. Cell Biol.* **129**, 1301–1310.

Gaertig, J., and Wloga, D. (2008). Ciliary tubulin and its post-translational modifications. *Curr. Top. Dev. Biol.* **85**, 83–113.

Greer, K., Maruta, H., L'Hernault, S.W., and Rosenbaum, J.L. (1985). Alpha-tubulin acetylase activity in isolated *Chlamydomonas* flagella. *J. Cell Biol.* **101**, 2081–2084.

Gundersen, G.G., Khawaja, S., and Bulinski, J.C. (1987). Postpolymerization detyrosination of alpha-tubulin: A mechanism for subcellular differentiation of microtubules. *J. Cell Biol.* **105**, 251–264.

Habermacher, G., and Sale, W.S. (1995). Regulation of dynein-driven microtubule sliding by an axonemal kinase and phosphatase in *Chlamydomonas* flagella. *Cell Motil. Cytoskeleton* **32**, 106–109.

Habermacher, G., and Sale, W.S. (1996). Regulation of flagellar dynein by an axonemal type-1 phosphatase in *Chlamydomonas. J. Cell Sci.* **109**(Pt. 7), 1899–1907.

Habermacher, G., and Sale, W.S. (1997). Regulation of flagellar dynein by phosphorylation of a 138-kD inner arm dynein intermediate chain. *J. Cell Biol.* **136**, 167–176.

Hammond, J.W., Cai, D., and Verhey, K.J. (2008). Tubulin modifications and their cellular functions. *Curr. Opin. Cell Biol.* **20**, 71–76.

Hubbert, C., Guardiola, A., Shao, R., Kawaguchi, Y., Ito, A., Nixon, A., Yoshida, M., Wang, X.F., and Yao, T.P. (2002). HDAC6 is a microtubule-associated deacetylase. *Nature* **417**, 455–458.

Hughes, R.M., and Waters, M.L. (2006). Arginine methylation in a beta-hairpin peptide: Implications for Arg-pi interactions, DeltaCp(o), and the cold denatured state. *J. Am. Chem. Soc.* **128**, 12735–12742.

Johnson, K.A. (1998). The axonemal microtubules of the *Chlamydomonas* flagellum differ in tubulin isoform content. *J. Cell Sci.* **111**(Pt. 3), 313–320.

Kilmartin, J.V. (1981). Purification of yeast tubulin by self-assembly in vitro. *Biochemistry* **20**, 3629–3633.

King, S.M., and Witman, G.B. (1994). Multiple sites of phosphorylation within the alpha heavy chain of *Chlamydomonas* outer arm dynein. *J. Biol Chem.* **269**, 5452–5457.

Kozminski, K.G., Diener, D.R., and Rosenbaum, J.L. (1993). High level expression of nonacetylatable alpha-tubulin in *Chlamydomonas reinhardtii. Cell Motil. Cytoskeleton* **25**, 158–170.

L'Hernault, S.W., and Rosenbaum, J.L. (1985a). *Chlamydomonas* alpha-tubulin is posttranslationally modified by acetylation on the epsilon-amino group of a lysine. *Biochemistry* **24**, 473–478.

L'Hernault, S.W., and Rosenbaum, J.L. (1985b). Reversal of the posttranslational modification on *Chlamydomonas* flagellar alpha-tubulin occurs during flagellar resorption. *J. Cell Biol.* **100**, 457–462.

Mahjoub, M.R., Qasim Rasi, M., and Quarmby, L.M. (2004). A NIMA-related kinase, Fa2p, localizes to a novel site in the proximal cilia of *Chlamydomonas* and mouse kidney cells. *Mol. Biol. Cell.* **15**, 5172–5186.

Marshall, W.F., and Rosenbaum, J.L. (2001). Intraflagellar transport balances continuous turnover of outer doublet microtubules: Implications for flagellar length control. *J. Cell Biol.* **155**, 405–414.

Maruta, H., Greer, K., and Rosenbaum, J.L. (1986). The acetylation of alpha-tubulin and its relationship to the assembly and disassembly of microtubules. *J. Cell Biol.* **103**, 571–579.

Matsuyama, A., Shimazu, T., Sumida, Y., Saito, A., Yoshimatsu, Y., Seigneurin-Berny, D., Osada, H., Komatsu, Y., Nishino, N., Khochbin, S., Horinouchi, S., and Yoshida, M. (2002). In vivo destabilization of dynamic microtubules by HDAC6-mediated deacetylation. *EMBO J.* **21**, 6820–6831.

McBride, A.E., and Silver, P.A. (2001). State of the arg: Protein methylation at arginine comes of age. *Cell* **106**, 5–8.

Moores, C.A., Perderiset, M., Francis, F., Chelly, J., Houdusse, A., and Milligan, R.A. (2004). Mechanism of microtubule stabilization by doublecortin. *Mol. Cell* **14**, 833–839.

Multigner, L., Gagnon, J., Van Dorsselaer, A., and Job, D. (1992). Stabilization of sea urchin flagellar microtubules by histone H1. *Nature* **360**, 33–39.

Mumby, M. (2001). A new role for protein methylation: Switching partners at the phosphatase ball. *Sci. STKE* **2001**, PE1.

North, B.J., Marshall, B.L., Borra, M.T., Denu, J.M., and Verdin, E. (2003). The human Sir2 ortholog, SIRT2, is an NAD+-dependent tubulin deacetylase. *Mol. Cell* **11**, 437–444.

Ohkura, H., Garcia, M.A., and Toda, T. (2001). Dis1/TOG universal microtubule adaptors—one MAP for all? *J. Cell Sci.* **114**, 3805–3812.

Palazzo, A., Ackerman, B., and Gundersen, G.G. (2003). Cell biology: Tubulin acetylation and cell motility. *Nature* **421**, 230.

Pan, J., and Snell, W.J. (2003). Kinesin II and regulated intraflagellar transport of *Chlamydomonas* aurora protein kinase. *J. Cell Sci.* **116**, 2179–2186.

Pan, J., and Snell, W.J. (2005). *Chlamydomonas* shortens its flagella by activating axonemal disassembly, stimulating IFT particle trafficking, and blocking anterograde cargo loading. *Dev. Cell* **9**, 431–438.

Pan, J., Wang, Q., and Snell, W.J. (2004). An aurora kinase is essential for flagellar disassembly in *Chlamydomonas. Dev. Cell* **6**, 445–451.

Panda, D., Miller, H.P., Banerjee, A., Luduena, R.F., and Wilson, L. (1994). Microtubule dynamics in vitro are regulated by the tubulin isotype composition. *Proc. Natl. Acad. Sci. USA* **91**, 11358–11362.

Pathak, N., Obara, T., Mangos, S., Liu, Y., and Drummond, I.A. (2007). The zebrafish fleer gene encodes an essential regulator of cilia tubulin polyglutamylation. *Mol. Biol. Cell* **18**, 4353–4364.

Pazour, G.J., Agrin, N., Leszyk, J., and Witman, G.B. (2005). Proteomic analysis of a eukaryotic cilium. *J. Cell Biol.* **170**, 103–113.

Pedersen, L.B., Geimer, S., and Rosenbaum, J.L. (2006). Dissecting the molecular mechanisms of intraflagellar transport in *Chlamydomonas. Curr. Biol.* **16**, 450–459.

Piao, T., Luo, M., Wang, L., Guo, Y., Li, D., Li, P., Snell, W.J., and Pan, J. (2009). A microtubule depolymerizing kinesin functions during both flagellar disassembly and flagellar assembly in *Chlamydomonas. Proc. Natl. Acad. Sci. USA* **106**, 4713–4718.

Piperno, G., Huang, B., Ramanis, Z., and Luck, D.J. (1981). Radial spokes of *Chlamydomonas* flagella: Polypeptide composition and phosphorylation of stalk components. *J. Cell Biol.* **88**, 73–79.

Piperno, G., and Luck, D.J. (1976). Phosphorylation of axonemal proteins in *Chlamydomonas reinhardtii. J. Biol. Chem.* **251**, 2161–2167.

Pugacheva, E.N., Jablonski, S.A., Hartman, T.R., Henske, E.P., and Golemis, E.A. (2007). HEF1-dependent Aurora A activation induces disassembly of the primary cilium. *Cell* **129**, 1351–1363.

Raff, E.C., Hutchens, J.A., Hoyle, H.D., Nielsen, M.G., and Turner, F.R. (2000). Conserved axoneme symmetry altered by a component beta-tubulin. *Curr. Biol.* **10**, 1391–1394.

Redeker, V., Levilliers, N., Schmitter, J.M., Le Caer, J.P., Rossier, J., Adoutte, A., and Bre, M.H. (1994). Polyglycylation of tubulin: A posttranslational modification in axonemal microtubules. *Science* **266**, 1688–1691.

Redeker, V., Levilliers, N., Vinolo, E., Rossier, J., Jaillard, D., Burnette, D., Gaertig, J., and Bre, M.H. (2005). Mutations of tubulin glycylation sites reveal cross-talk between the C termini of alpha- and beta-tubulin and affect the ciliary matrix in *Tetrahymena*. *J. Biol. Chem.* **280**, 596–606.

Rogowski, K., Juge, F., van Dijk, J., Wloga, D., Strub, J.M., Levilliers, N., Thomas, D., Bre, M.H., Van Dorsselaer, A., Gaertig, J., and Janke, C. (2009). Evolutionary divergence of enzymatic mechanisms for posttranslational polyglycylation. *Cell* **137**, 1076–1087.

Schneider, M.J., Ulland, M., and Sloboda, R.D. (2008). A protein methylation pathway in *Chlamydomonas* flagella is active during flagellar resorption. *Mol. Biol. Cell* **19**, 4319–4327.

Sloboda, R.D., and Howard, L. (2007). Localization of EB1, IFT polypeptides, and kinesin-2 in *Chlamydomonas* flagellar axonemes via immunogold scanning electron microscopy. *Cell Motil. Cytoskeleton* **64**, 446–460.

Sloboda, R.D., and Howard, L. (2009). Protein methylation in full length *Chlamydomonas* flagella. *Cell Motil. Cytoskeleton* **66**, 650–660.

Sloboda, R.D., Rudolph, S.A., Rosenbaum, J.L., and Greengard, P. (1975). CyclicAMP-dependent endogenous phosphorylation of a microtuble-associated protein (MAP). *Proc. Natl. Acad. Sci. USA* **72**, 177–181.

Song, Y.H., and Mandelkow, E. (1993). Recombinant kinesin motor domain binds to beta-tubulin and decorates microtubules with a B surface lattice. *Proc. Natl. Acad. Sci. USA* **90**, 1671–1675.

Szasz, J., Yaffe, M.B., Elzinga, M., Blank, G.S., and Sternlicht, H. (1986). Microtubule assembly is dependent on a cluster of basic residues in alpha-tubulin. *Biochemistry* **25**, 4572–4582.

Talian, J.C., Olmsted, J.B., and Goldman, R.D. (1983). A rapid procedure for preparing fluorescein-labeled specific antibodies from whole antiserum: Its use in analyzing cytoskeletal architecture. *J. Cell Biol.* **97**, 1277–1282.

Tam, L.W., Wilson, N.F., and Lefebvre, P.A. (2007). A CDK-related kinase regulates the length and assembly of flagella in *Chlamydomonas*. *J. Cell Biol.* **176**, 819–829.

Thoma, C.R., Frew, I.J., Hoerner, C.R., Montani, M., Moch, H., and Krek, W. (2007). pVHL and GSK3beta are components of a primary cilium-maintenance signalling network. *Nat. Cell Biol.* **9**, 588–595.

Verhey, K.J., and Gaertig, J. (2007). The tubulin code. *Cell Cycle* **6**, 2152–2160.

Vitre, B., Coquelle, F.M., Heichette, C., Garnier, C., Chretien, D., and Arnal, I. (2008). EB1 regulates microtubule dynamics and tubulin sheet closure in vitro. *Nat. Cell Biol.* **10**, 415–421.

Wargo, M.J., Dymek, E.E., and Smith, E.F. (2005). Calmodulin and PF6 are components of a complex that localizes to the C1 microtubule of the flagellar central apparatus. *J. Cell Sci.* **118**, 4655–4665.

Wargo, M.J., and Smith, E.F. (2003). Asymmetry of the central apparatus defines the location of active microtubule sliding in *Chlamydomonas* flagella. *Proc. Natl. Acad. Sci. USA* **100**, 137–142.

Webster, D.R., Wehland, J., Weber, K., and Borisy, G.G. (1990). Detyrosination of alpha tubulin does not stabilize microtubules in vivo. *J. Cell Biol.* **111**, 113–122.

Wilson, N.F., and Lefebvre, P.A. (2004). Regulation of flagellar assembly by glycogen synthase kinase 3 in *Chlamydomonas reinhardtii*. *Eukaryot Cell* **3**, 1307–1319.

Wloga, D., Webster, D.M., Rogowski, K., Bre, M.H., Levilliers, N., Jerka-Dziadosz, M., Janke, C., Dougan, S.T., and Gaertig, J. (2009). TTLL3 is a tubulin glycine ligase that regulates the assembly of cilia. *Dev. Cell* **16**, 867–876.

Xia, L., Hai, B., Gao, Y., Burnette, D., Thazhath, R., Duan, J., Bre, M.H., Levilliers, N., Gorovsky, M.A., and Gaertig, J. (2000). Polyglycylation of tubulin is essential and affects cell motility and division in *Tetrahymena thermophila*. *J. Cell Biol.* **149**, 1097–1106.

INDEX

G

H

P

Q

R

S

X

Y

Z

VOLUMES IN SERIES

Founding Series Editor
DAVID M. PRESCOTT

Volume 1 (1964)
Methods in Cell Physiology
Edited by David M. Prescott

Volume 2 (1966)
Methods in Cell Physiology
Edited by David M. Prescott

Volume 3 (1968)
Methods in Cell Physiology
Edited by David M. Prescott

Volume 4 (1970)
Methods in Cell Physiology
Edited by David M. Prescott

Volume 5 (1972)
Methods in Cell Physiology
Edited by David M. Prescott

Volume 6 (1973)
Methods in Cell Physiology
Edited by David M. Prescott

Volume 7 (1973)
Methods in Cell Biology
Edited by David M. Prescott

Volume 8 (1974)
Methods in Cell Biology
Edited by David M. Prescott

Volume 9 (1975)
Methods in Cell Biology
Edited by David M. Prescott

Volume 10 (1975)
Methods in Cell Biology
Edited by David M. Prescott

Volume 11 (1975)
Yeast Cells
Edited by David M. Prescott

Advisory Board Chairman
KEITH R. PORTER

Volume 23 (1981)
Basic Mechanisms of Cellular Secretion
Edited by Arthur R. Hand and Constance Oliver

Volume 24 (1982)
The Cytoskeleton, Part A: Cytoskeletal Proteins, Isolation and Characterization
Edited by Leslie Wilson

Volume 25 (1982)
The Cytoskeleton, Part B: Biological Systems and *In Vitro* Models
Edited by Leslie Wilson

Volume 26 (1982)
Prenatal Diagnosis: Cell Biological Approaches
Edited by Samuel A. Latt and Gretchen J. Darlington

Series Editor

LESLIE WILSON

Volume 27 (1986)
Echinoderm Gametes and Embryos
Edited by Thomas E. Schroeder

Volume 28 (1987)
***Dictyostelium discoideum:* Molecular Approaches to Cell Biology**
Edited by James A. Spudich

Volume 29 (1989)
Fluorescence Microscopy of Living Cells in Culture, Part A: Fluorescent Analogs, Labeling Cells, and Basic Microscopy
Edited by Yu-Li Wang and D. Lansing Taylor

Volume 30 (1989)
Fluorescence Microscopy of Living Cells in Culture, Part B: Quantitative Fluorescence Microscopy—Imaging and Spectroscopy
Edited by D. Lansing Taylor and Yu-Li Wang

Volume 31 (1989)
Vesicular Transport, Part A
Edited by Alan M. Tartakoff

Volume 32 (1989)
Vesicular Transport, Part B
Edited by Alan M. Tartakoff

Volume 33 (1990)
Flow Cytometry
Edited by Zbigniew Darzynkiewicz and Harry A. Crissman

Volume 34 (1991)
Vectorial Transport of Proteins into and across Membranes
Edited by Alan M. Tartakoff

Series Editors

LESLIE WILSON AND PAUL MATSUDAIRA

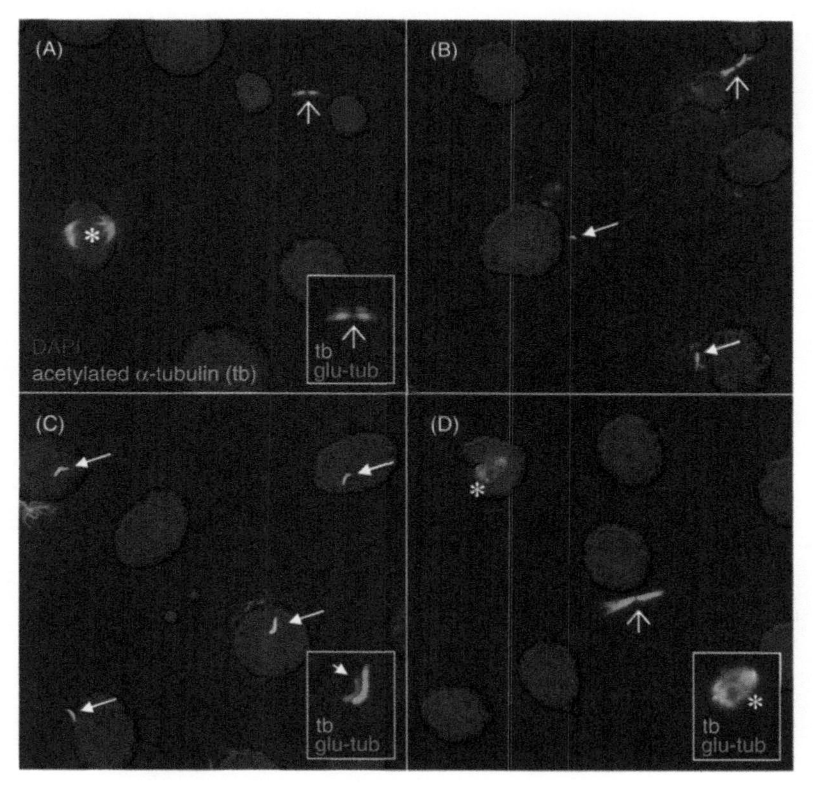

Plate 1 **(Figure 1 on page 74 of this volume)**

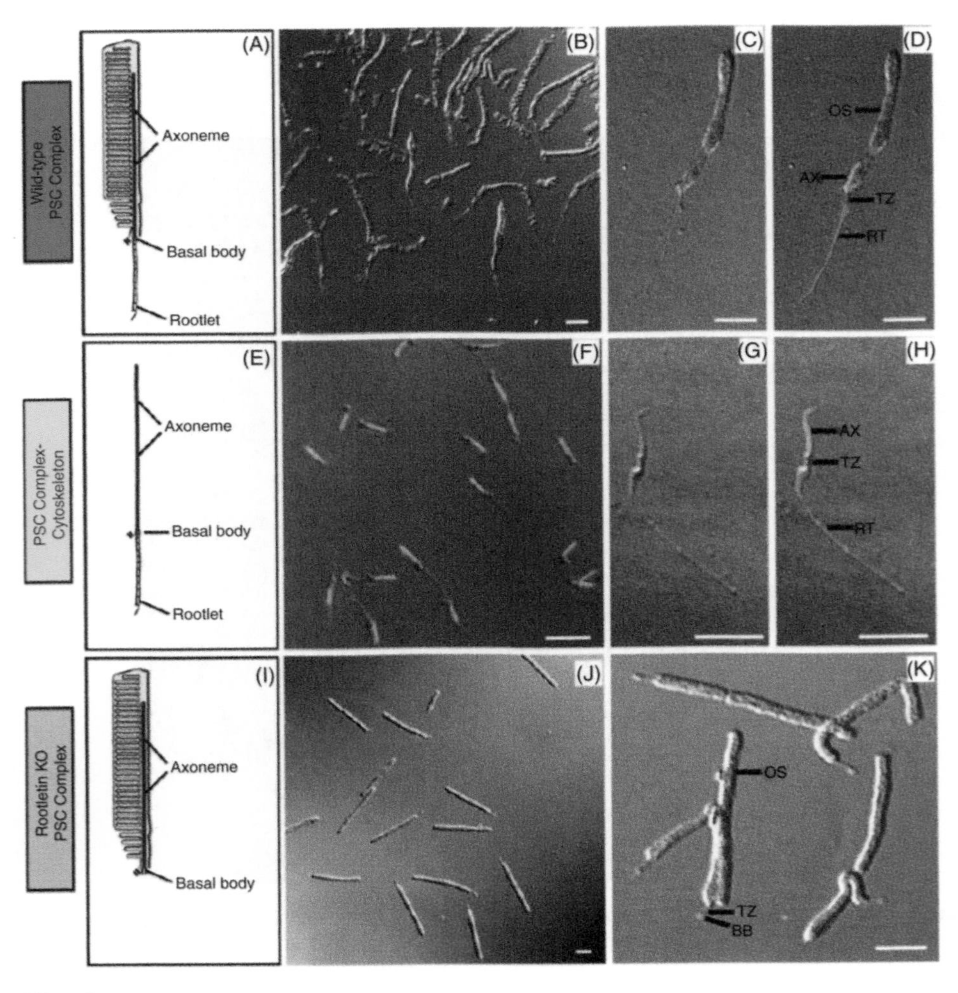

Plate 2 (Figure 1 on page 89 of this volume)

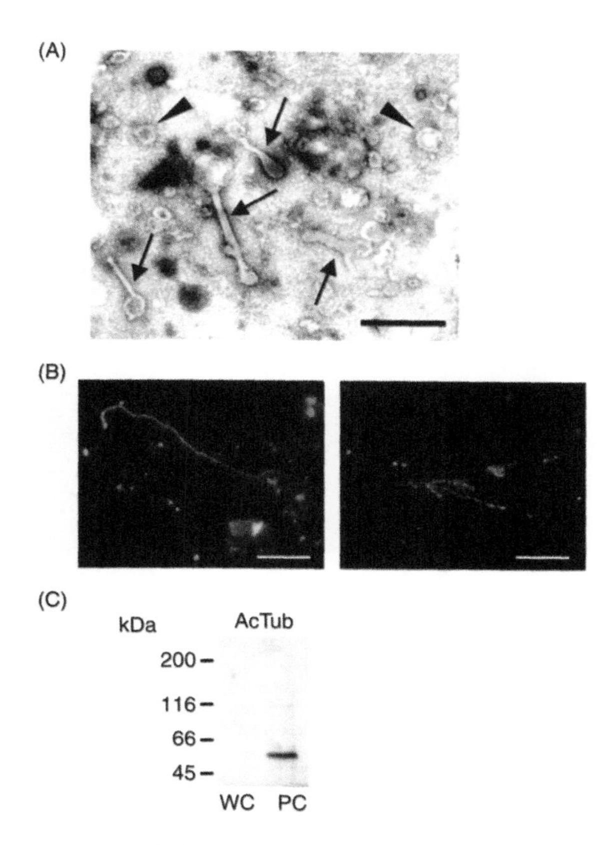

Plate 3 **(Figure 2 on page 94 of this volume)**

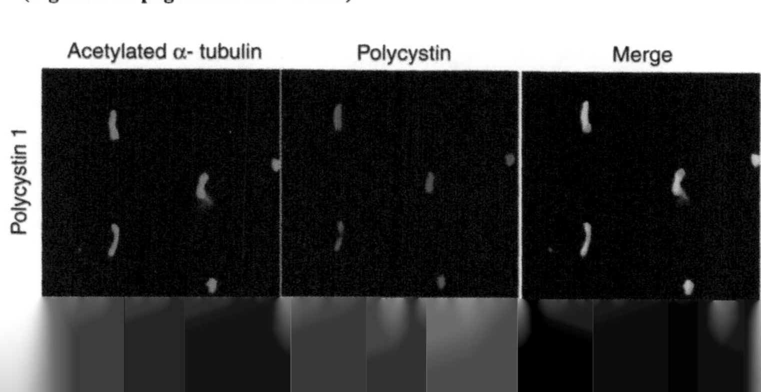

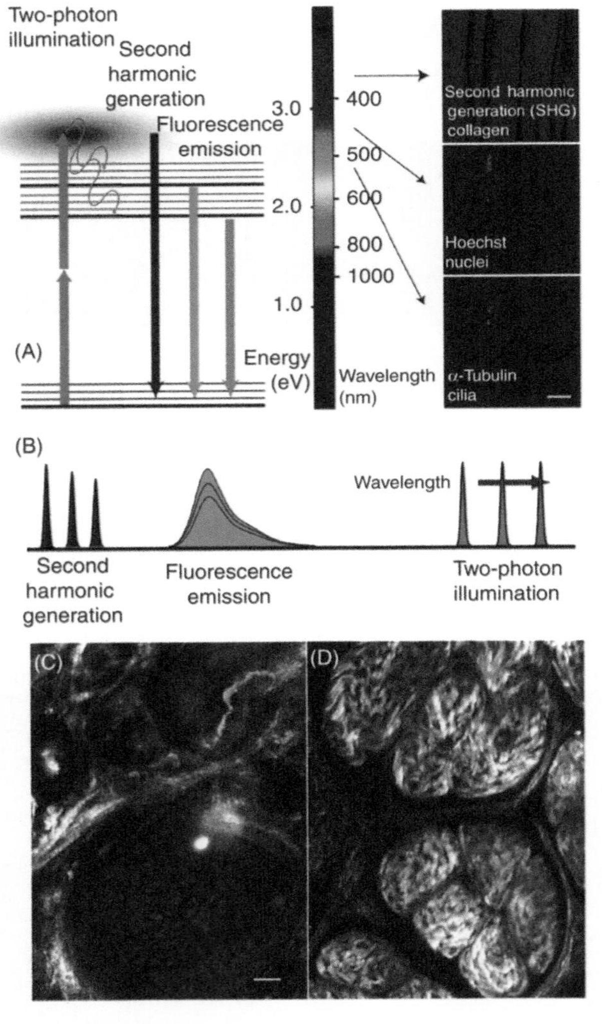

Plate 5 (Figure 3 on page 124 of this volume)

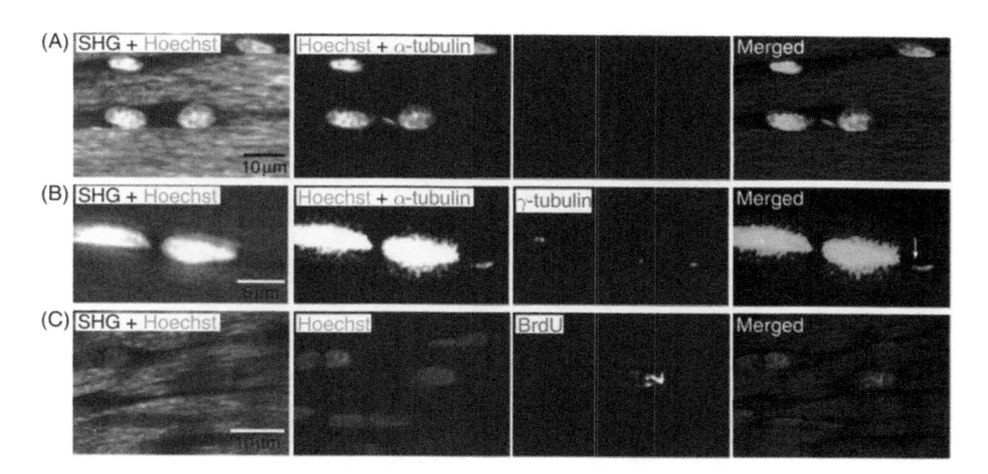

Plate 6 **(Figure 7 on page 130 of this volume)**

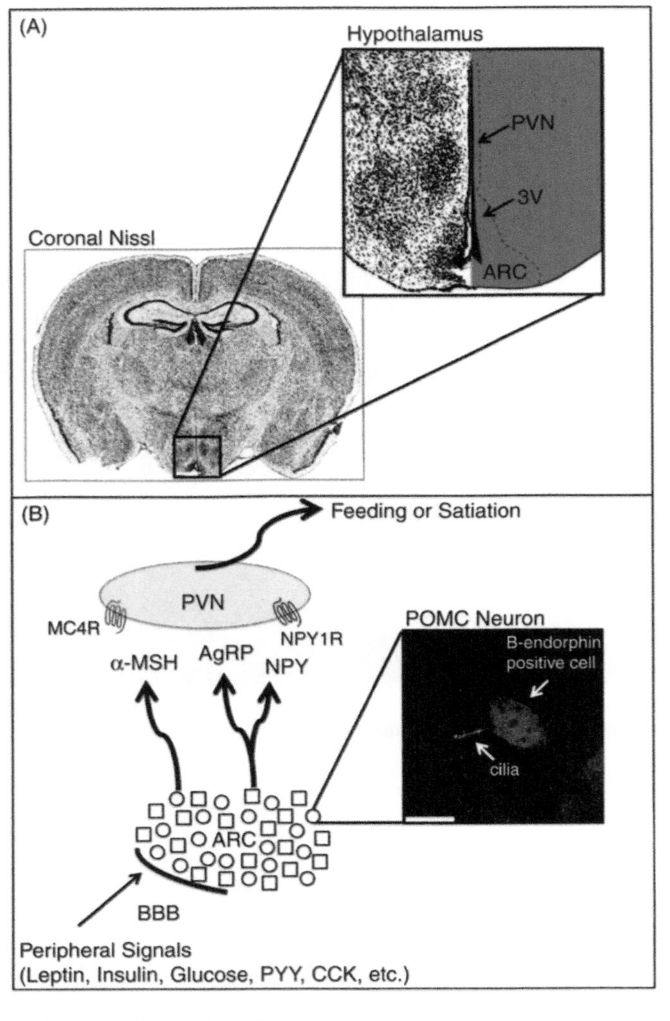

Plate 7 (Figure 1 on page 166 of this volume)

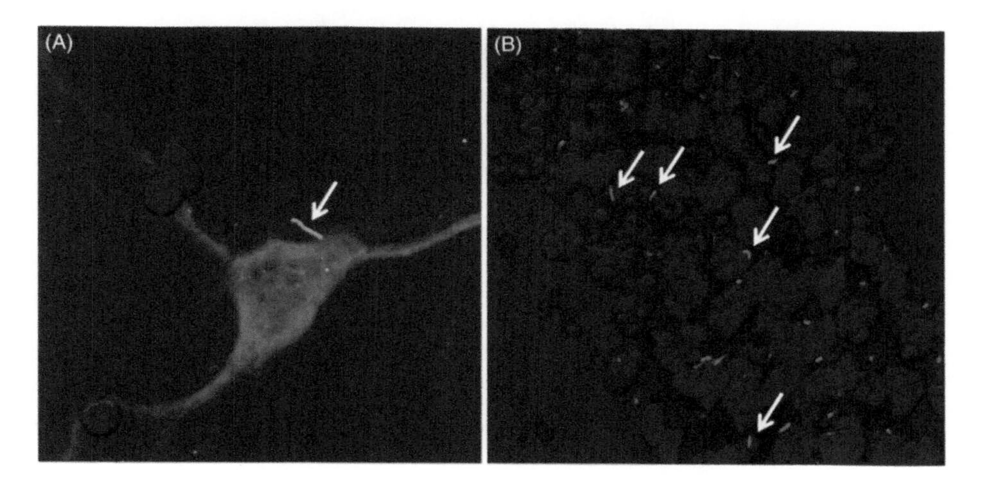

Plate 8 **(Figure 3 on page 171 of this volume)**

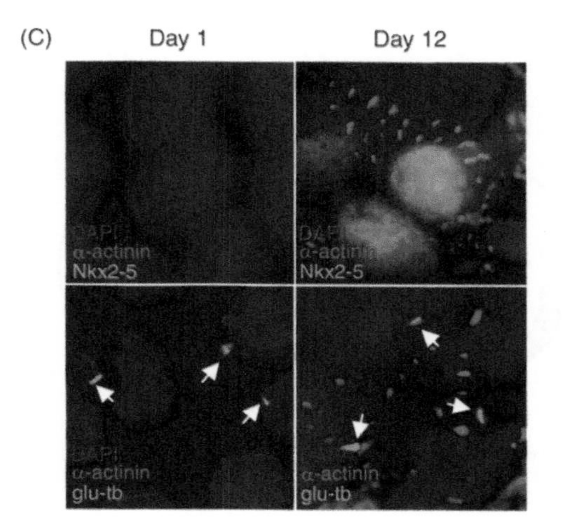

Plate 9 **(Figure 1 on page 191 of this volume)**

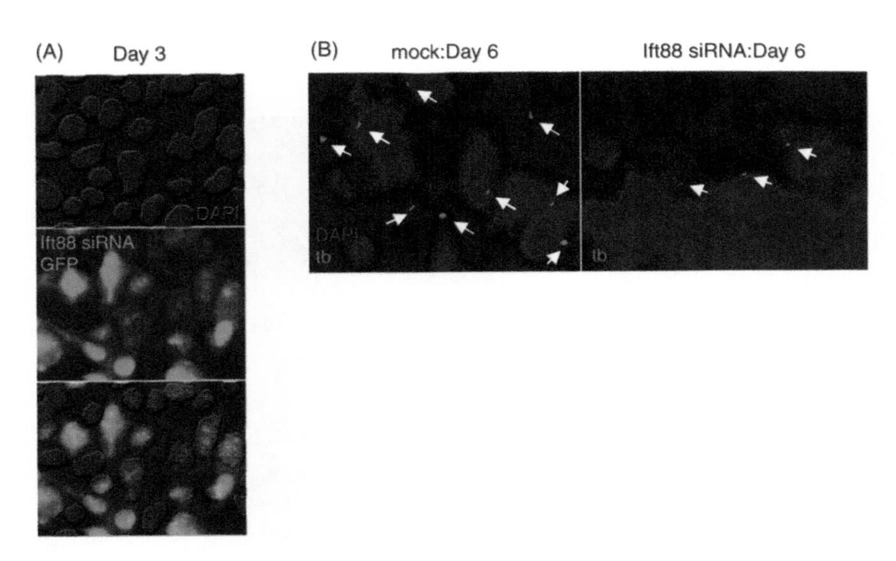

Plate 10 (Figure 2 on page 193 of this volume)

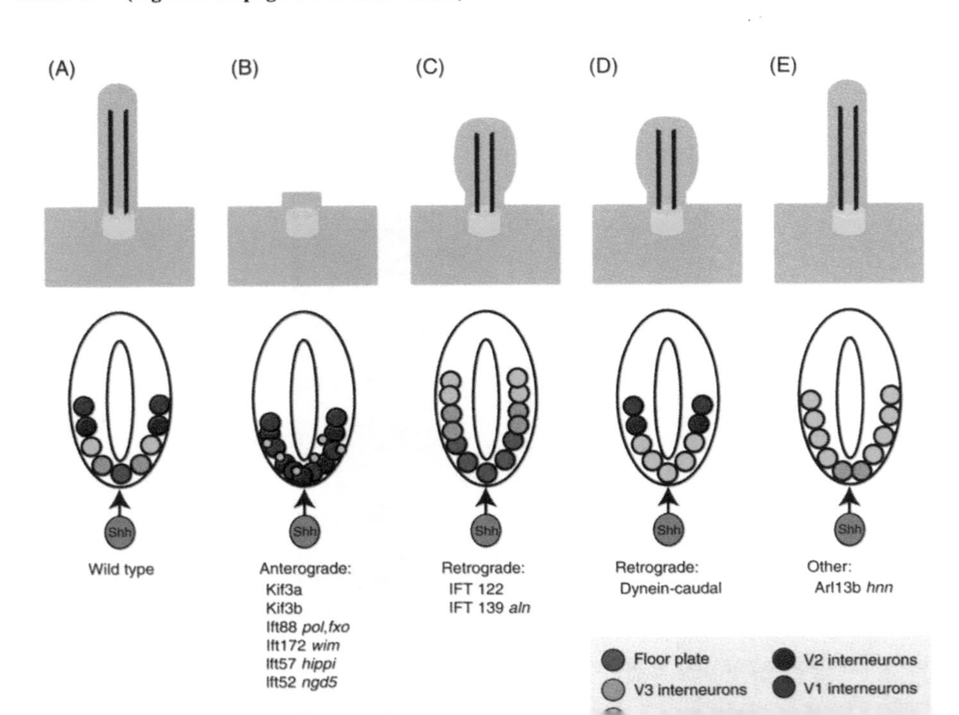

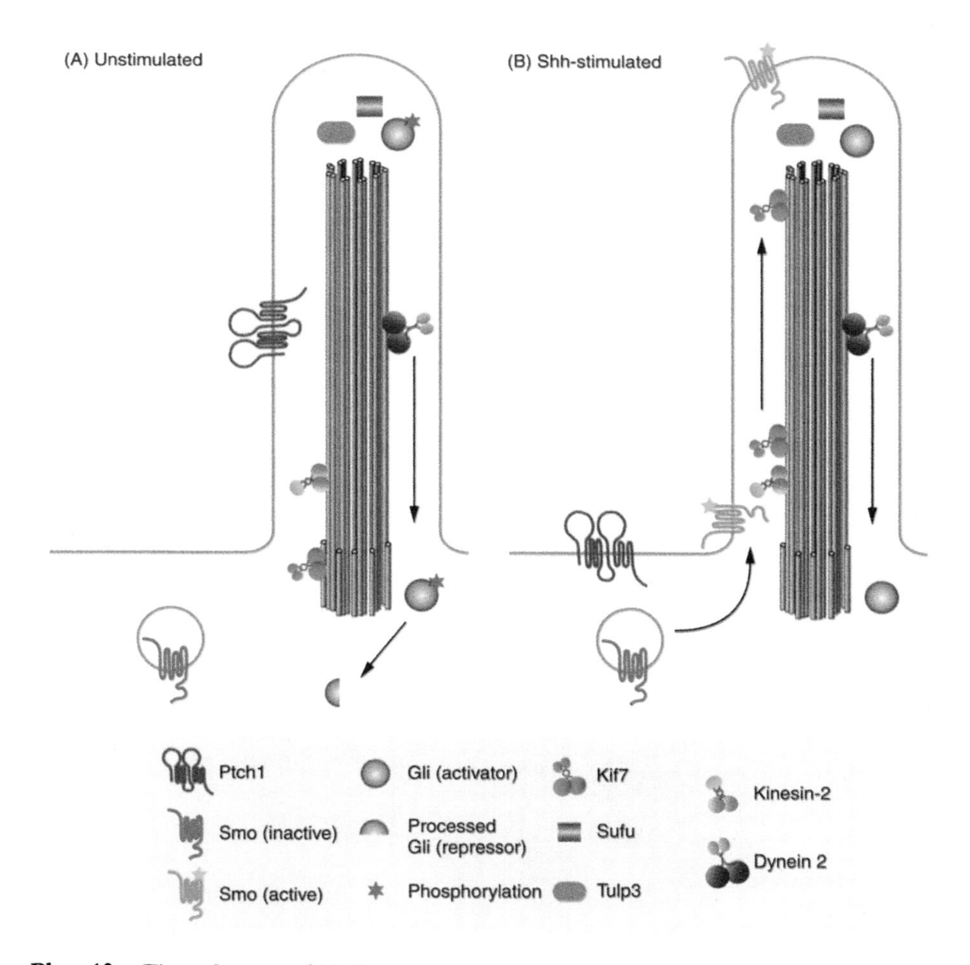

Plate 12 (Figure 3 on page 205 of this volume)

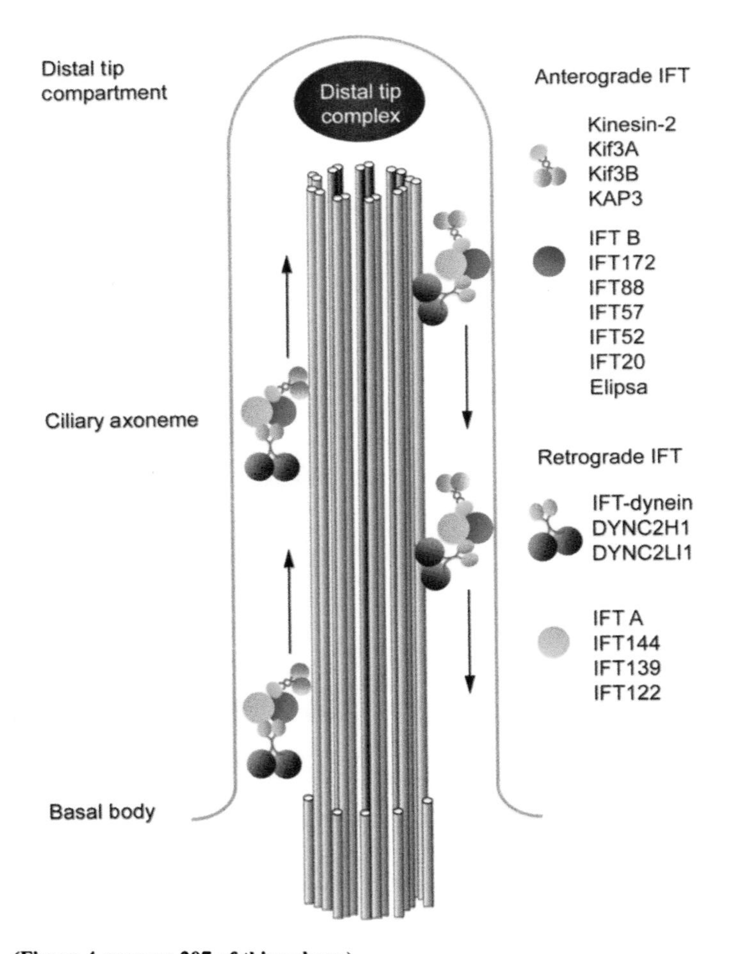

Plate 13 **(Figure 4 on page 207 of this volume)**

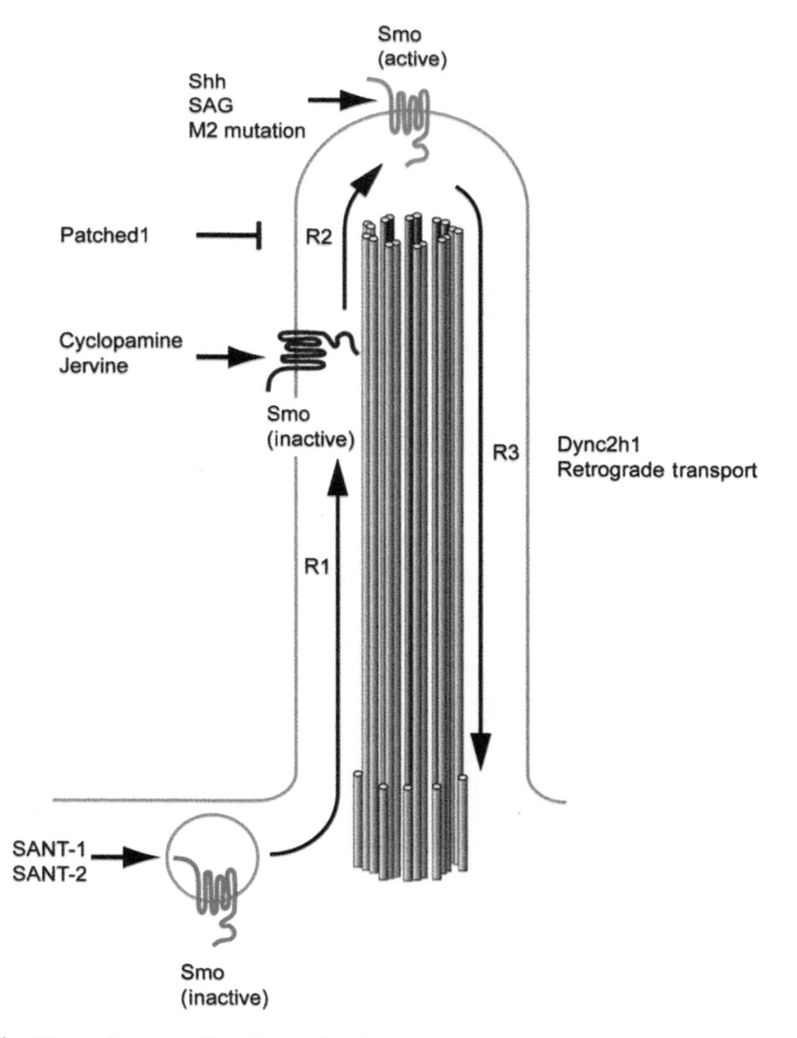

Plate 14 **(Figure 5 on page 214 of this volume)**

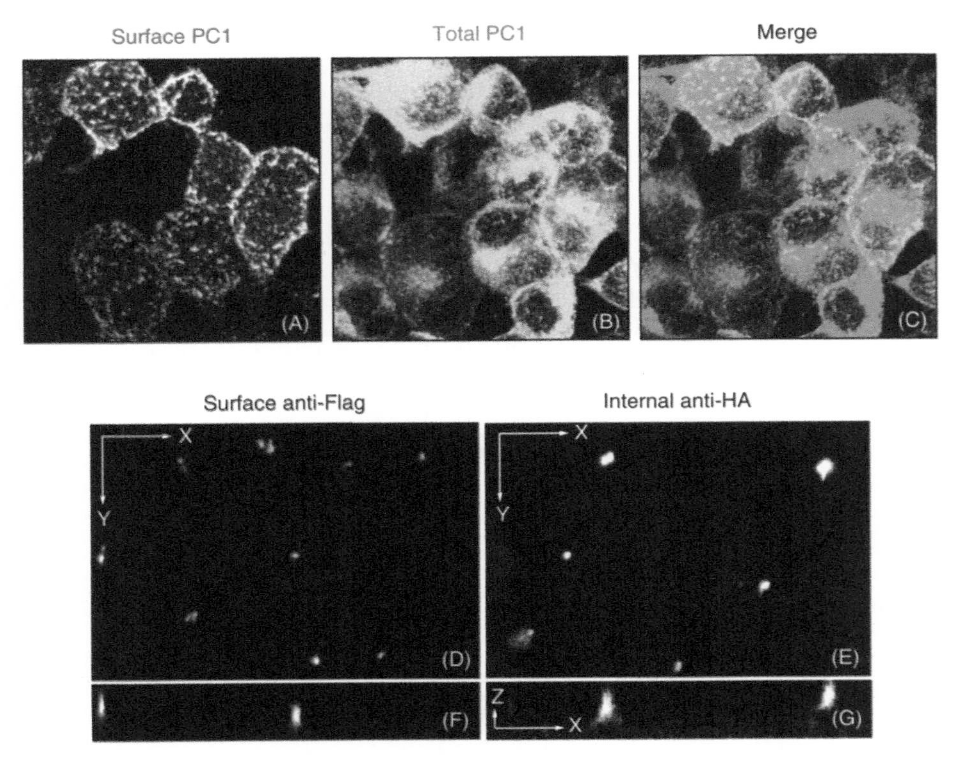

Plate 15 (Figure 3 on page 231 of this volume)

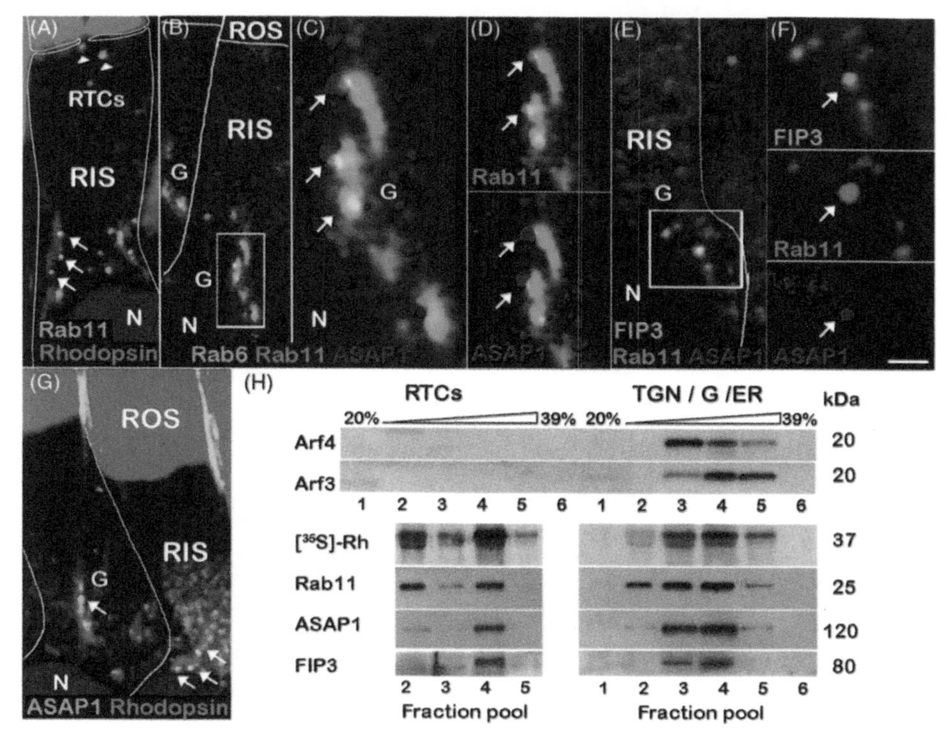

Plate 16 (Figure 4 on page 254 of this volume)

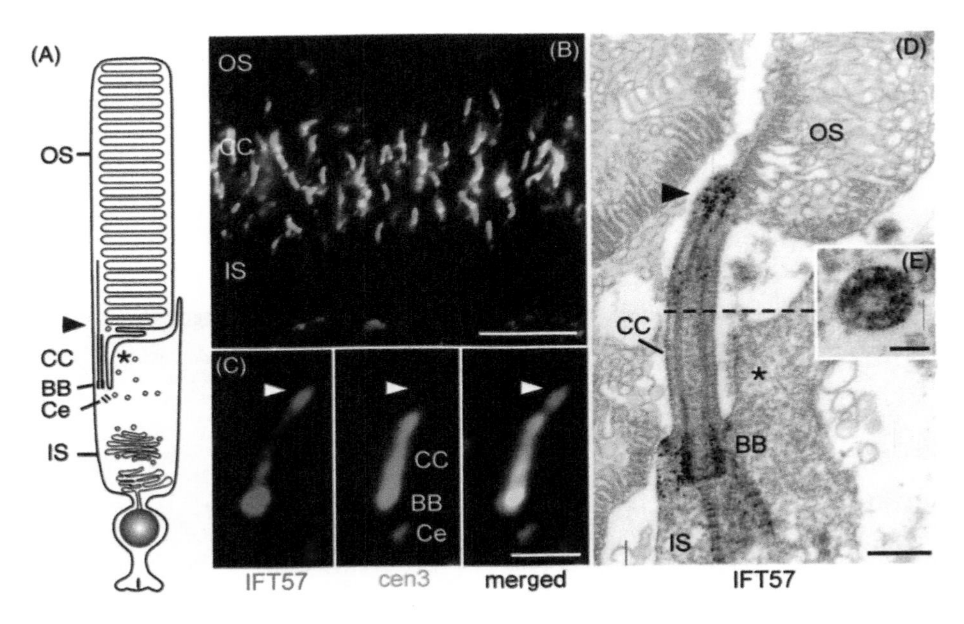

Plate 17 (Figure 3 on page 268 of this volume)

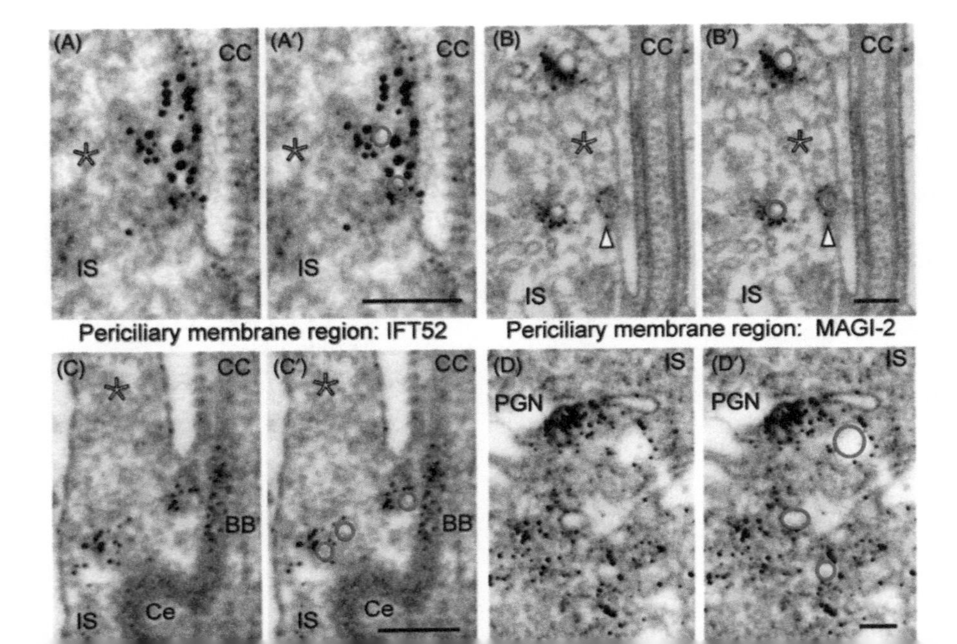

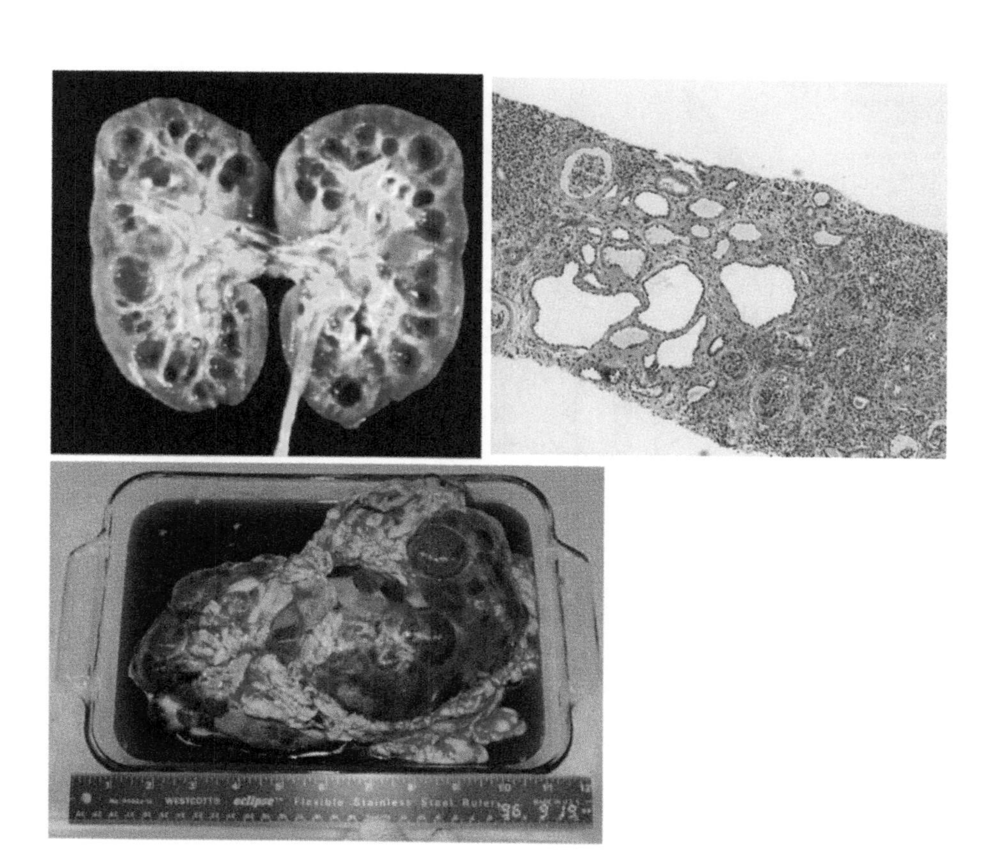

Plate 19 **(Figure 1 on page 275 of this volume)**

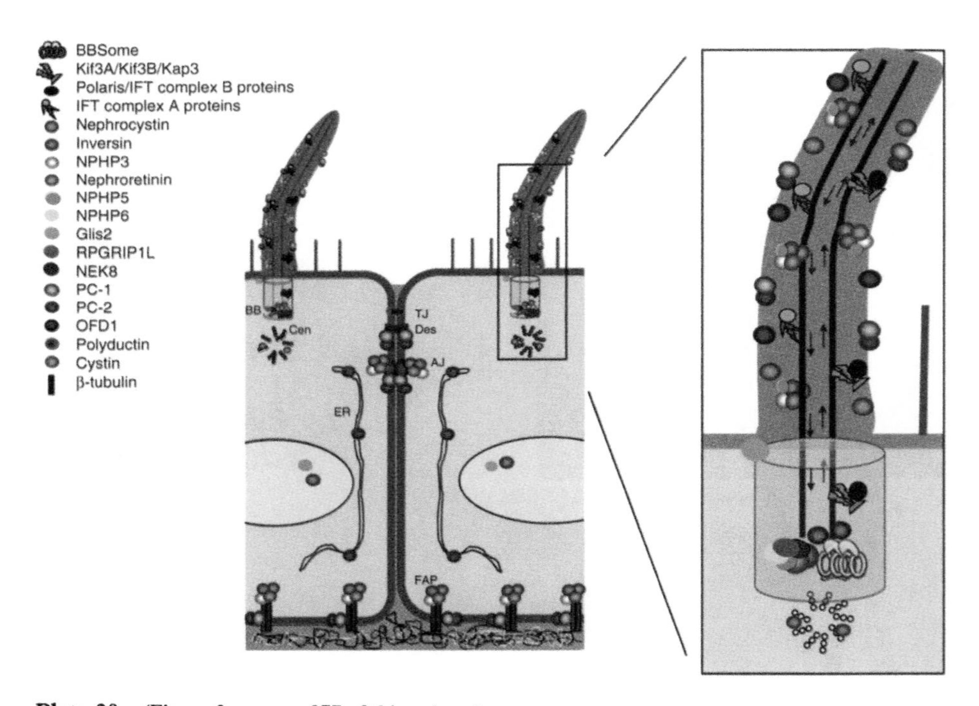

Plate 20 (Figure 2 on page 277 of this volume)

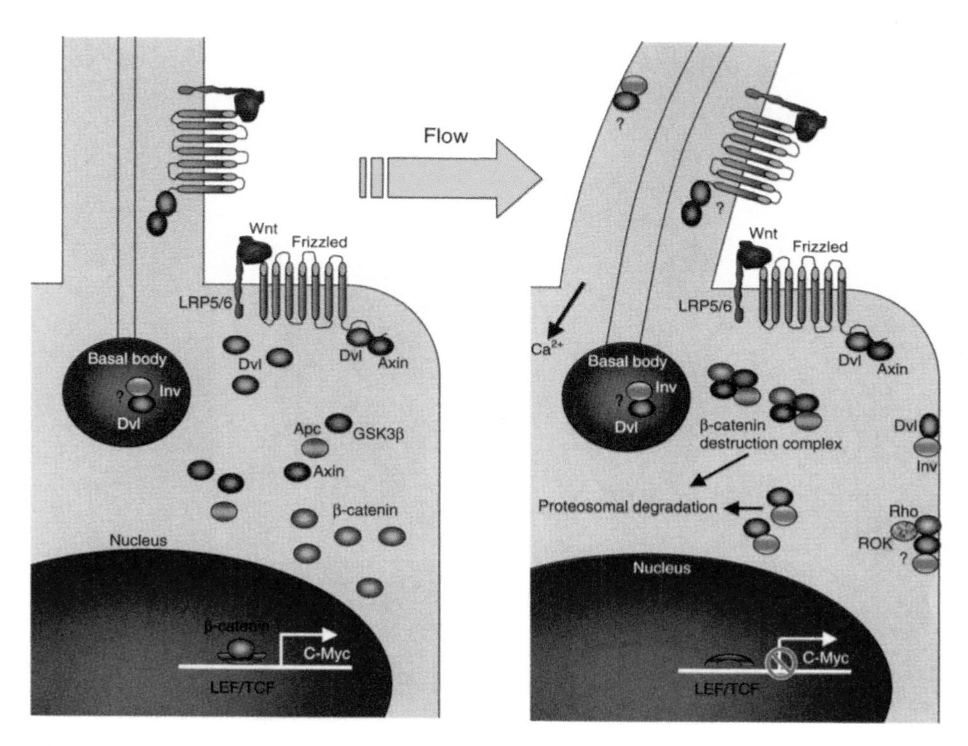

Plate 21 (Figure 3 on page 281 of this volume)

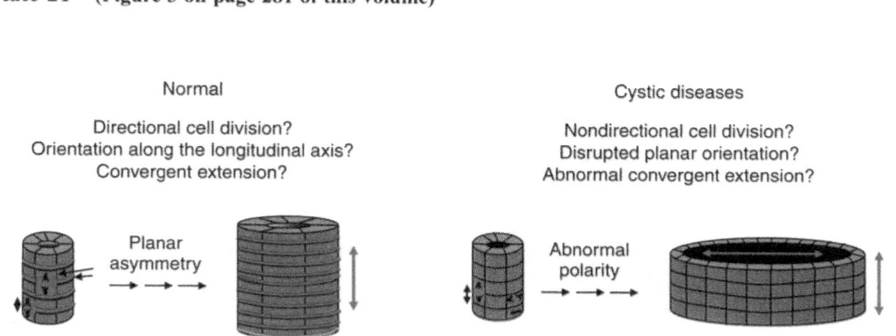

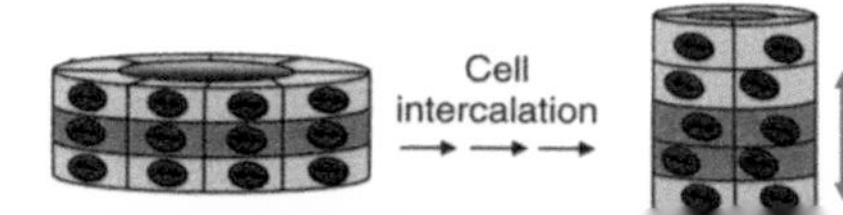

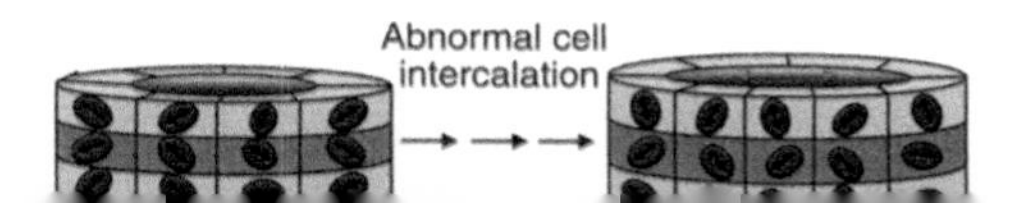

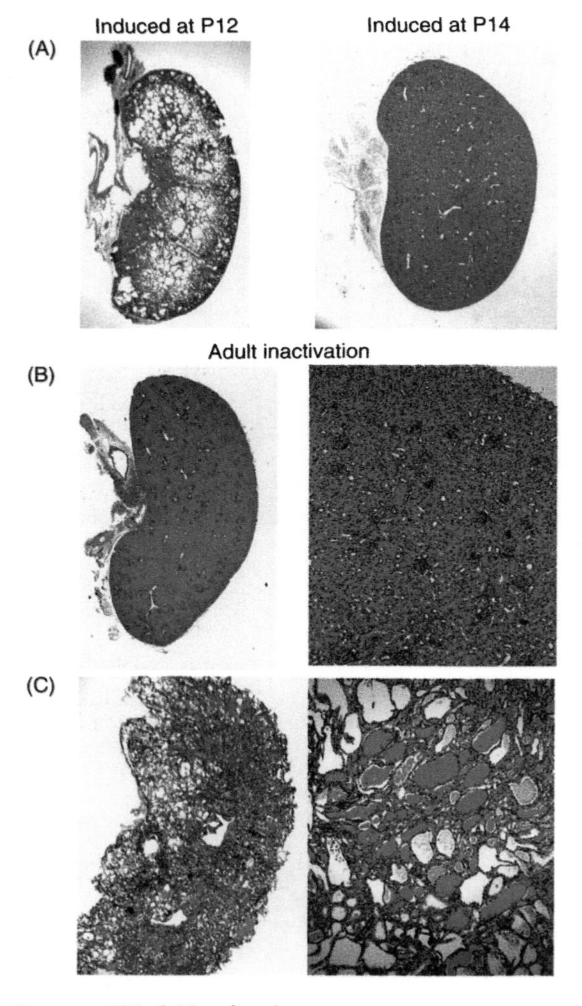

Plate 23 (Figure 5 on page 286 of this volume)

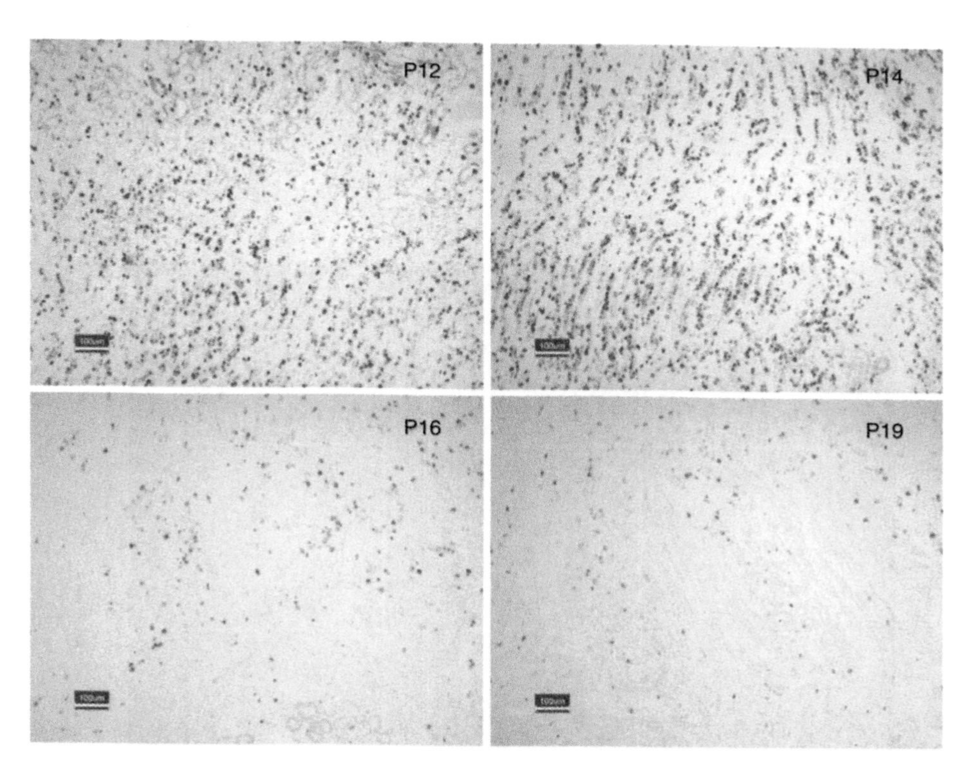

Plate 24 **(Figure 6 on page 287 of this volume)**

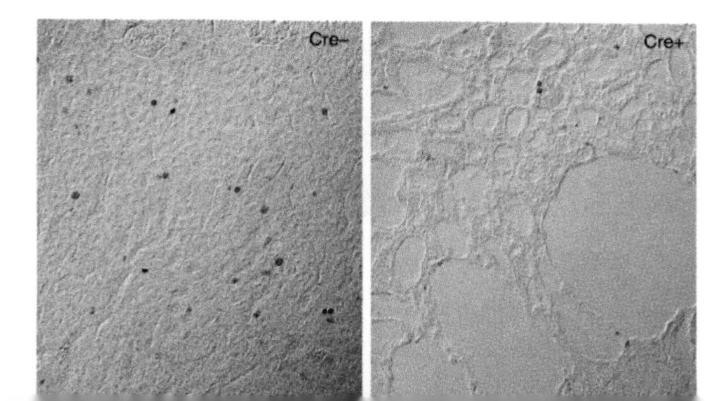

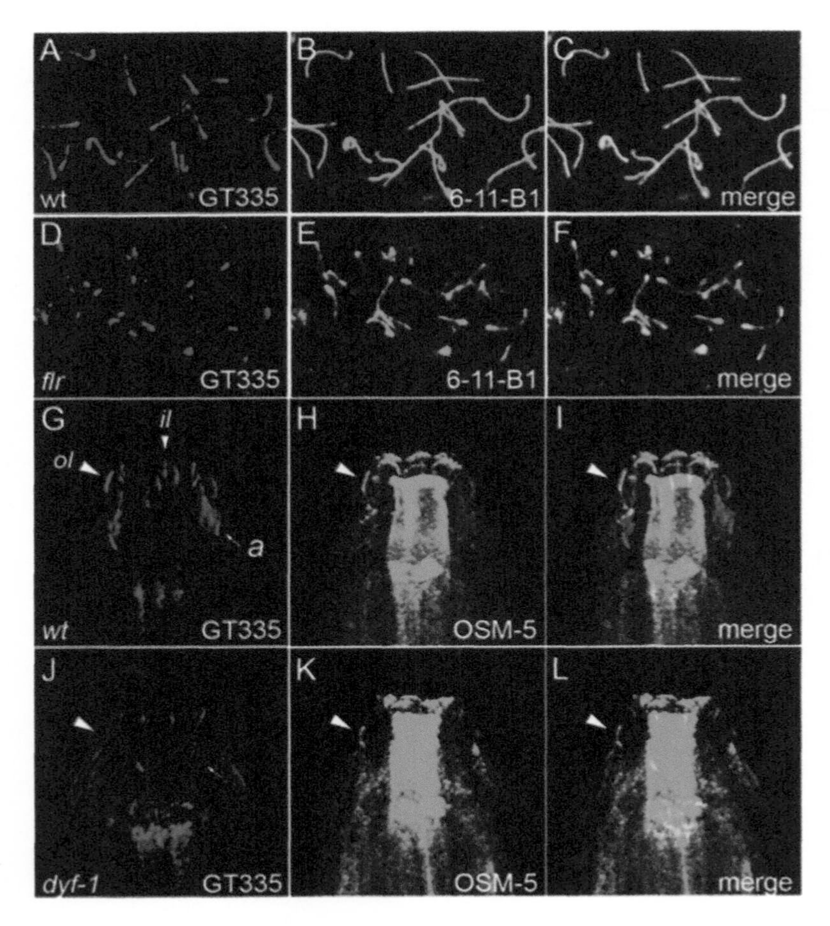

Plate 26 **(Figure 3 on page 323 of this volume)**

Printed and bound by CPI Group (UK) Ltd, Croydon, CR0 4YY

08/05/2025

01864953-0006